Systematic POMOLOGY

Volume - 2

Dr. Om Prakash Pareek
Former Director
Central Institute for Arid Horticulture (ICAR)
Bikaner - 334 006 (Rajasthan)

Dr. Suneel Sharma
Prof. of Horticulture
Dept. of Horticulture,
CCS Haryana Agricultural University
Hisar (Haryana)

SCIENTIFIC
PUBLISHERS

Published by
SCIENTIFIC PUBLISHERS (INDIA)

Jodhpur –

5 A, New Pali Road
P.O. Box 91
Jodhpur - 342 001 INDIA

Delhi –

4806/24, Ansari Road
Daryaganj
New Delhi - 110 002
INDIA

Print 2026

ISBN: 978-93-6346-067-6 (SET) Vol 1-2
eISBN: 978-93-6346-068-3 (SET) Vol. 1-2

Visit the Scientific Publishers (India) website
at http://www.scientificpuonline.com

Printed in India

PREFACE

Systematic Pomology is an integral part of the science of fruits. Its knowledge is a prerequisite for improvement as well as sustainable and profitable production of fruit crops. The nomenclature and classification of fruit plants had been traditionally done on the basis of the Bentham and Hooker's natural system of classification published in *Genera Plantarum* (1862-1883). Now after the development of the ordinal classification, using new systematic data especially on DNA sequences, for the families of flowering plants by the Angiosperm Phylogeny Group (APG), updated as APG III in 2009, adoption of the concept of phylogenetic nomenclature and classification in fruit species is more appropriate. Also, the latest Code of nomenclature for the cultivated plants should be adopted. Therefore the authors considered it necessary to put the fruit crops into the more systematic framework of the APG system of nomenclature and classification. This requirement has been felt ever since the authors taught the Systemaic Pomology course to postgraduate students. In fact, this interest has its roots when one of us (OPP) drew inspiration from the lectures of Prof. Ranjit Singh, an authority on citrus taxonomy, who taught the subject in the Horticulture Division of the Indian Agriculture Research Institute, New Delhi during 1962-63, and from the earlier publications from USA by Prof. F A Waugh (1916) and Prof. Q B Zeilinski (1955). The intention was also to consolidate and update the scattered information for those interested in the subject of fruit science.

With continued work on this task the present book has been brought out in two volumes. The subject matter of the book, having 40 chapters, has been presented in three parts. Part A, with 4 chapters, gives general information on methods for (i) description of plant characters, (ii) identification, (iii) nomenclature, and (iv) classification. In Part B and C, nearly 370 species belonging to 72 families have been grouped based on the Phylogenetic System of Classification of the APG. Part B has Gymnosperm fruits in 3 chapters and Part C has fruits of Angiosperms in 33 chapters. In each chapter, the fruits have been arranged giving their taxonomical status from family to the cultivar level with description of important pomological traits. Comparatively brief narrative is given on the characters at the levels of class, subclass, angiosperm clade, order and family. More emphasis has been laid on the description of characters at species level with considerably exhaustive description at the cultivar level. The book is richly illustrated with figures and photographs. Taxonomic keys have also been given for some species groups. This first volume contains 27 chapters; all the chapters of Part A and Part B and 20 chapters of Part C.

The remaining 13 chapters of Part C along with a references list, fairly exhaustive glossary and an index are continued in the second volume.

The book is intended to serve as a reference material for researchers, teachers, students, fruit growers and all concerned with fruit flora particularly those interested in their genetic resources, breeding and bioinformatics. The knowledge will be useful to a gene bank curator to enable him to systematically maintain fruit germ plasm *in-situ* or *ex-situ* and in herbaria including their proper documentation, to the taxonomists for use not only in preparation of keys and correct identification but also in classifying the fruit genotypes into categories and in removing confusion arising from synonimity in nomenclature. Correct knowledge of origin and evolution of the genotypes will be useful in their systematic use in scientific studies based on the information on their relationships and differences (taxonomic proximities) and also in selecting genetic material (considering specific characters) for use in improvement programmes. In addition, the information would be useful to the commercial fruit growers as a guide book for systematic planting and culture of fruits by selection of the right genotype/cultivar considering their suitability for use as rootstock and as scion material to obtain high productivity and quality of fruits in the given environment.

The authors have drawn information from number of published texts, monographs, books and periodicals for compilation of this work. The authors gratefully thank the organizations Biodiversity International, formaly IPGRI Regional Office for Asia, Rome, Italy, ProMUSA, TFNet, Selangor, Malaysia and IITA, Nigeria for using line diagrams/drawings of plant species from their published work. Our special thanks are also due to Dr S.N. Pandey, ex-ADG, ICAR, Dr. Sanjay Singh and Sh Sanjay Patil, CHES, Bikaner, Ms. Monica Lisa, CCS HAU, Hisar, Dr Sunil Pareek MPUAT, Udaipur, Dr. Murlidharan, Date Palm Research Station, Mundra, for providing photographs of cultivars of fruit species used in this publication. Thanks are also due to Dr R.A. Sharma and Dr S. Kathju, CAZRI, Jodhpur for valuable suggestions on portions of the text. Authors are grateful to the Vice Chancellor CCS Haryana Agricultural University, Hisar, and the Director CIAH, Bikaner for the kind encouragement to undertake this task. The task would not have been accomplished without the inspiration, encouragement and support provided by the family members especially Mrs. Kamla Pareek and Mrs. Deepali Sharma.

O.P. Pareek
Suneel Sharma

CONTENTS

[MALVIDS : EUROSID II]

Chapter 28

MYRTALES

The plants are trees, shrubs or herbs. Intraxylary (internal) phloem is always present in woody stems. Vessels are mostly with simple or rarely with scalariform perforations (Geissolomataceae and some Myrtaceae) and with vestured and bordered pits. Lateral pitting is almost always alternate. Fibres are with simple or bordered pits, septate or non-septate. Rays are heterogeneous to homogeneous. Axial parenchyma is most typically scanty paratracheal, but in addition it may be apotracheally diffused. Sieve element plastids are of Ss-type. Nodes are unilacunar or sometimes trilacunar. Leaves are opposite or less often alternate or verticillate, simple, mostly entire, usually with vestigial stipules and rarely exstipulate. Stomata are of various types, mostly anomocytic. Flowers are borne in cymose or racemose inflorescences or sometimes solitary in the axils, bisexual or rarely unisexual, actinomorphic or sometimes more or less zygomorphic (usually slightly zygomorphic), mostly 4 or 5-merous, sometimes apetalous (some Lythraceae and all Alzateaceae and Penaeaceae). Sepals are connate, sometimes much reduced or obsolete. Stamens are alternisepalous, numerous, rarely reduced to one. Filaments are bent inward in the bud or rarely erect (as in *Psiloxylon*) and anthers are 2-locular (Heteropyxidoideae in Myrtaceae, Rhynchocalycaceae) or 4-locular at anthesis opening longitudinally by apical pores (some Myrtaceae and the majority of Melastomataceae). Tapetum is secretory. Microsporogenesis is simultaneous. Pollen grains are 2-celled or less often 3-celled, in monads or sometimes (as in Onagraceae) in tetrads, basically 3-colporate, but sometimes colpate or porate, basically tectate and frequently characterized by pseudocolpi. Intrastaminal annular nectary disc is often present. Gynoecium is of 2 to many (often 4 or 5) united carpels with stylodia connate into a style (very short in *Psiloxylon*) and simple or lobed stigma. Ovary is superior to inferior, plurilocular to 1-locular, usually with (1)2 to many ovules per carpel. Ovules are anatropous or in various taxa hemitropous, amphitropous or campylotropous, usually bitegmic or very rarely unitegmic and crassinucellate (in some Myrtaceae). Endothelium is absent or rarely present (*Axinandra* in the Crypteroniaceae). Female gametophyte is of various types. Endosperm is nuclear. Fruits are of various types. Seeds are mostly with very little or no endosperm. The tissues contain flavonols, myricetin and methylated ellagic acid.

Family Combretaceae

It contains 13-20 genera and 500 species (Heywood *et al.*, 2007) distributed throughout the subtropical and warm temperate regions.

Genus *Punica* L. (n = 7, 8)

The genus has two species, *P. granatum* L. (pomegranate), native to the semi-arid eastern Mediterranean region, and *P. protopunica* Balf. f., indigenous to the island of Socotra in the Democratic Republic of Yemen.

The plants are glabrous shrubs or small trees having branches often terminating as spines. Leaves are small and shining. Flowers are borne terminally or axillary as solitary or in clusters of 1-5. These are campanulate, leathery, red or yellow, 2-3 cm long having 5-8 persistent sepals, showy, crumpled, red or white petals and many stamens covering the inner floral tube surface from the rim to the multilocular ovary. Carpels are in one whorl or in 2-3 superposed layers, the lower with axile placentation and the upper with apparent parietal placentation, a result of differential asymmetric growth. Fruit is a leathery berry crowned by persistent and irregularly dehiscent sepals. Seeds are many with translucent sarcotesta. Cotyledons are spirally rolled.

Pomegranate

(*Punica granatum* L.) x=8; 2n=16

The pomegranate (*Punica granatum* L.) probably originated in the region of Iran, Afghanistan and Baluchistan. *P. granatum* has been classified into two subspecies, *chlorocarpa* and *porphyrocarpa. Chlorocarpa* is found in trans-Caucasus region and *porphyrocarpa* in central Asia.

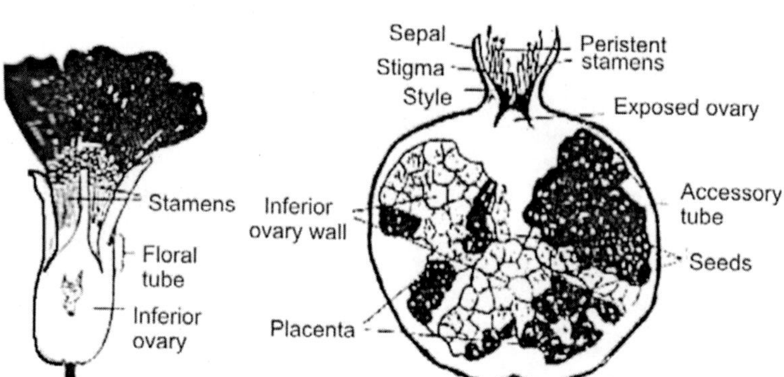

Fig. 28.1. Flower and fruit of pomegranate (Source: Ryugo, 2008).

Pomegranate is an evergreen deciduous shrub or small tree with a smooth, dark gray bark and spiny and angled branchlets. The leaves are opposite, short-petioled, 2-8 cm long, oblong or obovate, glossy, bright green and glabrous. Flowers are typically of two types: (i) large fertile ones with anthers and stigma at the same height emerging on old wood and (ii) the smaller sterile flowers with the stigma much below the anthers on new wood and having defective pollen that do not usually form fruits (Morton, 1987; Verheij and Coronel, 1991). The flowers are bisexual, 4-6 cm in diameter, borne axillary, solitary or in small clusters towards the ends of the branchlets. The calyx is tubular, persistent and 5-7 lobed. The 5-7 petals are brilliant orange red inserted in the calyx lobes. Stamens are numerous. Ovary is inferior with several locules in 2 series, one above the other (Fig. 28.1). The fruit is globose, smooth, coriaceous and woody berry, 6-8 cm in diameter, which becomes brownish yellow to red on ripening. Seeds are many sided, 1.3 cm long consisting of thin and transparent fleshy testa (axil) containing reddish juice surrounding the elongated angular seed kernel.

Cultivars

The cultivars show variation in shape and colour of fruit, rind thickness and taste and colour of aril. These have been classified on the following bases:

1. Flower characters-
   a. Single-flowered- White flowered (seeds are not soft), red or vermillion flowered (the seeds are soft and edible).
   b. Double-flowered – It has high sterility. The seeds are not soft.
2. Colour of aril- White, pink and red (e.g. cv. Kandhari).
3. Sweetness of aril- Acidic (important for making *anardana*), subacid (e.g. cv. Wonderful) and sweet.

Some of the important cultivars are Paper Shell, Spanish and Ruby. Important cultivars of Iraq are Ahmar, Aswa and Halwa and of USA are Wonderful and Grenada. The cultivar Japanese Dwarf (*P. granatum* L. var. *nana* Persoon.) is hardy and ornamental and can be grown in pots. Other ornamental double flowered forms bear white to red and purple flowers, e.g., Multiplex and Variegata. Four groups of pomegranate types based on colour are dark red, yellow green, black violet and white. Mangulati is an important cultivar of Saudi Arabia. Wonderful and Red Loufani are often grown in Jewish sector of Israel while Malissi and Ras el Baghal are favoured in the Arab sector. The local cultivars are light coloured but are sweet. The Afghan cultivars are dark red but are acidic.

Ganesh: it is a selection from open pollinated seedlings of Alandi and is very prolific yielder. The plants are medium tall. Fruit is greenish yellow with red tinge, of medium size having 234 g weight, 0.33 mm thick rind, soft seeds, 371 pink arils per 100 g, 78% juice, 15.5°B TSS and 0.5% acidity with pleasant agreeable flavour and taste.

Kandhari: The plants are medium tall, vigorous and deciduous shrubs having attractive foliage. The flowers are long having deep scarlet calyx and 6-7 deep crimson petals, darker than in Dholka. The fruit is large, angular in outline and deep red in colour. The rind is thin and crimson to deep red in colour. The fleshy testa or aril is blood red or deep pink with sweet and slightly acidic juice having pleasant flavour. The seeds are hard and crimson-coloured.

Dholka: The plants are evergreen, moderately vigorous and spreading shrubs. The stem is non-branching. Spines are sparse. The flowers are campanulate and light crimson. Petals are 5-6 and of vermillion colour. The fruits are large in size, somewhat greenish overlain by light yellow and very faint pink blush, yellowish crimson having patches of different colours, thick skin and are faintly ridged. The arils are white with light pink spots. The juice has 9°B TSS, 0.4% acidity and mild flavour. The quality is good. It is a soft seeded cultivar.

Alandi or Vadki: It is a good cultivar, bears regularly and heavily, but has very hard seeds. The fruits are of medium size. The aril is deep pink to intense red (blood red), juicy, sweet and slightly acidic in taste.

Muskat White: The plants are evergreen, tall and vigorous shrubs. The spines are sparse. The flowers are tubular having scarlet calyx and 5-6 light crimson petals. The fruit is greenish yellow with pink patches, conspicuous ridges and thin rind. The seeds are creamy white with pink dots. The TSS of juice is high and acidity is low. The fruit quality is good.

Muskat Red: Fruit is small to medium in size and rind is quite thick. The aril or juicy testa is red and moderately sweet. The seeds are not very hard.

Spanish Ruby: The cultivar is good cropper. Fruit is small to medium in size, rind is smooth, glossy and thin and the fleshy testa is red or rose coloured. The seeds are soft.

Kabul: The fruit is large in size, rind is thick and deep red mixed with pale yellow. The fleshy testa is dark red with slightly bitter juice.

Paper Shell: Fruit is medium in size, rind is thick and the fleshy testa is reddish to pink with sweet juice. The seeds are soft.

Jyothi: It is a selection from open pollinated seedlings of Bassein Seedless and Dholka. Fruit is medium to large in size with yellowish red colour and fleshy, sweet and pink arils and soft seeds. It is a moderate yielder.

Ruby: It has been developed from a three-way cross (Ganesh x Kabuli x Yercaud x Ganesh x Gulsha Rose Pink)). Fruit is reddish with green streaks, large in size having 270 g weight, 0.24 mm thick rind, 37.2 g/100 g dark pink arils, 88.7% juice with 16°B TSS and soft seeds. It is a high yielding cultivar giving 16-18 tonnes fruits per hectare.

Bhagwa: It is a clonal selection of Ganesh x Gul-e-Shah Red. The cultivar is also known by names such as Shendari, Ashtagandha, Mastani, Jai Maharashtra and Red Diana. It is a heavy yielder (30 kg fruits/tree) and matures fruits in 180-190 days. The fruit is large in size having sweet, bold and attractive dark red arils. The rind is 0.4 mm thick which makes it suitable for transporting to distant markets. The surface is glossy, very attractive and saffron coloured. The arils have 15.4°B TSS and 0.37% acidity.

Phule Arakta (Ganesh x Gul-e-Shah Red): It is a heavy yielder and produces large sized, sweet and soft seeded fruits. Fruit size is medium of 183 g weight having 0.32 mm thick rind, dark red arils, 16°B TSS and 0.45% acidity in the juice. Fruit yield is 30 kg per tree.

G-137: It is a clonal selection from Ganesh. Fruit is yellowish pink, large in size having 290 g weight, 0.45 mm thick rind, 238 dark pink arils per 100 g, 88.7% juice, 16°B TSS and 0.39% acidity. Fruit yield is 12 kg per tree.

Mridula (Ganesh x Gul-e-Shah Red): The plant is dwarf. Fruit is medium in size, of 178 g weight having 0.29 mm thick rind, blood red arils, 15°B TSS and 0.46% acidity. Fruit yield is 14 kg per tree.

CO 1: It is a clonal selection. Fruit is 340 g in weight having soft seeds, pink juicy arils and 16°B TSS.

Yercaud 1 (Acc. 455): It is a clonal selection. Fruit has 350 g weight, soft seeds, deep purple juicy arils and 15.2°B TSS.

Amildana (Ganesh x *P. nama*): The trees are of short stature. Fruit weighs 120 g having highly acidic (4.8%) arils suitable for making anardana.

Goma Khatta: It is a selection made at Godhra Research Station of CIAH, Bikaner.

White Flowered: The plants are evergreen. The fruits are pale yellow or lemon coloured. The arils are large and lack acidity. The seeds are large and hard.

P. protopunica Balf. f. from Socotra has larger leaves, smaller flowers and fruits compared to *P. granatum* and a 3-locular and semi-inferior ovary.

P. granatum* var. *nana AGM (dwarf pomegranate): The flower colour shows variation.

Genus *Trapa* L. (x=12; secondary basic number x=24)

The plants are aquatic annual herbs, rooted or floating. Leaves are dimorphic with deeply cleft stipules. Submerged leaves are approximately opposite, sessile, linear and caducous. Floating leaves are alternate to spiral, forming a rosette having elongated petioles, inflated near the middle, rhombic blades, the distal half of the margin coarsely dentate, pinnately dissected to filiform. Adventitious leaf-like organs emerge at submerged nodes. Flowers are solitary emerging from the upper axils, actinomorphic, 4-merous with floral tube. The vines are woody with hooks. Petiole is not glanduliferous and lamina is simple and unlobed having pinnate venation and small stipules. Flowers are in cymose inflorescences with terminal flowers on main branches replaced by a hook having small bracts and narrow floral cup. Flowers are solitary, actinomorphic, 4-merous with short floral tube, perigynous to partly epigynous, 4 sepals, 2-4 persistent as hardened spiny horns of the fruit, 4 white petals and 4 stamens. Ovary is inferior, bilocular having 1 pendulous ovule per locule, surrounded at the base by coronary disc. The fruit is 2-4-horned, dehiscent, top-shaped drupe with one seed. The genus may have one to 45 or more species including many infraspecific taxa differing parimarily in the number of horns and shape and size of the fruit, occurring in eastern North America. *Trapa natans* L. is a well known water chestnut (Kubitzki, 2007).

Water chestnut

(***Trapa natans*** L. syn. ***T. bicornis;*** singhara; water caltrops; paniphal)

It has wide distribution throughout Europe, Asia and North Africa. It is cultivated in Sri Lanka, India, Africa, South East Asia and USA. The var. *natans* is naturalized in North America and Australia and var. *bispinosa* is widely scattered from India to China and Japan and in tropical Africa. The var. *africana* is endemic to Lake Victoria (Brenan, 1963).

It is an aquatic annual herb having dimorphic, submerged, opposite and pinnatipartite leaves with filiform segments. Leaves appear as rosette, rhomboidal, 5 x 8 cm in size with lower surface reddish purple to green and the upper surface green, often variegated with long swollen petioles. Its root anchors in the mud, the stem remains submerged and the leaves float on the water surface. The flowers are axillary, solitary, white and open above the surface of water in the afternoon. After pollination, the pedicels bend down submerging the flowers. The fruit is bony, one celled, nearly 3 cm long and broad, indehiscent and 1-seeded drupe with a short cylindrical beak at the top through which the radicle is protruded. The fruit has four spines of which two develop and the others are often obsolete. The seeds are white and starchy.

The cultivars can be divided into 3 groups according to their maturity period (Singh and Hoda, 1996; Anon., 1976):

a. Early - Bari Gajree, Singree and Barkatia.

b. Mid season - Bara Chikanna, Kanpuria Lal, Murwa and Surkha.

c. Late - Gulaba, Pesilal, Saheranpursa and Kadema.

In Bihar, four important types are cultivated. These are Kanpur, Patna, Deshi Big and Deshi Small and differ in shape and colour of leaf and fruit. The fruits of Kanpuri and Patna are small and are eaten after boiling. The Basmati, Kangar and Dogru types are cultivated in Kashmir. In Agra (Uttar Pradesh), *buriya-ke-tal-ke-singare* is a very popular type. Kota Sudhar has soft, thin-skinned and sweeter nuts. It has been selected from the local heterogeneous population of Nagza cultivar in Kota (Rajasthan). It bears fruits 20 days earlier giving 6200 kg/ha yield compared to 4800 kg/ha in the local Nagza. The nuts of Kota Sudhar become available in first week of September and can be eaten green whereas those of the local type can be consumed only after cooking.

Family Myrtaceae (myrtle family)

The family contains 140 genera and about 5800 species, mainly of the tropics of Australia and America (Heywood *et al.,* 2007). This is one of the largest and distinctive families and includes the common fruits like guava, feijoa, rose apple, pimenta or allspice, cloves and some of our important ornamentals like myrtle, *Eucalyptus*, bottle brush (*Callistemon* and *Melaleuca)*, *Eugenia hookeri* and tea tree (*Leptospermum laevigatum*). The family has 2 subfamilies Psiloxyloideae with 2 tribes, Psiloxyleae and Heteropyxideae each with one genus, and Myrtoideae with 15 tribes and having the genera of economic importance.

The plants are aromatic trees and shrubs with flaky bark and contain terpenes. Hairs are unicellular or bicellular. The leaves are simple, usually opposite, evergreen, thick and mostly entire having short petioles, pinnate venation, more or less glandular punctate with scattered pellucid dots (spherical secretory cavities containing terpenoids and other aromatic, spicy resinous compounds) with or without stipules. The flowers are showy, regular, bisexual and polygamous by abortion, borne solitary in axils or in corymbs or racemes. Receptacle or calyx tube is more or less adnate to the ovary and is often elongated. Sepals are mostly 4-5, greatly reduced, usually free and persistent on the fruit. The petals are imbricate or connivant (coming in contact), rarely lacking. The stamens are numerous and conspicuous, often in fascicles opposite the petals, inserted at the margin of the disc. Anthers are small, opening by slits or apical pores. The ovary is inferior having one to many cells each with one to many ovules, mostly on axile placentae. Nectariferous tissue is atop ovary or lining the inner surface of hypanthium. The fruit is a berry, drupe, capsule or nut. Seeds are with no or very little endosperm. Embryo is with large or small cotyledons.

Genus *Psidium* L. (2n = 22, 44, 55, 77, 88)

It is a genus of 70 or more species, growing from Mexico and the Caribbean to northern Argentina. *P. guajava* L., the guava, is widely cultivated and naturalized. The plants are tropical and subtropical. It is native of tropical America.

The plants are small trees and shrubs. Branchlets are glabrous or pubescent with simple hairs. Inflorescence is a monad, triad or dichasium (rarely a bracteate shoot). The flowers are large, whitish and axillary, usually 5-merous having calyx not divided into lobes and persistent, about 8 petals, single style, usually capitate stigma, ±inferior ovary having 2-7 locules and few to numerous ovules in each locule on an axile and peltate placenta. The fruit is ovoid or globose berry crowned by the calyx lobes, remnants of the calyx or by a circular scar. Seeds are few to numerous having bony seed coat with a dull or rough surface, C-shaped embryo and very small cotyledons.

Guava

(***Psidium guajava*** L. syn. ***Psidium cujavillis*** Burm.f.) (2n=22, 33)

The plant is shrub or small evergreen tree, generally 3-10 m high having many branches, crooked stems and light to reddish brown, thin, smooth and continuously flaking bark. Leaves are opposite, simple with no stipules, 3-10 mm long short petiole with oblong elliptic, 5-15 x 4-6 cm blade having obtuse to bluntly acuminate apex, rounded to subcuneate base, entire margins, somewhat thick and leathery, dull gray to yellow green above, slightly downy below, prominent veins dotted with glands. Young shoots are quadrangular. In the subtropics, the flower emerges from the new growth produced twice a year. Inflorescence is axillary, 1 to 3-flowered with about 2 cm long pedicels and 2 linear bracts. Calyx splits irregularly into 2-4 lobes that are whitish and sparsely hairy within. The petals are 4-5, white, linear ovate, 2 cm long and delicate. The stamens are numerous with pale white, about 12 mm long, erect or spreading filaments and straw coloured anther. The ovary is inferior having numerous ovules with about 10 cm long style and green, capitate stigma. Fruit is an ovoid or pear-shaped berry, 4-12 cm long, weighing up to 500 g having yellow skin when ripe, sometimes flushed with red, juicy, creamy white or creamy yellow to pink or red pulp. The exterior of the fruit is fleshy, and the centre consists of a seedy pulp. The mesocarp is thick and edible. The soft pulp envelops numerous, cream to brown, kidney-shaped or flattened seeds. There are two botanical varieties; *P. guajava* var. *pyriferum* and *P. guajava* var. *pomiferum*.

Cultivars

Cultivars have been selected from the great variability of the seed propagated progeny and have been classified on the following bases :

i. Fruit shape- Pomiferum (apple-shaped), Pyriferum (pear shaped), Round and Karela.

ii. Colour of fruit skin- Red (Apple guava), Yellow (Ordinary) Lalit and Chittidar.

iii. Colour of flesh- Red-fleshed, Pink-fleshed, Yellow-fleshed and White-fleshed.

iv. Location of origin- Allahabad Safeda, Nasik, Dharwar and Dholka.

Cultivars of Puerto Rico

Some clones have been identified and evaluated for processing characters and the others have been tested and rated for resistance to *Glomerella* disease. Few named cultivars are Corozal Mixta, Corriente and Seedling 57-6-79.

Cayenne: The white-fleshed type was identified in Trinidad.

Elizabeth: It has large, round, very acidic and pink flesh fruits good for processing.

Red x Supreme x Ruby: It has large, ovoid fruit with deep pink flesh, suitable for eating fresh.

Large White: It has large, round, white and astringent flesh fruits with low sugar content. It can be used as filler in preserves.

Acid Speer: It has large, round fruit with pale yellow acid flesh. It is recommended as a source of pectin.

Red x Supreme x Ruby x White: It has large to very large, pear-shaped fruit with creamy white flesh. It is good for eating fresh and for juice and nectar.

Pink Indian: It has medium size, red-fleshed fruit having agreeable acidity. It is good for eating fresh and for processing.

Red Hybrid: It has medium, subovoid and red-fleshed fruit of medium quality.

Supreme x Ruby: It has medium, subovoid, white-fleshed fruit and has high productivity.

Stone: It has small, ovoid fruit with deep pink flesh having attractive and agreeable flavour for eating fresh.

Supreme: It has small, ovoid fruit with pale yellow, pink-tinged and sweet flesh, good for sherbet and paste. It is very productive.

Patricia: It has very small, ovoid, salmon-fleshed attractive fruit good to eat fresh but quickly loses its distinct strawberry flavour. It is good for syrup. It is very productive.

Malaysian cultivars

Kampuchea: This seeded and high yielding clone is extensively grown. It begins fruiting about 6-9 months after planting giving 60 tonnes/ha fruit yield during the peak production period. There are 2-3 flowering and fruiting peaks in a year. The tree is medium in height with flat, ovate and large leaves of 12 x 6.5 cm size with smooth leaf margin. The fruit is roundish, pear-shaped weighing about 400-600 g, sometimes as high as one kilogram. The skin is light yellow, smooth with inconspicuous dent at fruit stalk. The flesh is white, thick, crunchy and sweet with 6.5-7°Brix TSS and 55 mg ascorbic acid per 100 g of edible flesh. The fruit contains few seeds and is excellent for dessert and juice making and even for canning. This cultivar is resistant to root nematodes, anthracnose and *Phomopsis psidii* (end ring rot).

Taiwan: The fruit is large and pear-shaped. There is a deep recess where the fruit is attached to the stalk. Each fruit weighs about 300-500 g. The flesh is white, thicker than in the Kampuchea guava having sweet-sour taste, crunchy and enclosed by a rough peel. The fruit contains very few seeds and has good keeping quality. The tree is resistant to anthracnose, end ring rot and root nematodes.

Glom Toon Klau: The tree begins to bear 10-12 months after planting. The yield is medium to high of about 40-50 tonnes per hectare during peak periods (2-3 times/year). The fruit is roundish tapering slightly towards the stalk, rough skinned, slightly dented where the fruit is joined to the stalk. The fruit weighs 400-500 g and is white-fleshed, sweet, crunchy having large core, few seeds and TSS of 8-10°Brix. The shelf life of fruit is short (2-4 days).

Glom Sali: The tree is spreading and short to medium in height. The yield is about the same as Kampuchea guava. The green leaves are oblong with 15-18 pairs of lateral veins which are prominent on the lower surface. Leaf margins are smooth and slightly wavy. The fruit is roundish (globose) with a rough skin, light green when firm ripe and weighs about 200-400 g. The fruit has characteristic prominent ridges at the stalk end that is often grooved. The calyx falls off the ripe fruit. The flesh is thick, white, crunchy, and sweet and has a pleasant flavour and distinct aroma. The seed cavity is small and contains few to moderate number of seeds. The TSS is 6-7°Brix. The cultivar is susceptible to anthracnose.

Thai Seedless: This seedless cultivar has low yield of around 20-25 tons/ha/year because flower and fruit drop can be more than 50%. It fruits 14-18 months after planting and has one peak fruiting in a year. The fruit weighs 150-300 g. The fruit is round but asymmetrical and has a hollow cavity that can occupy one third the length of the fruit. It has a smooth, green and thin skin. The flesh is white, crunchy and sweet.

Bangkok Apple: This seedless triploid clone bears fruits about 10 months after planting. It suffers from more than 50% fruit abortion and yields around 15-20 tons fruit/ha/year. The tree is short to medium in height and has large, dark green leaves with a simple but wavy leaf margin. The leaves are ovate and have 25-27 lateral veins that are distinct on the lower surface. The fruit is apple-shaped, round (globose) or slightly flattened at the ends (oblate), often ridged, misshapen, and asymmetrical. The fruit weighs around 100-250 g. When firm ripe the skin colour turns light green and the flesh is white with a very faint tinge of pink. It is crunchy with a hollow carpellary cavity instead of a seed core. The TSS is 10-11°Brix. The cultivar is susceptible to anthracnose and root knot nematode. It bears excellent dessert fruits.

MAHA 65 (GU5): It has a medium size tree and a longer period of fruiting. The fruit is large and weighs 220-250 g. It is round to oblong, has a smooth skin with pinkish red, soft flesh and lot of seeds. The fruit is sweet-acid and has a grainy texture because of the presence of stone cells. The fruit is suitable for processing.

Beaumont: This clone is from Hawaii. It was selected from a seedling population derived from fruits found in Halemanu, Oahu, Hawaii. The tree is vigorous and spreading and yields 24 tons fruit/ha/year. The fruit is medium to large, oblong to roundish in shape, mildly acidic in flavour and weighs 230-250 g. The TSS is 9.9°Brix. The flesh is pink, mildly acidic and seedy. The pulp in the locule has a tendency to be loose causing puffiness. The fruit appears to be susceptible to fruit rot. The fruit is suitable for processing with a recovery of 77-79% puree.

Burma Red: This is a processing cultivar believed to be from Myanmar but is not widely cultivated. It has a medium sized spreading tree that flowers and fruits profusely. It has a short ripening period of 2-3 weeks and has round fruit that weighs 150-200 g having smooth skin. The fruit has pink flesh that is sweet-acid and moderately seedy. Its trees begin fruiting 10-12 months after planting. The maximum yield attained at 3-4 years is 30-35 tons per hectare with two peak seasons in a year.

In Colombia, the cultivars grown are Puerto Rico, Rojo Africano and Agrio. All these cultivars yield over 2200 fruits annually. Other high yielding cultivars are White, Red, D-13, D-14 and Trujillo 2.

Hawaiian cultivars

A large number of cultivars imported from other countries have been evaluated at the Waimanalo Experimental Farm. Four sweet, white-fleshed, thick-walled cultivars were rated as commercially desirable. These are Indonesian White, Indonesian Seedless, Lucknow 49 and No. 6363 (a Ruby x Supreme hybrid from Florida). The other promising ones are Apple (too musky and seedy), Allahabad Safeda (too bumpy on surface), Burma (too seedy) and Hong Kong White (too seedy). Of the sweet, pink-fleshed, thick-walled cultivars, Hong Kong Pink was preferred with second choice to No. 6362 (a seedling of a 'Ruby' x 'Supreme' cross) whereas No. 7199, a seedling of a 'Stone Acid' x 'Ruby' cross was considered too musky in Florida.

Among the acid, non-musky, thick-walled guavas, Beaumont, a Hawaiian selection, is large and pink-fleshed. Pink Acid (#7198), from a cross of Speer and Stone Acid, has dark pink flesh and few seeds.

Ka Hua Kula: It was selected from seedlings of Beaumont and released and recommended for guava puree. The fruit is large with thick, deep pink flesh and fewer seeds than in Beaumont and is less acidic and heavier bearer.

Egyption cultivars

Bassateen El Sabahia: It has long been the standard commercial guava cultivar in Egypt.

Bassateen Edfina: This produces pear-shaped fruit of medium size, sometimes pink-blushed, with thick, white flesh, few seeds, good flavour and higher ascorbic acid content than in the parent Bassateen El Sabahia. It bears over a long season.

Indian cultivars

Allahabad: It has large, white-fleshed fruits with few, medium-sized, fairly hard seeds.

Allahabad Safeda: The cultivar resembles L-49 and is most popular commercially grown cultivar in North India. The trees are vigorous, medium in height having profuse branching, long shoots, compact and rounded crown and dense foliage. Crown is broad and compact, often dome-shaped and rarely loose. The leaves are light green, long and elliptic to oblong. The fruits are medium to big in size, round and smooth having yellow surface. The fruit has bumpy surface, 173 g weight, 7.5 cm diameter, 5.0 cm cavity diameter, 3.2% seeds, 12.5°B TSS and 0.6% acidity (Nakasone *et al.*, 1967). The flesh is white, soft and sweet in taste with good flavour, less seeds and good keeping quality.

Behat Coconut: It has large fruits with thick white flesh, few seeds, poor for canning.

Lucknow 42: The fruit is of medium size, roundish, with creamy white, soft flesh, sweet, pleasant flavour, very few seeds and good quality and keeps fairly well but is not suitable for canning. The tree bears heavily.

Sardar (Lucknow-49): It is a selection made at Pune. The tree is semi-dwarf with flat crown. The leaves are elliptic ovate to oblong in shape. Fruit is large, very sweet and tasty, roundish ovate in shape with primrose yellow surface and excellent keeping quality. Pulp is cream white having high TSS and vitamin C and high pectin good for making jelly. The halves are good for canning. The fruit is medium in size, thick-fleshed and has few seeds, 227 g weight, 7.3 cm diameter, 4.4 cm cavity diameter, 3.3% seeds, 11°B TSS and 0.4% acidity (Nakasone *et al.*, 1967). It resembles Allahabad Safeda but the fruit skin is not as smooth.

Banarsi: It is one of the few very sweet guava cultivars commercially grown in India. The tree is of medium size and somewhat tall having spreading branches and broad crown. The leaves are light green, elliptic oblong with short pointed apex and rounded obtuse base. The fruits are roundish and yellow in colour having white and very sweet flesh. Keeping quality is not very good.

Chittidar: The tree characters resemble those of Allahabad Safeda. The tree is tall growing with spreading branches and large and rounded crown. Leaves are light green, pubescent underneath, elliptic ovate to oblong in form with obtuse leaf base and shortly pointed apex. Fruit is medium to large, subglobose having light straw yellow colour with conspicuous scattered red dots on the smooth skin surface. Keeping quality is good. The flesh is white, sweet in taste having acid-sweet flavour and small and soft seeds. It has higher TSS than Allahabad Safeda and Lucknow 49 but lower vitamin C content. The fruit keeps well and is good for canning.

Allahabad Surkha: It is a seedling selection having sweet and pink fruits with deep pink flesh and strong flavour and few seeds. Trees are vigorous, dome-shaped and compact. The plant produces up to 120 kg fruits in the sixth year of fruiting. Fruits are sweet and strongly flavoured and have few seeds.

Apple Colour: The tree is of medium size with broad crown and spreading growth and is heavy bearer. The fruit is medium in size, spherical in shape, pink in colour with deep minute dots on surface. The flesh is very sweet having good keeping quality, good for canning. The flesh has 0.34-2.12% acid and 9 to 11.36% sugar.

Apple: The cultivar has sweet, seedy, musky fruit with 136 g weight, 6.35 cm diameter, 4.6 cm cavity diameter, 6.3% seeds and white flesh having 8.8°B TSS and 0.4% acidity (Nakasone *et al.*, 1967).

Red Fleshed: The tree is tall and vigorous with spreading branches and open crown. The leaves are elliptic oblong, light green and pubescent below. Fruits are roundish ovate, saffron yellow in colour having few conspicuous scattered red dots on the smooth and glossy surface. The flesh colour is dark pink. Quality wise it is inferior to yellow green guavas. Keeping quality of the fruits is poor.

Red Lady (Taiwan 786): This is a gynodioecious cultivar having blood red colour flesh with good taste

Other red fleshed cultivars are Paluma, Ankapalle, Super Acid, Hybrid Red Supreme, Africa Florida, Florida Seedling, Kothrud, Patillo, China, Surkha and Portugal. Most of these cultivars have fruits lower in vitamin C content and poorer in quality to those of the white-fleshed types.

Seedless: It is a triploid. The tree is tall with long trunk and upright branches. The leaves are light green, elliptic oblong and slightly pubescent below. Fruits are globose or short oblong and small sized having straw yellow, smooth and shiny skin surface, small core and thick, sweet and creamy white flesh and few seeds. The fruits withstand long distance transportation and the cultivar is known for its excellent keeping quality. The fruit bearing poor.

Safed Jam (Allahabad Safeda x Kohir): Trees are medium vigorous. Fruits are round, large and yellowish green having thin peel, sweet taste and few soft seeds located at the core. Fruits have good keeping quality.

Kohir Safeda (Kohir Selection x Allahabad Safeda): Trees are large. Fruits are white-fleshed, large and have few soft seeds. The yield and fruit quality are better than those of the parents.

Arka Mridula: It is a selection from open pollinated seedlings of Allahabad Safeda. The tree is semi-vigorous and heavy yielder. Fruits are medium sized (180-200 g), round having yellow and smooth skin, white flesh and soft seeds. Pulp is white and sweet with good pectin content and 12°B TSS. The fruits are medium in size, have good keeping quality and are suitable for jelly making.

Arka Amulya (Allahabad Safeda x Seedless): The trees are semi-vigorous, spreading and heavy yielding. The fruits are round in shape, have smooth skin, green to yellow colour, few soft seeds, white pulp and 12°B TSS.

Arka Kiran (Kamsari x Purple Local): Fruits are yellow, medium in size (180-200 g) having soft seeds, deep pink pulp with high lycopene content and 12°B TSS.

Hisar Safeda (Allahabad Safeda x Seedless): It is a hybrid developed at Hisar. The tree growth is upright with compact crown and oblong leaves. Fruits are round in shape having creamy white pulp with few soft seeds having 13.4°B TSS and 185 mg/100 g ascorbic acid.

Hisar Surkha (Apple Colour x Banarasi Surkha): The tree is medium in height with broad to compact crown and oblong leaves. The fruits are medium in size, round in shape having yellowish green surface. The pulp is pink having 13.6°B TSS, 0.48% acidity and 169 mg/100 g^{-} ascorbic acid.

Pant Prabhat: Plants are upright with broad leaves. The fruits are round having smooth and light yellow skin. Fruit size is 150-175 g. The seeds are smaller and softer than those of Sardar. The flesh has 10-13°B TSS. Fruit yield is 100-125 kg per tree.

Lalit: It is a selection from half sib population of Apple Colour. It has wide adaptability in different agro-ecological zones of India. Fruit weighs 185-200 g and is saffron yellow with red blush, firm, and pink flesh having good blend of sugar and acid. The flesh has 15°B TSS and 250 mg/100 g vitamin C. It is suitable for both table purpose and processing. The pink colour of beverage remains stable in storage for more than one year. The fruit yield is 100 kg/plant at the age of 6 years.

Shweta: It is a selection from half sib population of Apple Colour. Trees are semi-vigorous with medium height and prolific bearer. Fruits are globose, medium sized weighing 225 g with creamy white exocarp having attractive red spots or blush, snow white flesh, 12.5-13.2°B TSS and 300 mg/100 g vitamin C. Fruits have few soft seeds and good keeping quality. The fruit yield is 90 kg/plant at the age of 6 years. It is a high yielder.

CISH-G-1: It is an open pollinated seedling selection from Apple Colour. The fruits have deep red peel rich in anthocyanin pigment and long shelf life.

CISH-GS-35: It is an open pollinated seedling selection from Allahabad Safeda. It is heavy bearing having round and medium to large (200-250g) and white fruits, sweet pulp (13.4°B TSS) and few seeds.

MPUAT S-1: It is a seedling selection having fruits with pink peel on yellow background.

MPUAT S-2: It is a seedling selection.

Hybrid 1 (Sardar x Allahabad Safeda): It is high yielder having fruits with soft seeds.

Hybrid 2 (Chittidar x Allahabad Safeda): It is high yielder having fruits with soft seeds.

TRY (G) 1: Plants are high yielding (41 kg/tree), tolerant to drought and soil sodicity and have off-season bearing. The fruits are shiny greenish yellow having desirable aroma and 10°B TSS.

SRD Hybrid-1 (Safed Jam x Red Fleshed): Trees are semi-vigorous and heavy yielder bearing medium to large (200-230 g) fruits having red and sweet pulp (11-12°B TSS).

SRD Hybrid-4 (Safed Jam x Red Fleshed): Trees are semi-vigorous bearing ovate and medium large (160-250 g) fruits having greenish white pulp.

Pusa Srijan (Seedless (triploid) x Allahabad Safeda (diploid)): A tetrasomic (2n+2). It is suitable as a dwarfing rootstock.

Habshi: The fruit is of medium size with thick, white flesh, few seeds giving halves good for canning.

Hafsi: The fruits are round, smooth and are red fleshed. The fruits are good in flavour but are not sweet.

Karela (Karelia): The fruit is medium large, pear-shaped, furrowed, rough-skinned, with soft, granular white flesh and sweet, rich and pleasant flavour. It is poor bearer and not popular.

Dholka: It is from Maharashtra, India.

Dhareedar: Trees are vigorous with erect and upright branching. Fruits are medium to large in size, roundish ovate in shape, greenish in colour and soft-seeded with 5-7 raised lines on the surface of mature fruits.

Nagpur Seedless: The fruit is small to medium, often irregular in shape and white-fleshed.

Safeda: The fruit is of medium size with very thin skin, thick white flesh, few seeds and outstanding quality for canning. It is susceptible to wilt and branches are brittle and break readily.

Smooth Green: The fruits are of medium size with thick white flesh and few small and hard seeds. Halves are firm and good for canning.

Anakapalle: It has small fruits with thin, red flesh and many seeds. It is not suitable for canning.

Florida Seedling: It has small fruits with thin, red, acid flesh and many seeds. It is not suitable for canning.

Hapi: It has medium to large fruits with red flesh.

Hybrid Red Supreme: It has large fruits with thin, red, acid flesh and moderate amount of seeds. It is not suitable for canning.

Kothrud: It has fruits of medium size with medium thick, red flesh and moderate amount of seeds. It is not suitable for canning.

The other Indian cultivars are Dharwar, Mirzapuri, Nasik, Sindh, White Supreme x Ruby (white-fleshed), and Banaras, Dholka, Hasijka, Kaffree and Wickramasekara (red-fleshed).

In Hawaii, Campbell (1963) described cultivars Blitch and Patillo and recommended that Blitch is suitable for jelly making while Patillo is good for cooking. The cultivar Rica and Paluma have been selected in Brazil. Rica is an early and high yielding cultivar and its fruits can withstand transportation and are suitable for processing while Paluma is late and high yielding having large fruits having red pulp suitable for both fresh consumption and processing (Pereira, 1984).

Nakasone *et al.* (1967) have described fruit characters of following guava cultivars:

Burma: Origin is Burma. Fruit is sweet and seedy and has white flesh, 210 g weight, 6.4 cm diameter, 4.9 cm cavity diameter, 4.5% seeds, 10.7°B TSS and 1.3% acidity.

Hong Kong Pink: Origin is Hong Kong. Fruit is sweet and thick-fleshed with few seeds and has pink flesh, 218 g weight, 7.6 cm diameter, 4.0 cm cavity diameter, 1.8% seeds, 10.0°B TSS and 0.2% acidity.

Hong Kong White: Origin is Hong Kong. Fruit is sweet and seedy and has white flesh, 181 g weight, 6.7 cm diameter, 4.7 cm cavity diameter, 4.2% seeds, 12.5ºB TSS and 0.4% acidity.

Indonesian White: Origin is Indonesia. Fruit is sweet with few seeds and has white flesh, 105 g weight, 6.3 cm diameter, 4.1 cm cavity diameter, 1.4% seeds, 10.5ºB TSS and 0.3% acidity.

Indonesian Seedless: Origin is Indonesia. Fruit is sweet, thick-fleshed and with no or few seeds and has white flesh, 176 g weight, 7.3 cm diameter, 4.3 cm cavity diameter, 0.7% seeds, 12.5ºB TSS and 0.4% acidity.

Patillo (7197): Origin is Florida. Fruit is highly acidic and has pink flesh, 113 g weight, 5.8 cm diameter, 4.0 cm cavity diameter, 3.3% seeds, 11.9ºB TSS and 1.7% acidity. The cultivar is good for processing.

Pink Acid (7198): Origin is Florida. Fruit is highly acidic and has dark pink flesh, 167 g weight, 6.9 cm diameter, 4.7 cm cavity diameter, 2.7% seeds, 11.7ºB TSS and 1.7% acidity. The cultivar is good for processing.

6362: Origin is Florida. Fruit is sweet with few seeds and has pink flesh, 176 g weight, 6.7 cm diameter, 4.5 cm cavity diameter, 2.0% seeds, 10.5ºB TSS and 0.4% acidity.

6363: Origin is Florida. Fruit is sweet, large-fruited with few seeds and has white flesh, 250 g weight, 7.1 cm diameter, 4.6 cm cavity diameter, 2.1% seeds, 11.4ºB TSS and 0.3% acidity.

7199: Origin is Florida. Fruit is sweet, thin-fleshed with few seeds and slight muskiness and has white flesh, 153 g weight, 6.9 cm diameter, 5.1 cm cavity diameter, 1.9% seeds, 11.3ºB TSS and 0.4% acidity.

Blitch: A seedling which originated in West Palm Beach and was planted at Homestead. Its fruit has strong odour, medium size and oval shape having light pink flesh, numerous small seeds, tart and pleasant flavour. The fruits are good for jelly.

Miami Red and **Miami White:** The fruit is large, nearly odourless and thick-fleshed. These were released by the University of Miami.

Redland: It has very large fruits with little odour, white flesh and relatively few seeds. Because of its excessively mild flavour, low ascorbic acid content and susceptibility to algal spotting, it was abandoned.

Red Indian: Its fruits have strong odour, medium to large size, round but slightly flattened at the base and apex, yellow skin often with pink blush with medium thick, red flesh of sweet flavour, numerous but small seeds, good for eating fresh. Cultivar is fairly productive.

Ruby: The fruits have pungent odour, medium to large size, ovate with thick, red flesh, sweet flavour and relatively few seeds and is excellent for eating fresh and for canning. Cultivar is fairly productive.

Hart: It has fairly large fruits with pale yellow flesh and 8% sugar content.

Rolfs: It has fruits of medium size with pink flesh, of good quality and contains 9% sugar.

Webber (Riverside): It has fruits of medium large size, pale yellowish flesh, good flavour and 9.5% sugar.

Cattley guava

(***Psidium cattleianum*** Sabine syn. ***P. littorale*** Raddi, ***P. littorale*** Raddi var. ***longipes*** (Berg) Fosb.; *P. chinense* Hort.; strawberry guava; Chinese guava; Calcutta guava, yellow cattley guava)

It is native of South America (Martin *et al.,* 1987). It is believed to have originated in the lowlands of eastern Brazil (Morton, 1987). It is cultivated to a limited extent in South and Central America, West Indies, Bermuda, Bahamas and southern California (USA). It is grown occasionally in subtropical Africa, in the highlands of Philippines at elevation of 1500 m and in Sri Lanka and Malaysia (Morton, 1987; Verheij and Coronel, 1991). It can withstand cold.

It is a fairly slow growing shrub or small tree growing up to 2-4 m in height. The yellow fruited form may attain 12 m height. Both the forms have slender, smooth and brown stems and branches. The young shoots are smooth and rounded. The leaves are evergreen, alternate, obovate, dark, smooth, glossy, leathery and 3.4-12 x 1.6-6 cm in size. The fragrant white flowers are 1.5-6 cm wide with prominent 2 cm long stamens and are borne singly or in 3's in the leaf axils. The fruit is round or obovoid, 2.5-4 cm long, tipped with 4-5-parted, thin-skinned, dark red or purple red and protruding calyx. Red skinned fruits have faintly yellowish flesh. The flesh is aromatic and 3 mm thick surrounding the central juicy and translucent pulp filled with hard, flattened, triangular, 2.5 mm long seeds. The flavour is somewhat strawberry-like, spicy and subacid in contrast to the muskiness of the common guava (Morton, 1987; Verheij and Coronel, 1991).

No named cultivars are known but there is considerable variability from which promising types can be selected. Red and yellow forms exist. Types with pubescent foliage are found in tropical America. The botanical variety *lucidum* is a distinct yellow fruited form.

Yellow strawberry guava (*P. cattleianum* var. *lucidum*).

Lemon guava (*Psidium cattleianum* var. *acre*).

Psidium cujavillis Burm: This species originated from the tropical America. The characteristics of growth, flowering, leaf development, fruit and plant cytology are quite similar to that of *P. guineense*. The fruit is small globose, ovoid with white flesh and sour taste.

Costa Rican guava (*Psidium friedrichsthalianum* (Berg) Niedenzu; cas): The species originated from Central America. The tree is small to medium with drooping stem and branches. The bark is smooth, dark brown and shiny. The leaves are obovate, dark green, shiny and covered with tiny hairs. Flowers are small and scented. The fruit is small, globose to oval and yellowish with thin peel having soft and white flesh. Fruits are acidic and are used to make jelly and jam.

Puerto Rican guava (*Psidium microphyllum* Britt., ***P. sartorianum*** Niedenzu): The tree is 15 m high. Fruit is spherical, 1-1.5 cm in diameter, borne solitary on new growth. The external colour of fruit is pale yellow and internal is whitish. Pulp is eaten fresh or in jellies. Flavour is sweetish and has low appeal.

Psidium guineense Sw. syn. ***Psidium araca*** Raddi, ***P. molle*** Bentol; ***P. schiedeanum*** Berg; ***P. policarpum*** Lamb; Brazilian guava: The species originated from the dry highlands of Brazil. It is known as "Brazil guava" or nicknamed as Arace Do Campo. The tree is small and drooping. Leaves are obovate, long and oval in shape, not tapered and pale green in colour. The fruits are ovoid and small in size filled with tiny seeds. Fruits are edible, sour to subacid with pleasant flavour and are processed into jelly.

Psidium montanum Sw.: The species originated from the mountains of Jamaica. The branches are quadrangular and small with multiple flowers per cyme. The oval leaves are without hair, tapering towards the tips. The fruit is globose or almost globose.

Psidium guayabita A. Rich.: The species grows wild in West Cuba bearing small edible fruits which are regarded valueless nutritionally.

Calycolpus goetheanus (Mart. ex DC.) O. Berg syn. ***P. fluviatile*** Miq.: It comes from Brazil and Guyana. The plant is dwarf or small shrub planted only in certain areas for fresh consumption.

Genus *Acca* O. Berg

It is a genus of 3 species, two of Andes (Peru) and one of eastern temperate South America (Kubitzki, 2011). The pineapple guava or feijoa is cultivated for its edible fruits. The plants are trees or shrubs with simple hairs. Inflorescence is unimerous. Flowers are 4-merous having red to pink and open calyx, reddish to pink petals, fleshy when mature, numerous stamens, stiff and red, more or less straight and erect in bud, 3-4-locular ovary, basically axile placenta but sometimes partially parietal and numerous ovules in each locule. Fruit is crowned by the calyx lobes. Seeds are numerous, small, lentil-shaped having hard or subcrustaceous seed coat, spiral embryo, strap-like cotyledons, as long as the hypocotyl.

The feijoa

(***Acca sellowiana*** Burret.; syn. ***Feijoa sellowiana*** O. Berg; pineapple guava)

It is a native of the upland warm temperate area of southern Brazil and adjacent areas of Uruguay, Paraguay and northern Argentina. It is cultivated in Uruguay and mountainous areas of Bahia province of Brazil and on limited scale in California and New Zealand (FAO, 1982; Martin *et al.*, 1987; Morton, 1987). It is heat tolerant, evergreen and subtropical.

The plants are large bushy shrubs or small evergreen trees growing up to 6 m in height. The leaves are small, 3-7 cm long with dark green upper surface and silver gray underside. The flowers are large, purple on the inside and whitish on the outside. The hermaphrodite flowers are bright, showy, attractive and about 30 mm in diameter and are borne solitary in the leaf axils of current season growth. The four petals are fleshy, edible, white outside and purplish within. The numerous stamens are red in colour. The fruit is a round, ovoid or oblong berry surrounded by persistent calyx lobes, 4-7 cm long, having dark green and thin skin with a whitish bloom, cream coloured gelatinous pulp containing 20-30 minute soft seeds (FAO, 1982; Page, 1984; Morton, 1987).

Cultivars

The main cultivars are Triumph (large, oval fruit with smooth skin and heavy yield), Mammoth (large, oval fruit with smooth skin and ripening by mid April in New Zealand), Andre (self fertile), Besson (rich aromatic flavour), Coolidge (self fertile, variable fruit size and shape and heavy cropper), Choiseana (good flavour), Superba (self incompatible), David and Roundjon (Morton, 1987). The cultivars Mammoth and Triumph show some self incompatibility and should be interplanted to ensure cross pollination (Page, 1984). The cultivars Coolidge, Superba, Large Oval and Chapman have fruits of inferior quality. The cultivars Apollo and Gemini are the selections made in New Zealand besides the other selections B55, 9-37, J3, T27, T31, Unique, Candy and K85.

Genus *Syzygium* Gaertn. (2n = 22, 44, 66, 110)

It has been separated from the genus *Eugenia* since it has two distinct cotyledons. The seed coat is rough and is adhering the pericarp. It contains species originated from the Old World. It contains over 500 species that occur in tropical and subtropical regions of the Old World (Kubitzki, 2011). The plants are trees or rarely shrubs and are usually glabrous. Inflorescence is terminal or axillary (occasionally ramiflorous or cauliflorous), usually a panicle. Flowers are 4(5)-merous having occasionally calyptrate calyx, distinct petals or occasionally forming calyptras, numerous stamens, dorsifixed and versatile anthers with parallel anther sacs, ±inferior, 2(3)-locular ovary having several ovules on an axile placenta, style about as long as the stamens and small stigma. Fruit is succulent having firm or spongy flesh, usually purple to red and usually persistent calyx. Seed is usually solitary, rarely 2 or more having soft, leathery or papery testa and thick and distinct cotyledons.

Key to the classification of *Syzygium*

A. Petals are united to form a calyptra, chromosomes 2n = 66 (x = 11) ....*S. cumini*

AA. Petals are free

B. Flowers are large, chromosomes 2n = 46 (x = 11)..........................*S. jambos*

BB. Flowers are small.. *S. aromaticum*

Jamun

(***Syzygium cumini*** (L.) Skeels syn. ***S. jambolanum*** DC., ***Eugenia cumini*** Druce., ***E. jambolana*** Lam., ***Myrtus cumini*** L., ***Calyptranthes jambolana*** Willd.; Java plum or jambolan, Portuguese plum, Malabar plum, black plum, Indian black berry, jaman, jambu, jambul and duhat), 2n=66.

It originated in India or in the East Indies. The jamun is fast growing, glabrous, tall and handsome evergreen tree having 15-24 m height. The bark is rough and cracked, flaking on the lower part of tree. The numerous white branches slightly droop at the ends and form a dense head. The leaves are thick, leathery, glossy, opposite, broadly oblong or oval, 5-25 cm long, 2-10 cm wide and have conspicuous yellowish midribs with lateral veins very close to each other. Panicle inflorescences are usually borne on leafless branches, which are solitary or fascicled, dense, and many-flowered. Flowers are greenish white, small, up to 0.8 cm in diameter, fragrant with widely campanulate, 4-6 m long calyx, yellow disk, 4 free, orbicular, grayish white to pink petals, 4-7 mm long and white, stamens, 2-3 celled ovary and 6-7 mm long and white style. The fruit clusters have 10 to 40 round or oblong fruit berries, 1 to 5 cm in length, which usually turn from green to light magenta and then dark purple or nearly black on ripening. The fruit skin is thin, smooth, glossy and adherent. The pulp is purple or white, very juicy and normally encloses a single, oblong, green or brown seed, which is 1-4 cm long. Some fruits may be seedless. The fruit is usually astringent having acid to fairly sweet flavour (FAO, 1982; Morton, 1987; Verheij and Coronel, 1991).

Cultivars

CISH J-42: Fruits are unique, seedless with of 97-98% pulp content and 14-15°B TSS .

CISH J-37: Fruits are bold with high pulp quality (90-92% pulp and 14-16-17°B TSS).

Goma Priyanka: The plants are semi-dwarf with spreading growth, dense foliage and drooping branches. It is early, precocious and suitable for high density planting. Flowering starts in March ripening fruits in the end of May. Average fruit weight is 19.86 g with 85% pulp and 16-80°B TSS.

Konkan Bahadoli: Fruits are bold sized having good keeping quality and high juice content.

Ra Jamun: This is a large fruited type having oblong, sweet and juicy fruits of deep purple colour known as Paras in Gujarat. The fruits have small seeds. It ripens during June-July.

Narendra Jamun: It is a selection. The fruits are seedless.

Dhoopdal: The cultivar from Belgaum has large and sweet fruits.

Water apple

(***Syzygium aqueum*** Burm. f. (Alston) syn. ***Eugenia aqueum*** (Burm. f.) Alston; watery rose apple; bellfruit)

It is native to South East Asia. The tree grows up to 3-10 m in height and develops a short and crooked trunk, which is 30-50 cm in diameter and often branched near the base making an irregular canopy. The leaves are opposite, elliptic cordate to obovate oblong, glossy, dark green, 7-25 x 2.5-16 cm in size having 0.5-1.5 mm long petioles. Inflorescences are terminal as well as axillary and 3-7 flowered. The flower is 2.5-3.5 cm in diameter having 5-7 mm long calyx tube, 4 spathulate petals which are up to 7 mm long and yellow white, numerous, 0.75-2 cm long stamens and 17 mm long styles. The fruit is a turbinate berry, pyriform in shape and 1.5-2 x 2.5-3.5 mm in size crowned by the fleshy calyx segment and is white to red and glossy. The flesh is very juicy and watery having 1-2 rounded and small seeds (Morton, 1987; Verheij and Coronel, 1991).

Java rose apple

(*Syzygium samarangense* (Blume) Merr. and Perry syn. ***Eugenia javanica*** Lam.; wax apple; java apple; wax jambu; samarang rose apple; jamrul) (2n=33, 42, 44, 60)

It is a native of South East Asia and is distributed in the Andaman Islands and is also grown in parts of South India and West Bengal. Java rose apple is widely grown throughout the tropics. These often occur in home gardens with limited commercial production in Malaysia, Australia and Central and South America. Taiwan has a more extensive plantation around Pingtung at the southern end of the island (Nakasone and Paull, 1998).

The related species that produce edible fruits are found growing in tropical Africa (*S. guineense* A.DC., water berry), the Philippines (*S. calubcob* Merr., kalubkob; *S. cordatilimbum* Merr., kara; *S. polycephaloides* Merr., lipote; *S. simile* Merr., malaruhat; and *S. xanthophyllum* Merr., malatampoi), and Indonesia (*S. curranii* Merr., gewok; *S. pycnanthum* Merr. and Perry. syn. *S. densiflora* (Blume) Duthie, kelampok) (Martin *et al*., 1987; Morton, 1987; Verheij and Coronel, 1991).

It can grow in hot and humid climatic conditions and up to 1200 m elevation. It does not tolerate frost and does the best in areas with a fairly long dry season. However, it requires assured water supply. The trees prefer heavy, deep fertile soils with an easy access to water but not light soils having deep water table (Morton, 1987; Nakasone and Paull, 1998).

It is a 5-15 m tall evergreen tree with wide irregular canopy, short and crooked trunk, 25-50 cm in diameter, often branched near the base. The leaves are opposite, elliptic to elliptic oblong, 10-25 x 5-12 cm in size, coriaceous, emit strong aroma when bruised and have thick and 3-5 mm long petiole and pellucid, dotted and thin margin. Inflorescences having 3-30 flowers are borne in the axils of fallen leaves. The white or pinkish flowers are self compatible, 3-4 cm in diameter having 1.5 cm long calyx tube and 3-5 mm long lobes, 4 orbicular to spathulate, 10-15 mm long petals, numerous 3 cm long, yellow white stamens and pistil with 3 cm long style. The fruits of water apple and wax jambu are very similar and cannot be easily distinguished. Fruit is a broadly pyriform berry, 3.5-5.5 x 4.5-5.5 cm in size and bright red to white in colour, crowned by the fleshy calyx with incurved lobes and has a wax like glossy skin. The flesh is white, spongy, juicy, aromatic and sweet sour in taste having 0-4, mostly suppressed, globose seeds which are up to 8 mm in diameter (Morton, 1987; Verheij and Coronel, 1991; Nakasone and Paull, 1998). The fruit is eaten fresh usually with salt or sauce, used in fruit salads (rujak) and preserved as pickle (asinan). Leaves are used for the treatment of ulceration of the tongue in the Philippines and of fevers in Cambodia. The wood is reddish and hard and is used in construction. The tree can be grown in home gardens and also along the driveways (Morton, 1987; Verheij and Coronel, 1991).

Cultivars

Forms having red, pink and white fruits exist in Indonesia and India. In West Bengal, the white form, known as Albes, bears seedless fruits and the pink form contains one or two seeds. Percy-Lancaster and Bose described a hybrid of Alba form of *E. javanica* with rose apple (*E. jambos*), which has prolific fruiting habit of

Alba and fragrance and sweetness of rose apple but the fruits are seeded and larger than of either parents. In Thailand, wax jambu clones have been recognized as cultivars, e.g. green fruited Khiew Savoey (Morton, 1987; Verheij and Coronel, 1991). Taiwan has clones having pink, deep red, light red, white and green with pink fruit skin colour, which are grown on large extent (Nakasone and Paull, 1998). The Institute of Plant Breeding, Los Banos has several clones of wax jambu (Verheij and Coronel, 1991). Some of the cultivars are Green Red (light green with red tinge at distal end and oval), Kong White (white oval), Thai Green (light green oblong oval), Black Diamond (sparkling red oval)

CHESM-1: The tree is 4.5 m in height, upright, branched with pyramidal canopy. The leaves are opposite, elliptic lanceolate, 20- 22 cm long and 8.5-9 cm wide, dark green with short petioles and glossy upper surface. The flowers are white, 2.5 cm in diameter, quadrupetalate with numerous stamens. Flowering takes place in February-March and fruits ripen in May-June. The fruits are borne in clusters of 5-7. The fruit is bell shaped, fleshy, glossy, pinkish red and 40-45 g in weight and is almost seedless and contains a lot of water having 6.7°Brix TSS, 0.14% acidity and 20-25mg/100 g ascorbic acid. The fruit yield is 2400-2500 per tree.

Rose apple

(***Syzygium jambos*** (L.) Alston. syn. ***Eugenia jambos*** L., ***Jambosa vulgaris*** DC; malabar plum; gulab jamun)

It is found in South India. Rose apple is an attractive evergreen tree and grows up to 9 m high. The young leaves are dark green, thick, glossy, oblong lanceolate, 10-20 cm long, acuminate, thick and short-petioled having secondary veins running parallel. The flowers are greenish white, 6-8 cm in size and appear in short terminal corymbs. The calyx has 4, suborbicular sepals of 10 x 7 mm size. Four petals are sub-orbicular, up to 10 x 7 mm in size and white to greenish white in colour. The stamens are numerous, 400 in number, conspicuous and 4 cm long; the style is also up to 4 cm long and the pedicel is up to 1.5 cm long. The fruit is a round or oval berry, 2-5 cm in diameter, greenish to pale yellow and flushed with yellow with a taste like that of apricot and an aroma of rose. The flesh is somewhat dry and spongy. Seeds are gray, 1-2 and lie loose in the cavity of a succulent pericarp and are polyembryonic producing 1-2 plants (FAO, 1982; IBPGR, 1986b; Morton, 1987; Verheij and Coronel, 1991).

Genus *Eugenia* L. (2n=22, 33, 44, 66)

The *Eugenia* differs from *Syzygium* in having cotyledons that are usually united, smooth seed coat free from the pericarp and generally a raceme inflorescence having flowers with pedicels. It contains 550 species growing from Mexico and the Caribbean to northern Argentina. A number of species also occur in the Pacific; 60 in New Caledonia, 11 in Philippines and 120 in Africa and neighbouring islands (Kubitzki, 2011). The plants are trees or shrubs having simple or two-armed hairs. Inflorescence is uniflorous, a bracteate shoot, rarely a dichasium, the bracteate shoots sometimes with an abbreviated axis bearing 2 to many flowers. Flowers are always 4-merous having free sepals, 2-locular ovary, 2 many ovules on pad-like to axile placentas. Fruit is crowned by the calyx lobes. Seeds are 1-2 having membranous or crusty seed coat and embryo with fused cotyledons.

Malay rose apple

(***Eugenia malaccensis*** L. syn. ***E. domestica*** Bailey, ***Jambosa malaccensia*** (L.) DC., ***Syzygium malaccense*** (L.) Merrill and Perry; Malay apple; mountain apple)

It is known as Ohia in Hawaii. It is native of South East Asia and is grown in the tropics and subtropics (IBPGR, 1986b; Verheij and Coronel, 1991). It is an evergreen tree of small to medium size, 15-20 m in height, straight trunk, branched near the base with broadly ovoid and dense crown. Leaves are dark green, opposite, shortly stalked, thick, coriaceous, elliptic oblong or obovate oblong, 15-30 x 7-20 cm in size with entire margin and 0.5-1.5 cm long petioles, which are thick and red when young. Inflorescences are borne exclusively on defoliated twig parts and are short, dense and 1-12-flowered. Flowers appear in clusters of 3-5 on older branches, which are very short-stalked, brilliant red, 5-7 cm in diameter. Calyx is oblong campanulate or turbinate and the calyx tube is 1.5-2 cm long with 4-8 cm long and broad lobes. Four petals are dark wine red, 12-20 mm long and oblong ovate or orbicular ovate. Stamens are numerous, up to 2 cm long and with red filaments. The style is 3-4.5 cm long and red. The ovary is oblong and red. Fruit is ellipsoid globose or pear-shaped 1-seeded berry, usually 5-8 cm long and 5-6 cm in diameter crowned by fleshy calyx segments which on ripening become dark red, pink or pale yellow with purple streaks, rarely uniformly yellowish white. The flesh is 0.5-2.5 cm thick, juicy, white with fragrant savoury. Seed is globose, brown, 2.5-3.5 cm in diameter and polyembryonic (FAO, 1982; IBPGR, 1986b; Morton, 1987; Verheij and Cornonel, 1991).

The types identified from the rich variability are red fruited, light red fruited, dark red fruited having longitudinal stripes, white oblong to pear-shaped fruits and white fruits with 90% seedlessness. Kingston Pride has large fruit, small seeds and superior flavour.

Pitomba

(***Eugenia luschnathiana*** Klotzsch ex Berg syn. ***Phyllocalyx luschnathianus*** Berg.; Uvalha do Campo)

It is cultivated to a limited extent in Bahia State of Brazil. It is a very rare fruit outside Brazil (Martin *et al.*, 1987; Morton, 1987). Pitomba is a good home garden fruit (Martin *et al.*, 1987). The tree is attractive evergreen and slow growing, of 10 m height. Leaves are opposite, short petioled, oblong lanceolate, 2.5-7.5 cm long, glossy and dark green on the upper surface and paler beneath. Solitary flowers are long stalked (2-3 cm pedicel), 4-petalled, white and borne in the leaf axils. Fruit is broad ovate, faintly 4-lobed, 2.5-3.2 cm long, bright orange yellow with 4 or 5 green sepals, 1.25 cm long, protruding from the apex. The skin is thin, tender and the pulp is golden yellow, apricot-like texture, soft, melting, juicy, aromatic and slightly acidic and with faintly resinous flavour. There may be one round seed or 2 to 4 irregular angular seeds of light tan colour and 1-1.6 cm diameter (Morton, 1987). Lot of variability occurs in nature with respect to fruit size (Morton, 1987).

Araca-boi

(***Eugenia stipitata*** McVaugh; pichi; araza; araza-buey)

It is native of western Amazonia and is found growing wild in Peruvian Amazon, Ecuador, Bolivia and Colombia (FAO, 1986). It is cultivated in western Amazonia especially around Iquitos, Peru (Leaky and Newton, 1989). Araca-boi is a shrub of 3-5 m height. Leaves are simple, opposite, elliptical oval in shape and 8-20 x 3-10 cm in size. Inflorescence is racemose, 4-10 mm in width bearing 3-8 flowers. The flower has 4 white petals, 75-150 stamens and one pistil. The ovary is multi-locular. The fruit is globose depressed, 50-100 g in weight having 5-10 cm longitudinal diameter, yellowish colour and velvety rind on ripening. The pulp is aromatic, light yellowish in colour, succulent and slightly fibrous. Each fruit contains around 6 to 12 seeds, oblong in shape and 2.5 cm in length.

There are two subspecies of *E. stipitata:*

i. *E. stipitata* subsp. *stipitata* McVaugh is called pichi in Peru and araca-boi in Brazil. Its tree grows up to 12-15 m tall with short-petioled, opposite, broad elliptic leaves, acuminate at the apex, rounded at the base, 7.5 - 18 x 3.4 - 8.25 cm in size with entire margin and 6-10 pairs of impressed lateral veins on the upper surface, densely hairy on the underside and faintly dotted with oil glands on both sides. Inflorescence is racemose, axillary with 4-10 mm long axis bearing 2-5 pairs of flowers, which are white, hairy, 2 cm wide with numerous prominent stamens. Terminal flower is absent. Pedicels are 1-2 mm long, calyx lobes are rounded, broader than long overlapping in the bud. Five petals are white, obovate and 7-10 x 4 mm in size. Ovary is 4-locular, each locule having 5-8 ovules. Style is 5-8 mm long. Fruit is an oblate or spherical berry like a small guava. It is very aromatic. In wild trees, the fruit is 1.5 cm wide and less than 56 g in weight. In cultivated types, the fruit is 12 cm wide weighing 420 to 800 g with thin and delicate skin, soft, juicy and very acid pulp, containing 8 to 10 irregular oblong or kidney shaped seeds which are 12.5 cm long and 1.5 cm wide. The fruit is light green at first but turns pale or orange yellow on ripening with a thin velvety skin enclosing a juicy and thick pulp which accounts for 60 per cent of the fresh fruit (FAO, 1986).

ii. *Eugenia stipitata* subsp. *sororia* is called rupina caspi in Peru. It is a shrub or a small tree of 3 m height with elliptic leaves, 9-12.5 cm x 2.5-4.5 cm size, barely visible veins, minutely hairy or hairless beneath when fully mature having a few dark dots. The flowers are 1.25 cm wide with 75 stamens (FAO, 1986; Morton, 1987).

Grumichama

(***Eugenia dombeyi*** (Sprengel) Skeels syn. ***E. brasiliensis*** Lam., ***Myrtus dombeyi*** Spre.; Brazil cherry)

It is a native of Brazil. The tree is small, evergreen and grows up to 7-10 m tall with narrow and compact crown and short trunk having gray bark and suberect branches. The leaves are opposite, ovate to obovate elliptic, 10-12 x 5-6 cm in size, leathery, glossy with recurved margin and 3-4 mm long petioles. Flowers are solitary, axillary, 2.5 cm wide, white and slightly fragrant and having 4-8 mm long and green sepals and 4-15 mm long white petals and 100 white stamens. Fruit is a subglobose to oblate berry hanging on long and slender pedicel, up to 3-5 cm in diameter with persistent sepals at the apex, bright red, dark purple to nearly black or white in colour with thin and delicate skin and red or white, juicy, soft and sweet pulp. Seeds are one to several, globose to angular, about 1 cm in diameter, hard and light brown to gray green in colour (Morton, 1987; Verheij and Coronel, 1991).

Three botanical varieties have been distinguished based on fruit colour. *E. dombeyi* var. *iocarpus* Berg. has deep violet fruits, *E. dombeyi* var. *erythrocarpus* Berg. has red fruits and *E. dombeyi* var. *leucocarpus* Berg. has white fruits (Morton, 1987; Verheij and Coronel, 1991).

Surinam cherry

(***Eugenia uniflora*** L. syn. ***E. michelli*** Lam.; Brazil cherry; pitanga cherry)

It is native of Brazil. It is evergreen shrub or tree growing up to 7 m tall with spreading branches. The leaves are opposite, ovate to lanceolate, 2.5-6 x 1.5 cm in size with rounded or slightly cordate base, obtuse to shortly acuminate apex, glabrous, glossy, bronze when young and dark green when mature and turn red in cold dry weather. The flowers are 1 cm in diameter, creamy white, fragrant and are borne 1-4 together in leaf axils having a 4-lobed tubular calyx, 7-11 mm long white petals and 50-60 stamens. The fruit is a pendulous berry, depressed globose in outline, 1-4 cm in diameter, 7-8 ribbed, greenish when young and turn orange and then bright red to blackish when mature. The skin is thin and the flesh is orange red, juicy and acidic to sweet and slightly resinous having a single large round seed. The fruit may occasionally contain 2-3 small and flattened seeds (IBPGR, 1986b; Morton, 1987; Brücher, 1989; Verheij and Coronel, 1991).

Cultivars

Two distinct forms of Surinam cherry are recognized, one with bright cherry red fruits and red tinged leaves and the other with deep crimson to almost black fruits and black tinted leaves. Seedlings have variable fruit size and quality. The black-fruited type is rare and tends to be sweeter and less resinous (Morton, 1987; Verheij and Coronel, 1991).

Genus *Myrciaria* O. Berg (2n = 22)

It has 22 species growing from Mexico and the Caribbean to northern Argentina. The plants are trees or shrubs having simple hairs. Bracteate inflorescence shoot is often reduced to a glomerule. Flowers are 4-merous, the hypanthium extending well beyond the ovary, circumscissile at the summit of the ovary and mostly falling with the calyx as a unit after anthesis, having open calyx, 2-locular ovary with usually 2 ovules per locule. Fruit is crowned by circular scar. Seeds are 1-2 having membranous seed coat and embryo as solid mass or with 2 free, piano-convex cotyledons (Kubitzki, 2011).

The important species having edible fruits are *M. dubia* (Kunth.) McVaugh (camu camu), *Myrciaria cauliflora* (DC) Berg. (jaboticaba), *M. edulis* (Vell.) Skeels.,

M. floribunda (West ex Willd.) Berg. (guava berry), *M. paraensis* Berg., *M. vexator* Mc Vaugh (blue grape), and *M. trunciflora* O. Berg., which are found growing in South America (Page, 1984; FAO, 1986; Morton, 1987; Brücher, 1989).

Camu-camu
(*Myrciaria dubia* McVaugh syn. *M. paraensis* Berg, *M. spruceans* Berg)

It is native of western Amazon basin and grows wild in swamps along the rivers and lakes, especially in the Amazon River basin and upper Orinoco basin especially in Rio Mazan near Iquitos, Peru, Colombia, Brazil and Venezuela (FAO, 1986; Morton, 1987).

It is a shrub or small tree growing up to 8-13 m high with thin, pale to bronzy brown bark, much branched stem with branches arising lower down the main stem. Leaves are opposite, simple, broadly ovate or elliptic, 45-10 x 1.5-4.5 cm in size with acuminate apex, rounded to sub-cuneate base, entire margin, dark green above and dull and paler below with 20 pairs of obscure lateral nerves and 3-5 mm long petiole. Inflorescence is axillary with 1-1.5 mm long axis usually bearing 4 subsessile flowers in 2 opposite pairs. Fragrant flowers are regular and bisexual. Calyx lobes are broadly rounded and 2.2 mm wide. Petals are ovate, 3-4 mm long and ciliate. Stamens are 125, 9-10 mm long with 0.5 - 0.7 mm long anthers. Ovary is inferior and style is 10-11 mm long. Fruit is a reddish brown to purple black globose berry, 1-3 cm in diameter with a circular hypanthial scar at the apex and fleshy and soft pulp at maturity enclosing 2-3 seeds (FAO, 1986; Morton, 1987). Sweet and flavoured types are available in nature from where selections can be made.

Jaboticaba
(*Myrciaria cauliflora* (DC) Berg; Brazilian grape tree; Yba-puru)

It is native of southern Brazil. The trees grow wild on river banks in northern Argentina, Paraguay and South Brazil. It is widely distributed throughout the tropics but there are no extensive and regular orchards because its early growth is extremely slow. It has also been introduced into California (USA), South Africa and Australia. The other related species having edible fruits are *M. edulis* (Vell.) Skeels, *M. floribunda* (West ex Willd.) Berg. (guava berry), *M. paraensis* Berg., *M. vexator* Mc Vaugh (blue grape), *M. dubia* (Kunth.) Mc Vaugh and *M. trunciflora* O. Berg., which are found growing in South America (Page, 1984; FAO, 1986; Morton, 1987; Brücher, 1989).

Jaboticaba is a handsome slow growing evergreen bushy tree growing up to 12 m in height. The tree develops an attractive symmetrical crown of pale green foliage under favourable conditions. The leaves are simple, 2-5 cm in length, opposite, lanceolate and have acute to acuminate apex. The yellow-white flowers are small, borne directly on the trunk and larger branches and have 4 sepals, 4 white petals and numerous stamens, which are produced either singly or in clusters. The hypanthium is tubular and circumscissile. Cultivated species are often polyembryonic. The fruits are globose, 2-3 cm in diameter and resemble large muscadine grapes with black or purple skin, but with persistent green calyx and tough skin unlike that in grape. The pulp is white or pale pink, sweet and juicy and usually contains one seed but occasionally as many as four (Page, 1984; Morton, 1987; Brücher, 1989). Cultivated species are often polyembryonic.

Cultivars

In Brazil, white or pink fleshed and small or large fruited forms are known. There is need to establish not only their species status but also the forms collectively known as jaboticabas with a view to identifying and naming superior cultivars (Morton, 1989). Important cultivars of jaboticaba are Branca, Paulista, Rajada, Sabara and Ponhema (CRFG, 1996b).

Rumberry
(***Myrciaria floribunda*** Berg. syn. ***M. protracta*** Berg., ***Eugenia floribunda*** West ex Willd.)

It is native of Mexico, West Indies and South America (Martin *et al*., 1987). It is also known as guava berry and mirto or murto. The rumberry occurs wild over a broad territory in Cuba, Hispaniola, Jamaica, Puerto Rico, the Virgin Islands, St. Martin, St. Kitts, Guadeloupe, Martinique, Trinidad, South Mexico, Belize, Guatemala, El Salvador to northern Colombia and also in Guyana, Surinam and French Guiana and eastern Brazil. It has been cultivated in Burmuda. It has been introduced into Philippines, Hawaii and Miami, USA (Morton, 1987).

It is an attractive evergreen shrub or slender tree reaching 10-15 m in height having reddish brown branchlets with flaking bark but are downy when young. Leaves are ovate elliptic or oblong lanceolate, pointed at the apex, 2.5-8 x 0.8-3 cm in size, wide, glossy, slightly leathery and minutely dotted with oil glands. The flowers borne in small, axillary or lateral clusters are white, silky hairy having about 75 prominent white stamens. The fruit is round or oblate, 8-16 mm in diameter, dark red or yellow orange, highly aromatic and of bitter sweet, balsam-like flavour having one globular seed. There are 3 or 4 cultivars in the dry hills of St. Croix. They vary in size and colour but all are intensely aromatic (Morton, 1987).

Genus *Rhodomyrtus* (DC.) Rehb.

It is a genus of 24 species of shrubs and trees. The genus is native to Asia, Malesia, Melanesia and Australia. Its greatest diversity is found in New Guinea and northeastern Australia.

Downy myrtle
(***Rhodomyrtus tomentosa*** Wight.; hill gooseberry; rose myrtle)

It is native of tropical Asia. It grows wild and is also cultivated in South East Asia, India, Sri Lanka and southern China. Downy myrtle requires cool tropical or warm subtropical climate and can tolerate temperature as low as -6°C. It thrives on degraded sandy sites along the shores and on riverbanks. It can tolerate strong sun and flooding. Moist and acid soils are generally preferred. The plant is not well adapted to limestone soils (Martin *et al*., 1987; Verheij and Coronel, 1991).

Two important botanical varieties are found. The var. *tomentosa* (*Myrtus canescens* Lour.) is generally found in waterlogged areas up to elevations of 300 m but rarely up to 1300 m in South East Asia, southern China and Indochina and is characterized by white tomentose leaves. The var. *parviflora* (*Rhodomyrtus parviflora* Alston.) occurs in mountainous woodlands and grasslands at 1800-2700 m altitudes in India and Sri Lanka and have cream or yellowish tomentose leaves (Verheij and Coronel, 1991).

It is an evergreen shrub or a small tree growing up to 3-4 m in height. The young leaves and inflorescences are densely white or yellowish tomentose. The leaves are elliptic to oblong elliptic, 4.5-8 x 2.3-4 cm in size, opposite, coriaceous with 3 conspicuous longitudinal veins, glossy and glabrous upper surface and whitish or yellowish tomentose lower surface. The flowers are solitary or occur in 3-flowered dichasia in the axils of upper leaves with up to 1 cm long peduncles and 0.5-2.5 cm long pedicels. The bracts are elliptic, leaf-like and 6-12 mm long. Bracteoles are elliptic or ovate, 2-3 mm long and persistent. Calyx is campanulate, 5-7 mm long, tomentose, 5-10 ribbed, 5-lobed and persistent. Petals are 5, broadly obovate, 15-18 x 9-13 mm in size and red or pink. Stamens are numerous, 10-15 mm long with pink filaments. Style is 13-15 mm long and ovary is 3-10-locular. The fruit is oblongoid berry, 10-15 x 8-10 mm in size, puplish black, tomentose and crowned with calyx lobes. The pulp is sweet having many seeds in 6 pseudo-locules and divided by thin and false septa (Verheij and Coronel, 1991).

Genus *Ugni* Turcz

The genus has 4 species that occur from Mexico to Chile. The plants are often densely branched shrubs having simple hairs. Flowers are white, 5-merous, nodding with persistent prophylls, free sepals, petals forming an ericoid, lampshade like structure, numerous stamens as long as the petals, sagittate anther, (2-)3 locular ovary with few to many 2-4-seriate ovules. Fruit is crowned by the calyx lobes. Seeds are small, numerous having hard, shiny seed coat and C-shaped embryo and hypocotyl as long as the cotyledons (Kubitzki, 2011).

Besides *Ugni molinae* Turcz. (Chilean cranberry), the three related species, *U. candollei* (Berneoud) Berg, found in continental Chile, *U. selkirkii* (Hook. et Arn.) Berg, endemic to the Juan Fernandez Island, and *U. myrcioides* (HBK) Berg, found at altitudes above 2500 m in the Andes from Bolivia to Venezuela on the steppes of Guyana highlands and in the mountains of Central America to Oaxaca, Mexico. These species, however, produce poor quality fruits (Landrum *et al.,* 1990).

Chilean cranberry

(***Ugni molinae*** Turcz. syn. ***Eugenia ugni*** W. J. Hook. et Arnett.; murtilla; Chilean guava)

It is native of Chile. It is a slow growing evergreen tree and attains a height up to 2 m. The young twigs are whitish and pubescent but the old branches are smooth and gray. The leaves are green, elliptic to ovate, 1.4-3.6 x 0.8-2 cm in size, glossy above and whitish underneath. Solitary flowers are bell shaped and have pink tinge with 1-2.5 cm long peduncles borne in the axils of leaves. Calyx is (4) 5-lobed, glabrous to very sparsely pubescent. Petals are sub-orbicular, slightly fleshy, 5-8 mm long, 0.3-0.4 mm thick at base and are pink when fresh. The disk is 2-4 mm across, glabrous and the style is 4-5 mm long. Stamens are 40-60 and 2-4 mm long. The ovary is 3-loculed with 13-30 ovules per locule. The fruit is subglobose, 1 cm in diameter having 1.5-2 mm long seeds (Anon, 1987; Landrum *et al.,* 1990).

Cultivars

Chilean cranberry is of three types depending on size, colour and flavour of the fruits (Landrum *et al.,* 1990):

i. Rosada type produces dark pink or red and aromatic fruits with an average weight of 0.5 and 10.2 mm diameter.

ii. Blanca type produces very soft, aromatic and sweet fruits of light yellow colour with pink spots and average weight of 0.4 g and 10.4 mm diameter.

iii. Garnate type has garnet coloured fruits, harder than of the other two types, which are sour and aromatic with an average weight of 0.34 g and 9.4 mm diameter.

Genus *Pimenta* Lindl.

The genus has 15 species that are mostly Caribbean but with one species in Brazil. The plants are trees and shrubs having simple hairs. Inflorescence is a dichasium or a panicle of 3-15 flowers. Flowers are 4-merous having open calyx, 2-locular ovary, 3-6 ovules per locule, on a sub-apical placenta. Fruit is crowned by the calyx lobes. Seeds are usually 1 or 2 having membranous or hard and shiny seed coat with spiral embryo, swollen hypocotyls and very small cotyledons (Kubitzki, 2011).

Allspice

(***Pimenta dioica*** (L.) Merill.; pimento; Jamaican pepper)

The tree is well known for its berries called Pimento used as an important spice since time immemorial and for its culinary and medicinal qualities. It is also known as Allspice due to its intricate aroma which is a medley of aroma from spices such as clove, nutmeg and cinnamon. In India, its leaves are used to flavour rice to give a typical aroma. The dried berries are used to marinate meat in traditional culinary practice.

Allspice is a small evergreen tree up to 15 m tall having a pale brown bark. Leaves are simple, opposite, entire, oblong elliptical, 6-20 cm long and punctuate with pellucid glands that give off the odour of allspice. The flowers having a peculiar aroma, present in groups of cymes, are small and whitish. These are structurally hermaphrodite but functionally dioecious. Male trees having above 100 stamens bear no fruit but the female trees having around 50 stamens bear fruits. The receptacle has four cream coloured calyx lobes, spread at anthesis and is persistent in the fruit. Petals are 4, whitish and quickly deciduous. The style is white with a yellow stigma. In female trees, the style is slightly shorter and the stigma is longer than in the barren male trees. The ovary is inferior and 2-celled, usually with one ovule in each cell. The plants flower during March-June and the berries mature 3-4 months later. For spice purpose, picking is done when these are fully developed but still green. The fruit has two kidney shaped seeds.

The dried green mature fruit is a commercial flavourant and curing agent. Its extract is an essential component in the classic liqueurs Chartreuse and Benedictine and in the local Jamaican drink Pimento Dram. Ground fruits are preferred in desserts, relishes, sausages and preserves. Traditionally, a water extract of the berries is used to treat flatulence and diarrhea while the powdered fruit is used for corns, neuralgia and rheumatism. Young woody shoots of pimento are used to make popular walking sticks and umbrella handles. In India, it is used as an aromatic stimulant in digestive troubles, as an adjuvant to tonics and purgatives and as an anodyne against rheumatism and neuralgia. In Turkey, it is taken along with honey as an aphrodisiac. The essential oils of *P. dioica* leaves and fruits are utilized in food industry mainly meat and tanning industries as well as in perfumery and cosmetic products. An oleoresin from the pimento berries is produced in small quantities. The therapeutic properties of the essential oils are anesthetic, analgesic, antimicrobial, antioxidant, antiseptic, acaricidal, carminative, muscle relaxant, rubefacient, stimulant and tonic. Pimento oil can be helpful for the digestive system, for cramp, flatulence, indigestion and nausea. It can help in cases of depression, nervous exhaustion, tension, neuralgia and stress and is used as natural repellent. The leaf and fruit oil is also used in perfumes, aftershaves and commercial food flavouring. Pimento seed is reported to be used by Haitaian immigrants and their descendants in the province of Camaguey, Cuba to treat abdominal pain when taken triturated and ingested with rum and sugar.

⑧ ⑧ ⑧

Chapter 29

SAPINDALES

The plants are trees, shrubs, woody lianas or sometimes subshrubs or herbs. Woody plants contain silica. Vessels are with simple perforations or less often perforations are scalariform. Lateral pitting is alternate or sometimes opposite. Fibres are with bordered or simple pits, often septate. Rays are heterogeneous or homogeneous. Axial parenchyma is mostly paratracheal, sometimes scanty or even wanting. Sieve element plastids are of Ss-type. Nodes are usually trilacunar. Leaves are opposite or less often alternate, simple or more often compound, exstipulate or sometimes with small caducous stipules. Stomata are of diverse types. Flowers are in various types of inflorescences, rarely solitary, bisexual, more often unisexual, actinomorphic and less often more or less zygomorphic (strongly zygomorphic in Hippocastanaceae), usually 5-4-merous. Sepals are free or more or less connate, imbricate or rarely valvate (some Sapindaceae). Petals are free, equal or unequal, imbricate or rarely valvate (some Sapindaceae) or seldom wanting. Nectary disc of receptacular origin is commonly present, extrastaminal or less often intrastaminal (Aceraceae). Stamens are (4)5-10, rarely more with free filaments, tetrasporangiate, introrse, basifixed, dorsifixed anthers opening longitudinally. Tapetum is secretory. Microsporogenesis is simultaneous. Pollen grains are 2 or 3-celled and mostly 3-colporate. Gynoecium is of 3-2(4-5) united carpels with free or more or less connate stylodia and superior ovary with (1)2 ovules per locule. Ovules are epitropous or apotropous, anatropous or less often hemitropous, campylotropous or orthotropous, bitegmic and crassinucellate. Ovules when pendulous, the raphe is outside the axis and when erect the raphe is towards the axis. Female gametophyte is of *Polygonum*-type. Endosperm is nuclear. Fruits are of various types. Seeds are exotestal, mesotestal or rarely with fibrous exotegmen. Embryo is mostly curved. Endosperm is copious, very scanty or lacking. The order contains 9 families and 5800 species. Biflavonoids and bitter terpenes are present in the leaves.

Family Burseraceae (frankincense and myrrh family)

It is found in the tropics of Africa, Asia, America, West Indies and Australia, extending also into subtropical and arid areas of North Africa. It contains 18 genera and 700 species (Heywood *et al.*, 2007).

The trees are large to small, less commonly shrubs, rarely epiphytes and are often aromatic with resinous plant parts, characteristically with flaky bark, sometimes shedding in sheets (*Bursera* spp.). The leaves are alternate, spirally arranged, usually crowded at twig tips, imparipinnate, less frequently unifoliate and usually exstipulate or with pseudostipules (*Garuga*). The flowers are small, usually greenish, cream coloured or pinkish, actinomorphic, unisexual (plants are dioecious or polygamodioecious), often with vestigial remains of the other sex or hermaphrodite (*Garuga, Dacryodes*), usually hypogynous (perigynous in some *Bursera* and *Commiphora* spp.) and borne in axillary or terminal to sub-terminal, paniculate or racemose inflorescences. The sepals are 3-5(6), connate below and lobed, imbricate or valvate. The petals are 3-5, free to partly connate, imbricate or valvate. The stamens are as many as or twice as many as petals and the filaments are usually free or sometimes connate at the base (e.g., *Canarium* spp.). Staminodes are present in female flowers. The ovary is superior with 2-5 carpels and locules having 2 ovules per locule and the style is simple. The fruits show considerable diversity but are basically compound, drupaceous or non-fleshy, capsular (pseudocapsules), with the pyrenes free or partly connate and dehiscent or remaining fused together. The dehiscent fruits in the tribes Protieae and Bursereae have as many valves as the developed locules and when fleshy, the exposed pyrenes are covered partially or almost completely by a pseudo-aril. In the tribe Canarieae, the fruits are indehiscent and consist of a compound drupe with 2-3 locules having a thin exocarp, an oily mesocarp and a tough or bony endocarp.

Genus *Canarium* Stickm.

About 120 species occur mostly in low lying and wet semi-deciduous pleotropical forests principally in Malesia and Madagaskar (Kubitzki, 2011). The plants are small to large, usually dioecious trees, rarely shrubby or scandent often with buttresses having smooth, rough, scaly or scalloped, often fissured and sometimes densely lenticellate bark. Resin is translucent, watery or sticky, white or dark brown or blackish. The leaves are imparipinnate, evergreen, usually with linear to foliose stipules on petioles toward or at base or less often on branch near the petiole base. Sometimes pseudostipules are present and pulvinus is usually present on terminal and lateral petiolules. Leaflets are entire to serrate or dentate with tertiary veins usually mixed, opposite to alternate percurrent. Inflorescence is axillary or (pseudo)-terminal, sometimes pseudo-spicate, pseudo-racemose or fasciculate. Flowers are 3-parted, unisexual. Rarely the plants are monoecious having cupular to funnel-shaped calyx and receptacle. Calyx is synsepalous but variously divided, valvate, usually persistent and enlarged in fruit. Petals are distinct, valvate at apex but imbricate in the middle having inflexed apiculate apex. Stamens are in two subequal or less often unequal series, rarely the antepetalous series are obsolete and filaments are sometimes partly or mostly connate to form a staminal tube and stamens are inserted at base of disk, sometimes on the disk near the base, rarely on the rim. Disk is usually annular but variously shaped, sometimes pubescent, rarely adnate to a shallow hypanthium in male flowers. Pistillode is usually ±reduced, sometimes fused with disk to form an ovariodisk, rarely the disk, stamens and ovary are borne on a short androgynophore. Staminodes are usually only slightly reduced. Ovary is 3-locular with short style, subcapitate or 3-lobed stigma or 3 separate subsessile stigmas. Fruit is an indehiscent compound drupe. Seeds are 1-2 (3).

The species that bear edible nuts are *C. ovatum* Engler (pili nut), *C. pimela* C. Koenig. (Chinese black olive), *C. album* Reauch. (Chinese white olive), *C. indicum*

L. (canary nut or Java almond) and *C. schwemfurthii* Engler (incense tree). These have originated in Asia and South East Asia. The species that are found in Australia are *C. australeanum* (in Aruhem land), *C. harveyl* Seemann (around Atherton Tablelands) and *C. haileyanum* (Lismore through Queensland).

Java almond

(***Canarium indicum*** L. syn. ***C. amboinense*** Hochr.; Canary nut)

It is native of Malaysia to Melanesia. The tree is tall growing up to 40 m in height with buttressed trunk. Leaves are compound, glabrous, with persistent ovate to oblong stipules having 3-8 pairs of leaflets that are oblong lanceolate and 7-35 x 3-16 cm in size. The blades are herbaceous to coriaceous. The tree is dioecious. Inflorescence is terminal, many flowered, 15 to 40 cm long and minutely tomentose. Male flowers are sub-sessile and female flowers are short stalked. The calyx is 5-7 mm long in male flowers and 7-10 mm in the female flowers. The 3 petals are cream coloured. The stamens are free in male flowers and adnate to the disk in female flowers. The pistil is minute or none in male flowers. Fruit is a drupe, ovoid round in shape, slightly triangular in cross section, 3.5 to 6 cm long and 2 to 4 cm in diameter, green in colour and contains a white kernel (3.5 x 2 cm in size) covered with a thick brown skin or membrane (Duke, 1989; Verheij and Coronel, 1991).

Cultivars

Several races are cultivated in Melanesia, varying in form and size of fruit. Two botanical varieties are recognized. *C. indicum* var. *indicum* has branchlets up to 13 mm thick, stipules up to 6 x 5 cm in size, dentate and herbaceous leaflets up to 28 x 11 cm in size and the fruit is 6 x 3 cm in size. It is more widespread and cultivated. *C. indicum* var. *platycerioideum* Leenhouts has branchlets up to 2.5 cm thick with stipules sometimes inserted on the bases of the petiole only and leaflets are 25-35 x 13-16 cm in size and the fruits of 6 x 3.5-4 cm size. It is found only in New Guinea up to 30 m altitude (Duke, 1989).

Pili nut

(***Canarium ovatum*** Engler; Philippine nut)

It is native of the Philippines and Bicol region is the centre of diversity. It is distributed throughout the tropics and is cultivated in Malaysia, Hawaii, Brazil and Honduras.

It is a fairly large evergreen tree growing to a height of about 20 m with a trunk diameter of 40 cm. Leaves are alternate, compound with opposite, smooth leaflets rounded at the base and pointed at the apex. The leaflets are 10 to 20 cm long. The plants are dioecious. The flowers are borne in large terminal clusters. Flowers are 3-merous, unisexual and sub-sessile having cupular calyx, 3 petals and 6 stamens. The ovary contains 3 locules each with 2 ovules but nearly always only one of the 6 ovules develops. The fruit consists of a single, 5-7 cm long, nearly cylindrical, oily, pleasantly flavoured kernel, hard and thick endocarp and a thin layer of soft pulp. It develops a glossy dark purple skin on ripening. The pulp is edible but is not tasty. Oil can be extracted from the pulp (Verheij and Coronel, 1991).

Dabai
(*Canarium odontophyllum* Miq.)

Dabai is an indigenous fruit enjoyed by the communities of Sarawak (Borneo), Malaysia. The fruit is sought after for its delicious creamy taste and pleasant aroma and high protein, fat and minerals. The tree is resiniferous, 8-25 m tall with a girth of 15-60 cm. The trunk is straight with erect, semi-erect or horizontal branches. Leaves are odd-pinnate with 7-17 leaflets, distinctly acuminate apex and are hairy below and on the midrib above. The fruit is ovoid to ellipsoid in shape with slightly triangular cross section. The skin is dark purple covering a light yellow flesh. The seed is covered by a hard shell and can be eaten as a nut. It is a rich source of protein, fat, carbohydrates, sodium, calcium, iron and antioxidants. To prepare for eating, the fruit is steeped in warm water (about 60°C) for 10-15 minutes. This softens the flesh giving it a smooth avocado-like texture. The fruit can be seasoned with salt, sugar or soya sauce to enhance the taste. The tree is relatively free from pests and diseases.

Genus *Dacryodes* Vahl

About 70 species occur in the moist tropics, usually lowlands, some in montane areas (Kubitzki, 2011). The plants are dioecious or hermaphrodite trees (one species is scandent), sometimes with plank buttresses. The bark is smooth or rough due to scales or raised lenticels, relatively thin that shed in irregular plates. Resin is clear or somewhat milky. Stellate hairs are sometimes present. Leaves are evergreen, exstipulate with pulvinus usually apparent on terminal and lateral petiolules. Entire leaflet margin and intersecondary veins are usually present. Inflorescences are axillary or sometimes (pseudo)-terminal. Flowers are unisexual or less often bisexual having 3-parted perianth, (shallowly) cupular to urceolate, truncate or 3-lobed calyx, sometimes divided nearly to base, distinct petals, valvate at apex, valvate on sides (American species) or slightly imbricate with an inflexed apiculum. Stamens are in two sub-equal series, inserted at base of disk or on the side or rim. Anthers are continuous with the filaments or sagittate. Pistillode is either reduced but provided with locules and ovules or a variously shaped parenchymatous structure or fused with the disk to form an ovariodisk, otherwise the disk is annular and pistil is 2-3-locular, short and thick with 2-3-lobed style. Fruit is an indehiscent, compound drupe, oblique ovoid to ellipsoid, rarely globose having thin and drying wrinkled exocarp, thin and oily resinous mesocarp, (thinly) cartilaginous, smooth, not lobed endocarp, 1(2)-seeded pyrenes, 1-2 aborted and strongly compressed locules forming a separable plate on periphery of pyrene. Testa is smooth and cotyledons are palmate or up to 13-palmátifid, folded or contortuplicate.

Sofou
(*Dacryodes edulis* (G.Don) H.J. Lam. syn. *Pachylobus edulis* G. Don.; African pear; African plum tree; bush butter; safoutier)

Sofou or African plum tree is a native of tropical West Africa and is distributed in tropical Africa and Malaysia (Martin *et al.,* 1987). The other related edible species are *D. buettneri, D. igaganga, D. klaineana* and *D. macrophylla.*

It is an evergreen tree and grows up to 8-12 m in height with a straight and cylindrical trunk. The leaves are compound imparipinnate with 4-12 pairs of leaflets. The tree is dioecious. Male and hermaphrodite flowers are borne on axillary or terminal 8-25 cm long inflorescences. Female inflorescence is 5-15 cm long and produces more fruits than the hermaphrodite ones. The ovary is bilocular, each locule having 2 ovules. After fertilization they fail to develop normally and only one seed develops. Fruit is an ellipsoid to globular or conical drupe, 4-12 cm long and 3-6 cm in diameter. The drupes are rose coloured when young turning deep blue at maturity. The pericarp consists of a very thin, waxy and coloured epicarp and pulpy edible mesocarp. The light rose, green or whitish mesocarp is 0.5 cm thick and has varied flavour. The endocarp is thin, smooth and plated extremely with a barely visible escutcheon. The seed is light green and consists of two fleshy and five lobed cotyledons (FAO, 1982; Ngatchou and Kengue, 1989).

Family Anacardiaceae (cashew family; cashew, mango, sumacs and poison ivy)

The family contains 82 genera and more than 700 species (Heywood *et al.*, 2007). It is largely tropical and subtropical family and occurs in both the eastern and western hemispheres with a more or less equal representation in South America, Africa and Asia. It has diversified in Central and South America, southern and central Africa, Madagascar, Indochina and Malaysia. Several genera are native to temperate North America and Eurasia (*Rhus, Pistacia*, and *Cotinus*). The largest genus *Rhus* with 200 species grows in temperate and tropical regions of both the hemispheres. *Semecarpus* with 60 species is Indo-Malaysian in distribution.

The plants are trees and shrubs, occasionally scandent shrubs, herbaceous climbers or lianas having resin canals and clear to milky sap that may turn black on exposure to air and is often poisonous. It causes severe dermatitis. The irritant chemicals may be distributed throughout the plant body (vertical resin canals in the bark are associated with the larger veins of the leaves, often also in the parenchymatous tissues) or concentrated in particular organs, e.g. fruit wall of cashew. The leaves are alternate, rarely opposite (*Abrahamia, Blepharocarya, Bouea*), exstipulate, often pinnate and sometimes simple (*Cotinus, Anacardium, Mangifera*). The flowers are actinomorphic and bisexual or sometimes unisexual (in dioecious plants), small and inconspicuous borne in terminal or axillary panicles, thyrses, spikes, racemes, fascicles or are solitary. Perianth parts are in fives. Sepals are (3-)5(-7), usually connate at the base but free above and toothed. Petals are (3-) 5(-7), rarely absent and free. Stamens are 5 to 10 or more with the filaments distinct or rarely connate at the base forming a tube. An intrastaminal fleshy receptacle (disk) is usually present. Staminodes are present in female flowers. The ovary is usually superior (inferior in *Drimycarpus*) and comprises 1 or 3 to 5 united, very rarely free carpels with 1 to 5 locules, each containing a single ovule. There are 1 to 3 styles, sometimes none. Intraseminal nectar disk is present. The fruit is usually indehiscent and drupaceous but a diversity of unusual fruit types also occurs such as samaras encircled by a marginal wing (*Campylopetalum, Dobinea* and *Laurophyllus*), samaroid with a single wing (*Faguetia, Loxopterygium*), dry syncarps (*Amphipterygium, Orthopterygium*) or achenial (*Apterokarpos*). In *Anacardium* and *Semecarpus*, the hypocarp subtending the drupe is enlarged and edible. The solitary seed may have thin or no endosperm and fleshy cotyledons.

The family has been divided into 2 subfamilies; subfam. Anacardioideae having the tribes Ancardieae, Dabineae, Rhoeae and Semecarpae and subfamily Spondiaeae having the tribe Spondiaeae and few members of the Rhoeae. The family includes fruits such as the mango. The other distant relatives of mango are cashew (*Anacardium occidentale*), gandaria (*Bouea gandaria*), pistachio (*Pistacia vera*), marula (*Sclerocarya birrea*), ambarella (*Spondias cytherea*), yellow mombin (*S. mombin*), red mombin (*S. purpurea*), imbu (*S. tuberosa*), dragon plums (*Dracontomelum* spp.), kaffir plum (*Harpephyllum caffrum*), etc. The ornamentals include members of genus *Rhus* (e.g. sumacs of California) and the well known pepper tree (*Schinus molle*). The family also has species that yield valuable wood, gums and resins, wax and varnishes and tanning materials. Some of its members are well known for causing dermal irritation. Some *Mangifera* spp. have resinous sap that causes allergic reaction.

Genus *Mangifera* L. (n=20 or 30, polyploidy common) (2n=40)

The trees are andromonoecious and have exudate that causes contact dermatitis. Leaves are evergreen, alternate, simple, petiolate, entire and lanceolate with acute to acuminate apex with cataphylls sometimes present. Inflorescences are terminal and/or axillary thyrsoids. Flowers are pedicellate and articulate having 5-parted imbricate perianth, white to pink corolla, often with yellow or pink glandular ridges on the adaxial surface, 5(-10) stamens of which only 1-2 (3-5 or all) are fertile and the others are much reduced. Filaments are distinct or connate at the base with dorsifixed anthers, reduced pistillode, glabrous, extrastaminal (intrastáminal), papillose or glabrous, 4-5-lobed disk, pseudomonomerous gynoecium and lateral style, sometimes approaching gynobasic, punctiform stigma and basal ovule. The fruit is subglobose, sigmoid, sub-reniform or ovoid, 1-locular drupe having green, yellow, orange, red to brown or black exocarp, fleshy mesocarp and woody and fibrous endocarp. Seed is oblong ovoid and compressed having usually piano-convex cotyledons (sometimes lobed and the ascendant radicle), rarely the cotyledons and the inner seed coat forms a labyrinth, chartaceous testa and curved embryo.

The genus *Mangifera* consists of 69 species mostly restricted to tropical Asia. The highest diversity occurs in Malaysia, particularly in peninsular Malaya, Borneo and Sumatra representing heart of the distribution range. The natural occurrence of all the *Mangifera* species extends as far north as 27° latitude and as far east as the Caroline Islands (Bompard and Schnell, 1997). Wild mangoes occur in India, Sri Lanka, Bangladesh, Myanmar, Sikkim, Thailand, Kampuchea, Vietnam, Laos, southern China, Malaysia, Singapore, Indonesia, Brunei, the Philippines, Papua New Guinea and the Solomon and Caroline Islands. Maximum species diversity exists in western Malesia where about 28 species are found. These are mostly distributed below 300 m altitude but can occur at 600-1900 m above sea level and are found as scattered individuals in tropical lowland rainforests. Most of the species (44) are found on well drained soils. *M. indica* is cultivated pantropically (Kubitzki, 2011). The species like *M. gedebe, M. griffithii* and *M. parvifolia* occur in certain type of swamp forests. *M. bompardii*, *M. dongnaiensis* and *M. orophila* are mainly found in submontane forests above 1000 m and occasionally up to 1500-1700 m above sea level whereas few species like *M. caloneura, M. collina, M. timorensis* and *M.*

zeylanica are acclimatized to seasonally dry climates in deciduous or semi-deciduous forests. *M. sylvatica* and wild *M. indica* can be found in Sikkim and southern China at altitudes of 600-1900 m above sea level.

Mukherjee (1949) reported that out of about 41 species 9 are found in the Archipelago. In wild types, the petiole is much longer. He divided the species on the basis of the type of disc:

i. Prominent disc – *M. indica, M. sylvatica, M. khasiana, M. pentandra, M. coloneura* and *M. siamensis.*

ii. Narrow disc – *M. odorata, M. foetida* and *M. caesia.*

Kostermans and Bompard (1993) raised the sections to the rank of subgenera:

Subgenus *Limus* (Marchand) Kosterm: The plants have the disc narrower than the base of the ovary or stalk like or even lacking. It consists of 11 species. The subgenus has been divided into the following two sections:

i. Section *Deciduae:* The section contains the deciduous species *M. kemanga* BI., *M. caesia* Jack., *M. pajang* Kosterm., *M. superba* Hooker f. and possibly *M. blommesteinii* Kosterm., *M. decandra* Ding Hou and *M. lagenifera* Griff.

ii. Section *Perennes*: The section contains the non-deciduous species *M. foetida* Lour., *M. leschenaultii* Marchand, *M. macrocarpa* BI., *M. caesia* Jack and *M. odorata* Griff.

Subgenus *Mangifera* (Ding Hou) Kosterm: The plants have the disc broader than the base of the ovary, which is cushion-like, often divided in four or five lobes. It contains 47 species under the following five sections:

i. Section *Marchandora* Pierre: It has one species, *M. gebede* Miq.

ii. Section *Euantherae* Pierre: It has 3 species, *M. caloneura* Kurz (syn. *M. duperreana* Pierre), *M. cochinchinensis* Engler and *M. pentandra* Hooker f.

iii. Section *Rawa* Kosterm: It has 9 species, *M. gracilipes* Hooker f., *M. griffithii* Hooker f., *M. microphylla* Griff. ex Hooker f., *M. paludosa* Kosterm., *M. parvifolia* Boerl. & Koorders, *M. merrillii* Mukherji, *M. andamanica* King, *M. nicobarica* Kosterm. and *M. minutifolia* Evard.

iv. Section *Mangifera* Ding Hou: It has more than 30 species : *M. altissima* Blanco, *M. mucronulata* Bl., *M. applanata* Kosterm, *M. oblongifolia* Hooker *f.*, *M. austro-indica* Kosterm, *M. orophila* Kosterm, *M. austro yunanensis* Hu, *M. pedicellata* Kosterm, *M. casturi* Kosterm, *M. pseudo-indica* Kosterm, *M. collina* Kosterm, *M. quadrifida* Jack, *M. dewildei* Kosterm, *M. rigida* Bl., *M. dongnaiensis* Pierre, *M. rubropetala* Kosterm, *M. flava* Evard, *M. rufocostata* Kosterm, *M. indica* L., *M. similes* Bl., *M. lalijwa* Kosterm, *M. sulauesiana* Kosterm, *M. laurina* Bl., *M. sumbawaensis* Kosterm, *M. linearifolia* (Mukherji) Kosterm, *M. sylvatica* Roxb., *M. longipetiolata* King, *M. swintonioides* Kosterm, *M. magnifica* Kochummen, *M. timorensis* Bl., *M. minor* Bl., *M. torquenda* Kosterm, *M. monandra* Merr. and *M. zeylanica* (Bl.) Hooker f.

15 species have tetramerous flowers (e.g. *M. altissima, M. torquenda, M. magifica, M. quadrifida*) and 4 have tetra-and pentamerous flowers (e.g. *M. casturi, M. quadrifida, M. rubropetala* and *M. indica*).

Key for identification of *Mangifera indica* (Kostermans and Bompard, 1993)

SUBGENUS *MANGIFERA* (section *Mangifera*)

Fertile stamen 1, the other four staminodial, very minute or lacking, sometimes with empty anthers. Flowers 5-merous (5 sepals, 5 petals)

1. Leaves spathulate to obovate-oblong to oblanceolate, obtuse to submarginate, rigidly coriaceous, 1.5 x 45 x 16 cm .... *M. zeylanica*

1. Leaves lanceolate to oblong, rarely subovate-oblong, acutish to acute, chartaceous to coriaceous

2. Lower surface of leaf smooth, no reticulation

3. Leaves oblong to elliptic oblong, up to 13 x 32 cm. Petiole 4-9 cm. Panicle minutely, sparingly puberulous, lax. Fruit roundish, smooth, yellow, green, pulp yellow .... *M. oblongifolia*

3. Leaves subovate-oblong to oblong, 4 x 12---6 x 17 (10 x 30) cm. Petiole 1-2 cm. Panicle glabrous, branchlets racemoid. Fruit globose, dull, brownish, roughish, pulp white, something like that of *M. foetida,* extremely acid .... *M. rufocostata*

2. Lower leaf surface with distinct reticulation

4. Panicles puberulous

5. Leaves underneath sparingly long pilose .... *austro-yunnanensis*

5. Leaves glabrous

6. Panicles densely, minutely puberulous. Flowers glomerulate ....***M. indica***

6. Inflorescences sparingly or densely puberulous, lax. Flowers not glomerulate

7. Petals ovate, c. 2.5 mm.. *M. collina*

7. Petals linear-oblong, 4- 5.5 mm ....*M. laurina*

4. Panicles glabrous

8. Pedicels 5- 6 cm long..*M. pedicellata*

8. Pedicels 1-2 cm long

9. Petals up to 11 mm long... *M. flava*

9. Petals up to 6 mm long

10. Flowers glomerulate ....*M. austro-indica*

10. Panicles lax, flowers not glomerulate

11. Leaves chartaceous. Apical part of fruit narrowed and curved like a parrot beak

12. Pedicels very short on a short supporting branchlet... *M. sylvatica*

12. Pedicels very long on a long supporting branchlet...*M. minor*

11. Leaves coriaceous

13. Tips of petal ridges free of the surface. Young fruit dark green, sweet.................... *M. latijiwa*

13. Ridges not free from the petal surface

14. Pedicels 1 mm. Petals elliptic spathulate, 3 mm, ridges ....*M. pseudo-indica*

14. Pedicels 2-4 mm. Petals narrowly oblong, 3-4 mm, ridges very faint................*M. orophila*

v. Species of uncertain toxonomic position: This group contains 11 species: *M. acutigemma* Kosterm, *M. persiciformis* Wu & Ming, *M. bompardii* Kosterm, *M. subsessifolia* Kosterm, *M. bullata* Kosterm, *M. taipa* Buch-Hamilton, *M. campospermoides* Kosterm, *M. transversalis* Kosterm, *M. hienalis* Liang Jian Ying, *M. utana* Utana and *M. maingayii* Hooker f.

***M. sylvatica* Roxb.** is characterized by its long petioles and inedible fruits. *M. pentandra* Hooker.f has all the 5 stamens functional and edible fruits. Both *M. rufocostata* Kosterm. and *M. swintonioides* Kosterm show off season bearing. *M. altissima* Blanco, grown in the Philippines, shows resistance to hopper, tip and seed borers. *M. casturi* Kosterm has black fruit colour and *M. caloneura* Kurz., found in Myanmar, bears acidic sweet fruits.

***M. laurina* Bl.:** It is a species of lowland forests of Malaysia where it is also cultivated. It can be distinguished from the common mango by the lax and widely pyramidal, glabrous or sparingly puberulous panicles. The flowers are smaller and are not glomerulate. The petals have a different shape, texture and colour. The small fruit has orange yellow pulp and is generally consumed unripe. The species is highly resistant to anthracnose.

Mango
(*Mangifera indica* L.)

The plants are hardy, evergreen and tropical to subtropical trees, withstand frost but are susceptible in the younger age and require dry weather during blossoming and heat for ripening of the fruits. The tree is large, erect or spreading with a dense crown. There are several flushes of growth in a season. The bark of older trees has corky layer that splits but that of younger branches is smooth. The colour of young growth is green to coppery. The leaves are large, dark green, lanceolate to elliptic oblong, smooth, simple, alternate, coriaceous, glabrous and often wavy.

The flowers are yellow, small, numerous, borne in large terminal panicles. The ultimate branchlets of the inflorescence are 4-5–merous and cymose. Both male and hermaphrodite flowers are borne on the same panicle. The panicle is yellow to red, often pubescent. Large number of flowers are staminate having aborted pistils which are almost invisible. Only 6-11% are perfect flowers. These are only 1-3% in shy bearing types while in some 60-70% are perfect flowers. In general, out of 180 flowers only one fruit sets owing to degeneration of egg apparatus, anthracnose and damage by storms. Petals are 4-5. Sepals are 4-5. Half the length of petals is free and deciduous. The disc is fleshy and 5-lobed. Stamens are 4-5 but only one or 2 are functional. The ovary is single-celled having single ovule. The fruit weighs up to 2 kg (in cv. Tenneru). Pollination is carried out by insects. The number of maturing fruits is very small. Although more perfect flowers are at the base of panicle good set is usually more at the proximal end. The colour and flavour of fruit are variable but in general it is turpenoid. The stone is large, fibrous and has hard endocarp (Fig. 29.1). Generally the cultivars are monoembryonic. Polyembryony is found in the Philippines cultivars (e.g. Olour of Malabar).

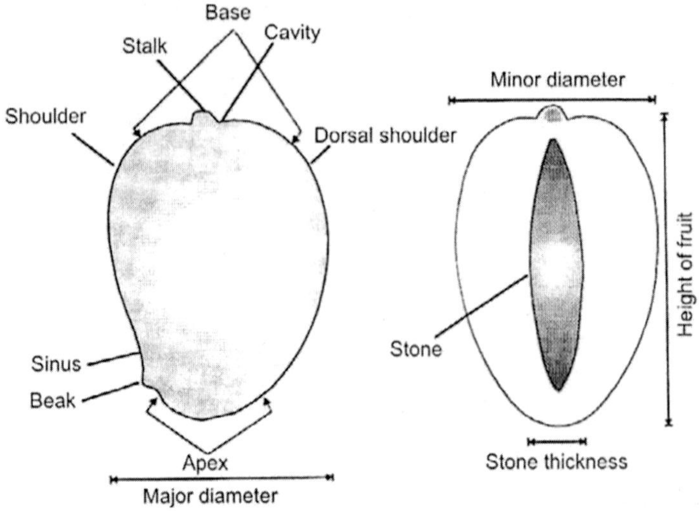

Fig. 29.1. Parts of mango fruit (Source: Sant Ram and Rajan, 2003).

Cultivars

Being heterozygous, seed progenies gave rise to enormous variability in mango. Innumerable cultivars developed as a result of fancy of people and business interests of nurserymen and were preserved in different regions by vegetative propagation. This has made the basis of their classification difficult owing to considerable confusion created by:

1. Varied names of the same cultivar, e.g. Suvarnrekha and Sinduri; Totapari, Sundershah, Bangalora and Collector; Baneshan, Banganpalli, Safeda and Chapta; Alphonse, Alphonso, Apus, Hafuz and Badami; Pairi and Peter; and Malda, Bombay Green, Sarauli and Sehroli (sometimes Bombay Yellow).

2. Same name for different cultivars, e.g. Safeda in Delhi (Banganpalli), Safeda Malihabad and Safeda Lucknow, and Malda Handle, Fazli and Bombay Green.

3. The names Fazli, Fazri and Fazri Zafrani appear confusing.

First systematic record of mango cultivars is available in Ain-i-Akbari (1590 AD), wherein cultivars of different regions having different size, shape, flavour and taste have been described (Singh 1960). Attempts have also been made to characterize and systematically classify the cultivars. Watt (1889) gave the earliest description of the Indian mango cultivars in a book, "Dictionary of Economic

Products". Maries (1902) described 500 cultivars from Bihar with botanical terminology, mainly fruit characters. Collins (1903) and Higgins (1906) also made attempts to classify the cultivars in Puerto Rico and Hawaii. Later, Woodhouse (1909) described the distinguishing characters of 40 cultivars from Bhagalpur (Bihar) mainly based on fruit characters and these cultivars were classified in three broad groups: (i) round fruit, (ii) Bombay type, and (iii) long oval or rectangular fruits, longer than broad. In Florida, Rolfs (1915) classified the cultivars based on the fruit and stone characters and growth habit of the tree into 9 groups: (i) Turpentine, (ii) Elcanor, (iii) Pineapple, (iv) No. 11, (v) Bombay, (vi) Cambodiana, (vii) Sandersha, (viii) Mulgoba, and (ix) Gola. Wester (1920) suggested that the system of classification should be comprehensive to include the cultivars not only from India but also from the other mango growing regions. He described important cultivars of the Philippines and India.

Burns and Prayag (1921) listed 89 cultivars from Bombay and divided them, based on the system given by Woodhouse, into 3 groups called 'cohorts': i) Round fruited, ii) Intermediate, and iii) Long fruited in "The Book of Mango". Hartless (1913), for the first time, used floral characters as criteria for classification. He classified the cultivars as small-panicled (10.2 to 20.3 cm), medium panicled (20.3 to 40.0 cm) and long-panicled (above 40 cm). Popenoe (1920, 1974) described 300 cultivars from all over the world and divided them into 4 groups: i) Mulgoba (mono-embryonic), ii) Alphonso (monoembryonic), iii) Cambodiana (polyembryonic), and iv) Sundershah (monoembryonic), using the characters such as i) colour of the flowering axis (panicle), ii) colour of the lateral branches, iii) pubescence on the panicle, iv) number of embryos/seeds, and v) fruit characters. This was the first attempt to include the characters of panicle for the purpose of classification. So far fruit and inflorescence characters were used for this purpose. Sturrock and Wolfe (1944) described 38 cultivars of Florida and gave the first key for their identification.

Mukherjee (1948) described 72 mango cultivars of Uttar Pradesh, Bihar and Bengal and also used vegetative characters for classification. The characters used were - i) tree growth habit, ii) colour of emerging leaves, iii) colour of mature leaves, iv) leaf margin, v) size of leaf, vi) colour of branches of the inflorescence, vii) pubescence on the panicle, viii) presence and length of bracts, ix) size of flower, x) ridges on the petals, xi) shape and colour of the ridges, and xii) fruit characters. Sturrock (1951) developed an artificial key based on fruit characters for the identification of mango cultivars grown in Florida. Gangolly *et al.* (1957) have described various fruit shapes in mango. Jintanawong *et al.* (1992) made a systematic approach in classifying 170 Thai mango cultivars based on the IBPGR Tropical Fruit Descriptor and divided them into 8 groups. Naik and Gangolly (1950) described 355 cultivars, based on a system that seems to be natural and divided them into 6 cohorts and further into 24 classes as given below:-

Primary characters (giving rise to 6 cohorts) :

i. Fruit shape: a) Round, b) Intermediate, and iii) Long (Fig. 29.2).

ii. Beak: a) Prominent beak, and b) Beak absent (Fig. 29.3).

Secondary characters (to divide each cohort into 4 classes giving rise to 24 classes):

i. Leaf apex: a) Acute and ii) Sub-acuminate.

ii. Folding of the leaves: a) Slight folding, and b) Strongly folding

Characters used at tertiary level:

i. Shape of the inflorescence.

ii. Shape of leaf.

iii. Venation on the stone.

iv. Amount of fibre on the stone.

v. Shoulder of the fruit (Fig. 29.5).

vi. Colour of the emerging leaves.

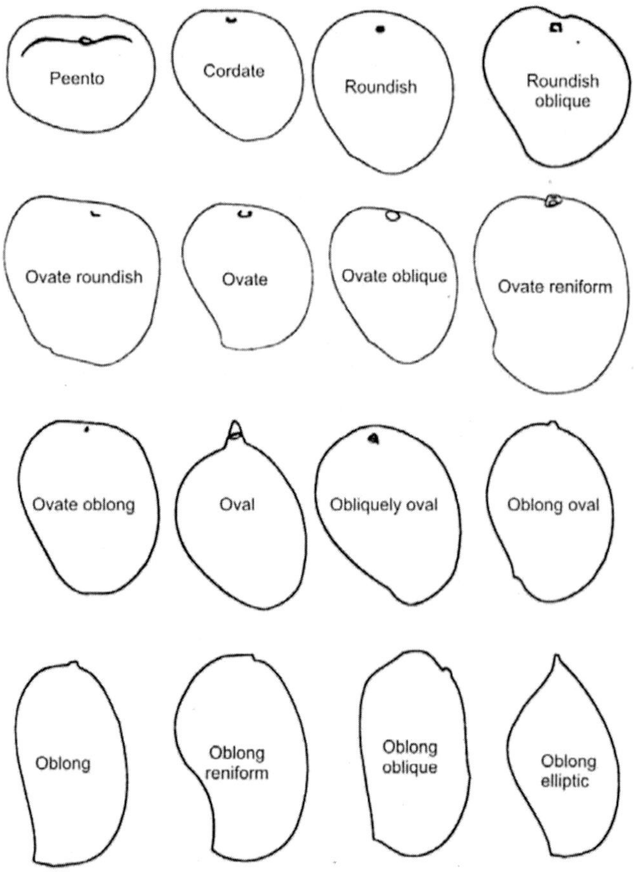

Fig. 29.2. Fruit shape in mango (Source: Sant Ram and Rajan, 2003).

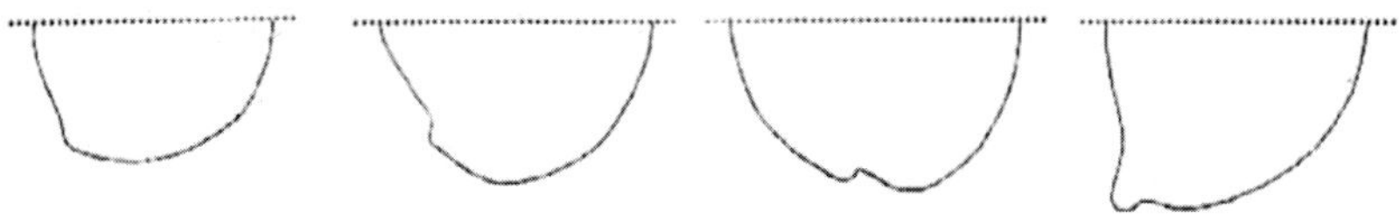

Fig. 29.3. Beak types in mango fruit (Source: IPGRI, 2006).

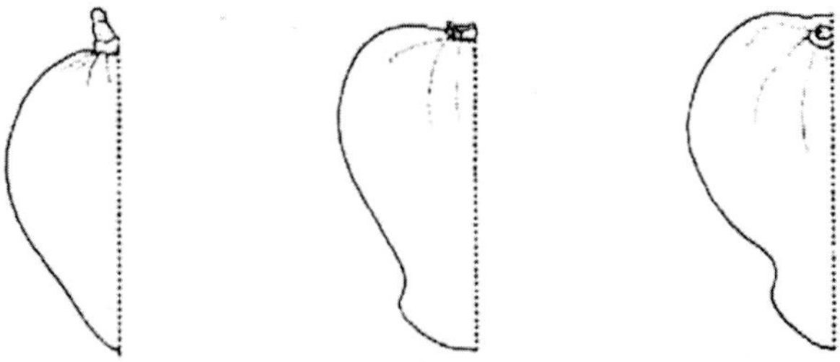

Fig. 29.4. Sinus types in mango fruit (Source: IPGRI, 2006).

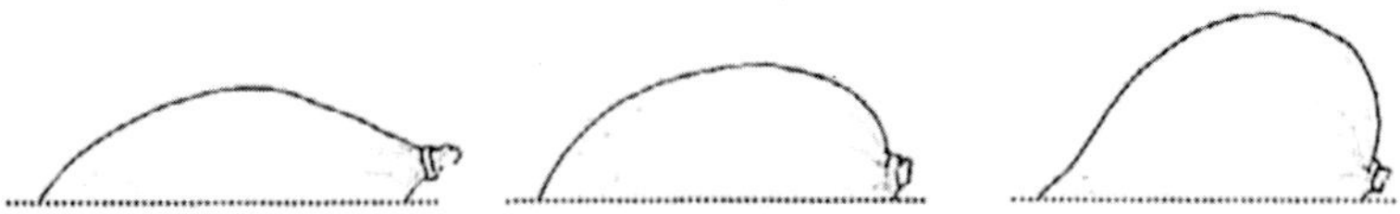

Fig. 29.5. Shoulder slope types in mango fruit (Source: IPGRI, 2006).

Singh and Singh (1956) from Saharanpur divided the mango cultivars into 6 cohorts and further into 72 classes as given below:

Primary characters (giving rise to 6 cohorts) :

i. Three fruit shapes

ii. Two types of beak

Secondary characters (to divide each cohort into 12 classes giving rise to 72 classes):

i. Venation on the stone: a) Depressed, b) Raised, and c) Level.

ii. Colour of the panicle: Four colours

Characters used at the tertiary level:

i. Apex of the leaf : a) Acute, b) Acuminate, and c) Sub-acuminate

ii. Folding of the leaf : a) Flat, and b) Folded.

Popenoe has divided the cultivars into 6 horticultural races:-

a. Polyembryonic

i. West Indian, e.g. cv. Carbao
ii. Hawaiian
iii. Cambodiana, e.g. cv. Cambodiana

b. Monoembryonic

iv. Mulgoba, e.g. Mulgoba, Haden
v. Alphonso, e.g. Alphonso, Pairi
vi. Sandershah, e.g. Totapuri, Bangalora (pulp hard, skin thick).

Except in subnormal or deformed fruits, the form seems to be the most prominent character for identification as it exhibits the least variation under different environmental conditions. Singh and Singh (1956) considered presence of fibre, its length and texture in the pulp, nature of venation on the stone, in depressed or raised form, leaf tip (apex) and folding of leaves as important characters for classification of the cultivars and that the characters such as colour of leaves and shape of inflorescence are not so important. Time and duration of flowering are of less importance. The colour of fruit skin and pulp can be of importance for identification of coloured cultivars but Rajan and Lal (1993) observed great variation in fruit colour even within Dashehari as a result of solar radiation. Accumulation of anthocyanin pigments in mango fruit peel and carotenoids in the pulp is dependent on exposure to sunlight and stress conditions. Although size and distribution of lenticels (dots) on the fruit is a useful character for distinguishing closely related cultivars but environmental interactions may reduce its usefulness. Jintanawong *et al.* (1992) observed that specific leaf and fruit characters were highly heritable.

Thus the most stable characters that emerge important for classification of mango cultivars are fruit shape and size, presence and form of beak, fruit base, shoulder, sinus, apex and cavity of stalk insertion, size and distribution of lenticels (dots) on the fruit, presence of fibre, its length and texture in the pulp, nature of venation on the stone, in depressed or raised form, colour of panicle axis, presence of pubescence on the inflorescence, panicle length, shape, presence of deciduous bracts, leaf tip or apex and folding of leaves.

A. Cultivars in India

I. Traditional cultivars

Bombay Green or Sarauli: It is commonly grown in North India due to its early ripening habit. The fruits mature by mid June. It is also called Malda in Northern India. The cultivar is susceptible to malformation and biennial bearing and has moderate bearing. The fruits are spinach green in colour when ripe, medium in size and somewhat irregular in shape from oblique to ovate. The flesh has good taste and quality but the keeping quality is not very good.

Rataul: The cultivar has fruit maturity is mid season (July). Fruit is small, oval. Peel is thin, lemon yellow. Pulp is firm having good quality, excellent flavour and luscious and delightful. The keeping quality is good.

Zardalu: It is an early to mid season cultivar developed as chance seedling in Murshidabad, West Bengal. Fruit is medium, oblong to obliquely oblong, golden yellow having firm to soft pulp. Fruit quality is very good.

Dashehari or Aman Dashehari: This cultivar derives its name from the village Dashehari near Lucknow. It is a leading commercial cultivar of North India and is one of the best among Indian cultivars. It matures fruits by June end and produces heavy crops but is alternate in bearing. The fruits are medium sized, elliptic oblong with attractive greenish yellow colour and smooth shiny surface. Fruit quality is excellent and keeping quality is good. The flesh is yellow, fibreless, sweet, juicy and excellent in taste and flavour. It is mainly used for table purpose and is suitable for canning.

Langra or Banarsi Langra: This cultivar is indigenous to Varanasi area of Uttar Pradesh and is extensively grown in northern India. The cultivar has also done well in southern and western India. The cultivar has tendency of biennial bearing but is heavy cropper. Fruits are larger in size, round to oblong oval in shape with very small beak having light green (lime green) to yellowish green, smooth and leathery surface. The flesh is yellow, more fibrous having very good sugar-acid blend and pleasant and aromatic flavour and taste. Fruit quality is good. It is sometimes rated better than Dashehari in quality. Keeping quality is medium.

Chausa (Samar Behist Chausa): The cultivar originated as a chance seedling in the orchard of a Talukadar of Sandila in Hardoi district of Uttar Pradesh and is commonly grown in northern India. The trees start bearing late, 15-20 years after planting. The fruits are large in size, ovate to oval oblique in shape having smooth and shiny yellow surface. The flesh is yellow, soft and juicy having pleasant flavour and very good quality. It is one of the cultivars producing very sweet fruits having very low or negligible acidity when fully ripe. The cultivar suffers from the maladies of mango malformation and alternate bearing. Keeping quality of the fruit is medium.

Ramkela or Ramgola: The cultivar is indigenous to North India. Fruit maturity is late. The fruit is small to medium, roundish oblique having medium thick peel, firm, primuline yellow and fibreless pulp. The fruit quality is medium to good and keeping quality is medium. The fruits are used for pickle making.

Bambai: It is a commercial cultivar from Bihar. It is also known as Malda in West Bengal and Bihar. Fruit size is medium having ovate oblique shape and yellow colour. Fruit quality and keeping quality are medium.

Himsagar: This cultivar is indigenous to West Bengal. This is one of the choicest cultivars of Bengal and has gained extensive popularity. Fruit is of medium size and ovate to ovate oblique in shape. Fruit colour is yellow. Both fruit and keeping quality are good.

Gulabkhas: The cultivar is indigenous to Bihar. Fruit maturity is mid season. The fruit is small to medium, oblong oblique and cadmiun yellow having medium thin peel, firm and fibreless pulp. Fruit quality and keeping quality are very good.

Krishna Bhog: This cultivar is indigenous to Murshidabad in West Bengal. Fruit size is medium having roundish oblique shape and yellow colour. Fruit quality and keeping quality are good.

Fazli or Fajri: The cultivar is commonly grown in Uttar Pradesh, Bihar and West Bengal. Fruit is very large and obliquely oval in shape and the colour is light chrome. Fruit quality and keeping quality are medium.

Sukul: the cultivar is indigenous to Bihar. Fruit maturity is late. The fruit is medium to large, oblong oblique and primuline yellow having thick peel, firm and fibreless pulp. The fruit quality is medium to good and keeping quality is good.

Pairi or Raspuri: It is an excellent cultivar adapted to western and southern parts of India. It is high yielding and produces quality fruits but suffers from the problem of biennial bearing. The fruits are medium sized and ovate having significant apex (beak) with very slight sinus, smooth and shiny surface, yellowish green colour with very brilliant red tinge on the shoulders.

Alphonso: The cultivar is known by different names in different regions, viz. Badami, Gundu, Khader, Appas, Happus and Kagdi Happus. It is rated very high in India and has high export value. The cultivar has specific climatic requirements and performs well in Ratnagiri, Maharashtra on west coast of India. It is a mid season and fairly regular cultivar but is medium cropper. The fruits are medium large, oval-ovate oblique with prominent ventral shoulder and glossy attractive yellowish with bright pinkish blush at the basal end. The flesh has yellow colour, pleasant flavour, very less fibre and an excellent blend of sugar and acid and superb taste. The fruits have excellent table, canning and processing qualities. Keeping quality is good.

Fernandin: It is a late cultivar and has good keeping quality. This is one of the oldest cultivars of Bombay believed to have originated in Goa. Fruit size is medium to large having oval to obliquely oval shape and yellow colour with a red blush on the shoulders. Fruit quality and keeping quality are medium. It is a late season cultivar mostly used for table purpose.

Mankurad: The cultivar is of commercial importance in Goa and in the neighbouring Ratnagiri district of Maharashtra. The cultivar develops black spots on the skin in rainy season. Fruit is medium in size, ovate in shape and yellow in colour. Fruit quality is very good. Keeping quality is poor. It is a mid season cultivar.

Kesar: This is a leading cultivar of Gujarat. Fruit size is medium and oblong shape having red blush on the shoulders. Keeping quality is good. It is an early cultivar.

Vanraj: It is a highly prized cultivar of Vadodra district of Gujarat and fetches good returns. Fruit is medium in size, ovate oblong in shape and deep chrome colour with a blush of jasper red on the shoulders. Fruit quality and keeping quality are good. It is a mid season cultivar.

Jamadar: It is a famous cultivar of Gujarat. Bearing is heavy and regular. Fruit is of medium size and good quality. Pulp is firm, yellow and fibreless. Flavour is pleasant. Keeping quality is good.

Rajapuri: It is a commercial cultivar of Gujarat. Maturity is early to mid season. Fruit is medium to large, ovate to ovate oblong having thick peel, deep chrome colour and firm and fibreless pulp. Fruit quality is good and keeping quality is medium.

Bangalora or Totapuri: It is a very popular mid season cultivar of South India. It bears regularly and heavily. The fruit quality is not very good but on account of its regular cropping and good processing traits, it is preferred by processing industry. The fruit is medium to large, oblong with a neck at base and conspicuous beak and sinus at apex and is attractive in appearance with bright apricot yellow colour overlain with red blush. The skin is thick. The flesh is yellow to orange, firm, fibreless and somewhat dry with pleasant flavour and good sugar-acid blend. The fruits have very good storage quality. It is a mid season cultivar.

Neelum or Neelam: This cultivar is indigenous to Tamil Nadu. It is a very regular and high yielding cultivar grown commercially in South India. It has wide range of adaptability. The fruits have fairly good storage ability and therefore, are ideal for transporting to distant places. The fruit is medium in size, ovate oblique in form having somewhat rounded beak, prominent sinus and saffron yellow, smooth and shiny surface at ripening. Fruit quality is good, better than that of Bangalora. The flesh is firm having little fibre, pleasant flavour and good taste with acidic blend. It is a late season cultivar.

Rumani: It is an apple shaped mango. It is more round with a basal cavity. It has good keeping quality. Fruits become hopeless if harvested when fully mature.

Banganpalli: It is a commercial cultivar of Andhra Pradesh and Tamil Nadu and is also known as Chapta, Safeda, Baneshan and Chaptai. Fruit is large in size and obliquely oval in shape. The colour of the fruit is golden yellow. The fruit pulp is without fibre and is of good quality. Keeping quality is good. It is a mid season cultivar and is good for canning.

Suvarnarekha: This is a commercial cultivar of Visakhapatnam district of Andhra Pradesh. Other names of this cultivar are Sundari and Lal Sundari. Fruit is medium in size and ovate oblong in shape. Colour of the fruit is light cadmium with a blush of jasper red. Fruit quality is medium and keeping quality is good. It is an early cultivar.

Mulgoba or Mulgoa: This is a popular commercial cultivar of southern India. It does well in dry regions and not in the coastal areas. Fruit is large in size, roundish oblique in shape and yellow in colour. Fruit quality is very good. Keeping quality is good. It is a late season cultivar.

Raspoonia: It is a sucking cultivar.

The other cultivars are Bombay yellow, Gopal Bhog, Bharat Bhog, Fazri Zafrani, Khasulkhas, Khajuri, Taimuriya, Mithua, Bathua and Sepia Shah Pasand.

II. Cultivars based on fruit size and peel colour

1. Very large fruits (over 500 g): Hathijhool, Pansera (4.6 kg) Fazli (800-1200 g).
2. Large (300-500 g): Arka Aruna (450 g), Mallika (300 g).
3. Medium large (150-299 g): Dashehari (150-260 g), Langra (175-240 g), Sammar Behist Chausa (200-275 g), Bombay Green (150-200 g).
4. Small (100-149 g): Amrapali (130 g) Rataul (107 g).

5. Very small (99 g and below): Bhowani Chowras (70 g), Malviya Bhog (81.5 g).
6. Various shades and amount of green and yellow peel colour.
7. Various shades and amount of red peel colour.

III. Cultivars based on use

1. Dessert cultivars: Dashehari, Langra , Sammar Behist Chausa.
2. Processing uses: Ramkela (pickle), Amrapali (beverage and nectar), Bangalora (pulp), Sukul (juice), Mallika and Dashehari (canning).
3. Rootstock use: Kerala Dwarf (dwarfing), vigorous (Muvandan), salt tolerant (Kurkkan and 13-1).

IV. Promising Selections

Dashehari 51: This is a regular bearing clonal selection from Dashehari.

Menaka: This is a seedling selection from Gulabkhas. It is a late maturing cultivar having good yield, attractive fruit colour (red blush on basal side of fruit), medium size (300 g) with moderate storage life. The pulp content is 75% and is deep yellow, less fibrous, firm and sweet (20ºB TSS).

Subhash: This is a chance seedling from Zardalu. The cultivar is free from serious pests and diseases. The fruit has attractive bright yellow colour of Zardalu with shape similar to Langra. The fruit is medium in size (220 g), very sweet (24ºB TSS) pulp having 0.294% acidity.

Pant Chandra: This is a suitable cultivar for hills and valleys. It is a mid season cultivar having tall and erect trees. Fruit is green, medium in size having reddish yellow and sweet (18ºB TSS) pulp.

Pant Sinduri: The tree has medium height and round canopy. It is an alternate bearer. Fruit has medium size (200 g), yellow peel with pink shoulder and yellow and sweet (16-18ºB TSS) pulp.

Alphonso Clone 900: It is a highly productive and early maturing cultivar having large fruits suitable for processing.

Paiyur-1: This is a clonal selection from Neelum made at Karukkanchavadi, Dharmapuri. The trees are dwarf, low spreading and suitable for high density planting. It is a regular bearer giving mean yield of 22.3 kg/tree (8929 kg/ha) at 9 year age. Fruits are medium long, ovate with yellow skin, excellent taste and good keeping quality.

Pusa Surya: This is a selection from Eldon. It is a regular bearer, semi-vigorous cultivar and is suitable for close planting giving high yield of good quality fruits having attractive peel colour.

Mahmooda Vikarabad: The dwarf cultivar is suited for high density planting. The tree is small in size. The fruit is medium having thin and yellowish green skin and moderately firm and fibreless flesh. The fruit quality is very good to high. It is mid season to late cultivar regularly bearing heavy crop and tolerates wind but is susceptible to hoppers. The keeping quality of fruit is good.

V. Hybrid cultivars

Al Fazali (Alphonso x Fazali): Developed at Bihar Agricultural College, it is a good quality and late maturing cultivar, free from malformation and fruit fly.

Amrapali (Dashehari x Neelum): Bred at IARI, New Delhi, it is a dwarf, regular bearing and late maturing cultivar. It is suitable for high density planting. About 1600 plants can be planted in one hectare. It yields 16 tonnes fruits per hectare, on an average.

Mallika (Neelum x Dashehari): The fruit is large in size, oblong elliptical in shape and cadmium yellow in colour. Fruit and keeping quality are good. It is a mid season cultivar.

Manjeera (Rumani x Neelum): Bred at Sangareddy, Andhra Pradesh, it is a regular cultivar, heavy bearing in bunches and suitable for high density planting.

Pusa Arunima (Amrapali x Sensation): The cultivar is regular bearer. It gives good fruit yield of high quality fruits having attractive colour.

Pusa Lalima (Dashehari x Sensation): The tree is vigorous, regular and highly productive. Fruit is medium in size having attractive red peel with some green portion, sweet, fibreless and firm flesh having pleasant flavour.

Pusa Pratibha (Amrapali x Sensation): It is a regular bearing cultivar. The fruits have long shelf life and excellent quality.

Pusa Shreshtha (Amrapali x Sensation): It is a regular bearing cultivar with long shelf life giving excellent quality fruits.

Pusa Pitamber (Amrapali X Lal Sundari): Fruits have attractive bright yellow and are sweet having good sugar/acid blend.

Ambika (Amrapali x Janardan Prasad): Bred at Lucknow, the trees are semi-vigorous. The fruit skin is attractive and bright yellow having dark red bluish and firm flesh with scanty fibre and 21°B TSS. It is a late cultivar.

Arunika (Amrapali x Van Raj): The trees are semi-vigorous. It is a good quality late cultivar, free from malformation and fruit fly. Fruits are large (460 g), sweet (18.5°B TSS) and have better taste and lesser fibre than Fazli.

Sunder Langra (Langra x Sunder Prasad): It is a moderate and regular bearing cultivar. The fruits are medium in size (240 g) and very sweet (23°B TSS). The shape of fruit is similar to Langra with red blush on the peel.

Safari (Gulabkhas x Bombai): It is an early maturing cultivar having medium size (153 g), greenish yellow fruits with red blush on the base and very sweet (22.5°B TSS) and firm flesh. The fruits are suitable for canning.

Jawahar (Gulabkhas x Mahmud Bahar): It is a semi-dwarf, precocious and regular bearing cultivar having medium-sized (215 g), greenish yellow and highly sweet (22.5°B TSS) fruits.

Sai-Sugandh: Fruit colour is golden yellow on ripening and has 380 g weight, 22°B TSS, pleasant aroma and sweet taste, and is free from spongy tissue and stone weevil. It is suitable for table purpose. It is a monoembryonic cultivar.

Konkan Ruchi (Neelum x Alphonso): It is a regular bearer, high yielding cultivar having fruits suitable for pickle making. The fruits are large (430 g) with 78% pulp content, 8ºB TSS and 3.2% acidity before ripening at maturity.

Suvarna (Alphonso Clone 900 x Neelum 287): The tree is vigorous and regular and cluster bearing. Fruits are of medium size, ovate oblique with yellow peel having orange blush, thin peel, 78% pulp, 22.5ºB TSS and 4800 IU Vitamin A.

Konkan Raja (Bangalora x Himayuddin): This is a regular bearing cultivar. Fruits are large (616 g) having thick stalk, yellow peel and high pulp content (83%). This is suitable for use as salad at maturity (before ripening).

Arka Aruna (Banganpalli x Alphonso): Bred at IIHR, Bengaluru, It is a dwarf, regular bearing and precocious cultivar suitable for high density plantation. Fruits are large having attractive skin colour with red blush and are free from spongy tissue.

Arka Puneet (Alphonso x Banganpalli): Bred at IIHR, Bengaluru, it is regular and prolific bearer. Fruits are medium sized having excellent keeping quality and attractive skin colour with red blush and are free from spongy tissue.

Arka Anmol (Alphonso x Janardhan Prasad): Bred at IIHR, Bengaluru, it is a regular bearer and good yielder. Fruits are medium sized having uniform yellow peel colour, excellent keeping quality and free from spongy tissue.

Arka Neelkiran (Alphonso x Neelum): Bred at IIHR, Bengaluru, it is regular bearer and late season cultivar with medium sized fruits having attractive red blush, free from spongy tissue.

PKM-1 (Chinnasuvarnarekha x Neelum): It is a regular bearing and high yielding cultivar and bears in clusters. The fruits are medium sized (250-300 g), long, juicy and suitable for pulp making.

PKM-2 (Neelum x Mulgoa): It is a regular bearer having excellent fruit quality.

Neelphonso (Neelum x Alphonso): It is a regular bearer, dwarf, late maturing cultivar suitable for table purpose and juice making.

Neeleshan (Neelum x Baneshan): It is a dwarf and late maturing cultivar with fruits having good keeping quality. The fruits are large (318 g), oval and sweet (16-20ºB TSS).

Nileshan Gujarat (Neelum x Baneshan): The tree is dome shaped having moderate growth, medium-sized fruits suitable for table use. The fruit has 16.26ºB TSS, 11.51% total sugar, 0.12% acidity and 16 mg/100g pulp vitamin C having good keeping quality. The fruit yield is 48 kg/tree.

Neeleshwari (Neelum x Dashehari): It is a dwarf and late maturing cultivar with fruits having non-fibrous, firm texture pulp of very good quality.

Neelgoa (Neelum x Pera Mulgoa): The trees are regular bearer producing fibreless and moderately juicy fruits.

Neeluddin (Neelum x Himayuddin): The trees are regular bearer producing fibreless, medium sized, flavoured and juicy fruits.

Prabhasankar (Bombai x Kalapadi): The trees are regular bearer producing medium size fruits that resemble Bombai but have comparatively late maturing habit.

Ratna (Neelum x Alphonso): The tree is vigorous and precocious. The fruits are medium sized, attractive in colour and free from spongy tissue. It is a regular bearer.

Sindhu (Ratna x Alphonso): It is a regular bearer. The fruits are medium sized, seedless, early maturing and free from spongy tissue with high pulp to stone ratio and very thin and small stone.

Au Rumani (Rumani x Mulgoa): It is precocious, heavy and regular bearing cultivar giving large fruits having yellow cadmium skin colour.

AU Rumani (Rumani x Neelum): The cultivar bears heavy yield of fibreless large fruits.

Other released hybrid cultivar is Sabri.

B. Cultivars grown in USA (Florida and Hawaii)

Aloha: The cultivar originated in San Diego and Jerry Staedeli from Hawaiian seed. The tree is spreading and light bearer depending on rootstock affinity. The fruit is large (397-510 g), dull yellow and covered with red. It is an early (Oct-Nov) cultivar. It is susceptible to anthracnose.

Brooks: The cultivar originated in Miami from seedling of Sandersha. The tree is somewhat dwarf. Fruit is medium to large (283-566 g), kidney-shaped, green with yellow shoulder and rather fibrous. It is a very late cultivar and is resistant to anthracnose.

Cambodiana: The cultivar originated in Miami from the seedling of Saigon. It is a Philippine type. Fruit is small to medium, elongated ovate, yellow green and juicy having acidic flavour. It is an early cultivar.

Carabao: The cultivar originated in Philippines. This is early to mid season cultivar. It is polyembryonic. The tree is vigorous forming a large and dense canopy. The fruits are produced in clusters. The fruit is medium (220-311 g), oblong to elongated, kidney-shaped, 13 x 7.5 cm, light green to yellow blushed with few small green dots (lenticels) having rounded to slightly flattened base. The skin is thin and medium tough. The flesh is stringy, tender and melting with little fibre, lemon yellow, spicy, juicy, acidic and sweet with good to excellent eating quality. It is an early to mid season cultivar. The seed is very large. The cultivar has fair/good resistance to anthracnose and powdery mildew but is susceptible to fruit fly and alternate bearing.

Carrie: The cultivar originated in Delray Beach, Florida from seedling of Sophie Fry. The tree is dwarf. Fruit is small, 340 g, regular, ovate, green yellow and fibreless having high flavour. It is an early cultivar.

Cooper syn. Cooper No. 1 or 3: The cultivar originated in Hollywood. The tree is spreading and dense. The fruit is large (454-566 g), long and green. The flesh is of high quality. It is a late cultivar.

Costa Rica: The cultivar originated in East Los Angeles from the seed from Costa Rica. Fruit is small, 283 g, elongated, flat, pale green and juicy. It is a very early cultivar.

Doubikin: The cultivar originated in Kelmscott, West Africa. The tree is dwarf, rounded, slow growing and fruits in two years even from seed. It is polyembryonic. The fruits are round and large (340-454 g). It is a mid season cultivar.

Earlygold: The cultivar originated in Pine Island, Florida. The tree is upright. Fruit is small, 340 g, obliquely round, orange having red blush and fibreless. The seed is often abortive. It is a very early cultivar and is resistant to anthracnose.

Edgehill: The cultivar originated in Vista, California. It is an Indian type. The tree is upright, hardy and vigorous. It is monoembryonic. The tree blooms early and produces small (227-340 g), almost fibreless fruits, green in colour with red blush. It resists mildew but is subject to soft nose. It is a mid season (November-December) cultivar.

Edward (Haden x Carabao): The cultivar was developed at Miami. It is intermediate between Indian and Philippine forms. The tree is dense and compact. The fruit is medium to large, elongated ovate with apex often oblique and yellow green with red blush and excellent flavour. Seed is very small and can be easily removed. It is an early cultivar.

Fascell: The cultivar originated in Miami from seedling of Brooks. The tree is open and slow growing. Fruit is medium to large, elongated flattened and yellow with pink blush and acidic flesh. It is an early cultivar.

Gouveia: The cultivar originated in Honolulu. The tree is upright and open. The fruit is medium to large (283-566 g), long ovate and green becoming bright red. It is sweet, juicy with no fibre. It is a late cultivar having uneven ripening.

Haden: The cultivar originated in Florida from a seedling of Mulgoba (Indian type). The tree becomes quite large and spreading. It is alternate bearing. The fruit has very attractive appearance and excellent quality. It is medium to large-sized (431-680 g), regular ovate, 10 x 8 cm, and plump with a rounded base and depressed beak, bright yellow with deep crimson or red blush and numerous large whitish/ yellow glands. The skin is thick and tough. The flesh is yellowish orange, firm, very juicy with a pleasant aroma and little fibre only close to the seed which is monoembryonic and covered in a medium thick woody shell. It has good storability. It is an early cultivar, susceptible to anthracnose and only moderately resistant to powdery mildew.

Irwin: The cultivar originated in Miami. It is a seedling of Lippens and is Florida's leading local cultivar. The tree is somewhat dwarf to medium-sized, moderately vigorous with an open canopy and produces consistently good yields. It is a mid season cultivar and matures fruits from late January until March. Fruits are often produced in clusters. Fruit is medium (372 g), elongated, ovate regular in form, 12.3 x 8.5 cm, orange yellow with a crimson or dark red blush and numerous large white dots. The flesh is lemon yellow, soft, tender, melting, juicy, sweet with a pleasant aroma and only slightly fibrous. It has short storage life. The seed is mono embryonic and embedded in a relatively small and thin stone shell.

Julie: The cultivar originated in Trinidad . The tree is dwarf and slow growing. Fruit is small (170-284 g), flat oblong and obliquely almost two-nosed, orange, rather fibrous, juicy and sweet. It is a late cultivar.

Keitt: The cultivar originated in Homestead and is probably a seedling of Mulgoba. Fruit is large (566-737 g), ovate with slightly oblique apex, green having rich flesh and fibre only around the seed. It resists mildew. It is a late cultivar. In Florida, fruiting takes place during July-August and sometimes up to September.

MacPherson: The cultivar originated in Encinitas. The tree is dense and low branching. Fruit is small (170-227 g) and yellow green with red blush having fairly good flesh. It is a mid season cultivar for coastal region.

Manila: The cultivar originated in Mexico as a strain from Hawaii and is common in Veracruz state. It is Philippine type. The tree is dwarf and dense. The fruit is small to 284 g, long, flat and yellow having a sharp flavour. It is susceptible to anthracnose. It is an early season (October-December) cultivar but is best when the fruits are picked late.

Mulgoba: The cultivar originated in Bombay and is distinct from the ancient Mulgoa. The fruit is medium (454 g). The cultivar is suitable for greenhouse.

Ott: The cultivar originated in La Habra heights. It is a seedling of Saigon. The tree is dwarf. Fruit is medium to 15 cm in size and orange yellow with pink blush. It is a very early cultivar.

Pina syn. Pineapple: The cultivar originated in Mexico as a seedling strain. It is a Philippine type. The tree is upright. Fruit is small to 340 g, of ovoid shape and orange yellow colour. Its flavour suggests that of pineapple. It is an early season (November-December) cultivar.

Pirie (syn. Paheri): The cultivar originated in India. The tree is broad and spreading. Fruit is small (227-284 g), almost round with oblique apex, yellow with red blush, juicy and fiberless having rich flavour. The cultivar has alternate bearing and blooms every 18 months. It is an early to mid season cultivar, suitable for greenhouse.

Reliable: The cultivar originated in San Diego, California from seedling of Sensation. The tree is broad, dense and slow growing. The fruit size varies from 284 to 567 g, shape is oblong and is yellow blushed red. It rarely misses a crop. It has long ripening season (October-February). It is susceptible to anthracnose and soft nose.

T1: The cultivar originated in Vista from seedlings of Edgehill. The tree is low and spreading and vigour depends upon the rootstock. Fruit is medium to large, 15-20 cm, of broad oval shape, green with red blush and fibreless. It is susceptible to anthracnose and resists mildew and soft nose. It is a late season (December-January) cultivar and is very late on the coast (January-February).

Thomson syn. Thomson Large Seedling: The cultivar originated in Vista, Manila from seedlings. It is polyembryonic. The tree is spreading, vigour depending upon the rootstock. Fruit is small to medium (170-340 g), yellow and flat-shaped. It resists mildew. The fruit has high fibre when chemical fertilizer is used. It is an early cultivar (September-November) and the fruits ripen well indoors if picked prematurely.

Tommy Atkins: The cultivar is of unknown parentage, originated from a seed planted at Fort Lauderdale, Florida. It is commercially grown for export in Florida. The tree is vigorous and large with a rounded dense canopy and produces consistently heavy and regular good crop. It is an early to mid season cultivar and ripens well if picked immature. Fruit is medium to large (522 g), ovate to oblong, 12.6 x 9.9 cm having orange yellow and smooth, tough and thick skin covered with red and heavy purple bloom and numerous white lenticels and a broadly rounded base. The flesh is yellow to deep yellow, firm, juicy with a strong pleasant aroma and medium fibre. It is of fair to good eating quality having excellent shipping and shelf life qualities. The seed is monoembryonic and covered in a thick, woody stone (6.6% of total fruit weight). The cultivar is highly resistant to anthracnose and powdery mildew.

Villasenor: The cultivar originated in Los Angeles. The tree is dwarf, spreading and strongly responds to rootstock. Fruit is medium to 340 g having ovate shape, greenish yellow colour with pink blush, mild flavour. It is a late season (Dec-Jan) cultivar.

Winters syn. M20222, Southland: The cultivar is a selection made in Miami, USDA from the seedlings of Ono. It is Philippine type and is polyembryonic. The tree is broad giving variable production. Fruit is medium to 397 g, smaller having half round shape, yellow colour and is blushed red. It is susceptible to anthracnose and resists soft nose. The parthenocarpic fruits attain full size. It is a mid season (Nov-Dec) cultivar and ripens well if picked immature.

Zill: It originated in Lake Worth, Florida from seedling of Haden. The tree becomes fairly large with an open, very spreading canopy. This is an early to mid season cultivar and produces well and fairly consistently. The fruit is small to medium, 225-345 g (average 285 g), yellow with an intense red or crimson blush. The fruit shape is almost round resembling that of van Dyke, 10.6 x 8.2 cm. The apex is rounded with a small beak. Lenticels are yellow/brown and the flesh is deep yellow, juicy, soft and without fibre. The flavour is rich and sweet and of good to excellent eating quality. The seed is monoembryonic and covered by a thick woody stone (8% of fruit weight). The cultivar has moderate resistance to anthracnose and powdery mildew, but does not withstand storage and shipping stress well.

Apple: The trees are large, vigorous of pyriform growth habit. It is an early cultivar. The fruits are nearly round, 9.7 x 11 cm, medium to large, 280-580 g (mean 397 g) having smooth and thin skin, yellow/orange to red colour when ripe. The flesh is fibreless, juicy and yellow of excellent flavour and melting texture. It is not a polyembryonic cultivar. The cultivar is susceptible to anthracnose and powdery mildew and is alternate bearing.

Arumanis: This cultivar originated from Indonesia and is called Harumanis in Malaysia. It is widely planted in humid parts where many cultivars of better quality fail to fruit. The tree is vigorous and tall with a slightly open canopy. It is polyembryonic. It is biennial bearing, poor yielding and has low to fair resistance to powdery mildew and anthracnose. The fruit is small, oval to oblong, yellow with large yellow white dots and a rounded base. The beak is inconspicuous and the skin is thin and tough. The flesh is firm and juicy with little fibre, lemon yellow, sweet, slightly insipid with a strong aroma, but with poor to fair eating quality.

Batawi: The trees are relatively small, round in shape and medium in productivity. It is polyembryonic. It does not have a distinct biennial bearing. The fruit is very large, 470-590 g (average 523 g), 10.7 x 10.6 cm, round having a rich olive green to purple maroon colour and a prominent beak. The flesh quality is good both in texture and flavour with pale orange colour and little fibre. The cultivar has good resistance to anthracnose but is susceptible to powdery mildew.

Boribo: The tree is large and vigorous. It is polyembryonic and is regular in bearing. The fruit is large, 430-640 g (mean 511 g), oblong, 11.5 x 7.8 cm, but not as slender as Ngowe. The shoulders are only slightly curved and the beak is obscure. The fruits are pale olive green with bloom and become yellow apricot when ripe. The flesh quality is good to excellent. It is of a deep orange colour, virtually free from fibre, juicy and of a very strong flavour. The cultivar has fair resistance to anthracnose but is susceptible to powdery mildew.

Gesine: The tree is small to medium in size and forms a dense canopy. It is an early and regular bearing cultivar.The fruits are medium-sized, 240–300 g (average 280 g), oblong, 11.7 x 7.8 cm, brightly coloured and attractive of very good quality. Fruits exposed to the sun are intense red, while those developing inside the canopy are apricot in colour with a reddish blush. Lenticels are white turning brownish green at maturity. The flesh is melting, juicy, deep orange, aromatic and relatively free from fibre. The cultivar is heavily attacked by anthracnose and to a lesser degree by powdery mildew.

Golek: The cultivar originated from Indonesia. The tree is moderately vigorous with an upright and open canopy. It is polyembryonic and mid season. The fruit is oblong with a rounded base, greenish yellow with an orange overlay and prominent white dots. The fruit size is 10.9 x 8.3 cm of 210 – 500 g (average 325 g). The skin is thin and easily separated. The flesh is deep yellow, sweet, insipid, soft and juicy with a mild aroma and has abundant fibre and a fair eating quality. The cultivar has good resistance to anthracnose.

Heart: The tree is vigorous with a large, spreading rounded canopy. The mid season and polyembryonic cultivar is regular producer. The fruit is small to medium in size, 8.4 x 8.0 cm of 140 to 255 g and of a bright lemon yellow colour with spicy, resinous aroma. It is heart-shaped with a flattened base and beak and a few large white dots. The skin is thick. The flesh is firm and juicy with a moderate amount of coarse fibre. Its eating quality is only fair. The cultivar has good resistance to anthracnose.

Kensington (Kensington Pride, Bowen): The cultivar was discovered in Bowen, Australia, thought to have originated from seed imported from India. The trees are vigorous and spreading and produce consistent, high yields. The cultivar is early to mid season. The fruit is of medium size, 319 g, almost round, 12.2 x 8.1 cm, with flattened base and slight beak. The skin is thick, yellow with a slightly orange/pink blush and is adherent to flesh. The flesh is yellow, sweet, soft and juicy with unique flavour and moderate to little fibre. The seed, covered by a moderately thick woody stone (7.8% of fruit weight) is polyembryonic. The fruit has good shelf life. The cultivar is moderately susceptible to anthracnose.

Kent : The cultivar originated in Miami, Florida from open pollinated seedling of cultivar Brooks. The tree is large and vigorous with a dense upright canopy. Fruit bearing may be alternate. It is similar to the cultivar Keitt. It is also a late cultivar but matures earlier than Keitt. The fruit is large, 545 g, regular ovate, 12.4 x 9.7 cm with a rounded base and often with two slight beaks. The skin is thick, tough and greenish yellow with a red or crimson blush on the shoulder and numerous small yellow lenticels. The flesh is juicy, melting, deep yellow, fibreless and of a rich flavour. The seed, embedded in a thick, woody stone (8.5% of fruit weight) is monoembryonic. The fruits can ship well but are prone to storage diseases.

Madoe (Madu): The cultivar originated from Indonesia. The tree is moderately vigorous with a dense, rounded canopy. It is a mid season and polyembryonic cultivar. Its fruits resemble those of local Apple cultivar but are much more resistant to anthracnose. The fruits are medium to large in size, 310-450g (mean 380 g), oval/oblique in shape, 9.7 x 10.7 cm, with a rounded base and a slight beak. The skin colour is deep yellow/apricot with the shoulders showing a reddish flush. The flesh is yellow, soft, tender and juicy, almost fibreless having rich flavour and of outstanding quality.

Matthias: The cultivar is of unknown origin but most probably comes from West Africa. The tree is of medium to large size and forms a dense canopy. This is mid season, polyembryonic, regular bearer cultivar, moderately resistant to anthracnose and powdery mildew. The fruit is medium-sized with an average weight of 251 g, ovate in shape, 10.3 x 7.8 cm and with a slight beak. The lenticels are at first green and later turn yellow. The skin is deep yellow having red blushed shoulders. The flesh is firm, yellow, sweet, juicy and relatively free from fibre. There is a moderate resistance to anthracnose and powdery mildew.

Maya: The cultivar selected in Israel is of unknown parentage, resembles Haden but the fruits are much smaller. The tree is large and vigorous, tends to be alternate bearing and is very susceptible to anthracnose. The fruit is plump, yellow with a reddish blush, medium-sized, 250-380 g (mean 295g), ovate (10.3 x 7.8 cm) and a small rounded beak. Lenticels are white at first, changing to yellow/brown later. The flesh is firm, yellow, juicy and aromatic, virtually free from fibre and of high eating quality. The fairly large seed (9.2% of fruit weight) is monoembryonic.

Ngowe (Lamu): The cultivar is polyembryonic, susceptible to powdery mildew and may be alternate bearer. The trees are comparatively small and round in shape. The fruit is large, 425–600 g, oblong and slender, 14 cm x 9.5 cm, with a very prominent hook-like beak at the apex. The skin colour is initially pale green and becomes most attractive yellow to orange as the fruit develops. The flesh is deep yellow of excellent quality, virtually free from fibre, melting and carries no turpentine taste.

Nimrod: This cultivar originated from Israel and resembles the Apple cultivar. The tree is vigorous, medium-sized with an upright and dense canopy. The fruit is large, 340 - 580 g, oval/oblique, 11 x 10 cm, deep yellow with a light red blush with numerous yellow lenticels when ripe. The base is flattened with a slight indication of a small rounded beak. The skin is thick and tough and separates easily. The flesh is yellow, soft and juicy with little fibre having mild, aromatic flavour and good eating quality. The seed is monoembryonic and embedded in a medium thick and woody stone (7.6% of fruit weight).

Parwin: The cultivar of unknown parentage originated from Bradenton, Florida. It resembles Haden but lacks the latter's bright red colour. The tree is vigorous with a slightly open habit and there is a remarkable resistance to anthracnose and powdery mildew. It is regular bearing. The fruits are often borne in clusters. The fruit is medium to large, 380-560 g with oblong to ovate shape, 10.8 x 8.5 cm. It tends to be plump, light yellow with a pink/red blush with numerous distinct lenticels. The flesh is yellow juicy, moderately sweet with a good flavour and relatively free from fibre. The stone is medium-sized (7.5% of fruit weight) enclosing the monoembryonic seed.

Peach: The trees are big. The fruits are roundish oblique, 9.3 x 8.1 cm, medium-sized, 241 g and fibrous having broadly rounded apex with a depression on the ventral side and a slight beak. The skin is smooth, thick and tough with white lenticels and an attractive yellow orange colour. The flesh is apricot yellow with a tender juicy texture having sweet flavour and a very slight turpentine taste. The eating quality is good. The seed is large (8.1% of fruit weight) and polyembryonic. The cultivar is suitable for higher elevations with fair resistance to diseases and good shelf life.

Sabine: The tree is moderately vigorous and upright with a dense canopy. The cultivar seems to have slight tendency of alternate bearing and is affected by anthracnose and powdery mildew. The fruits are medium to large-sized, 360 - 520 g, elongated, 14.2 x 6.6 cm, having rounded apex and small depressed beak and of very good quality. Those developing inside the canopy are deep yellow while those exposed to the sun are bright yellow with a dark red blush. The flesh is yellow, pleasantly sweet, juicy and fibreless with medium texture and mild aroma. The seed is monoembryonic and covered by a medium-sized woody stone (9.6% of fruit weight).

Sabre: The cultivar probably originated in South Africa. The tree is small to medium. It is polyembryonic, regular and heavy bearer and has fairly good resistance to anthracnose and powdery mildew. The fruits are small to medium sized, 180–290 g (average 233 g), broad oblong and kidney-shaped, 11.8 x 6.9 cm with broadly rounded apex curved into a prominent beak. The skin is smooth, tough and leathery, easily removed from the flesh and is yellow green, often with a reddish blush. The flesh is deep orange, sweet to insipid-flavoured with a melting texture having medium amount of fibre and fair eating quality normally giving a turpentine aftertaste. The seed is large, up to 9.4% of total fruit weight.

Sensation: This cultivar originated from Miami, Florida. The trees are moderately vigorous and develop into a broad rounded, symmetrical canopy. It is a late cultivar depending on location, susceptible to anthracnose and has alternate bearing. The fruit is medium-sized, 307 g in weight, oval oblique, 10.8 x 7.8 cm, deep yellow with a prominent dark red to purple blush covering most of its surface with a rounded apex and slight beak formation. Lenticels are numerous and pale yellow in colour. The skin is medium thick, tough and separates easily from the flesh. The flesh is deep yellow, fibreless, firm, juicy, sweet and having a distinctive mild flavour and good quality. The seed is monoembryonic covered in a thick woody stone (5.8% of fruit weight). The cultivar is severely susceptible to anthracnose and has alternate bearing and uneven fruit ripening.

Smith: This is an open pollinated seedling of Haden found in Honolulu, Hawaii. The trees are upright and vigorous. The cultivar is mid season, quite susceptible to anthracnose and its fruits must be left on the tree till they develop full colour and flavour. The fruits are large of 550 g, elongated, 13.9 x 7.5 cm, having orange yellow base colour combined with a deep crimson blush with broadly rounded apex and no beak. The skin is thick and tough covered with large white lenticels. The flesh is orange yellow, juicy, spicy, almost fibreless and of a firm texture. The fruit quality is rated as good and yields are moderate to heavy and regular. The seed is fairly large (7% of fruit weight), long, flat and monoembryonic.

Van Dyke: The cultivar originated from Homestead, Florida and is regular producer and has great resistance to anthracnose and powdery mildew. The trees are medium-sized with a large open canopy. The fruit is small to medium sized having 280 g average weight and ovate, 10.5 x 7.9 cm. It has very attractive bright yellow ground colour with a heavy crimson blush and prominent beak. The skin is thick, easily separating and covered with numerous white yellow lenticels. The flesh is quite firm, melting and juicy with little fibre, orange yellow, rich, spicy and sweet with a strong pleasant aroma. It is of good to excellent quality with good shelf life and shipping qualities. The seed is monoembryonic and covered by a medium-sized woody stone (7.1% of fruit weight).

Zillate: The origin of the cultivar is assumed to be in Florida. The tree is moderately vigorous forming an upright tight canopy. It is a late season and regular cultivar having moderate resistance to powdery mildew but is affected by anthracnose. The fruit is medium-sized of 260-350 g (average 291 g), oblong, 12.5 x 7.1 cm with a yellow ground colour and an intense red blush. There are numerous small white lenticels covering the thick and tough skin. The rounded apex carries an almost non-existent beak. The flesh is firm, juicy and yellow relatively free from fibre, aromatic and of good eating quality. The fairly large flat seed (7.8% of total fruit weight) is monoembryonic.

C. Cultivars grown in other countries

Australia: Kensington Pride, Banana, Earlygold, Glenn, Haden, Irwin, Keitt, Kent and Zill.

Bangladesh: Aswina, Fazli, Gopal Bhog, Himsagar, Khirsapati, Langra, Krishn Bhog, Kohinoor, Kua Pahari and Mohan Bhog.

Brazil: Bourbon, Carlota, Coracoa, Espada, Itamaraca, Maco, Magoada, Rosa and Tommy Atkins.

China: Baiyu, Guixiang, Huangpi, Huangyu, Macheco, Sannian and Yuexi No.1.

Costa Rica: Haden, Irwin, Keitt, Mora and Tommy Atkins.

Ecuador: Haden, Keitt, Kent and Tommy Atkins.

Egypt: Alphonso, Bullocks Heart, Hindi Be Sennara, Langra, Mabrouka, Pairie, Taimour and Zebda.

Guatemala: Haden, Kent and Tommy Atkins.

Haiti: Francine and Madame Francis.

Indonesia: Arumanis, Dodol, Gedong, Golek, Madu, Manalagi, Cengkir and Wangi.

Israel: Haden, Tommy Atkins, Keitt, Maya, Nimrod, Kent and Palmer.

Kenya: Boubo, Ngowe and Batawi.

Malaysia: Arumanis, Kuala Selangor 2, Golek, Apple Rumani, Malgoa, Apple Mango, Maha-65, Tok Boon, Chok Anan and Nam Dok Mai.

Mali: Amelie and Kent.

Mexico: Haden, Irwin, Kent, Manila, Palmer, Sensation, Tommy Atkins and Van Dyke.

Myanmar: Aug Din, Ma Chit Su, Sein Ta Lone and Shwe Hin Tha.

Pakistan: Anwar Ratol, Baganapalli, Chausa, Dashehari, Gulab Khas, Langra, Siroli, Sindhri, Suvarnarekha and Zafran.

Peru: Haden, Keitt, Kent and Tommy Atkins.

Philippines: Carabao, Manila Super, Pico, Binoboy, Dudul, Pahutan and Senorita.

South Africa: Fascell, Haden, Keitt, Kent, Sensation, Tommy Atkins and Zill.

Sri Lanka: Karutha Colomban, Willard, Vellai Colomban, Petti Amba, Malwana Amba, Parrot Mango, Peterpasand, Dapara and Hingurakgoda.

Thailand: Nam Dok Mai, Ngar Charn, Okrong, Rad, Chok Anan, Kao Keaw, Keow Savoey and Pimsenmum.

Venezuela: Haden, Keitt, Kent and Tommy Atkins.

Vietnam: Cambodiana.

Binjai

(***Mangifera caesia*** Jack. ex Wall.; kemang; belunu; binglu)

It is native of South East Asia (Martin *et al*., 1987) and is found growing in Sumatra, Borneo and peninsular Malaysia. It requires hot tropical lowlands below 300 m altitude having monsoon climate and fertile soils having good drainage (Verheij and Coronel, 1991).

The tree is large growing about 35 m tall with 50-80 cm stem diameter. Bark is grayish brown and superficially fissured and contains an irritant sap. Leaves are elliptic to lanceolate, 10-12 x 4-5.5 cm in size, medium green, coriaceous and shiny above. Panicle emerges at the terminal end of the shoot and is densely flowered, 15-25 cm long and much branched with stout rachis and branches. Pale pink and fragrant flowers are 5-merous with up to 10 mm long linear petals, one fertile stamen, 5 mm long filament and 4 teeth like staminodes. Disk is narrow, 1-1.5 mm long and pale green having reddish brown ovary, 8 mm long, excentric, obliquely globose and white style that becomes violet after anthesis. Fruit is an obovate oblong drupe, necked at base and 12-15 x 6-7 cm in size. Fruit skin is very thin and yellowish or pale brownish. Pulp is whitish, soft and juicy and fibrous with a peculiar sourish taste and strong smell at maturity. Stone is ellipsoid lanceolate, 7 x 3.5-4 cm in size, not flattened, thin walled and monoembryonic. Endocarp is not woody and is composed of matted coarse fibres (Verheij and Coronel, 1991).

Several forms of binjai have been recognized based on fruit characters. The "Wani" form has ellipsoid, rounded fruits, 9-11 x 6.5-7 cm in size, which is glossy and pale green at maturity with milky white flesh. Many fibreless forms with sweet pleasant taste are also available (Verheij and Coronel, 1991). The fruit has an excellent and unique flavour (Verheij and Coronel, 1991).

Horse mango
(***Mangifera foetida*** Lour.; bachang; machang; ma-mut; xoaihoi)

Horse mango is a native of South East Asia. It occurs wild in dipterocarp forests of peninsular Malaysia, Thailand, Sumatra and Borneo. It is apparently wild in Java. It grows up to 30-35 m tall without buttresses. Bark is light brown to dark grayish brown, shallowly fissured containing irriatant whitish sap turning black on exposure. Leaves are elliptic oblong, 15-40 x 9-15 cm in size, stiffly coriaceous, dark green above and clear green below with 1.5-8.0 cm long and stout petioles, very swollen at the base. Panicles are sub-terminal, upright, pyramidal, 10-40 cm long, sparsely 5-merous bearing scentless flowers. Sepals are obovate, lanceolate and 6-9 x 1.5-2.5 cm in size. Of the 5 stamens, one is fertile and each has 8 mm long pinkish purple filament and dark violet anthers. Ovary is subglobose and yellow and style in excentric. Fruit is variable in size and shape and is an obliquely ovoid oblong or almost globose drupe, 9-14 x 7-12 cm in size, dirty dark olive green or yellowish green, smooth and dull with brown lenticels and 5 mm thick skin. Flesh is pale orange yellow, fibrous and juicy with strong smell and turpentine taste. Stone is plump, 6-5 x 3 cm in size and coarsely fibrous. Seed is monoembryonic (Morton, 1987; Verheij and Coronel, 1991).

Cultivars

Three forms of horse mango are found; *Limus piit* having small, globose fruit found in West Java, *Local type* having large, oblong fruit found in Malaysia and *Limus tipung* having large, oblong, hardly fibrous and fine textured fruits. Several other forms with variable fruit characters are also found in Borneo particularly in South Kalimantan (Verheij and Coronel, 1991).

Kuwini
(***Mangifera odorata*** Griffith; kuini; manguier odorant; bembem; kaweni)

It is a native of South East Asia. The species possibly represents hybrid forms between *M. indica* and *M. foetida* (Ding Hou, 1978). It is commonly cultivated in Borneo, Sumatra and Java. It is also found in peninsular Thailand, South Sulawesi and the Philippines. The kuwini is a popular fruit having local economic significance in areas where mango cannot be grown satisfactorily.

The tree is medium to large growing up to 10-24 m in height with globose or broadly ovoid crown. The sap of its bark has a strong resinous smell. Leaves are oblong lanceolate, 12-35 x 4-10 cm in size, coriaceous, shortly acuminate with prominently reticulated veins especially on the lower surface. Flowers are borne on 10-50 cm long and densely flowered terminal panicles. Rachis is yellowish green and tinged reddish brown. Flowers are whitish to yellowish, pentamerous, 6 mm wide and fragrant. Sepals are ovate, 3-4 mm long, brown red or partly green. Petals are lanceolate, 5-6 x 1.2-2 mm in size, yellowish at the base and pale pinkish towards the apex. Stamens are 5-6, reflexed, 1.5-2 mm long, of which only one is fertile. Ovary is subglobose, yellowish with 3-5 mm long and dark red style. The fruits are attractive. Fruit is an obliquely ellipsoid oblong, hardly flattened drupe, 10-13 x 6-9 cm in size, 300-450 g in weight, green to yellowish green, sparingly spotted with dark brown lenticels having rather thick (3-4 mm) rind and orange yellow, firm, fibrous, sourish sweet and juicy flesh which has a pungent smell and turpentine taste. Stone is 8-10 x 4.5-5 x 2.5-3 cm in dimension and is covered with rather soft fibres. Seeds are frequently polyembryonic (Verheij and Coronel, 1991).

Cultivars

Kuwini is a polymorphic species. Several forms are distinguished in West Java. Bembem is an inferior form having fruits with a strong smell and turpentine taste, reminiscent of the fruit of *M. foetida*. Its less fibrous flesh having mild turpentine taste is palatable. Gandarassa is a rare form said to be superior to kuwini. It is less sweet but is more juicy and has less strong smell (Verheij and Coronel, 1991). Because of the flavour and firmness of its fruit and its ability to grow in moist areas, kuwini deserves attention for use in breeding programme (Verheij and Coronel, 1991).

Bambangan

(***Mangifera pajang*** Kostermans; asem paying; alim; hambawang)

Bambangan is native of Borneo where it occurs wild in lowland dipterocarp forest. It is found in cultivation in Dayak orchards in Sabah, Sarawak, Brunei and East Kalimantan. This attractive fruit tree remained confined to its native area for long and was introduced outside Borneo only in the 1980's.

Trees of bambangan grow up to 15-33 m tall with 30-70 cm trunk diameter. The deciduous tree has very stout twigs with leaves more or less aggregated at the apices. The bark is brown or gray and smooth or fissured. Leaves are elliptic oblong to obovate oblong, 28-45 x 10-15 cm in size, rigidly coriaceous with prominent nerves. The fragrant flowers are borne terminally or sub-terminally on erect, pyramidal, up to 30 cm long inflorescence. The flower has 5-lobed calyx, 5 petals, purple on the inner surface and pinkish white outside, 5 stamens of which 2 are fertile and white ovary. Fruit is a globose or ovoid drupe, 9.5-12 x 6.5-10 cm in size. The fruits are the largest in the genus (up to 20 cm in diameter). Its rough and potato brown rind that can be peeled as that of banana. The thick and fibrous flesh is bright deep yellow and sweet with a distinctive taste and pleasantly aromatic flavour. Stone is plump, 9x6.5x4.5 cm in dimension with woody endocarp and monoembryonic seed (Verheij and Coronel, 1991).

The other species of *Mangifera* having usefulness in breeding are:-

M. sylvatica Roxb.: It has long petiole. The fruit is not edible.

M. altissima Blanco: It is grown in the Philippines

M. pentandra Hooker f.: It has all the 5 stamens functional. Fruit is edible.

M. coloneura Kurz: It is found in Burma.

Genus *Bouea* Meisn.

Three or more species occur in South East Asia to western Malesia (Kubitzki, 2011). The trees are andromonoecious and have prominent terminal and axillary buds with bud scales. Leaves are evergreen, opposite decussate, simple and petiolate. Inflorescences are axillary (rarely also terminal) panicles. Flowers are pedicellate and non-articulate having 3-5-parted perianth, valvate calyx, imbricate, white, greenish, or yellow corolla, petals keeled along the midrib, androecium haplostemonous, filaments subulate and glabrous, anthers basifixed with an apiculate connective, very reduced pistillode, no staminodes, glabrous, small, flat or concave disk, pseudomonomerous gynoecium, short style, stigma rounded and flat, sometimes 2-3-grooved and basal ovule. The fruit is subglobose to ellipsoid and 1-locular drupe with yellow, orange or red exocarp, fleshy, edible mesocarp and fibrous endocarp. Seed has straight embryo.

The gandaria

(***Bouea gandaria*** Blume syn. ***B. macrophylla*** Griff.; maprang; ma-praang)

Gandaria is a native of South East Asia. It has now spread to tropical America and Honduras (Martin *et al.*, 1987). It is cultivated widely as a fruit tree in Sumatra, wetter parts of Java, Borneo and Ambon and in Thailand (Verheij and Coronel, 1991). It is an evergreen tree. It grows to the size and shape of a mango tree, 10-27 m in height with a straight, cylindrical, low branched trunk and a thick, dense and broadly ovoid crown. The young drooping branches are four angled. The dark green, oblong lanceolate leaves are opposite, short petioled, thick, leathery, 10-35 cm long, 3-12 cm broad, shining above and dull light green beneath, glabrous with prominent midrib on both surfaces having acute or rounded base and entire margins. The panicles are borne in the axils of leaves. The andromonoecious tree has greater proportion of small male flowers having 3-5-merous parts. The staminate flowers are borne on short puberulent yellow stalks and the perfect flowers are sessile. The sepals are erect, ovate, yellow and externally pubescent. The petals are erect or spreading obliquely, obovate ovate, light yellow and 0.15-0.2 cm long. The stamens alternate with the petals and have very short filaments. The ovary in the perfect flowers is minute and glabrous. The oblong ovoid or ellipsoid fruits are glabrous, obtuse or rounded at both the ends and turn yellow or orange on ripening. These are 3-5 cm long, 3-4 cm thick with a thin rind and contain thick, juicy, pulpy, yellow or orange and subacid flesh (mesocarp). The endocarp is thinly coriaceous and woolly fibrous externally and smooth within. The seed is 1.5-3 x 1.5-2 cm in size with a thin light brown testa adnate to the endocarp. The cotyledons are thick planoconvex, blue violet in colour and astringent (Martin *et al.*, 1987; Morton, 1987; Verheij and Coronel, 1991). Cultivars have been recognized based on taste of the fruits. Hintalu is a very sour cultivar. Ramania Pipit and Ramania Tembaga are red fleshed sweet cultivars (Verheij and Coronel, 1991).

Genus *Anacardium* L. (n =12, 29)

Eleven or more species occur in Honduras south to Paraguay, Brazil and Bolivia. The species *A. giganteum* and *A. nanum* can be subdivided based on the colour (yellow or red) and shape (round, pear shaped or elongated) of apple (Nair, 2014). *Anacardium occidentale* is cultivated pantropically (Kubitzki, 2011).

The plants are andromonoecious subshrubs or trees with contact dermatitis causing exudate. Leaves are generally evergreen, alternate, simple, sessile to petiolate, entire, chartaceous to coriaceous, domatia is present in secondary vein axils abaxially. Inflorescences are terminal and/or axillary thyrsoids. The flowers are polygamous, numerous and borne in panicles. These are pedicellate and articulate having 5-parted perianth, irnbricate calyx, corolla of reflexed petals, 6 fertile stamens, 1(-4) much larger and exerted, filaments basally connate into a tube, very reduced pistillode, no staminodes or reduced in bisexual and male flowers, no disk, pseudomonomerous gynoecium having straight to sigmoid (sometimes approaching gynobasic) style, apical or lateral, punctiform stigma and obscure and basal ovule. The fruit is a kidney-shaped nut borne on fleshy receptacle called the one-seeded accessory nut. The disc provides the fleshy portion. The drupe is reniform, 1-locular subtended by hypocarp, sigmoid to pyriform (rarely 0), white, green, yellow or red having woody pericarp with large rectangular cavities containing caustic oils and greenish to brown to black exocarp. Seed is with reniform embryo.

Cashew

(***Anacardium occidentale*** L.; monkey nut)

It originated in northeastern Brazil (Morton, 1987) and is spread over South and Central America. In India, it is grown in Kerala and Tamil Nadu. It is hardy evergreen tropical plant and requires heat for maturity of the fruit. The tree grows up to 10-12 m high with a wide dome shaped crown. The leaves are simple, alternate, obovate to obovate oblong, 20 x15 cm in size, leathery, red brown when young, swollen at base and flattened on upper surface. Inflorescence is a lax, up to 25 cm long, terminal, drooping and many-flowered panicle with fragrant male and hermaphrodite flowers. The flower has 5, lanceolate to oblong ovate sepals that are 4-15 x 1-2 mm in size having reflexed whitish anthers that later turn pinkish red. Male flower has 7-9 stamens of 4 mm length and 1-3 stamens of 6-10 mm length. Hermaphrodite flower usually has 5 petals, 5 sepals, 9 short and one long stamen and a superior, reniform and monocarpellate ovary. The ovary is obovoid with the style attached on one side. The pistil is oblique. Long stamens produce viable pollen. Style is simple, 12 mm long, exerted from corolla to the length of long stamens. Fruit is a kidney shaped nut, about 3 x 1.2 cm in size with gray brown and resinous hard pericarp and much enlarged, fleshy and swollen pedicel forming the fruit-like cashew apple, which is pear shaped, 10-20 cm long, 4-8 cm broad, shiny, red-yellow, astringent, soft and juicy (Fig. 29.6). The seed is kidney shaped with reddish brown testa (Fig. 29.7). Two large white cotyledons and a small embryo together constitute the kernel (FAO, 1982; IBPGR, 1986a; Morton, 1987; Brücher, 1989; Verheij and Coronel, 1991). The nut is not edible until roasted.

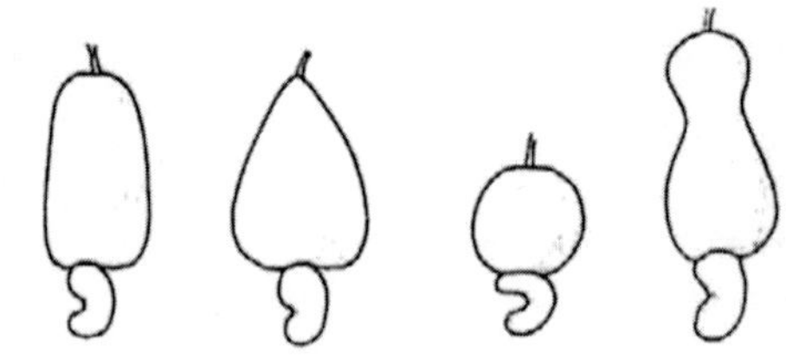

Fig. 29.6. Shape of cashew apple (Source: IBPGR, 1986a).

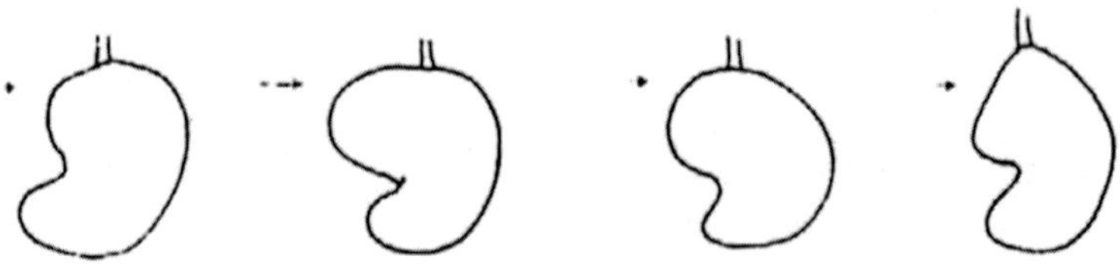

Fig. 29.7. Shape of base of cahew nuts (Source: IBPGR, 1986a).

Barros (1988) classified the natural diversity of cashew into two types (Table 29.1).

Table 29.1. Characters of the common and the precocious dwarf cashew types (Barros, 1988).

Characters	Common type	Dwarf type
Stature	High (8—15 m)	Low (<4 m)
Canopy width	10-20 m	5-6.5 m
First flowering	3-5 years after planting	6-18 months after planting
Hermaphrodite flowers	7.9%	3.9%
Nut weight	3-33 g	3-10 g
Apple weight	20-500 g	20-160 g
Production/plant/season	<10 to > 230 kg nuts	up to 43 kg nuts

1. The common type originated in northeastern Brazil is the most diffused. It has 8-15 m plant height and up to 20 m canopy width. The branch distribution and the canopy frame vary greatly, ranging from erect and compact to sprawling. Its productive capacity varies considerably. The nut weight ranges from less than 3 to 33 g and cashew apple weight ranges from 20 to over 500 g. The production from the tree stabilizes after 8 years, frequently at 12-14 years age.

2. The precocious dwarf cashew, also known as the "six-month variety", probably originated in the Amazon region. It is 4 m high with a homogeneous 5-6.5 m wide canopy. The flowering starts 6-18 months after planting. Among natural

populations, the nut weight ranges from less than 3 to 10 g and the apple weight from 20 to 160 g. The productive capacity is also lower, the highest yield observed being 43 kg nuts per plant (Barros 1988). The variability for the main agronomic traits is lesser than that observed in the common type.

Desai and Singh (2011) classified the cashew genotypes into the following 2 groups based on apple to nut ratio:-

1. Genotypes having apple/nut ratio of 6-12: Tiswadi-7, Red Local, GB-2, Dhave-1, GB-3, Dhave-2, Dhave-3, Valpoi-1, Valpoi-7, Balli-1, Balli-2, Tudal-3, Loutolim-2, Ashley-1, Raja-1, Bardez-3, Ganje-1, Ganje-2, Bardez-8, Bardez-9 and Mayem-1.

2. Genotypes having apple/nut ratio of >12: Tiswadi-3, Valpoi-3, KN2/98, Khola-2 and Tudal-1.

Cultivars

The species *A. occidentale* has been divided into two botanical varieties on the basis of nut/apple ratio that generally varies from 1:6 to 1:10 (Salem, 2007). The cultivars under each of these two types have red or yellow coloured apple. The type *A. giganteum* Hanc. ex Engle (wild cashew) has very large apple and edible but small nut.

Promosing cultivars of Brazil are Amarelo, Gigente and Manteiga (Brücher, 1989). In Kenya, Selection A 81 has shown high yielding ability. Of the 16 clones selected in Malaysia, C 11 yields 30% higher than C 21 and 2-6 times more than the other clones. In Thailand, Selections SK 60-1, SK 60-2, SK-A and Sirichai 25 are recommended. Cashew Balli 2 has been identified for high yield and better nut and apple characteristics (Desai, 2009). Some selections made at the Central Plantation Crops Research Institute, Kasaragod, India are M 10/4, M 6/1, M 14/3 and M 76/1. The hybrids evolved at Cashew Research Station, Vengurla, Maharashtra are 2/11, 2/12 and V7. High yielding cultivars of cashew are Amrutha (H 1697), Sulabha (K 10-2), Vengrula 2 and Vengurla 7 (Singh *et al.,* 2003). Cultivars suitable for different regions of India (Nambiar *et al*., 1990; Singh *et al.,* 2003) are given in Table 29.2.

Table 29.2. Cultivars of cashew in different regions of India

Regions	Cultivars
Karnataka	BLA-39-4, VR-1, VR-2, Ullal-1 and Ullal-2
Kerala	BLA-39-4, K-22/1, NDR-2-1
Andhra Pradesh	BPP-4, BPP-6, VRI-2, Ullal 1, Chintamani 1
Maharashtra	V-1, V-4, VRI-2
Orissa	WBDC-V, T-40, BPP-2, BPP-6, BPP-8, Ullai-1
Tamil Nadu	BR1-1, VRI-2, VRI -3
Goa	V-1, V-4, VRI-2
West Bengal	BLA 39-4, Jhargram 1, BPP 8

BPP-1 (Tree No 1 x Tree No 273): The cashew apple is medium in size giving juice recovery of 68 per cent and the nut of medium size (5 g weight) and about 28% shelling. The cultivar has about 13% perfect flowers and nearly 6-8% fruit set per panicle. The tree produces 15-17 kg nuts at about 25 year age.

BPP-2: The hybrid has the same parents as of BPP-1. The tree produces smaller nut of 4 g having nearly 20% protein content and yield of 18-20 kg nuts at an age of 25 years giving 24-26% shelling. The cultivar has nearly 7% perfect flowers and 7-10% fruit set per panicle.

BPP- 8 (H 2/16; Tree No 1 x Tree No 39): The nut yield is 14.5 kg/tree, nut weight is 8.2 g and gives 29% shelling of W 210 grade.

Ullal-2: It is a late flowering cultivar. It produces 18 kg nuts per tree at an age 25 years in small harvest duration of 80-85 days. The nut is bigger weighing 6 g giving shelling of nearly 30%. The cultivar has about 7% perfect flowers.

Vengurla-1: The tree is compact with profuse branching and has very widely differing forms. The cultivar produces about 8% perfect flowers per panicle. The nut weight is 6 g and shelling percentage is nearly 31. The cashew apple is yellow coloured.

Vengurla-4 (Midnapore Red x Vetore 56): Each nut is of about 8 g weight and bright red colour giving high shelling of nearly 30-32%. The cultivar has high sex ratio with nearly 35% bisexual flowers per panicle and 5-6% fruit set. It is cluster bearing and yields17.2 kg nuts of W 210 grade per tree.

Vengurla-6 (Vetore-56 x Ansur-1): It yields 13.8 kg nuts of W 210 grade per tree. The nut weight is 8 g and 28% shelling.

Vengurla-7 (VRI-1; Vengurla-3 x M 10/4): It yields 18.5 kg nuts of W 180 grade per tree. The nut weight is 10 g and 30.5% shelling with 2.9 g kernel weight.

Dhana (H-1598; ALGD-1 x K-30-1): It yields 17.5 kg nuts per tree of W 210 grade. It is cluster bearing and the nut weight is 9.5 g and 28% shelling.

Kanaka (H-1598; BLA-139-1 x H-3-13): It yields 19 kg nuts of W 210 grade per tree. The nut weight is 6.8 g with 31% shelling and kernel weight is 2.2 g.

Priyanka (H- 1591; BLA-139-1 x H-3-13): It yields 16.9 kg nuts of W 180 grade per tree. The nut weight is 10.8 g with 26.5% shelling.

Amrutha (H- 1597; BLA-139-1 x H-3-13): It yields 18.4 kg nuts of W 210 grade per tree. The nut weight is 7.2 g with 31.6% shelling and 2.2 g kernel weight.

Genus *Pistacia* L. (n = 12, 14, 15)

Based on phylogeny assessment using five molecular sets, sequences of nuclear ribosomal ITS, the third intron of the nuclear nitrate reductase gene (*NIA-i*3), and the plastid *ndhF*, *trnL-F*, and *trnC-trnD.* Yi *et al.* (2008) recognized 12 species in the genus (*Pistacia atlantica*, *P. chinensis*, *P. integerrima*, *P. khinjuk*, *P. lentiscus*, *P. mexicana*, *P. palaestina*, *P. saportae*, *P. terebinthus*, *P. texana*, *P. vera* and *P. weinmannifolia*). The species occur in Texas, USA, south to Nicaragua, Mediterranean Europe and North and East Africa, southwest and central Asia (former Soviet Republics) and east to Afghanistan and temperate central and southern China, to peninsular Malaysia and the Philippines (Kubitzki, 2011). Al-Saghir and Porter (2012) divided the genus into 2 sections, *Pistacia* and *Lentiscella,* and produced a key to the recognized 14 taxa.

Key to the species of *Pistacia*

1. Leaves large, 8-23 cm long, 5.2-23 cm wide.
  2. Leaf rachis not winged.
    3. Leaves coriaceous or becoming coriaceous
      4. Petiole flattened; leaflets 3-5, obtuse or mucronulate .... *P. vera*
      4. Petiole rounded, rarely flattened; leaflets (3-)6-11, mucronate .... *P. terebinthus*
    3. Leaves membranaceous.
      5. Petiole angled or rounded; leaflets 1-9, acuminate .... *P. khinjuk*
      5. Petiole angled or flattened; leaflets 6-15, attenuate.
        6. Leaflets paripinnate.
          7. Leaflets broadly lanceolate, 6-11.5 cm long, 2-4 cm wide, average ratio 3.4:1 ....*P. chinensis* subsp. *integerrima*
          7. Leaflets lanceolate, 4-10 cm long, 9-24 mm wide, average ratio 4.1:1 ....*P. chinensis* subsp. *chinensis*
  2. Leaf rachis winged.
    8. Leaves coriaceous, 10 cm long, evergreen ....*P.* × *saportae*
    8. Leaves membranaceous. 8-18.2 cm long, deciduous
      9. Leaflets (1-3-)5-7, 4.5-6.5 cm long, 1.5-3.8 cm wide, average ratio 2.3:1 ....*P. eurycarpa*
      9. Leaflets (5-)7-11, 3-7 cm long, 5-20 mm wide, average ratio 3.5:1 ....*P. atlantica*
1. Leaves small, 2-15.1 cm long, 1.8-10 cm wide.
  10. Leaves coriaceous.
    11. Leaflets 4-10, mucronate, lacking large lacking large distinct veins .... *P. lentiscus* subsp. *lentiscus*
    11. Leaflets 6-16, emarginate, with large distinct veins ....*P. lentiscus* subsp. *emarginata*
  10. Leaves membranaceous.
    12. Leaves paripinnate; leaflets (4-6)12-20, often alternate..................................... *P. weinmannifolia*
    12. Leaves imparipinnate; leaflets 10-30, never alternate...*P. mexicana*

Members of the genus are tropical hardy shrubs or trees with well developed vertical resin canals having clear exudate. They can, however, withstand frost and are also heat resistant. Leaves are alternate, deciduous or persistent, pinnately compound, but sometimes trifoliolate or unifoliolate, paripinnate or imparipinnate, memberanous or leathery, petiolate with rachis, sometimes alate. The plants are dioecious.

Inflorescences are determinate and terminal or axillary. Flowers are almost always unisexual, radial and small with well developed staminodes or carpellodes. Sepals are usually 5, distinct to slightly connate; stamens are 4-5, filaments are very short, usually glabrous and distinct with basifixed anthers, tricolporate or triporate pollen grains, carpels 3 variously connate with only 1 fertile and well developed carpel, ovary superior. Gynoecium is asymmetrical and unilocular with apical placentation, 1 ovule pendulous from a basal funicle, bilobed or simple and recurved capitate stigma. Nectariferous disk is present and is usually intrastaminal. Fruit is a flattened asymmetrical (globose or ovoid) drupe with chartaceous and red to purplish exocarp, thin mesocarp and woody endocarp. Embryo is curved and endosperm is scanty or lacking. The gunus includes the pistachio nut, which was previously put under the genus *Terebinthus*.

Pistachio or Pistache

(***Pistacia vera*** L.; green almond) (x = 15; 2n = 30)

It is a native of Iran, Afghanistan and central Asia from Turkemenia to Pamir-Alai and Tienshan (Wickens, 1995) and is cultivated worldwide in dry warm climates. The plants are small deciduous and resinous trees with straggling growth of 7.5-9.0 m height and with spread and stout gray twigs. The branches arise from the ground level. In seedling trees, the trunk rarely develops. The buds have several outer scales. The leaves are large, alternate, pinnate, gray green and shining with often reddish raised veins on both the surfaces. Each leaf has 3-7 leaflets that are ovate, sessile and leathery, the terminal one is often of twice the area of that of the others. The tree is dioecious and require cross pollination. The emergence of flowering shoots is accompanied by a marked increase in shoot diameter when the bark becomes gray rather than brown in colour. Fertile buds start growing during September and the flowers become receptive about three weeks later. Vegetative buds begin to develop about 7 days later so that the opening flowers remain partly protected from frost by the young expanding leaves. Flowers are minute, unisexual, apetalous or naked, clustered together in short lateral spikes. Male flower has 2 bractlets at base, 1-2 sepals and 2.5-3.0 mm long 5-6 stamens inserted on the disc. Female flower consists of 2 bractlets, 2-5 sepals, 3 mm long pistil, subglobose or ovoid 1-celled superior ovary with one ovule and 3-parted short style. Pollination is carried out by wind and not by bees. The receptive period of female flower lasts about 4 days. The male flowers are functional when the red anthers burst open showing bright yellow pollen. Protandry exists. Parthenocarpy is observed but the seeds are not viable.

The fruit is dry narrowly obliquely ovoid to oblong drupe, 1-2 x 0.6-1.2 cm in size having a dehiscent coat and a single large seed inside a thin and hard bivalved shell surrounded by a fleshy hull. After pollination, the fruit enlarges rapidly and attains full size within 6-8 weeks with a firm but nearly empty shell. During this period, the embryo itself develops little although the stalk (funicle) joining the embryo to the seed wall enlarges and may be mistaken for the embryo. Subsequently, the embryo enlarges rapidly and the seed becomes full sized within 16 weeks. It loses water and ripens 4-6 weeks later (IBPGR, 1986b). The endocarp is hard and yellowish red. The mesocarp has resinous gum. The endocarp splits along the ventral suture. The kernel is rich in oil.

Cultivars

Kermen: The cultivar is widely cultivated in California and is the backbone of pistachio industry there. The trees are vigorous having upright branches and give high yield of excellent quality nuts. The kernel is white, crisp and firm. The cultivar suffers from biennial bearing and has blank nuts and low natural splitting.

Joley: The cultivar is a selection made from open pollinated seedlings in California. Flowering and nut maturity are earlier than in Kermen. Nuts are of small size and the tendency of blank nuts is low and nut splitting is very good.

Sfax: The cultivar was selected in Tunisia. It is suitable for warm temperate region as its chilling requirement to break winter dormancy is lower compared to that for Kermen. It is very early maturing. The nuts are smaller and of good quality having quite high shell splitting.

Lassen: The cultivar was obtained at California by selection from plants raised from the seeds brought from Rafsinjan, Iran. Being late flowering, it escapes frost damage. The cultivar is fairly high cropper. The nuts are bigger in size and shell splitting is high and can be very easily opened by hand. The quality of roasted kernels is very good. Being cross pollinated some pollinizers such as Peters and Chico were selected for this cultivar.

Peters: It is a well established and universal pistachio pollinizer. It produces large quantities of fertile pollen with the added advantage of its flowering coinciding with a large number of pistillate cultivars including Kerman.

Chico: This cultivar was obtained at California from the plants raised from the seeds of *P. vera* brought from Syria. Like Peters it produces pollen in abundance. As it flowers early in the season it is a desirable pollinizer for the cultivars Bronte, Red Aleppo and Trabonella but is too early for Kerman, Chico and Peters. It covers the flowering periods of both early and late blooming cultivars.

***Pistacia terebinthus* L.** (Cyprus turpentine, terebinth pistache): It is found growing in the Mediterrnean region and central Asia. The fruit is edible.

***Pistacia atlantica* Desf.** (wild pistache): The fruit is edible. The tree is grown for producing the chewing gum used as a breath sweetner.

***Pistacia chinensis* Bunge:** The trees are large, 25 m in height. The shoots and leaves are downy at first. The leaflets are 10-12. The fruit is globose and obovoid.

Genus *Spondias* L. (n = 16)

It contains 16 or more species; 8 from South Mexico to southeastern Brazil and Bolivia, 8 from India and Sri Lanka to tropical China and South Pacific Islands, neotropical species naturalized in West Africa, the West Indies and South East Asia. Several species are cultivated pantropically (Kubitzki, 2011). These are *Spondias dulcis* Soland ex Parkinson syn. *S. cytherea* Sonn. (ambarella), *S. purpurea* L. (red mombin or Spanish plum), grown for its brilliant red coloured vitamin A and C rich fruits, *S. mombin* L. syn. S. *lutea* L. (golden mombin, yellow mombin, hog plum), *S. tuberosa* Arruda (imbu) and *S. venulosa* Mart. Geographical distribution of the *Spondias* species is given in Table 29.3.

The plants are polygamodioecious or hermaphrodite (rarely dioecious), strongly protandrous trees (very rarely semi-epiphytes) having exudate that causes contact dermatitis. The leaves are deciduous, alternate, imparipinnate (bipinnate) and petiolate. The leaflets are opposite, sub-opposite or alternate, sessile to petiolate. Inflorescences are terminal and/or axillary panicles (racemes), often appearing before the leaves or along with young leaves. Flowers are pedicellate and articulate, calyx slightly imbricate or apert, (4)5- lobed, corolla valvate, (4)5(6)-parted and cucullate having white, cream, purple or red colour, diplosteminous androecium, sometimes in two whorls of unequal length, filiform or subulate filaments, dorsifixed anthers, reduced pistillodes and staminodes, glabrous or papillose, annular and lobed disk, (3-)5 carpels, (3-)5 stylodia, capitate to spathulate stigmas and apical ovules. Fruit is globose, obovoid, oblong or ellipsoid, (1-)5-locular drupe having yellow orange, red purple or greenish exocarp, fleshy mesocarp, bony endocarp usually with a fibrous outer layer (very rarely spiny) projecting into the fleshy mesocarp. Seed has curved embryo.

Table 29.3. *Spondias* species and their geographic distribution.

Species	Distribution
S. acida BI.	Malay Peninsula
S. acuminata Roxb.	India, Burma, Thailand
S. bipinnata Airy Shaw & Forman	Thailand (limestone hills)
S. bivenomarginalis KM. Feng, P.Y. Mao & P. Me.	Yunnan, China
S. dulcis Parkinson	Asian; cultivated in New World, Brazil, Caribbean
S. macrocarpa Engl.	Eastern Brazil
S. malayana Kosterm	Malesia, Philippines
S. mombin L. var. *mombin*	Mexico to Bolivia and eastern Brazil, Central America
S. mombin L. var. *glohosa*	Colombia, Venezuela, Ecuador, Peru, Bolivia
S. novoguineensis Kosterm	New Guinea, Solomon Islands, etc.
S. pinnata (Koenig ex Linn. F.) Kurz	India, Himalayas, Burma, Sri Lanka
S. purpurea L.	Mexico, Central America, Ecuador
S. radlkoferi Donn. Sm.	Mexico, Central America, northwest Venezuela, West Ecuador
S. testudinis J.D. Mitch & Daly	Southwest Amazonia
S. tuberosa Arruda	Northeast Brazil
S. venulosa Mart. ex Engl.	Eastern Brazil
S. tonkinensis Kosterm	Tonkin, Langson Province
S. xerophila Kosterm	Sri Lanka

Ambarella

(***Spondias dulcis*** Soland ex Parkinson syn. ***S. cytherea*** Sonn.; great hog plum; jew plum; Tahitian apple; Polynesian plum; hevi; vi apple; golden apple; otahetite apple)

Ambarella is native of South East Asia and the south pacific region. It is cultivated on a small scale in Myanmar, Cuba, India, Indonesia, Jamaica, Malaysia, the Philippines, Sri Lanka, Reunion and USA (Puerto Rico) principally for its edible fruits (IBPGR, 1986b; Morton, 1987; Verheij and Coronel, 1991).

The deciduous trees grow rapidly attaining a height of 10-25 m with a straight, gray and smooth trunk. The leaves are alternate, crowded at the ends of the branchlets, pinnate, 30-60 cm in length, usually with 4-12 pairs of lateral elliptic leaflets, 4-14 x 2.5-5 cm in size and a terminal leaflet. At the beginning of dry and cool season, the leaves turn bright yellow and fall, but the tree with its nearly smooth, light gray brown bark and graceful rounded branches remains attractive. During the spring, small inconspicuous, whitish and pentamerous flowers appear on large (up to 50 cm long), widely branched terminal panicles. Flowers are either male or female or hermaphrodite borne in a cluster. Calyx is small, 2.5 cm in diameter having 5 reflexed yellowish white petals, 10 stamens, shorter than the petals and sessile and 5-celled ovary. Fruits are ellipsoid and glabrous drupes, yellow or orange when ripe, 5-10 x 4-8 cm in size with thick, fleshy and white pulp. Stones are comparatively thin consisting of one layer with prominent ridges from which long tough fibres radiate in all directions and are 5-celled. Woody core contains 1-5 flat seeds (IBPGR, 1986b; Morton, 1987; Verheij and Coronel, 1991).

Yellow mombin

(***Spondias mombin*** L. syn. ***S. lutea*** L.;
tapereba; hog plum; Spanish plum; gully plum; ashanti plum)

It is a native of tropical America (Martin *et al*., 1987). The yellow mombin is an erect deciduous tree up to 20 m tall with 60-70 cm trunk diameter having up to 2 cm long buttressed spines or knobs. The leaves are alternate, pinnate, 20-45 cm long, hairy, often with pinkish petioles and 9 to 19 sub-opposite, ovate or lanceolate and pointed leaflets which are 5-15 cm long and oblique at the base. Inflorescence is 15-30 cm long terminal panicles having small, polygamous and fragrant flowers that emerge after the new leaves. The fruits are aromatic, ovoid or oblong, 3.2-4 x 1 cm in size having thin and tough skin and medium yellow, translucent, fibrous, very juicy, musky and very acid pulp often with turpentine flavour and attached to the white, fibrous or corky stone (Morton, 1987; Brücher, 1989; Verheij and Coronel, 1991).

Amra

(***Spondias pinnata*** (Linn. f.) Kurz; wild mango; hog plum)

It is a native of tropical Asia. It grows wild from the Himalayas in the north to the Andaman Islands (Morton, 1987). It is also found in Myanmar, Thailand and Sri Lanka. It requires hot tropical lowland climatic conditions. It is a deciduous medium sized tree growing up to a height of 8-10 m and has slender but strong woody stem having branches covered with gray and aromatic bark and alternate and pinnate leaves. The leaf has 5-11 leaflets having mango like smell. Its polygamous flowers are white or greenish white and are borne in large terminal panicles. The fruits are ovoid, medium sized (4.5 x 3.2 cm) with a depression at the basal end. The fruit is olive green when unripe but turns yellow at maturity. The pulp is fairly juicy, orange coloured having pleasant flavour. The oblong elliptic seeds are 1-3 (Anon., 1976).

Purple mombin
(*Spondias purpurea* L. syn. *S. dulcis* Blanco; Spanish plum)

Red mombin is native of tropical America. It is found wild and also cultivated from southern Mexico through northern Peru and Brazil particularly in the arid zones. *S. tuberosa* Arruda ex Kost is a related species that yields juice. It has been introduced into other tropical and warm subtropical countries and has gained some economic importance as a fruit crop in southern Mexico, Peru, Brazil and Philippines (Morton, 1987; Verheij and Coronel, 1991; Macia and Barfod, 2000).

The plants are small deciduous trees and grow up to 10 m in height with 30 cm trunk diameter. The branches are thick and brittle. Leaves are alternate, pinnately compound having 4-12 pairs of leaflets, 6-12 cm long rachis and 2.5-4 cm long petioles. The leaflets are obliquely elliptic to elliptic oblong, 2-5.5 x 1-2.5 cm in size having entire or obscurely crenulate margin. Inflorescences consisting of unisexual and bisexual flowers appear before the leaves. These are axillary, paniculate or racemiform and up to 4 cm long. The flowers are reddish or purplish with triangular calyx lobes, 4-5 ovate oblong petals of 3-4 x 1.5-2 mm size, 8 or 10 stamens and 4 or 5 styles. Fruit is a oblongoid to ovoid drupe, 2.5-5 cm long, 30-30 g in weight, purple red and dark purple in colour having yellow, aromatic and juicy flesh and oblongoid stone which is up to 2 cm long, rough, fibrous and hard containing up to 5, usually abortive seeds (Morton, 1987; Brücher, 1989; Verheij and Coronel, 1991).

The other species like *S. tuberosa* Arruda, and *S. venulosa* Mart. are native to the New World.

Genus *Sclerocarya* Hochst (n = 13)

Two to three species occur in sub-Saharan Africa and Madagascar (Kubitzki, 2011). The trees are dioecious (very rarely gynodioecious). Leaves are deciduous, alternate, imparipinnately unifoliate and petiolate having opposite or occasionally sub-opposite and petiolate (sessile) leaflets. Inflorescences are terminal and/or axillary. Female inflorescence often reduced to a solitary flower and male flowers are arranged in spicate racemes. Flowers are pedicellate, non-articulate having 4-5-parted and imbricate perianth, yellow to red calyx, yellow to purplish pink to red corolla, (8-10- 15-25(-30) stamens, subulate filaments, basi-dorsifixed anthers, barely evident or no pistillode in male flowers, many (21-26) staminodes, glabrous, patelliform disk, (1)2(-4) carpels, 2-3, short and subapical stylodia, capitate stigmas, pendulous ovules, apical with very long funicle. The fruit is obovoid subglobose, 2-3(4)-locular drupe having yellow to orange exocarp, fleshy and thick mesocarp, bony endocarp with (1)2-3(4) apical opercula and stylodia often persistent. Seed is obclavate and compressed with chartaceous and brownish testa.

Marula nut

(*Sclerocarya birrea* (A. Rich.) Hochst. syn. ***Sclerocarya caffra*** Sond.; kaffir plum; cat thorn; marvala plum) (2n=30)

It is a native of eastern and southern Africa (Martin *et al.,* 1987) and is distributed throughout the tropical Africa and Madagascar (Martin *et al.,* 1987). It has a vast geographic range incorporating three subspecies; *caffra* in south, *birrea* further north and in West Africa and *multifoliolata* only in comparatively small areas where the ranges of the other two subspecies overlap (Hall *et al*., 2002). Marula nut has been introduced in USA (Florida), Israel, India, Oman and Australia.

Marula nut is a small to large, much branched and deciduous tree, up to 20 m tall with a spread of 10 m. Bark is pale, nearly smooth and peeling in disk shaped flakes. Leaves are alternate, imparipinnate, crowded towards the apex of the stem, up to 30 cm long with 7-13 leaflets that are petiolate, ovate to elliptic, blue green having serrate margins in juvenile phase but smooth in older plants. The tree is dioecious. Flowers are unisexual. Male flowers are borne on terminal reddish spikes or racemes. The flower has 4, dark crimson sepals, 4 pinkish petals and 12 to 15 stamens inserted around the fleshy and depressed disk. Female flowers are long peduncled borne singly or 2-3 together at the ends of young shoots having subglobose and 2 to 3 locular ovary. Fruit is a fleshy, obovoid subglobose, of 3-5 cm diameter, 2 to 3 celled and yellow drupe containing a seed in each cell and with an eye through which embryo can grow out of the shell. Seed or stone is hard, light brown, smooth, oval, 3-4 g in weight and about 2.5x1.5 cm in size (Arnold *et al*., 1985; Kokwaro, 1986; Hall *et al.,* 2002). Kernels are valuable but are difficult to extract from the stones.

Genus *Buchanania* Spreng. (n = 11)

Twenty five to thirty or more species occur in tropical Asia, Malesia, Australia, Micronesia, Melanesia and Samoa with the greatest diversity in Malesia (Kubitzki, 2011). The trees are hermaphrodite. Leaves are alternate, simple and sessile to petiolate, entire and sub-coriaceous to coriaceous sometimes with cataphylls present. Inflorescences are terminal and/or axillary panicles. Flowers are hermaphrodite, pedicellate, articulate or non-particulate having (4)5(6)-parted perianth, imbricate corolla, greenish to white, diplostemonous androecium, filaments articulated in some species, anthers basifixed and usually sagittate, glabrous disk, only one fertile, the fertile one sometimes glabrous, cupular and crenulate. Carpels are 4-6, basally pubescent. Stylodia are 4-6 and short. Stigmas are oblique and sterile carpels are with functional stigmas, fertile carpel is without functional stigma, 4-6, basal or sub-lateral ovules, only one fertile. Fruit is lentiform, 1-locular drupe with persistent or caducous calyx, red or brown exocarp and bony endocarp splitting vertically in half at maturity in some species.

The genus *Buchanania* consists of 20 species of trees and shrubs, 6 of which occur in India. The other related species *B. angustifolia* Roxb. is grown for the edible nuts.

Chironji

(***Buchanania lanzan*** Spreng. syn. ***B. latifolia*** Roxb.; cuddapa almond; almondette; hamilton mombin; charoli)

Cuddaph almond is native of India and Myanmar (Martin *et al*., 1987). The tree is found on the hillsides in dry deciduous forests of Maharashtra, Gujarat, Karnataka, Andhra Pradesh, Madhya Pradesh, Bihar and Orissa in India, Malaysia and Myanmar.

It is a medium sized evergreen tree attaining a height of 8-15 m. The dark gray bark is thick and resembles crocodile hides. The leaves are thick, alternate, simple, oblong and 12-25 cm long, leathery and softly hairy beneath. The flowers are sessile, small, greenish white and are borne in dense, many flowered panicles. Calyx is short, persistent having ciliate lobes. Petals are 4-5, 2.5 mm long, oblong and recurved. Stamens are 8-10, free and inserted at the base of disk. The fruit is a black, single seeded drupe, 8-12 mm in diameter. The stone is hard and 2-valved, containing one kernel. The kernels are pear-shaped and 10 cm long (Anon., 1990; Verheij and Coronel, 1991). The cultivar Thar Priya has been released by CIAH, Bikaner.

Family Sapindaceae (litchi, maple, horse chestnut family)

It is a large family consisting of 145 genera and 1900 species (Heywood *et al*., 2007). It is widely distributed principally in the tropical and subtropical regions and some are in the warm temperate regions of North America. Several species are useful for their fruits, drugs, dyes and timber. The most noted is the litchi of China. The genera *Sapindus, Melicocca* and *Euphoria* are also grown for their fruits.

The plants are trees, herbaceous bushes or lianas. The leaves are usually spirally alternate, sometimes opposite (in *Acer, Aesculus* and a few other genera). They are most often pinnately compound, sometimes palmately or just palmate (*Acer*, *Aesculus*) with petiole lacking stipules but having a swollen base. The plants may be dioecious or monoecious. Flowers are small and unisexual or functionally unisexual. They are usually borne in cymes in panicles. They most often have 4 or 5 petals and sepals (petals are absent in *Dodonaea*). The number of stamens range from 4 to 10, usually on a nectar disc between the petals and stamens often with hairy filaments. Most frequently the number is 8, in two rings of 4. The gynoecium contains 2 or 3 carpels, sometimes up to 6. There is usually only one style with a lobed stigma. The fruits are fleshy (litchi) or dry may be nuts, berries, drupes, schizoczrps, capsules (*Bridgesia*), or samaras (*Acer*). The embryos are bent or coiled, without endosperm in the seed, but are frequently with arils.

Four subfamilies are recognized, Dodanaeoideae with 38 pantropical genera including the genus *Dodonaea,* Sapindoideae with 111 pantropical genera including the genus *Paullinia,* Hippocastanoideae with 3 genera of northern warm temperate region including the genus *Aesculus,* Aceroideae with 2 genera of north-temperate region including the genus *Acer.*

Genus *Litchi* Sonn (2n = 28, 30)

The plants are duodichogamous trees having indument of 2-branched hairs. Leaves are alternate and paripinnate with distal rudimentary leaflets. Inflorescences are terminal and axillary thyrses. Flowers are actinomorphic, bisexual or functionally unisexual having cup-shaped calyx with 4-5 equal minute lobes, without petals, annular disk, (6)7 (-11) stamens, colporate, striate pollen, 2-carpellate ovary with a single ovule per carpel and 2 elongated stigmas with spreading or coiled lobes. Fruit is 1-coccate (one coccus rudimentary) and indehiscent with leathery and muricate pericarp. Seeds are partly or completely covered by a translucent, fleshy and convolute arillode (Kubitzki, 2011). The genus has single species, *L. chinensis* Sonn., from southeastern China, Indochina, Malay Peninsula, Java, Borneo and the Philippines. It is widely cultivated in subtropical regions.

Litchi

(*Litchi chinensis* Sonn syn. ***Nephelium litchi*** Cambess., ***Euphoria litchi*** Desf., ***E. litchi*** Juss., ***Dimocarpus litchi*** Willd., ***Sapindus litchi*** Roxb.; lychee)
(n=14, 15, 16, 17) (2n = 28, 30 or 32)

It is a native of China. In India, it is grown at the foothills of the Himalayas and Nilgiris. The trees are tender, subtropical and evergreen. Frost and dry winds are undesirable for its growth and development. The sex undergoes changes in different flushes on the same tree. Usually the flowers emerging earlier are male. The flowers are small, borne in terminal panicles and are unisexual or bisexual. The seedless types are self or cross sterile. The fruits are borne in clusters. The fruit is 2.5-3.75 cm in diameter, pinkish in colour and turns brown on ripening. The fruit surface is tuberculate. The tubercles become rounded when the fruit ripens. The pericarp is brittle and rough with the tubercles. The arils are free. The fruit is a berry.

Leenhouts (1978), Menzel (1991) and Tindall (1994) considered that the species *Litchi chinensis* has three clearly defined subspecies based mainly on twig thickness, flower arrangement, stamen number and fruit characters:

i. *L. chinensis* ssp. *chinensis* (syn. *Dimocarpus litchi* Lour., *L. sinense* J. Gmelin and *Nephelium litchi* Cambess.) has slender twigs, flowers in lax cymules, stamens usually six and fruits smooth or with pyramidal warts up to 1 mm high. This subspecies is widely grown in China. It also grows wild in northern Vietnam and Cambodia.

ii. *L. chinensis* ssp. *philippinensis* (Radlk) Leenh. (syn. *Euphoria didyma* Blanco, *L. philippinensis* Radlk) has slender twigs, flowers in lax cymules, stamens 6 to 7 and fruits with pyramidal warts up to 3 mm high. This subspecies is known only in the Philippines where it is widely distributed but is rarely cultivated.

iii. *L. chinensis* ssp. *javensis* Leenh. syn. *L. chinensis* Sonn. has thick twigs, flowers in sessile clusters, stamens 7-11 and fruits smooth or with pyramidal warts up to 1 mm high. This subspecies is known only in cultivation in west Java and southern Indochina.

Cultivars

In China, litchi germplasm is divided into 3 types, viz. Sharp splintered type, Humped type and Smooth type on the basis of the pattern of skin cracking in the middle when the fruit ripens. The other important characteristics are fruit size, shape and taste. The major horticultural traits based on which litchi cultivars are classified and valued are maturity time (early, medium, late), single fruit weight, fruit shape, skin colour, edible (pulp) portion, flesh texture, flavour, TSS and seed size (bold/chicken tongued). The shape of skin segments and protuberances are the reliable and stable genetic characteristics.

Key to identify Chinese cultivars

I. Protuberances protruding and hard

1. Protuberances relatively fine, dense and sharp-pointed, skin segments small and irregularly arranged .... Tai So type
2. Protuberances large, sharp and short-pointed, skin segments small and regularly arranged..Kwai May type
3. Protuberances relatively blunt and short, skin segments relatively large and irregular in size and arrangement; often small segments among the normal skin segments... Jin Feng type

II. Protuberances hair-like or sparse, fine and sharp pointed

1. Skin segments irregular in size and arrangement; fruit shoulder is extremely wide and pronounced, stalk thick and strong .... Sum Yee Hong type
2. Skin segments regular in size and arrangement; fruit shoulder flat.. Haak Yip type

III. Protuberances smooth or not evident

1. Skin segments obviously protruding usually long, narrow shaped and arranged in longitudinal rows .... No Mai Chee type
2. Skin segments smooth, slightly protruding, usually near round in shape and irregular in arrangement................. .... Wai Chee type

About 40 cultivars are known but only about 18 are commercially important.

Sanyehong (Sum Yee Hong, Sun Yueh Hong and Third Month Red): It has a history of 700 years in China and is widely cultivated in Guangdong, Guangxi, Hainan and Taiwan. Fruit matures in May in Guangdong, the third month of the Chinese lunar calendar (earlier in Hainan and Guangxi). The tree is open, spreading and of medium size with long, thin and fragile branches. The leaflets are long, narrow, shiny dark green and much thicker than in other cultivars. The fruit weighs 37-42 g and is heart shaped or uneven heart shaped with thick skin and flat and unevenly sized segments. The flesh is waxy, white, juicy and sweet acid (17ºB TSS and 0.3% acidity). Seeds are large. Flesh recovery from the fruits is 65%. Although the fruit has average eating quality but fetches good market price because it is one of the earliest available.

Baitangying (White Sugar Jar): It is indigenous to western Guangdong. The tree is open with dark brown branches and oval and light green leaflets. Fruit weighs 25 g and is uneven heart shaped with smooth and bright red skin. The flesh accounts for about 72% of fruit weight and is waxy, white, crisp and very sweet (18ºB TSS and 0.05% acidity). Seeds are normally large giving a flesh recovery of 65%. Fruits are harvested in late May in Guangdong when their quality is rated superior.

Baila (Bah Lup, Dianbaibaila): It is grown widely in western Guangdong, southern Hainan and eastern Guangxi. The tree is medium in vigour and dome-shaped and is reasonably productive. Leaflets are long, narrow, dark glossy green with short-pointed tips. Fruits are available in late May in Guangdong and are nearly heart-shaped, 25 g, with thin, soft, smooth and brilliant purple red skin. The flesh is waxy, white, juicy and delicately sweet (18°B TSS and 0.12% acidity). The fruit contains a large oval seed, giving a flesh recovery of 72%. Fruits are rated better than Sanyuehong but not as good as Baitangying. The cropping is biennial in many orchards.

Shuidong (Souey Tung, Yuanzhi): The tree tolerates wet soils and is commonly planted along water ways in Guangdong. It has been introduced into Australia, but is not widely planted there. Productivity is average in northern Australia and poor in southern districts. The tree is relatively low with thin, long, open and spreading branches that hang down. Leaflets are large, flat, dark glossy green and pointed. The inflorescences are large compared to those of many other cultivars. Fruit weighs 25 g and is heart shaped with distinct uneven shoulders. The skin is thin, smooth, and dull purple red. The fruit tip is obtuse or blunt. The flesh is soft, juicy and sweet. Seeds are mostly of medium size giving a flesh recovery of 72%. The fruit ripens from late May to early June in Guangzhou and between November and December in northern Queensland.

Feizixiao (Fay Zee Siu, Smile of the Imperial Concubine): It has a long history in China and is reported to have delighted the Emperor and his concubines. The fruit is described as, 'having the colour of amber, size and shape of a goose egg and sweetness of honey'. It is widely cultivated in Guangdong, Guangxi, Hainan and Taiwan and has also been introduced into Australia. Productivity is variable in Australia even in northern districts. The tree is vigorous having long, sparse and fragile branches. Leaflets are large, narrow and deep glossy green. Fruit weighs 30 g and is round to oval. The skin is thin with dense sharp pointed protuberances. Fruits are yellow red at maturity. The flesh is firm, sweet and slightly fragrant (18°B TSS and 0.2% acidity). Seed size is variable. The flesh recovery is about 80%. There are many small seeds. Fruit matures in early June in Guangzhou and in October in northern Queensland.

Dazao (Tai So, Big Crop, Mauritius, Hong Huay): It has a history of 500 years in Guangdong and is also found in most of the litchi growing areas of the world. A similar cultivar, Mauritius, selected from seedlings in Mauritius during 1870s (Ledin, 1957), forms the basis of industries in southern Africa, Israel and Florida. Practically all the trees in South Africa are descendants of a few trees imported from Mauritius. Hong Huay in Thailand is also very similar to Dazao (Tai So). It was the most popular cultivar in Australia, but has now lost favour because it bears erratically in most districts. There are also problems of tree splitting and browning or cracking of fruits before harvest. Trees are vigorous and spreading with an open crown having branches that tend to split from the trunk. Leaflets are large, glossy dark green and almost canoe shaped. Fruit weighs 24 g and is oblong with uneven shoulders. The bright red skin is thin with small sharp pointed protuberances. The flesh is milky white, slightly tough having 15°B TSS and 0.3% acidity. Flavour is sweet acid when immature, sweet when fully ripe and bland when overripe. Seeds are generally large giving a flesh recovery of 75%. Fruit matures during early June in Guangdong and early December in northern Queensland.

Heiye (Haak Yip, Hak lp, Blac Leaf and O-Hia): The name of the cultivar refers to the 'very dark glossy green leaves'. It has a history of over 600 years in China and is the most widely grown cultivar in southern districts. It is also popular in Thailand, where it is known as O-Hia. There are small plantings in Australia, although production is erratic and fruit quality is average. The tree and fruit resemble Shuidong (Souey Tung). The trees are vigorous and dome shaped with sparse foliage and long branches. The leaflets are long, narrow pointed and slightly curled at the tip. The fruit weighs 19 g and is uneven heart-shaped or oval with tough, smooth and dark red skin. The flesh is milky white, juicy, sweet and slightly fragrant (18ºB TSS and 0.4% acidity). Seeds are of medium size giving a flesh recovery of 65%. Quality is rated 'above average' in China. Fruits are harvested in late June in Guangdong and in late December in northern Queensland. It is the main cultivar used for canning in South East Asia.

Nuomici (No Mai Chee, No Mai Tsi and Glutinous Rice Cake): It has a history of more than 200 years in Guangdong and is widely grown in the Pearl River Delta as well as in various districts of Guangxi, Fujian and Taiwan. It is one of the most highly prized cultivars in China and usually sells for three to four times the price received for other cultivars. Fruits are suitable for fresh market and for drying. The tree is umbrella shaped with a dense canopy and slim branches that hang down. The leaflets are small, soft and thin with wavy edges. The fruit weighs 25 g and nearly all are small seeded giving a flesh recovery of 85%. The fruits are heart-shaped and bright red at maturity. The skin is thin and smooth and the flesh is juicy with a distinctive sweet fragrant flavour. Fruits appear in the market in late June in Guangzhou. Cropping is irregular and the fruit is susceptible to cracking. However, the high price received guarantees a good return. The cultivar has been introduced into Australia and Israel, but is not widely grown or cropped as much as in China.

Guiwei (Kwai May Red, Kwai Mi): It is native to Luogan in Guangzhou and is widely grown in Guangdong, Guangxi and southern Taiwan. It is another highly prized cultivar. It was introduced into Australia where it is known as Kwai May Red, but it is not widely planted. A similar cultivar, Kwai May Pink, forms the basis of the industry there. In Australia, trees resemble those of Kwai May Pink but are more spreading. Leaflets are slightly larger than Kwai May Pink and are flatter. The tree is dome shaped with sparse, slim and hard branches. Leaflets are short, pointed at the tip and curl upward from the midrib. The fruit weighs 18 g and is round with thin and bright red skin. The segments are swollen and the protuberances are sharp pointed. The flesh is milky white, crisp, sweet and fragrant (19ºB TSS and 0.18% acidity). Seeds are small in most seasons giving flesh recovery up to 80%. The fruit matures in late June in Guangdong, the same season as Nuomici. In China, it is generally rated superior. In Australia, fruits are almost identical to those of Kwai May Pink except that it has red rather than pink orange skin, firmer flesh, higher proportion of small seeds (50-60%), higher flesh recovery (70-80%) and slightly better flavour.

Huaizhi (Wai Chee, Cherished Litchi Kim Cheng): It is one of the most common cultivars in Guangdong and Guangxi. It is also popular in Thailand (Firn Cheng) and Australia. It crops heavily in most areas in China although the trees are slow to establish after planting. Productivity is better in southern Queensland than in northern districts. Trees are low and dome shaped with thick and hard branches and compact foliage. The small leaflets are oval and curve upwards from the midrib and down along their length. The fruit weighs 20 g and is round and dark red. The skin is thick and tough with large and flat segments. The flesh is waxy, white, juicy and sweet (19ºB TSS and 0.25% acidity). The seeds are nearly all large giving a flesh recovery of 70%. Fruit quality in China is rated as 'above average'. Although fruits of Huaizhi have good flavour, their value in Asia is lower compared with Heiye (Haak Yip), Guiwei (Kwai May Red) and Nuomici (No Mai Chee) because its large seeds and soft flesh reduce eating quality. Fruits are available in early July in Guangdong and in early February in southern Queensland. The mature fruit can be left to hang on the tree for several days.

Chenzi (Chen Family Purple, Brewster): It is widely cultivated in Fujian and was taken to Florida by Rev. W. M. Brewster in 1903 (Groff, 1921) where it became the major cultivar but its production is very erratic and now Mauritius is more popular. It was also introduced into Australia but is not popular there. A similar cultivar, Floridian, was introduced into Israel from California (Goren *et al*., 2001). The tree is open, with a dome shaped crown and dense branches that hang down. The leaflets are long, oval and pointed at the tip. The fruit weighs 15 g and is heart shaped with dark purple red skin that has small segments and sharp protuberances. The flesh is waxy, white, juicy, fragrant and slightly sour at maturity (19ºB TSS, 0.22% acidity). There are many small seeds giving a flesh recovery of 75%. Fruit quality in China is rated as 'average'. Fruit matures in mid to late July in Fujian and in late December in southern Queensland.

Lanzhu (Difficult to Obtain): It has a history of 300 years in China and is one of the most popular cultivars in Fujian. The tree is vigorous and semi-spherical with a dense canopy and slim branches that hang down slightly. The leaflets are long and oval with straight edges and short pointed tips. The fruit weighs 26 g and is heart shaped to nearly round with pointed protuberances and yellow purple red. The skin is thin. The segments are large. The flesh is milky white, juicy and slightly sour (16ºB TSS). The seeds are generally aborted. Flesh recovery of 75% and fruit quality is rated as 'above average'. Fruits are available from late June to early July in Fujian, China.

Sah Keng: It was developed in Taiwan from a seedling of Haak Yip (Yen, 1984). It was introduced into Australia but is not grown commercially outside Taiwan. Sah Keng produces large fruits having small seeds. The trees show significant variation in the orchard. Trees are of medium height, dome shaped with short and fragile branches and pale green leaves. It is a mid season cultivar. Yields are heavy but irregular. Fruit weighs 30-35 g and is heart-shaped with purple red skin. The skin segments are swollen and protuberances are blunt. The flesh is soft and sweet. Seeds are variable, often small giving a flesh recovery of 75%. Fruit quality is rated as 'good' or 'better'.

Kom: The trees are vigorous and erect and have long and strong branches and dense foliage. Leaflets are narrow, pointed and medium in size and dark green. They are generally flat but curve downwards slightly towards the tip. The new growth flush is red changing to green with maturity. Fruit weighs 8-20 g and is long and heart shaped to nearly round. The very thick skin is blotchy and yellow purple red at maturity. Shoulders are flat or even and the fruit apex is obtuse. The skin segments are smooth at maturity and variable in size, shape and arrangement. The protuberances are sharp pointed. Fruits are borne in small loose clusters. The flesh is tough to fibrous and mild in taste becoming bland at maturity. Seed size is small in proportion with the small fruit having 'chicken tongue'. Flesh recovery ranges from 60 to 80%. Fruits are available early in the season.

O-Hia (Baidum, Heiye and Haak Yip): It resembles Haak Yip but does not match it in all characteristics. Trees are of medium size with dense foliage on long and thin branches (not as long as in Haak Yip). Leaflets are large, narrow, dark green and slightly curled upwards from the midrib. The new growth flush is red brown. Fruit weighs 21 g and is heart shaped. The skin changes from blotchy yellow to deep red with maturity. Skin segments are irregular in size, shape and arrangement having smooth to obtuse protuberances. The flesh is juicy and sweet. Seeds are plump giving a flesh recovery of 65%. It is a mid season cultivar.

Chakrapad (Chacapat): It produces very large fruits late in the season. Fruits are sweet and acceptable in Thailand but are often acidic in Australia. Fruit production is average in Australia. Trees are moderately vigorous, erect and have long branches and dense foliage. Leaflets are small, long, narrow, pointed and dark green. They curl upwards from the midribs and downwards along their length towards the tip. The new growth is green. Fruit weighs 30 g and is round to slightly heart shaped. The skin is thin, soft and deep red with yellow markings. Shoulders are flat and the fruit tip is round. Skin segments are swollen with obtuse protuberances. The flesh is moderately juicy remaining slightly acidic when fully ripe in Thailand and very acidic in Australia. Seeds are large giving a flesh recovery of 65%.

Luk Lai: The trees are medium in vigour and dome shaped with many short upright branches and dense foliage. The colour of the new growth is brown green changing to light green with advancing maturity. Leaflets are mid-green and broad oval in shape. The leaflets are curled up slightly from the midrib to the margin and down along their length towards the tip. Leaflets characteristically hang vertically from the branches. Fruit weighs 17 g and is almost round to slightly egg-shaped with flat shoulders and an obtuse apex. The skin is yellow red changing to deep red with maturity, The segments are smooth and irregular in size and arrangement. The skin is relatively thick and the protuberances are very sharp and strong. Fruits are normally produced in small tight clusters and have uniform maturity. The flesh is moderately crisp and separates easily from the seed. Flavour is subacid to sweet becoming bland once the fruits are fully mature. Fruit contains a medium to large and dark brown seed giving a flesh recovery of 60%. Flavour is acceptable if the fruits are picked at proper stage of maturity The large seeds limit commercial acceptance of the cultivar.

Kwai May Pink: It is believed to have originated in China and resembles Guiwei (Kwai May Red). It is popular in Australia where large number of trees have been planted. It is relatively unknown elsewhere. Trees are large, very erect and have long, slim branches that point upwards. The trees tolerate strong winds to some degree. The new growth flush is attractive red. Leaflets are narrow, long, oval and shiny light green. They curl upwards slightly from the midrib and downwards along the length. Its bearing ability is good possibly due to a long harvest period. It is a mid season cultivar. It develops acceptable sweetness and flavour well before the fruit matures. Fruit weighs 20 g and is round with very rough and thick skin. The skin colour changes from yellow to yellow pink to orange pink at maturity with some green on the shoulders. Flesh is crisp, sweet and aromatic. Seeds are variable, 70% with chicken tongues giving a flesh recovery of 70%. Fruits are exported to South East Asia from Australia.

Salathiel: The original tree was found growing in northern Queensland. Its parentage is unknown. Yields are variable in subtropical districts and light in tropical areas. Fruits are harvested late just before Wai Chteea (Huaizhi). Trees are small and sometimes produce long branches with undeveloped leaves. The new flush is red changing to green with maturity. Leaflets are small, broad and curve down slightly at the tips. The tip of the leaflet is round with a short, distinctive point. Fruit weighs 20 g and is egg to ball shaped and is borne in small loose clusters. The skin is thick and moderately rough with prominent markings. It changes from blotchy yellow to deep red at maturity. The fruit tip is obtuse to round. The flesh becomes thick, crisp and sweet long before the fruits are fully coloured. Most fruits have chicken tongue seeds giving a flesh recovery of 75%. Occasionally, fruits can be almost seedless, although these are very small and unmarketable. Fruits attract a high price in domestic markets and are also exported.

Hom Shiu Chi: Each mature fruit has an immature fruit attached.

Loh Mai Chi: The cultivar is said to be one of the best in the world. It is grown in the vicinity of Canton (China).

No Mai Chi: It is most recognised cultivar. The fruit protuberances are large, sharp and short pointed. The skin segments are small and regularly arranged.

Characteristics used in India to distinguish cultivars

i. Tree height: The cultivars are grouped as tall (>6 m high), medium tall (5-6 m) and dwarf (<5 m) types.

ii. Leaf characters (shape and size): Leaf of the cultivar Rose Scented is boat shaped, while the cultivar China has distinctive twist along the length curved upward from the midrib and down along the length. Small leaflets of Bedana are oval shaped.

iii. Panicle characters: Emergence of panicle and its shape differ in cultivars. Singh and Singh (1954) distinguished the cultivars based on panicle characteristics. Shahi has long panicle, while Bedana produces short and compact panicle.

iv. Fruit skin colour and texture: The skin colour varies from pinkish brown to dark red in cultivars although it is also influenced by the growing conditions. Bedana and China cultivars have very thick fruit skin while Rose Scented and Shahi have very thin skin. Skin surface at maturity may have smooth swelling or sharp pointed protuberances. Bedana has very smooth protuberance while Shahi has very distinct protuberance.

v. Seed characters: Presence and absence of seed as well as structure and size of seeds vary in cultivars. In Rose Scented and Bedana, high proportion of chicken tongued seeds are observed while China is bold seeded. Swarna Rupa cultivar has high proportion of small seeded fruits.

The major horticultural traits used to classify and evaluate cultivars are maturity time (early, medium, late), fruit weight and shape, skin colour, flesh texture, flavour, edible (pulp) portion, TSS and seed size (bold/chicken tongued).

There are as many as 30 cultivars in Bihar but only Shahi, China, Purbi or Purvi, Early Bedana and Late Bedana are commercially important. Many of them have almost similar fruits, e.g. Shahi resemble Rose Scented or Dehra Rose and China resembles Calcutta and Bombai. Ray (2007) observed physico-chemical characters of 14 important cultivars at Pusa (Bihar), India (Table 29.4).

Shahi (Muzaffarpur, Rose Scented): It is the most popular cultivar in Bihar and is also found in other parts of India as well as in Bangladesh. Fruiting is early in the season. The fruit is oval with crimson red skin, large in size (22 g) having excellent quality and aril with delicate rosy flavour. The flesh is juicy, sweet and fragrant, and the seeds are variable in size. Yields are heavy and regular. A large tree produces 100-150 kg fruit. The aril remains tight (firm) for longer duration and thus the fruits have good keeping quality during storage or long distance transport.

China (Purbi, Calcuttia, Bengalia and Bombaiya): It is an important cultivar in India. It ripens when most of the other cultivars have been harvested. Its origin has not been determined although there is a similar cultivar in Bangladesh known as China Number Three. Trees are relatively small and high yielding but have alternate bearing tendency. Fruit weighs 25 g and is orange red. The aril becomes loose during storage or long distance transport. The flesh is soft, juicy and very sweet but is inferior to Shahi. Seeds are normally small.

Early Bedana (Early Seedless): It is a popular early cultivar from Bihar, Uttar Pradesh, Punjab and Bangladesh. Trees are medium in height and are regular in fruit production. Fruit weighs 17 g and is oval or heart shaped with rough and deep red skin at maturity. The flesh is white, juicy and sweet. Overall fruit quality is good.

Late Bedana (Late Seedless): It is a late cultivar from northern India. Trees are vigorous and productive giving average yields of 60-80 kg from 10-year old trees. Fruits are of medium size having good flesh recovery. The flesh is creamy white, juicy and sweet. Seeds are usually small.

Bombai: It is an important early cultivar from West Bengal and Bangladesh. Trees are vigorous and productive. Ripe fruit has an attractive deep red colour with gray white, juicy and sweet flesh. It is similar to China grown in other areas.

Dehra Dun (Dehra Rose): It is an important cultivar from Uttar Pradesh and Punjab. Trees are of medium height and produce moderate to heavy yields. Fruits are bright pink red at harvest and are very attractive. They have small seeds but are susceptible to cracking.

Bengal (McLean's Red): It is a seedling of the Indian cultivar Purbi sent to Florida in 1929 (Loomis, 1955). Although second most important at one time after Tai So in Australia, the cultivar has now lost favour. A similar cultivar, McLean's Red, makes up 16% of plantings in South Africa (Milne, 1999).

Trees are vigorous and spreading with thin branches but are reasonably resistant to wind damage. Leaflets are large, light green and have a distinctive twist or curl along their length. The new growth flush is red brown. Average cropping is low although trees produce very high yields in 'on year'. The fruits are borne in large clusters of 50 or more. Fruit weighs 23-27 g. The fruits are round, egg shaped to lop-sided heart shaped with uneven shoulders and a distinctive pointed tip and thick and rough skin having an attractive bright red colour. The protuberances are sharp pointed to wedge shaped. Fruits are attractive and have pleasant taste but have large seeds, poor flesh recovery and ripen unevenly. The flesh is soft, sweet and moderately juicy. The flavour of the fruits deteriorates if they are left to hang on the tree. There are very few abortive seeds. Under drought, the aril remains undeveloped and does not cover the seed at the pointed end giving a flesh recovery of 50% or lower.

Swarna Roopa: It is a clonal selection made at Ranchi. Fruits mature a week earlier than the late cultivar China. Fruits are round bearing in clusters, medium large (20 g), sweet (19°B TSS) having attractive deep pink colour, small flat seed and high TSS:acid ratio. It is highly resistant to cracking.

CHES 2: It is a selection from Bombaiya. Fruits mature late and are borne inside the canopy and are free from sunburn and cracking.

Sabour Madhu (H-105) (Purbi x Bedana): Trees are tall and vigorous. Fruits are borne in clusters of 24 per panicle, mature 8 days later than in the late cultivar Kasba. These are conical, medium large (15.5 g), light red and have higher TSS (21°Brix TSS) and aril content than Purbi. Fruit shape resembles that of Purbi.

Sabour Priya (H-73) (Purbi x Bedana): Fruit shape has characters of both the parents. Fruit weight and quality are better than Purbi.

Rajendra Hybrid-235: Trees are high yielding. The fruit is ovate and medium-sized (21.8 g) having bright red peel, high sweetness (23.4°B TSS) and cream white aril of excellent quality.

Rajendra Sabour Bedana: It is highly productive late maturing cultivar bearing medium-sized (27 g) fruits having bright red peel, high content (82.7%) of creamy white and very sweet (24°B TSS) aril having pleasant flavour.

Saharanpur Selection: It is a seedling selection from the local cultivar. It is a late maturing cultivar. The fruit has 19.8°B TSS and very low cracking (2%).

Madhu: It is a hybrid developed by crossing Purbi x Bedana. It has higher number of fruits (24) per panicle and ripens 8 days later than the late maturing strain Kasba in Bhagalpur (UP). It has higher TSS and aril percentage than Purbi. Its fruit shape resembles that of Purbi.

Mandrazi: It is a popular cultivar in Bhagalpur (UP). Its fruit size and yield potential are comparable with Shahi and China.

Bedana: It produces the best quality fruits but due to sparse fruiting it is a low yielder.

The cultivars Rose Scented (Gulabi), Early Large Red, Kalkattia, Late Seedless, Early Seedless, Macleen and Piaji (tastes like boiled onion) are from Uttar Pradesh.

Table 29.4. Characters of litchi cultivars observed at Pusa (Bihar), India (Ray, 2007).

Cultivars	Fruit wt. (g)	Aril wt. (g)	Juice %	TSS ^{0}Brix	Acidity %	Seed wt. (g)	Peel wt. (g)	Fruit cracking %	Fruit yield/ tree kg
Rose Scented	21.2	13.2	57.2	19.3	0.47	4.27	3.70	10.6	90.7
Deshi	18.1	12.0	54.6	18.8	0.56	3.49	2.59	6.2	85.2
China	21.9	14.8	56.7	20.7	0.53	4.02	3.08	-	106.5
Shahi	22.1	14.9	58.9	21.2	0.42	3.86	3.28	15.3	118.6
Bombai	19.5	13.1	53.1	17.8	0.57	3.40	3.02	4.1	89.3
Late Bedana	21.4	15.6	62.3	20.3	0.46	2.41	3.39	5.7	56.7
Calcutta	18.5	11.7	53.7	19.0	.0.55	3.74	3.02	3.2	86.5
Mandraji	20.0	12.5	54.6	17.5	0.51	4.36	3.15	4.8	92.0
Early Bedana	18.7	13.9	60.5	20.8	0.48	1.78	3.00	6.3	38.2
Purbi	19.3	13.0	52.1	19.2	0.46	3.21	3.02	5.5	64.2
Longia	15.7	10.2	50.1	17.0	0.58	3.01	2.48	14.2	36.5
Kasba	18.7	12.3	54.2	18.0	0.53	3.36	3.04	7.3	55.7
Kaselia	16.0	10.7	51.2	17.0	0.59	3.07	2.23	13.6	60.0
Dehra Rose	18.2	12.1	57.6	18.3	0.47	3.27	2.80	10.1	82.0

Genus *Dimocarpus* Lour (2n = 30)

Six species occur in southern and southeastern Asia from Sri Lanka and India to eastern Malesia and Australia. The trees or shrubs are falsely polygamous and have indumentum of simple and stellate hairs. Leaves are alternate, paripinnate and rarely unifoliate having serrate or entire leaflets, the distal leaflet being rudimentary. Inflorescences are terminal or less often axillary thyrses. Flowers are actinomorphic, functionally unisexual having 5-6 sepals, imbricate and connate at the base or distinct, 5(6) or 0 petals, without appendages, annular, 5-lobed and pubescent disk, (6-) 8(-10) stamens, colporate, striate to perforate pollen, sometimes scabrate or regulate, 2(3)-carbellate ovary with a single ovule per carpel, filiform style with 2-3 stigmatic spreading lobes. Fruits have 1(2) indehiscent or tardily dehiscent, warty, smooth or spiny mericarps. Seeds have a thin, translucent white and fleshy arillode around the hilum (Kubitzki, 2011).

Longan
(*Dimocarpus longan* Lour. syn. *Euphoria longan* (Lour.) Steud., *Nephelium longan* (Lour.) Hook.; ashphal) (2n=30)

It grows wild on Western Ghats up to 1500 m altitudes and in Assam up to 1050 m altitudes. The tree is smaller than that of litchi. Longan tree is evergreen, large and symmetrical growing up to 12-15 m in height and has corky bark on the trunk which splits and peels off. The tree resembles litchi in growth habit and appearance but differs in bark characteristics. The compound leaves have 2 to 3 pairs of leaflets that are elliptic, ovate oblong and 10-20 x 3.5 cm in size and are dark glossy green on the upper surface and paler green on the underside. The young leaves are red brown turning light green with maturity. Flowers are borne on terminal and axillary upright panicles which emerge on new growth. They vary from male to female in succession along the panicle. The flowers are small, greenish yellow, unisexual and unattractive having confluent calyx lobes at base, 2-5 x 1-3 mm in size with 5 petals, velutinous disc, 8 stamens and sessile and bilobed ovary. The flowers are self incompatible and insects are required for pollination. The fruit ripens later than that of litchi. Fruits are somewhat similar to those of litchi borne in bunches and are round to ovoid, small (about 1-3 cm in diameter), brownish yellow at maturity and have a smooth to warty skin. The tubercles are absent. The aril is free, white, translucent and gelatinous with sweet salty flavour and contains a shiny dark brown and large seed with an eye-shaped marking (hence dragon's eye) (IBPGR, 1986b; Morton, 1987; Verheij and Coronel, 1991; Menzel *et al.*, 1990). Sugar content in the pulp is lower. The fruit is inferior but is good for canning and has some kind of smell.

Cultivars

Seedling trees are extremely variable in precocity, yield and fruit quality. There are 300 to 400 cultivars in southern China but only 30 to 40 are grown. The cultivars vary with respect to canopy density, size, colour and arrangement of leaf, bark characteristics, yield, size and quality of fruit and disease and wind resistance. Important cultivars of China are Fu Yan, Wu Yuan, Wu Long Ling, Black Ball and Shia Xia. The best quality cultivars of Thailand are Biew Kieu, Chompoo E Bure, E Dol, E Haw and Biew Kiew and those of Florida are Kohala, Ponyai, Homestead No.1 and Kay Sweeney. Most important cultivars from Hawaii are Fukho No. 2, Kohala, Hac, Wai and Carambo. Shek Kip is an important cultivar of Hongkong. Yang Tao Ye, Chian Oh Diao, Funkien Lungan (Funkugan) and Lungan are the cultivars of Taiwan. Most of the modern cultivars are selections suited to different climatic conditions (Morton, 1987; Verheij and Coronel, 1992). Regular bearing cultivars are required.

Kohala: It is a tropical mid season and regular bearing cultivar having spicy fruit of 15 g weight.

Sao-an-liao: It is a subtropical mid season cultivar having fruit of 15 g weight, 13°B TSS, dark brown skin, 62% pulp and small seed. Fruit bearing is irregular.

Duan ru: It is a mid season cultivar having fruit of 13 g weight and 14°B TSS. Its bearing is regular in the tropics but is irregular in the subtropical region.

Yang tao ye syn. Carambola Leaf: It is a mid season cultivar having spicy fruit of 16 g weight and 11°B TSS. Its bearing is regular in the tropics but is irregular in the subtropical region.

Fen ke: It is a subtropical mid season cultivar having crispy fruit of 11 g weight, dark yellow skin colour, 15°B TSS, 66% edible flesh, white colour and medium sized seed. Fruit bearing is irregular.

Daw: It is a subtropical mid season cultivar having fruit of 15 g weight. Fruit quality is good. Fruit bearing is irregular.

Chompoo: It is an excellent cultivar.

Biew Kiew: It is an excellent cultivar.

Fuyan syn. Lucky Eye: It is a subtropical late bearing cultivar having good quality fruit of 12 g weight, 15°B TSS, having dark brown skin colour, 67% of transparent crispy flesh and small seed. The trees are vigorous. Fruit bearing is regular.

Wuyan syn. Blackround: It is a subtropical mid season and regular bearing cultivar having good quality fruit of 17 g weight having 71% white flesh with 14°B TSS, large seed and light brown skin colour. The trees are vigorous.

Shi Xia syn. Shek Yip, Shi Yuan: It is a subtropical mid season and regular bearing cultivar having excellent quality fruit of 10 g weight having dark brown skin colour, 69% white and crispy flesh with 19°B TSS and small seed. The trees are vigorous.

Genus *Nephelium* L. (2n=22)

There are over 30 species of *Nephelium* having edible fruits, including 5 in Myanmar, Thailand and Indochina, 13 in Peninsular Malaysia, 16 in Borneo, 4 in the Philippines and 3 in western Java (Tindall, 1994). The fruit producing species are *N. chryseum* Blume (mountain rambutan growing wild), *N. ramboutan-ake* (Labill.) Leenh (pulasan), *N. glabrum* Noronha [Redan (Malay)], *N. topengii* (Hainan rambutan growing wild in Hainan), *N. macrophyllum, N. malaiense* Griff. (mata kucing), *N. maingayi, N. melanosmiscum, N. philippiense* (bulala), *N. obovatum* (the vine longan), *N. bassocense*, *N. glabrum*, *N. hypoleucum* (the white underleaf rambutan of Thailand), *N. rimosum*, *N. xerospermoides* (aluoa), *N. cuspidatum* Blume syn. *N. eriopetalum* var. *robustum* [Lotong (Malay] (Mendiola, 1941; Galang, 1955; Rodrigo, 1976). The other related species producing edible fruits are *N. eriopetalum* Miq. (lotong, found growing in South East Asia), *N. hypoleucum* Kurz. (kyet-monk, found growing in Indochina), *N. uncinatum* Radlk. ex Leenh. (namun, found growing in Indonesia), *N. aculeatum* Leenh. (distributed in Sabah, Borneo), *N. juglandifolium* Blume. (lungsir, found growing in peninsular Malaysia), *N. meduseum* Leenh. (distributed in Borneo), *N. melanomiscum* Radlk. (parih, found growing in Borneo and Philippines), *N. melliferum* Gagnepain (distributed in peninsular Malaysia, Thailand and Vietnam), *N. reticulatum* Radlk. (buah stagok, distributed in Borneo) and *N. maingayi* Hiern. (ridden) found growing in Indonesia and Malaysia (Morton, 1987). The major edible types in semi-wild and domesticated species exhibit characteristic aril thickness, sourness/sweetness, juiciness and flavour traits (Rodrigo, 1976). In semi-wild and primitive forms the aril or sarcotesta invariably adheres to the seed (Arora, 1998). Among these, pulasan is reputed to have sweeter and larger fruits with thicker and better quality aril.

The trees are falsely polygamous or dioecious or less often shrubs. Leaves are alternate and paripinnate having leaflets distinctly glaucous beneath, distal leaflet being rudimentary. Inflorescence is axillary, pseudo-terminal or terminal (in *N. cuspidatum* Blume ramiflorous and cauliflorous) thyrses. Flowers are actinomorphic having cup-shaped calyx crowned by 4-6 sub-equal, valvate lobes, 0 or 4-6, clawed petals with a bibbed appendage, annular disk, 4-10 stamens having colporate and striate pollen, (1)2 (-4)-carpellate ovary with a single ovule per carpel, elongated style and 2-lobed stigma. Fruit is 1(2)-coccate, tardily loculicidally dehiscent having warty to spiny, coriaceous or less often woody or corky pericarp. Seeds are completely covered by edible sarcotesta (Kubitzki, 2011).

Rambutan
(*Nephelium lappaceum* L. syn. *Euphoria nephelium* DC.) (2n=22)

Rambutan is a very large taxonomic group comprising trees, shrubs, vines and lianas. It is a native of South East Asia (Martin *et al.*, 1987). The species is distributed from southern China through Indochina, Malaysian Peninsula, Indonesia (Sumatra, Java, Borneo and Sulawesi) to the Philippines (Verheij and Coronel, 1991). Considerable genetic diversity exists in Vietnam, Indochina, Thailand, Malaysia, Indonesia and the Philippines as the species is cross pollinated and has been widely grown over a long period. In India, it grows in the Nilgiri hills on the lower elevations. Based on leaflet characters, 6 botanical varieties of *N. lappaceum* are *lappaceum, pallens* (Hiern) Leenh, and *xanthoides* (Radlk.) Leenh (Van Welzen and Verheij, 1991).

***N. lappaceum* var. *lappaceum*:** The leaflets are widest above the middle, the midrib is sparsely pilose below and the nerves are strongly curved. This is widely distributed and cultivated in Thailand, Malaysia, Indonesia, Borneo and the Philippines.

***N. lappaceum* var. *pallens* (Hiern) Leenh:** The leaflets are widest at or below the middle, the midrib is usually glabrous below and the nerves are slightly curved. Distribution is more in China, Thailand, Laos, Cambodia, Vietnam, Malaysia, Indonesia (Sumatra), Borneo and southern Philippines.

***N. lappaceum* var. *xanthoides* (Radllk.) Leenh:** The leaflets are widest at or below the middle, the midrib is shortly hairy below and the nerves are slightly curved. It is confined to Borneo.

The other 3 varieties are *N. lappaccum* var. *cuspidatum, N. lappaccum* var. *eriopetalum* and *N. lappaccum* var. *robustum.*

Rambutan is a large, round topped and much branched evergreen tree. The wild trees are fairly large (13 m in height) whereas the clonal trees are 4-7 m high and usually have spreading growth. Leaves are alternate, paripinnate having ovate to obovate 3-4 pairs of leaflets of 5-28 x 2-10.5 cm size, usually horizontal, glabrous above or sometimes slightly hairy on the midrib and variably hairy beneath. Both monoecious and dioecious tree types occur. The inflorescences are pseudo-terminal to terminal. The actinomorphic, whitish, yellowish or greenish flowers are either male (having well developed stamens or have stigma that fails to open) or hermaphrodite or female (having small stamens that lack anther dehiscence). The 4-5 sepals are nearly free more than halfway, connate and 0.7-2.1 mm long. Petals are usually absent or sometimes up to 4 develop but remain reduced in size not exceeding 1.6 mm. Disc is complete and hairy or glabrous. In male flower, 5-8 stamens are exerted with filaments having dense long hairs at least in the basal part and anthers dehiscing lengthwise. In hermaphrodite flower, the pistil is 2 or rarely 3-merous, densely hairy and well developed, the ovary is warty with well developed style and spreading stigma. According to Naik, the female flowers set fruit by self pollination but Ochse observed that the anthers were sterile in such trees.

The fruits are ellipsoid to subglobular, 6 x 3.5 cm in size, weighing 20-95 g and having short pedicels. The pericarp is red or yellow and is covered with smooth appendages that look like curved thorns but are soft and flexible. The sharp pointed tubercles are as in castor. The interior aril is smooth, white, translucent, juicy and sweet and united with the seed. The seed is ellipsoidal and 2-3 cm long (IBPGR, 1986; Morton, 1987; Verheij and Coronel, 1991).

Cultivars

The cultivars were developed by cloning superior plants selected from the wild variability. The cultivars are distinguished by characters such as thickness, colour, juice content and aroma of the flesh and its adherence to the seed (clingstone or freestone) and whether the papery skin of the seed comes off with the flesh. The wild varieties *pallens* and *xanthioides* could be used for cultivar improvement (Verheij and Coronel, 1991). There are well over 200 cultivars available throughout tropical Asia (Almeyda *et al.*, 1979; Tindall, 1994; Zee, 1993), many of which are from the ASEAN region (Indonesia, Malaysia, Singapore, the Philippines, Thailand and Brunei Darussalam). About 20 clones are cultivated in different countries of South East Asia that originated from 6 botanical varieties. The main cultivars are Lebakbulus, Binjai, Seematjan, Rapiah, Tankuweh, Seenjonja, Seetangkooweh, Seelengkeng and Seekonto from Indonesia; Labakbulus, Binjay, Tankue, Rapish and Simacan in Java; Simecen, Sinyonya, Maharlika and Seematjan introduced from Indonesia in the Philippines; R134, Peng Thing Cheng (R3), Muar Gading (R156), Khaw Tow Bak (R160), Lee Long (R161), Oh Heok (R 162), Deli Cheng (R 170), R 99 and R 191 in Malaysia; Chooi Ang, Peng Thing Bee, Ya Tow, Azimat and Ayer Mas in Malaya (Morton, 1987); Deli Cheng and Jit Lee in Singapore; and Rongrien, Seechohmpoo, Bangyeekhan, Chompoo, Bang Yi Khan, See Tong, Nam Tan Kruad, Seetong, Namtangruad and Jemong in Thailand. In northern Australia, the cultivars R3, R134, R156, R168, Rapiah, Rongrien and Jit Lee were found acceptable as these are relatively free from testa adherence (Watson, 1983). Promising cultivars selected in Hainan, China are B-R-1 and B-R-2 (Kay-Ming, 1990b).

Seematjan: The cultivar is the most popular and commercially grown all over the tropics, mostly in Malaysia, Thailand and Burma. The tree is medium tall, much branched, evergreen and has open crown and long and spreading branches. When fully ripe, the epicarp of the fruit becomes intense red in colour, bright attractive in appearance and the entire surface is covered with about 2 cm long tender and curved spines. The cultivar has the following two distinct strains:-

i. Besar: The aril is thick, white and adheres to the seed and is very juicy and sweet having pleasant flavour.

ii. Kooinbarg: It has small sized fruits. The epicarp or rind is thin and the spines are also not too many and are present thinly on the surface. The edible aril is soft and less sweet in taste and its adherence is not very strong.

Seelengkeng: The plants are dwarf. Its propagation by air layering is difficult. The flowers appear on both terminal and axillary inflorescences. The fruit is somewhat smaller (3x2 cm) and ovate and the surface is covered with bright red and soft spines and fetches high price in the market. The aril is very sweet and resembles litchi very closely in taste.

Seetangkoozoeh: The plants are large and tall having broad tree top, spreading branches and bright green and attractive foliage. The fruit is large (5x4 cm), ovoid but somewhat compressed laterally and having the surface covered with short (1 cm) tender spines. The flesh or aril is yellowish white when fully ripe and adheres tightly to the seed coat. It is juicy and sweet in taste with pleasant flavour and has good storage quality and can withstand long distance transportation.

Lebakboeloos: The tree is large and vigorous having broad crown, spreading branches and bright green foliage. When ripe, the fruit is very attractive, dark red in colour having sparsely placed short spines of about 1-1.5 cm length on the surface. The edible aril is whitish, about 0.5 cm thick and adheres firmly to the seed coat and is juicy and sweet acidic in taste and can withstand long distance transportation.

See-Cholimpoo: It is widely cultivated and popular cultivar of Thailand. The tree is medium in growth having broad and open crown and spreading branches. The fruit is medium in size, ovate and pink red in colour having tender and curved spines. The most significant features are small seed and thick aril. The flesh or aril is white, firm and tough in texture, juicy and sweet with a characteristic pleasant flavour.

Roling-Rain: This is an important cultivar grown extensively in Thailand. Trees are of medium size, crown is broad and the branches are spreading. Fruit is bright red in colour, shiny and attractive having many ovate spines, red colour while the tips remain green even until maturity. The flesh is white having firm and tough texture, juicy and sweet taste. The flavour is pleasant.

B-R-1 (Boting Rambutan No. 1): It is a cultivar selected at Hainan, China. The tree is large and fast growing having broad and open crown, profuse, long and spreading branches and shiny and evergreen foliage. The fruit is large in size (30 g or more), ovate in shape with dense and fine red spines having thick, sweet (18.2-21.7°B TSS), juicy, crisp and aril. The fruit has more edible portion because of the small seed and whitish thick aril.

B-R-2: It is a cultivar selected at Hainan, China. The fruit has yellowish skin and short spines and contains high vitamin C but lacks good keeping quality.

Jit Lee (Deli): The cultivar is from Singapore. The tree is of medium vigour. The fruit is large in size (30-55 g) having orange/red skin with green tips, 35% edible portion and 20-22°B TSS and has long postharvest life.

Binjai: The cultivar is from Indonesia. The tree is small to medium in vigour. The fruit is large in size (32-41 g) having orange/red skin with green tips, 41% edible portion and 18-21°B TSS with firm pulp and red peel.

Rongrien (Roengrean): The tree is medium to large in vigour. The fruit is large in size (40-50 g) having dark red skin with green tips, 41% edible portion and 18-21°B TSS.

Chompoo (Seechompoo Dwarf): The tree is dwarf. The fruit is 28-35 g in size having orange coloured skin, 40% edible portion and 18-20°B TSS.

R3 (Peng ting ching): The cultivar is from Malaysia. The tree is medium in vigour. The fruit is 22-33 g in size having red coloured skin, 46% edible portion and 22.5°B TSS.

R9 (Tau po cheng): The tree is medium to large in vigour. The fruit is large in size (40-50 g) having pink/crimson red skin, 36% edible portion and 20-23°B TSS.

R137 (Gajah Mati): The tree is medium to large in vigour. The fruit is large in size (40-45 g) having pink/crimson red skin, 49% edible portion and 22-23°B TSS.

R170 (Deli Cheng): It is a vigorous clone, growing to a height of 7 m. The fruit is large and oval weighing about 38 g having red skin with long spinterns. The aril is dry and sweet (20°B TSS) with recovery of about 40%. The aril can be usually removed from the seed without attachment to the testa.

R134: The cultivar is from Malaysia. The fruit is red, round, sweet (21°B TSS), firm and juicy and of medium size weighing about 30 g with green spinterns giving aril recovery of about 40%.

R156 (Muar Gading): It is moderately vigorous clone growing to a height of 5-6 m. The ivory yellow fruit is round and weighs about 37 g. The aril is thick, firm and juicy with about 22°B TSS giving recovery of 57%.

R162 (Ong Heok): The cultivar is from Malaysia selected from Penang. The tree is vigorous with round canopy and grows to a height of 7 m. The fruit is large (36 g), sweet (about 22°B TSS) and elongated with a yellow red orange peel. The aril is thick, firm, dry and sweet giving recovery of 54%.

R167 (Chai Tow Cheng): The cultivar is from Malaysia. The fruit is large, sweet and elongated with a red peel.

R191 (Rongrien, anak sekolah): It is the most popular clone in Thailand originated from a seedling in Malaysia. The tree is rather small. The fruit is small weighing about 30 g and is ovoid round with red skin and greenish spinterns. The aril is sweet (18°B TSS), firm and moderately juicy giving recovery of about 50%.

R193 (Deli Baling): The cultivar originated from Baling, Malaysia. The tree is vigorous attaining a height of 6 m. The fruit is round and medium-sized weighing about 35 g with red skin and long spinterns. The aril is sweet (20°B TSS) giving recovery of about 42%.

Arka Coorg Arun: This is an early maturing cultivar. The semi spreading trees are medium in size. It flowers during February–March and the fruits mature in September-October. The fruit is dark red and weighs 40 to 45 g. Fruits are freestone without attachment to the testa and aril can be removed from the seed. The aril is white, thick, firm, dry and sweet (20°B TSS) with recovery of about 42%. A ten year old tree produces 750 to 1000 fruits per year.

Arka Coorg Peetabh: This is a mid season cultivar. The trees are semi spreading and regular bearer. It flowers during February-March in Coorg and the fruits mature in October. The fruit is yellow and weighs 25-30 g. The aril is white,

Pomegranate (Monalisha Hota and IIHR, Bengaluru)

Bhagwa

Mridula

Arka Ruby

Ganesh

G-137

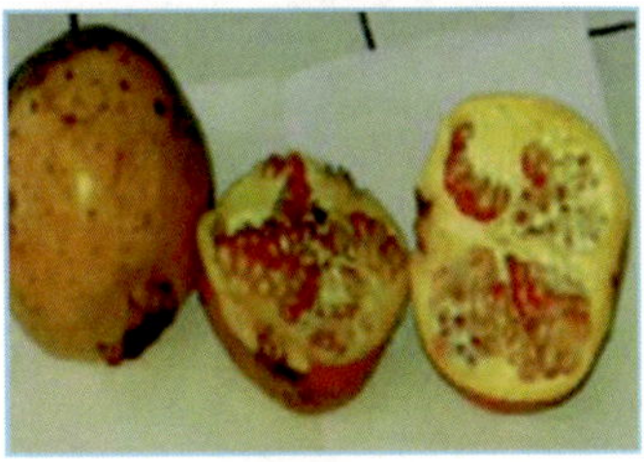
Wonderful

Guava (TFNet, 2008)

Allahabad Safeda

Lucknow 49

Apple Colour

Ruby Supreme

Arka Rashmi

Arka Mridula

Arka Kiran

Hisar Safeda

Wax apple (TFNet, 2008)

CHESM- 1

Green Red Wax Jambu

Kong White Wax Jambu

Black Diamond Wax Jambu

Java Wax Apple

Malay Rose Apple

Mango (Dr. S.N. Pandey; Griesbach, 2005; IIHR, Bangaluru)

Dashehari

Langra

Banganpalli

Mulgoba

Chausa

Totapari

Alphonso

Alphonso

Amrapali

Fazli

Mango (Dr. S.N. Pandey; Griesbach, 2005; IIHR, Bangaluru)

Gulabkhas

Arka Anmol

Arka Aruna

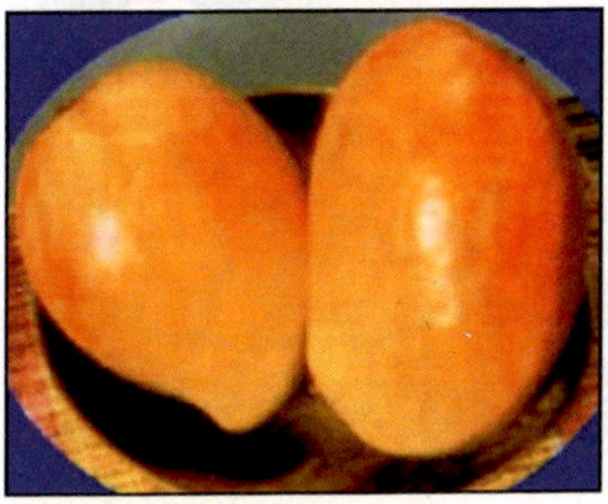

Arka Neelkiran

Arka Puneet

Anupam

Ambika

Kensington

Sensation

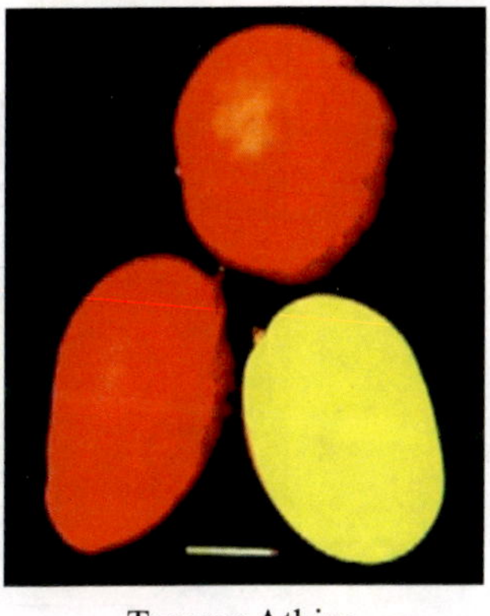

Tommy Atkins

Haden

Litchi

China

Kwai Mi

Purbi

Early Seedless

Rose

Swarna Roopa

Longan

Langsat

Nephelium ramboutan-ake (Labill.) Leenh

Rambutan (TFNet, 2008)

R3 (Gula Batu) from Malaysia

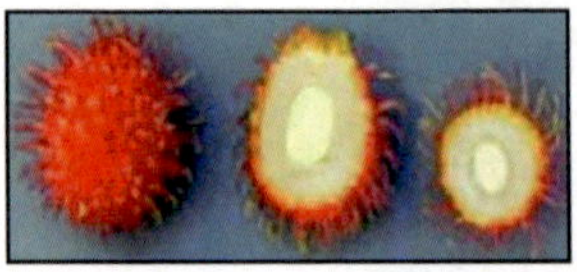

Seematjan

Binjai (from Indonesia)

Arka Coorg Arun

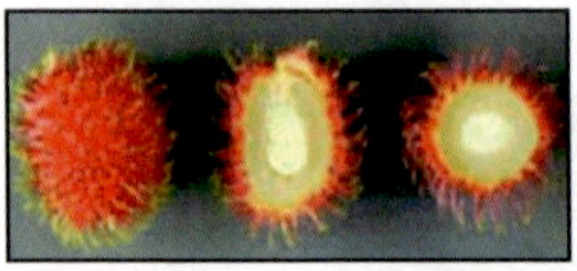

R167 ('Chai Tow Cheng') from Malaysia

CHES- 26

juicy, sweet (21°B TSS) with recovery of 41%. A ten year old tree produces 1200 to 1500 fruits per year.

CHES-14: The cultivar is late and regular bearer. The tree is semi-spreading. It flowers in February and the fruits ripen in October-November. Fruit is yellow and weighs 45- 50 g. The fruits are freestone, not attached to testa, and the aril is white and sweet (18.5°B TSS) having vitamin C content of 55 mg/100g pulp and aril recovery of 55%. The yield is 900-1200 fruits per tree.

CHES-26: It is a high yielding selection. The tree is of medium size, semi-spreading and early maturing. Flowering takes place during February and fruits ripen in September-October. The fruit is dark red, weighs 35- 40 g and is clingstone and the aril is attached to the testa. Aril is white, thick, juicy and sweet (17.5°B TSS) with 48 mg/100g vitamin C content giving recovery of 54%. Fruit yield is 2000-2500 per tree.

Pulasan

(***Nephelium ramboutan-ake*** (Labill.) Leenh. syn. ***Nephelium mutabile*** Bl.)

It is a native of West Malaysia, Borneo and Sumatra (Page, 1984) and is cultivated on large scale in the western part of Java with fewer plantings in rest of Indonesia, Thailand and Malaya. Pulasan has good potential for wider cultivation in hot tropics because of regular bearing (Martin *et al.,* 1987).

It is a medium-sized tree growing up to 10-15 m in height with a round, smooth, low-branched trunk, brown bark and irregular crown. The branchlets are thin, terete, densely rusty and pubescent when young and glabrescent. The leaves are alternate, petioled and even pinnate or spuriously odd pinnate by atrophy of one of the terminal leaflets. The leaflets are opposite, short stalked, oblong lanceolate or elliptic lanceolate, entire, coriaceous, dark green and glabrous above having short silky hairs below. The trees may be dioecious. The inflorescences are axillary and terminal, erect, clothed with short yellowish or brown pubescence and are 10-12 cm long. The flowers are polygamous or unisexual, apetalous, fascicled or solitary along branches of the inflorescence. The calyx is cup shaped, 4-5-lobed and membranous. There are 6-8 stamens in the male flower. The ovary is obcordate, two or three-lobed and coarsely hairy. The fruit consists of one well developed nutlet (drupe) with one or two abortive ones at its base. The fruits are oblong, dark red or occasionally yellow and are covered with coarse, short but pointed tubercles or soft spines. The pericarp is thick and shiny and yellowish white inside. The aril is clear, yellowish white and juicy. The seeds are ovoid oblong or ellipsoid, compressed laterally and 2-3.5 x 1.5-2 cm in size.

Cultivars

There are numerous types of pulasan in Java. These are divided into two main groups, one having dark red fruits with crowded tubercles and an aril which separates from the seed and the other having lighter coloured fruits with widely set tubercles and an aril which adheres firmly to the seed. The best forms of the first group have about 7 to 8 cm long and 4 to 5 cm broad fruits which have yellowish or light grayish white and about 1 cm thick aril. The fruits of the second group are usually smaller.

Around Bogor (Indonesia), there are three forms, characterized by the colour of the fruits, namely dark red, red and yellow. Superior cultivars are Asmerah Tjoplok and Kapoolasan Mera Tjoplok (Morton, 1987). Most widely planted pulasan cultivars are: P1, P4, P6, P8, P22, P28, P54, P63 and Lee in Malaysia; Merah, Poetih, Sibabad and Koeneng in Indonesia; and Sibabat in the Philippines (Page, 1984).

Genus *Blighia* Koenig

Three species from tropical Africa are well known. The plants are usually falsely polygamous dioecious trees or shrubs. Leaves are alternate and paripinnate having 1-5 leaflet pairs that are entire, distal leaflet being rudimentary are without stipules. Inflorescences are axillary racemose thyrses. Flowers are actinomorphic, functionally unisexual having 5 valvate sepals and 5 petals connate along the margins to form a pouch, 8-lobed annular disk, 8-10 stamens, colporate, striate pollen, 3(4)-carpellate ovary with single ovule per carpel and elongated conical style. Fruit is 3(4)-locular and loculicidal capsule having a basal arillode.

Akee

(***Blighia sapida*** Koenig; seso vegetal) (2n=32)

It is native of West Africa. It was introduced into the West Indies during the eighteenth century. It is also planted in West Indies, Ghana and Bahamas (Morton, 1987). It is widely cultivated in Jamaica for the edible arillodes and as an ornamental plant in Africa, the West Indies and areas of the Neotropics (Kubitzki, 2011).

It is a small evergreen tree growing up to 15 m in height and has 30 cm trunk diameter. The bark is gray and smooth to slightly rough. The inner of the bark is pinkish, gritty and slightly bitter. Twigs are light gray, stout, finely hairy having long and narrow ridges. The leaves are alternate and 15-20 cm long. Leaflets are in 3-5 pairs with short stalks. The blades are elliptic, mostly 7.5-20 cm long and 5-10 cm wide. The upper surface is shiny, yellow green and hairless with many sunken and parallel side veins. The lower surface is dull light green with raised and hairy side veins. The flowers are small, greenish white and fragrant borne on unbranched raceme of 7.5-18 cm length. Flowers are of two kinds, bisexual and male. The flower has calyx consisting of 5 pointed, brownish tinged sepals, 5 white petals, 8 stamens and pistil with 3-angled, 3-celled and 3-ovuled brownish ovary. The fruit is attractive, yellow to red, thick-walled, three valved pyriform capsule, 7-8 cm long and 4-5 cm in diameter borne in clusters of 3-10. The three globose, shiny black seeds are covered with firm and white nut-flavoured pulp (Morton, 1987).

Genus *Melicoccus* P. Browne (n = 32)

Ten species are native to South America and Dominican Republic. The trees are dioecious or monoecious. Leaves are alternate and paripinnate having 1-2 leaflet pairs, distal leaflet is rudimentary, absent or present exceptionally, usually with winged rachis and without stipules. Inflorescences are terminal panicles or racemes. Flowers are actinomorphic, functionally pistillate or staminate having cup-shaped calyx, 4(5) sepals, equal and imbricate, 4(5) erect or reflexed petals, without or with rudimentary appendages, annular and slightly lobed disk, 8 stamens, colporate and striate pollen, 2-carpellate and unilocular ovary, single ovule per carpel and obsolete style with stigmatic surface lobed and sub-capitate. Fruit is subglobose or ellipsoid and indehiscent having leathery pericarp. Seeds are 1(2) having edible sarcotesta (Kubitzki, 2011).

Mamoncillo

(***Melicoccus bijugatus*** Jacq. syn. ***Melicocca bijuga*** L.; Spanish lime; genip; Jamaica bullace; quenep; honey berry; kenep; mamoncillo)

Melicocco means honey berry owing to its typical flavour. It is a native of West Indies, central and southern America and has been introduced into the tropics. It is widely cultivated in the neotropics for its tasty fruits.

The plants are large, slow growing and upright trees with gray trunk and grows up to 20 m in height. The leaves are alternate, compound with 4 light green to blue green leaflets. The trees are dioecious. Male and female flowers are borne on separate trees, but occasionally hermaphrodite plants also occur. The flowers, borne on terminal panicles, are small, greenish white and unisexual. Owing to lack of pollination, isolated trees seldom produce fruit. Fruiting is good when the flowering time of staminate trees synchronizes with the receptive period of female trees. Fruits are borne in dense clusters which are small, green, round or oblong, 2-3.5 cm in diameter and weigh 10-25 g each. At maturity, the fruit has a thin, tough leathery pericarp surrounding a juicy white cream or light orange translucent pulp (aril) which encloses a single oval seed (Page, 1984; Morton, 1987; Brücher, 1989).

Cultivars

Some selections have been made in Florida and Puerto Rico which give heavy and regular yield of large fruits having thick aril, small seed, high sugar content and good flavour. These are Aree 1, Queen, Newcomb and Montgomery. Important cultivars of Puerto Rico are Puerto Rico 1, Puerto Rico 2, Puerto Rico 3 and Puerto Rico 4. In Florida, the cultivars Newcomb, Montgomery, Cuban No.1, Cuban No.2 and Queen are considered to be the best (Page, 1984; Morton, 1987).

Genus *Pometia* J. R. and G. Forst

Only one species under this genus occurs in Malesia.

Fijian longan

(***Pometia pinnata*** J. R. and G. Forst; Pacific lychee; taun; dawa; kasai)

It is a native of Malesia and originated in the region from Ceylon and Andaman islands through South East Asia including Papua New Guinea and the Pacific to Fiji and Samoa. It is most common in Papua New Guinea, Solomon Islands, western Samoa and Fiji. The tree is distributed in tropical areas within latitudes 20ºN and south of the equator and is mainly confined to the zones with a minimum of 2500 mm evenly distributed mean annual precipitation.

It is a large tree growing up to 40 m in height with a trunk diameter of 90 cm. The bark is dark gray and smooth but flakes off in round or irregular patches leaving scars on the trunk. The leaves are alternate, even pinnate with 4 to 9 pairs of leaflets. In young trees, up to 16 pairs of leaflets are borne on 100 cm long rachis. The broadly lanceolate or oblong leaflets are large, 12-18 x 4-7 cm in size, tomentose and light green but the new flushes are deep red. The small, cream white and unisexual flowers are borne on 50 cm long terminal racemes. The flower type is variable. Some trees may bear entirely functional male flowers while others may have predominantly functional females along with some functional male flowers in the same panicle. The fruits are borne in clusters and have hard, smooth, 4 mm thick pericarp, green or red in colour, round to oblong in shape, 40-80 g in weight and up to 7 cm in diameter. The aril is semi-transparent, white and juicy. The seed is round and about 25 mm in diameter (Page, 1984).

Cultivars

The forms of *P. pinnata* J. R. and G. Forster are *F. acuminata, F. alnifolia, F. cuspidata, F. glabra, F. macrocarpa, F. pinnata, F. repandra* and *F. tomentosa* (Walter and Sam, 2002). Few selections have been made on the basis of fruit quality characters. Three cultivars from Fiji and 8 forms based on fruit colour (yellow to purple) from Papua New Guinea are reported. The best form in Papua New Guinea is Sepik District (Page, 1984).

Genus *Talisia* Aublet

The genus has 52 species predominantly from South America and some from central America and southern Mexico. The plants are duodichogamous unbranched shrubs or small to large trees growing up to 20 m tall. Leaves are alternate, paripinnate or imparipinnate having entire leaflets with rudimentary distal leaflet. Inflorescences are axillary, terminal or cauliflorous thyrses. Flowers are 5-merous, actinomorphic, staminate or pistillate having usually cup shaped calyx, imbricate or valvate sepals, equal or unequal in size, reflexed or ascending petals with marginal appendages or with a basal, petaloid, sericeous appendage, annular, 5-8-lobed disk, 5-8 stamens, colporate, striate or brevicolporate pollen, psilate, 3-carpellate ovary with single ovule per carpel and subulate style crowned by a capitate to cylindrical stigma. Fruit is an oval indehiscent drupe, 2-4 cm long, 1(-3)-seeded surrounded by a translucent crisp, juicy layer of pulp (sarcotesta) and a thin orange or brown skin. It is leathery or less often having woody pericarp.

Pitombeira (pitomba)

(*Talisia esculenta* (St. Hil.) Radlk. syn. *Sapindus esculentus* St. Hil.)

It is native of South America (Martin *et al.*, 1987). It is distributed from Amazonas to Rio de Janeiro and Paraguay and Bolivia. It is found growing from northwest Argentina, Bolivia to Brazil and Mexico. The related edible species are *T. oliviformis* Radlk. (yellow genip), *T. intermedia* Radlk. and *T. guianensis* Aubl.

Pitomba is a medium sized tree growing up to 4-12 m in height with brown, thick and rough bark. Leaves are alternate, 1-pinnate, exstipulate, 2-4 sub-opposite to alternate pairs of leaflets, narrowly elliptic to oblong elliptic and 7-13 x 3-6 cm in size. Inflorescence is borne terminally on the branchlets. Flowers are bisexual with 5 elliptic, 3 mm long sepals, 5, white or pinkish, perfumed petals, 8 stamens, hairy filaments arising from within a glabrous disk at the base of ovary, ovoid ovary and apiculate berry, 2.5 cm in diameter, yellowish green with coriaceous exocarp having 1-2 seeds, enveloped by a white fleshy aril with thick and superposed cotyledons (FAO, 1986).

Yellow genip
(***Talisia oliviformis*** (Kunth) Radlk.)

The tree is medium sized, 15 m tall. The leaves are alternate and compound with 2-3 pair of leaflets. The plants are dioecious. The flowers are whitish and occur in terminal clusters. The fruit is ovoid, yellowish and about 2.5 cm long and has one seed. The pulp is whitish or cream coloured. The fruits are eaten fresh or are processed into juice or jam.

Genus *Paullinia* L. (2n=24)

About 190 species are native to the neotropics one of them extending into Africa. The plants are woody vines, usually producing milky sap. Cross section of stem shows a central and 3-5 peripheral vascular cylinder or a single vascular cylinder. Numerous species are used as fish poisons. Leaves are alternate, trifoliolate, pinnate (5-pinnate), bipinnate and biternate or variously dissected having minute to large stipules. Inflorescences are axillary thyrses with a pair of tendrils at base of the rachis or cauliflorous thyrses without tendrils. Flowers are zygómorphic and functionally unisexual having 4-5, unequal, imbricate sepals, 4 petals with a basally adnate hood shaped appendage, unilateral, 4-lobed disk, 8 stamens, porate and perforate pollen, 3-carpellate ovary with single ovule per carpel and style with 3 stigmatic branches. Fruit is 1-3-locular, septifragal marginicidal capsule, the locules dorsally winged or exalate and exceptionally spiny. Seeds are globose, oblong or ellipsoid with partial to nearly complete sarcotesta (Kubitzki, 2011).

Guarana
(***Paullinia cupana*** Kunth.)

The guarana is native of the lower Amazon region of South America. It is now widely cultivated in Brazil. The woody climber (vine) has irregularly compound leaves and coiled tendrils by which it supports itself. The inflorescence is borne in leaf axils. The inconspicuous yellow flowers having 5 mm long petals develop into small, bright red fruits containing shiny, dark brown seeds. The attractive rounded capsules (fruits) are 2-3 cm long and develop a red warty skin on ripening and contain 3 black seeds. It is the source of the economically important “guarans” used in the confection of stimulating drink or soft drinks in Brazil.

Genus *Aesculus* L.

The genus comprises 13–19 species of trees and shrubs native to the temperate northern hemisphere, with 6 species native to North America and 7–13 species native to Eurasia, besides several hybrids. Linnaeus named the genus *Aesculus* after the Roman name for an edible acorn. Common names for these trees include "buckeye" and "horse chestnut". Some are also called white chestnut or red chestnut. *Aesculus* seeds were traditionally eaten, after leaching. All parts of the buckeye or horse chestnut tree are extremely toxic, including the nut-like seeds. The toxin affects the nervous system causing paralysis and death. The toxicity is due to saponin aescin and glycocide aesculin, possibly with alkaloids. Native Americans used to crush the seeds and the resulting mash was thrown into still or sluggish water bodies to stun or kill fish. They would then boil and drain (leach) the fish at least three times in order to dilute the toxin's effect. New shoots from the seeds also have been known to kill grazing cattle.

Indian horse chestnut
(*Aesculus indica* (Wall. ex Camb.) Hook. f.)

Indian or Himalayan horse chestnut is an attractive tree growing to about 30 m in height with a spread of about 12 m. It is cold hardy and tolerates as low as -15°C temperature. The tree has a beautiful round canopy. The leaves are large, of 10-20 x 2-6 cm size and are ornamental. Flowering takes place from June to July. The flowers are hermaphrodite, white in colour and are pollinated by bees. The seeds ripen in October. The nuts are mildly poisonous to humans. The dried seeds are ground to make bitter flour. The bitterness is because of the saponins which can be rinsed out by thoroughly washing the flour. The flour is often mixed with wheat flour to prepare *chapattis* and *halwa* and sometimes porridge. It has been used in traditional medicine in India for the treatment of some skin diseases, rheumatism and as an acrid and narcotic astringent. The leaves are used as cattle fodder in parts of northern India.

Family Rutaceae

The Rutaceae is a family of approximately 158 genera and 1900 species with major centres of diversity in southern Africa and Australia (Mabberley, 2008). The species are distinctive and are often easily recognizable because of their often opposite, usually compound and pellucid punctate leaves, cymose inflorescences, often pachypetalous flowers and stamens with thick filaments often forming a ring and ovaries with a conspicuous disc at the base and an expanded stigma. Although the fruits are varied, most have a glandular punctate pericarp, some being fleshy and indehiscent while the others are dry and dehisce to reveal shiny black seeds.

The plants are mostly trees and shrubs and are tropical and subtropical, largely natives of eastern hemisphere. All the plant parts are characteristically dotted with glands that contain aromatic oil. The leaves are alternate or opposite without stipules, often thorny and simple or compound. The inflorescence is solitary or axillary cymose. The flowers are bisexual, usually regular and hypogynous. Sepals are 4-5 and united at the base or free and imbricate. The petals are 4-5 and free, imbricate or valvate. Stamens are equal to the number of petals or twice this number (genus *Citrus* has 4-8 times), free or united in bundles inserted at the base of thick annular disc. Pistils are 2-5 or one compound pistil, as in *Citrus* with 2-5 or more carpels and mounted on a fleshy disc. Styles are usually connate. Ovules are one to many per carpel. The ovary is superior. The fruits are of two distinct types: i) fleshy drupe or berry, and ii) small, dehiscent and dry in the form of a capsule or samara. The fruits are often very large. Seeds are with or without endosperm. Some members are polyembryonic and some are parthenocarpic.

Engler (1931) divided the family Rutaceae into 7 subfamilies; Rhabdodendroideae, Aurantioideae, Flindersioideae, Spathelioideae, Dictyolomatoideae, Rutoideae and Toddalioideae. Takhtajan (1997) excluded Rhabdodendroideae from Rutaceae; which is now recognized as a family under Caryophyllales (APG III) and retained the remaining six. Thorne (2000) recognized 3 subfamilies: Rutoideae (including Chloroxylon, Flindersia, Luvunga, Toddalioideae; 120 genera), Aurantioideae (30 genera), and Spathelioideae (including Dictyolomatoideae; 5 genera).

Subfamily Aurantioideae

The subfamily contains the citrus fruits. The artificial systems of classification proposed by Hooker (1875) and Engler (1896) were based on morphological characters and putative origin of species without considering the existence and importance of nucellar embryony. They also did not have access to biotechnological approaches to taxonomy. Later, Engler (1931) classified the subfamily Aurantioideae into one tribe, Aurantieae and two subtribes, Hesperethsinae and Citrinae. Tanaka (1936) classified it into 11 tribes, Aegleae, Atalantieae, Aurantieae (having 4 subtribes, Hesperethusinae, Citropsinae, Citrinae and Poncirinae), Clauseneae, Lavangeae, Balsamocitrinae, Feroniinae, Merrillnae (having a subtribe Swingleinae), Meropeae, Microcitreae and Micromeleae. Then Swingle (1948) and Swingle and Reece (1967) classified it into 2 tribes, Clauseneae (having 3 subtribes, Clauseninae, Merrilliinae and Micromelinae) and Citreae or Aurantieae (having 3 subtribes, Balasamocitrinae, Citrinae and Triphasiinae) (Table 29.5) and 9 subtribal groups and 33 genera. The subtribal group "true citrus fruit trees" of subtribe Citrinae contains the important genus *Citrus* (Table 29.6).

Table 29.5. Tribal and subtribal classification of subfamily Aurantioideae (Bayer *et al.*, 2009)

Engler, 1931	**Tanaka, 1936**	**Swingle and Reece, 1967**
Aurantieae, Hesperethsinae and Citrinae	Aegleae, Atalantieae, Aurantieae, Hesperethusinae, Citropsinae, Citrinae, Poncirinae, Clauseneae, Lavangeae, Balsamocitrinae, Feroniinae, Merrillnae, Swingleinae, Meropeae, Microcitreae and Micromeleae	Clauseneae, Clauseninae, Merrilliinae, Micromelinae, Citreae or Aurantieae, Balasamocitrinae, Citrinae and Triphasiinae

Results of AFLP (amplified fragment length polymorphism) analysis demonstrated that *Poncirus, Microcitrus*, and *Eremocitrus* of this subtribal group are distant from *Citrus* (Pang *et al.*, 2007). Based on test using nine cp-DNA sequence regions in representatives of all genera of the subfamily Aurantioideae (save *Limnocitrus*) and numerous species and hybrids referred to *Citrus* are confirmed as monophyletic (Bayer *et al.,* 2009). Within Aurantioideae, the genera in tribe Clauseneae are not monophyletic unless *Murraya* and *Merrillia* are removed to Aurantieae. Within the tribe Aurantieae, the three traditionally recognized subtribes are not monophyletic. Triphasinae is not monophyletic unless *Oxanthera* is returned to *Citrus* (*Citrinae*). Balsamocitrinae is polyphyletic. *Feroniella*, traditionally considered allied to *Limonia (=Feronia)*, is shown to be nested in *Citrus*. The proposed congenericity of *Severinia* and *Atlantia* is confirmed. The circumscriptions of *Citrus* are strongly supported by this analysis, with hybrids appearing with their putative maternal parents. Two clades are resolved: one consisting of the species from *Clymenia, Eremocitrus, Microcitrus, Oxanthera* and also *Citrus medica*; and the second of species from *Fortunella* and *Poncirus*.

Table 29.6. The Aurantioideae (orange subfamily) (Swingle and Reece, 1967)

Tribe	Subtribe	Subtribal group	Genus	Species
Clauseneae	Micromelinae (very remote citroid fruit trees)	-	*Micromelum* Blume	9
	Clauseninae (remote citroid fruit trees)	-	*Glycosmis* Correa	35
			Clausena Burm. f.	23
			Murraya Koenig ex. Linn.	11
	Merrillinae (large fruied citroid fruit trees)	-	*Merrillia* Swing.	1
Citreae	Triphasinae (minor citroid fruit trees)	Wenzelia group	*Wenzelia* Merr.	9
			Monanthocitrus Tan.	1
			Oxanthera Montrouzier	4
			Merope Roem.	1
		Triphasia group	*Triphasia* Lour.	3
			Pamburus Swingle	1
		Luvunga group	*Luvugna* Buch.-Ham.	12
			Paramignya Wight	15
	Citrinae (citrus fruit trees)	Primitive citrus fruit tree group	*Severinia* Tenore	6
			Pleiospermium (Engl.) Swing.	5
			Burkillanthus Swing.	1
			Limnocitrus Swing.	1
			Hesperethusa Roem.	1
		Near citrus fruit tree group	*Citropsis* (Engl.) Swing. & M. Kell.	11
			Atalantia Correa	11
		True citrus fruit tree group	*Fortunella* Swingle	4
			Eremocitrus Swingle	1
			Poncirus Raf.	1
			Clymenia Swingle	1
			Microcitrus Swingle	6
			Citrus L.	16

Balsamocitrinae (hard shelled citroid fruit trees)	Tabog group	*Swinglea* Merrill	1
	Bael-fruit group	*Aegle* Correa	1
		Afraegle (Swing.) Engler	4
		Aeglopsis Swingle	2
		Balsamocitrus Stapf	1
	Wood apple group	*Limonia* Correa (syn. *Feronia*)	1
		Feroniella Swingle	3

Tribe Clauseneae

Very remote citroid fruit trees (3 subtribes, 5 genera, 79 species and 20 varieties). It never has any spines. The odd-pinnate leaves have alternately attached leaflets. The fruits are usually small and carry semi-dry or juicy berries, except in *Merrillia*. The pulp vesicles are present.

Subtribe Micromelinae

Very remote citroid fruit trees (1 genus, 9 species and 4 varieties)

Genus *Micromelum* Blume

It has origin in South East Asia and Oceania (9 species, 4 varieties). The species are: *M. ceylanicum* Wight, *M. compressum* (Blanco) Merr., *M. compressum* var. *inodorum* (Blume) Tan., *M. diversifolium* Miq., *M. diversifolium* var. *cuneata* Miq., *M. falcatum* (Lour.) Tan., *M. hirsutum* Oliv., *M. integerrimum* (Buch.-Ham) Roem., *M. minutum* (Forst) Wt. and Arn., *M. minutum* var. *curranii* (Elm.) Tan., *M. minutum* var. *tomentosum* Tan., *M. pubescens* Blume and *M. scandens* Rechinger

Subtribe Clausenineae

Remote citroid fruit trees (3 genera, 69 species, 16 varieties).

Genus *Glycosmis* Correa

It has origin in South East Asia and Oceania (35 species). The species are: *G. angularis* Elmer, *G. bilocularis* Thwaites, *G. bonii* Guillaumin., *G. boreana* Narayana Swami, *G. citrifolia* (Willd.) Lind., *G. clemensii* Tanaka, *G. craibii* Tanaka, *G. crassifolia* Ridley, *G. chlorosperma* Blume, *G. cyanocarpa* (Blume) Spreng., G. *dinhensis* Pierre, *G. elata* Ridley, *G. erythrocarpa* Hayata, *G. esquirolii* (Lev.) Tanaka, *G. lindleyana* Swing., *G. longipes* (Craib) Tanaka, *G. macrantha* Merr. (syn. *G. oliveri* Stapf.), *G. ovoidea* Pierre, *G. macrocarpa* Wight, *G. macrophylla* (Bl.) Miquel, *G. malayana* Ridley, *G. mauritiana* (Lam.) Tanaka, *G. montana* Pierre, *G. monticola* Ridley, *G. oliveri* Stapf, *G. parkinsonii* Tanaka, *G. parva* Craib, *G. pentaphylla* (Retz.) Correa, *G. pierrei* Tanaka, *G. pseudoracemosa* (Guill.) Swingle, *G. puberula* Lindley, *G. rupestris* Ridley, *G. sapindoides* Lindley, *G. tomentella* Ridley and *G. wintii* Craib.

Genus *Clausena* Burm. f.

It has origin in South Asia and Oceania (23 species, 12 varieties). The species are: *Clausena abyssinica* Engl, *C. anisata* (Willd.) Hook. f., *C. anisata* var. *multijuga* Welw. ex. Hiern., *C. anisata* var. *mollis* Engl., *C. anisum-olens* (BI.) Merr., *C. brevistyla* Oliv., *C. brevistyla* var. *halmaheirae* (Miq.) Swing., *C. cambodiana* (Pierre) Guill., *C. dentata* (Willd.) Roem., *C. dentata* var. *dulcis* (Bedd.) Swing. , *C. dentata* var. *dunniana* (Lev.) Swing., *C. dentata* var. *henryi* Swing., *C. dentata* var. *longipes* (Craib) Tan., *C. dentata* var. *pubescens* (Wt. & Am.) Tan., *C. dentata* var. *robusta* Tan., *C. excavata* Burm. f., *C. excavata* var. *villosa* Hook. f., *C. guillauminii* Tan., *C. harmandiana* (Pierre) Guill., *C. harmandiana* var. *contracta* Tan., *C. heptaphylla* (Roxb.) Wt. & Arn., *C. heptaphylla* var. *engleri* (Tan.) Swing, *C. heptaphylla* var. *contracta* Tan., *C. inaequalis* (DC.) Benth., *C. indica* (Dalz.) Oliv., *C. lansium* (Lour.) Skeels, *C. laxiflora* Quis & Merr, *C. lenis* Drake., *C. luxurians* (Kurz) Swing., *C. macrophylla* Hook. f., *C. mollis* Merr., *C. papuana* Lauterb., *C. pentaphylla* (Roxb.) DC., *C. platyphylla* Merr., *C. suffruticosa* (Roxb.) Wt. & Arm., *C. todayensis* Elmer and *C. wallichii* Oliv.

Wampee
(***Clausena lansium*** (Lour.) Skeels.)

Wampee is native of South China (Martin *et al*., 1987). It is distributed mostly in tropical Asia, Africa and Australia. It is a highly esteemed fruit tree of southern China and Vietnam. It is also cultivated on a small scale in Nilgiri hills of India, South East Asia, Sri Lanka, Australia (Queensland) and USA (Florida and Hawaii) (Verheij and Coronel, 1991). The related species *C. dentata* (Willd.) Roemer is found growing in India and China and *C. indica* (Dalz.) Oliver is found growing in India and Sri Lanka. These also produce edible fruits.

It is a medium, much branched and round topped evergreen tree up to 3 m or more in height with gray brown bark. The spirally arranged 40 cm pinnate and leathery leaves have alternate and lanceolate leaflets, 15 x 7cm in size with obtuse apex. Flowers are greenish white, bisexual, pentamerous sub-sessile, sweet scented and whitish to yellowish and are borne on large terminal cymose inflorescence having less than 1 mm length, green, narrowly elliptic sepals of 5 x 2 mm size, 10 stamens and short and 5-celled gynophore. The fruit is ovoid or globose five-celled (1-2 seeded) berry about 1 cm in diameter. The fruit is dark green when unripe and turns lemon yellow at maturity. The fruit generally has five furrows on its surface with glandular dots. The pulp is translucent, pearly white and soft textured. Good blend of sugar and acid in its pulp makes it one of the most delicious fruits. Seeds are ellipsoid ovoid, 1.5 cm long, green with brownish chalazal cap and green cotyledons (Morton, 1987; Verheij and Coronel, 1991).

Genus *Murraya* Koenig ex Linn.

It originated in South Asia and Oceania (11 species, 4 varieties). The species are: *M. alata* Drake., *M. alata* var. *hainanensis* Swing., *M. altenans* (Kurz.) Swing., *M. crenulata* (Turaz.) Oliv., *M. euchrestifolia* Hayata , *M. glabra* (Guill.) Swing., *M. gleniei* Thwaites ex. Oliv., *M. koenigii* (L.) Spreng., *M. microphylla* (Merr. & Chun)

Swing, *M. paniculata* (L.) Jack, *M. paniculata* var. *omphalocarpa* (Hay.) Tan., *M. paniculata* var. *ovalifoliolata* Engl., *M. paniculata* var. *zollingeri* Tan., *M. siamensis* Craib. and *M. stenocarpa* (Drake) Swing.

Orange Jessamine
(***Murraya paniculata*** (L.) Jack.)

It is a tropical evergreen plant bearing small, white and scented flowers and is grown as an ornamental tree or hedge plant. *Murraya* is closely related to citrus and bears small orange to red fruits resembling kumquats although some are cultivated but do not set fruits.

Subtribe Merrilliinae

Large-fruited remote citroid trees (1 genus, 1 species)

Genus *Merrillia* Swing.

It originated in South East Asia (1 species), *Merrillia caloxylon* (Ridl.) Swing.

Tribe Citreae

Citrus and citroid fruit trees (3 subtribes, 9 subtribal groups, 28 genera, 124 species, 18 varieties). The plants are native of the Old World tropical region, from South China, South East Asia and South Asia to Australia. The trees or shrubs are evergreen except for 3 genera, viz., *Poncirus*, *Aegle* and *Feronia*. The simple leaves are compound, often odd pinnate, being strictly in opposite pairs. Nearly all the species develop axillary spines singly or in pairs. The fruit has 2-5 locules with one or two ovules in each or 6-18 locules with 4-18 ovules in each. Pulp vesicles are present. The peel of the fruit has well developed glandular layer on the outside.

Subtribe Triphasiinae

Minor citroid fruit trees (3 subtribal groups, 8 genera, 46 species, 3 varieties).

A. Wenzelia group (4 genera, 15 species, 1 variety)

Genus *Wenzelia* Merr.

It originated in Oceania (9 species, 1 variety). The species are: *W. archboldiana* Swing., *W. brevipes* Merr., *W. brevipes* var. *alabatensis* Swing., *W. dolichophylla* (Lauterb. & K. Schum) Tan., *W. grandiflora* (Lauterb.) Swing., *W. kambarae* Swing., *W. melanesica* Swing., *W. paludosa* (Lauterb.) Tan., *W. platysperma* Swing. and *W. tenuifolia* Swing.

Genus *Monathocitrus* Tan.

It originated in Oceania (1 species), *M. cornuta* (Lauterb.) Tanaka:

Genus *Oxanthera* Montrouzier (false oranges)

These originated in Oceania (4 species). The species are: *O. aurantium* Tan., *O. fragrans* Montr., *O. neo-caledonica* (Guill.) Tan., *O. undulata* (Guill.) Swing.

Genus *Merope* Roem.

It originated in South East Asia and Oceania (1 species); *Merope angulata* (Wild.) Swing.

B. Triphasia group (2 genera 4 species)

Genus *Triphasia* Lour.

It originated in South East Asia and Oceania (3 species). *T. brassii* (White) Swing., *T. grandifolia* Merr. and *T. trifolia* (Burm. f.) P. Wils.

Genus *Pamburus* Swingle

It originated in South and South East Asia (1 species). *P. missionis* (Wt.) Swingle.

C. Luvunga group (2 genera, 27 species, 2 varieties).

Genus *Luvunga* Buch.-Ham.

It originated in South, South East Asia and Oceania (12 species). The species are: *L. angustifolia* (Oliv.) Tan., *L. borneensis* Hochr., *L. calophylla* Kurz., *L. crassifolia* Tan., *L. eleutherandra* Dalz., *L. motleyi* Oliv., *L. nitida* Pierre , *L. papuana* Lauterb., *L. philippinensis* Merr., *L. sarmentosa* (Blume) Kurz., *L. scandens* (Roxb.) Buch. Ham. and *L. tavoyana* Lindl. ex. Oliv.

Genus *Paramigyna* Wight

It has origin in South and South East Asia and Oceania (15 species, 2 subsp.). The species are: *P. andamanica* (King) Tan., *P. armata* (Thwaites) Oliv., *P. beddomei* Tan., *P. citrifolia* (Roxb.) Oliv., *P. confertifolia* Swing., *P. cuspidata* (Ridl.) Swing., *P. grandiflora* Oliv., *P. lobata* Burkill, *P. longipedunculata* Merr., *P. mindanensis* Merr., *P. monophylla* Wight., *P. rectispinosa* Craib, *P. scandens* (Girff.) Craib, *P. scandens* subsp. *hispida* (Pierre) Swing., *P. scandens* subsp. *ridleyi* (Burkill) Swing., *P. surasiana* Craib and *P. trimera* (Oliv.) Burkill.

Subtribe Citrinae

Citrus fruit trees (3 subtribal groups, 13 genera, 65 species, 15 varieties). It contains 13 genera and 65 species of trees and shrubs that are almost exclusively native to the Indo-Pacific, including China and Australia. It includes fruits that are called citroid or citrous (=citrus like) fruits. The "true citrus fruit trees" of this subtribe are considered the most advanced genera based on morphological traits. Stamens are 2-4 times the number of petals or more. The rind of fruits is not hard. The fruit is large with sessile or stalked pulp vesicles. The juice sacs are present that are not found in any of the members of the family. The seeds are glabrous.

A. Primitive citrus fruit tree group (5 genera, 14 species)

Genus *Severinia* Tenore

It has origin in South China and South East Asia (6 species 1 form).

The species are: *S. buxifolia* (Poir.) Tenore, *S. buxifolia* brachytic form., *S. disticha* (Blanco) Swing., *S. lauterbachii* Swing., *S. linearis* (Blanco) Swing., *S. paniculata* (Warb.) Swing. and *S. retusa* (Merr.) Swing.

Box orange
(***Severinia buxifolia*** (Poir.) Tenore)

This dense, low branching and compact evergreen shrub has small, oval, glossy and dark green leaves closely spaced on slender and thorny branches. Growth of the tree is very slow. Small white, fragrant and orange-like blossoms emerge in spring and summer and develop into shining black and seedy berries.

Genus *Pleiospermium* (Engl.) Swing.

It originated in south Asia and Oceania (5 species). The species are: *P. alatum* (Wt. and Arn.) Swing., *P. dubium* (Bl.) Swing., *P. latialatum* Swing., *P. longisepalum* Swing. and *P. sumatranum* Swing.

Genus *Burkillanthus* Swing.

It originated in South East Asia and Oceania (1 species). *Burkillanthus malaccensis* (Ridl.) Swing.

Genus *Limocitrus* Swing.

It originated in South East Asia (1 species). *L. littoralis* (Miq.) Swing.

Genus *Hesperethusa* Roem.

It originated in South and South East Asia (1 species). *H. crenulata* (Roxb.) Roem.

B. Near citrus fruit trees group (2 genera, 22 species, 4 varieties).

Genus *Citropsis* (Engl.) Swing. & M. Kell.

It originated in Central Africa (11 species, 1 variety). The species are: *C. angolensis* Exell, *C. articulata* (Willd.) Swing. & M. Kell., *C. daweana* Swing. & M. Kell., *C. gabunensis* (Engl.) Swing. & M. Kell., *C. gabunensis* var. *lacourtiana* (De Wild.) Swing. & M. Kell., *C. gilletiana* Swing. & M. Kell., *C. latialata* (De Wild.) Swing. & M. Kell., *C. mirabilis* (A. Chev.) Swing. & M. Kell., *C. zenkeri* Engler, *C. presussii*. See *C. articulata*, *C. schweinfurthii* (Engl.) Swing. & M. Kell. and *C. tanakae* Swingle and M. Kell.

Genus *Atalantia* Correa

It originated in South and South East Asia (11 species, 3 varieties). Number of stamens is twice the number of petals. The species are: *A. ceylanica* (Arn.) Oliv., *A.*

citroides Pierre ex Guill., *A. guillaumini* Swinge., *A. hainanensis* Merr. & Chun. ex Swing., *A. macrophylla* (Oliv.) Kurz., *A. monophylla* DC., *A. racemosa* Wight, *A. racemosa* var. *henryi* Swing., *A. roxburghiana* Hook. f., *A. rotundifolia* (Thwaites) Tan., *A. roxburghiana* var. *kerri* Swing., *A. roxburghiana* var. *kwangtungensis* (Merr.) Swing., *A. simplicifolia* (Roxb.) Engl. and *A. wightii* Tan.

C. True citrus tree group (6 genera, 28 species and 14 varieties).

Genus *Fortunella* Swingle (Kumquats)

The kumquats originated in South China (4 species, 1 variety) and are indigenous to southeastern China and have been given differed names. These were considered part of the genus *Citrus* until Swingle separated it as *Fortunella* (kumquats). In Chinese language the word kumquat means 'gold orange' (Swingle, 1915).

It requires distinct winter and is hardy to cold conditions. The plants are shrubs or small trees. The plants have high temperature and water requirement. The tree is 2.4-4.5 m in height having twigs that are angled when young and may be thornless or with a few spines. The leaves are alternate and lanceolate in shape and 3.25-8.6 cm in size with finely toothed margin. These are blunt pointed with veins prominent on the upper surface only (as in mandarin) and the lower surface is lighter in colour and has raised glands. The petioles are narrowly winged. The flowers are white borne singly or in clusters of 2 to 4 in leaf axils and terminally (as in mandarin). Stamens are 16-20 adhering irregularly in bundles. The ovary has 3-7 locules with 2 ovules in each. The flower buds are conspicuous and angular (as in mandarin). Blossoming is late. The fruits are small, oval oblong or round, 1.6-4 cm in size having golden yellow to reddish orange rind and pale orange acidic pulp, but the rind is sweet. Seeds are small, smooth, plump and pointed and polyembryonic having deep green cotyledons.

Kumquats resemble calamondin and some small mandarins, but differ from citrus by the presence of few locules (3-5), usually 2 ovules per locule and being very small and sweet in taste with somewhat pulpy rind and angular flower buds. The fruits along with twigs and leaves are extensively used for decoration during Christmas time. Fruits are usually eaten fresh without removing the rind as the spicy aromatic flavour of rind and the acidic pulp makes a very delightful combination. Fruits are extensively used for making marmalades, jellies, preserves and candies as they contain very high amounts of pectin.

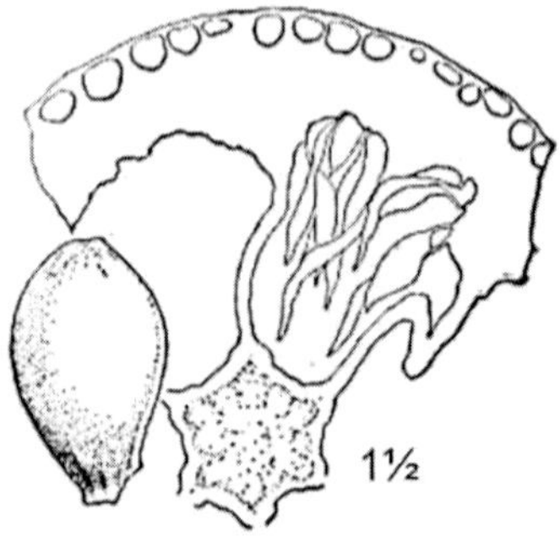

Fig. 29.8. *Fortunella crassifolia* (Source: Hume, 1957).

Large Round Kumquat
(***Fortunella crassifolia*** Swingle; meiwa)

It is a natural hybrid of oval and round kumquats. This is called the Ninpo, Neiha or Meiwa Kinkan of Japan. It is less cold hardy compared to Nagami but is considered to be the best for fresh eating. The characteristic features are that the fruit is large in size and round to short oblong in shape having nearly 7 segments. The rind is very thick. The pulp has sweet flavour and few or no seeds.

Marumi or Round Kumquat
(***Fortunella japonica*** (Thunb.) Swingle; marumi kinkan)

It is called Maru or Marumi Kinkan of Japan. Plants are branched shrubs with less vigour and none to very small spines. It is very handsome of all citrus fruits because of dwarf plants, much branched twigs and small and brightly coloured orange fruits. Leaves are light green in colour, broad and more slender and small and blunt and not sharp pointed apex. The fruits are small and rounded to slightly oblate obovate and orange in colour with no sign of rudimentary style. The number of segments is more (4-7) and seeds are small. Rind is thin and sweet.

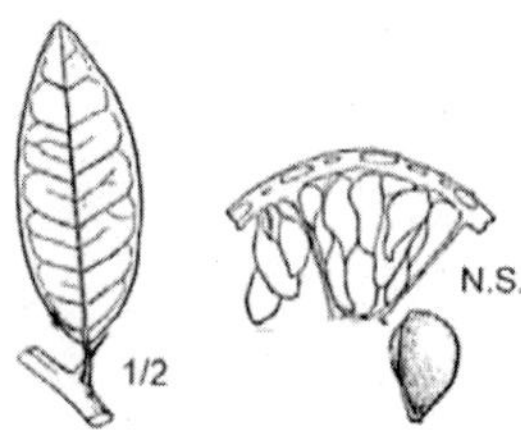

Fig. 29.9. *Fortunella japonica* (Source: Hume, 1957).

Nagami or Oval Kumquat
(***Fortunella margarita*** (Lour.) Swingle; nagami kinkan)

It is popularly called Naga or Nagami Kinkan of Japan. The characteristic features are that the fruits are oval, oblong or obovate in shape having less number of segments (usually 4-5) and both flesh and rind are deep orange and richly flavoured.

Plants are shrubs or small trees having dense branches. Early growth of both branches and branchlets are angular but becomes circular later. Spines are axillary, smaller than of *Poncirus* and are present along with the bud. Leaves are evergreen and short pointed, usually unifoliate having narrowly winged petiole, entire to slightly crenate margins, dark green above and lighter below with oil glands thickly dotted on the under surface. White coloured flowers, smaller than of the other citrus types, appear in pairs or in clusters from the axils of new or last year's growth. Flower stalk has 1-2 small bracts, 4-5 yellowish green calyx, 4-5 white oblong lanceolate corolla, 16-20 stamens with filaments of irregular length in a ring form at the base but free at top and the anthers attached at filament tips. Ovary is generally 3-6-celled, sometimes having 2 ovules per cell and usually small stigma. Fruit is small, globose or oblong having 4-5 segments with glands embedded in the thick fleshy rind, which is sweet and edible. Seed is small generally having greenish cotyledons.

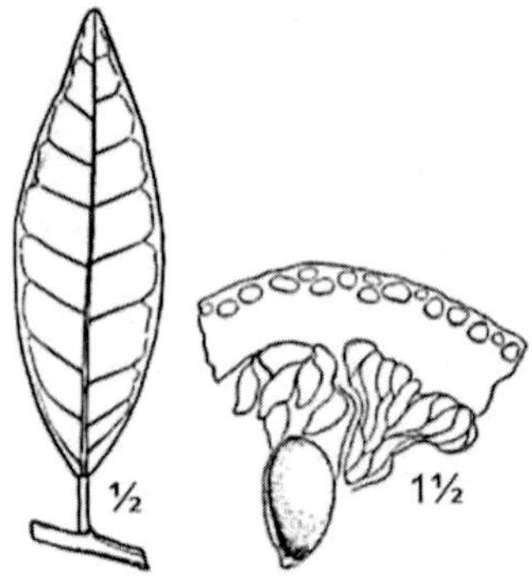

Fig. 29.10. *Fortunella margarita* (Source: Hume, 1957).

It is an unarmed glabrous bush having thinly coriaceous and unifoliate leaves having 16-19 mm long petioles, winged above. Few flowers borne usually 2 in an axil have 5 petals and 24 stamens. The fruit is globose and 4.5 cm in diameter with numerous large glands.

Hong Kong wild kumquat
(*Fortunella hindsii* (Champ.) Swing.)

It is indigenous to the hilly areas of Hong Kong, China and the nearby places. It is a spiny shrub or small tree having slender twigs that are angled when young, oval elliptical leaves tapering sharply at both ends, dark green above, pale beneath and faintly venose with winged petioles and short and broad flowers. The fruits are small, 1-1.5 cm in diameter, subglobose, bright orange or scarlet orange when ripe (the colour of a tangerine orange). The pulp vesicles are very few, small and fusiform. The seeds are thick, oval or ovate in shape having green cotyledons.

Dwarf kumquat
(***Fortunella hindsii*** var. ***chintou*** Swing. syn. ***Citrus japonica*** Thunb.)
2n = 18, 20, 36.

The plants are 5 m tall trees having 20 cm trunk diameter. Branchlets are numerous having variable spines, 5 cm long on young growth but some are only a few mm on the flowering shoots. Leaves are unifoliolate or sometimes mixed with simple leaves having 6–9 mm, narrowly winged petiole, elliptic to obovate elliptic leaf blade of 4–6 × 1.5–3 cm size with rounded to broadly cuneate base, dentate margin near the apex or rarely entire, rounded and mucronate apex. Flowers are solitary or fascicled and subsessile. Calyx is 4 or 5-lobed. Petals are 5, 5 mm or less in size. Stamens are 20 with filaments cohering into 4 or 5 bundles. Ovary is as long as the style, 3 or 4-loculed with 3 or 4 ovules per locule. Fruit is bright orange to red, globose to slightly oblate, 9–10 mm in diameter, smooth, 3 or 4-seeded having sweet and edible pericarp and acidic sarcocarp. Seeds are broadly ovoid with mucronate apex and smooth seed coat and sometimes numerous embryos having green cotyledons. Flowering occurs in April–May and fruiting in October–December.

Changshou kumquat
(*Fortunella obovata* Tan.)

Changshou kumquat is also called the Fukushu kumquat in Japan. The naturally small tree is very ornamental, has a spreading form and is thornless. The leaves are larger and broader than in other kumquat species. The fruit is also larger and slightly different in shape. Changshou is about 4 cm long and oval with a depressed apex. The rind is orange and thinner than Nagami or Meiwa. The flesh is orange, acidic and contains a few seeds.

***Eremocitrus* Swingle**

The genus resembles *Citrus* in the structure and appearance of the fruits. But its leaves are gray green, thick and leathery and markedly pellucid punctuate and both sides are nearly alike having 4 ventral and 2 dorsal layers of palisade cells, sunken stomates, an epidermis with a thick cuticle and scattered few celled appressed hairs on both the surfaces. The spines are usually long and slender, but are sometimes wanting, especially on fruiting branches of old trees. They occur singly in the axils of the leaves. The twigs are gray green, slender, very slightly angled when young with scattered stomates at the base of deep, narrow pits and two or more layers of palisade cells below the very thick walled epidermis. Besides, its 4 to 5-merous flowers have free stamens and 4 or 5-celled ovary with 2 ovules in each cell. The flowers occur singly or 2 or 3 together in the axils of the leaves and are borne on slender pedicels about as long or slightly longer than the petals. The calyx is 3 to 5-lobed. The 4 or 5 petals, rarely 3, are more or less narrowed at the base. The fruits are oval or somewhat pyriform and have a thin and fleshy peel like that of a lime, covered with oil glands. The pulp is vesicular, sour and juicy. The pulp vesicles are subglobose, separate easily in the ripe fruit and are borne on slender stalks. The seeds are small (5-6 x 3-4 mm), pointed, ovate and yellowish gray with a hard testa, irregularly verrucose and furrowed in a longitudinal direction. The cotyledons are plano-convex having hypogeous germination. The post-cotyledonary leaves are slender cataphylls.

Australian desert lime
(*Eremocitrus glauca* (Lindl.) Swingl.)

The species occurs over a large arid area of eastern Australia. It is very drought tolerant and can withstand both very hot (45°C) and very cold (-24°C) conditions. Of all species of Aurantioideae it takes the shortest time (about 8 weeks) from flowering to fruit maturity. The habitat of this species extends from Rockhampton to Longreach in Queensland, south to Dubbo in central New South Wales and west to Quorn in the Flinders Ranges of southern Australia. In the western part of this area it occurs in isolated clumps, of up to several hundred trees, that are located as much as 100 km apart.

It varies in size from 2–3 m high, thorny, multi-stemmed, shrubby thicket to 8 m high trees with rounded crowns and 2 m trunk and up to 12 m tall, narrow and upright trees with a well defined trunk. When a tree is seriously damaged it tends to sucker prolifically from the roots to form an impenetrable thorny thicket. The leaves are gray, long and narrow having thick cuticle and practically no petioles and no articulation. They differ from the true citrus fruits in having both upper and lower surfaces similar and having a palisade layer, sunken stomata and hairs parallel with the surface on young leaves. Trees vary in thorniness, some trees with single, short, stout and only a few cm long spines while the others have long and narrow and up to 8 cm long spines. Flowers are small, white having four or five petals and some with complete while others lacking a fully developed stigma and are only male functional. The stamens are free and are four times of the number of petals. The fruit resembles a small lime. It is round to oblate (occasionally pyriform) about 2 cm in diameter with a light yellow green and almost translucent skin with the obvious small oil cells. Fruits are often seedless but some have a few small plump seeds with a wrinkled seed coat. A gall wasp (*Eurytama*) parasitizes and develops large galls in place of the seeds.

The species has been hybridized with sweet and sour orange, mandarin and kumquat. It is also graft compatible with *Citrus* and *Microcitrus* species.

Genus *Poncirus* Raf.

It originated in North and Central China (1 species, 1 variety). It is a monotypic genus having only one species (*Poncirus trifoliata*). It is the only deciduous genus belonging to the warm temperate zone and having distinct chilling requirement. The plants are highly thorny. The spines are single, large and stout. The leaves are palmately trifoliate. The stem pith is discontinuous. The flowers are large and are borne singly. The petals are thin and papery and white in colour. Stamens are free and have 6-8 locules. The fruits are small, oblate or obovoid, dull lemon yellow in colour and pubescent and inedible. The vesicles are sessile. The seeds are large and polyembryonic.

Trifoliate Orange
(*Poncirus trifoliata* (L.) Raf.)

Plants are small trees with upright and densely branched growth. Young growth is smooth, angular, dark green and thorny and the old shoots are circular. Thorns are very stout, sharp, 4-5 cm long and flattened at base. The leaves are trifoliate, deciduous and obovate having notched central leaflet and crenate margin. Flower buds are formed in summer and emerge singly or in pairs in leaf axils covered by scales during spring before the leaves on naked shoots. Flowers are short pedicelled or nearly sessile. Calyx is small, 5, oval and 1-1.5 cm long. The corolla are 5, white, large, 3-4 cm across, obovate and clawed at the base. The stamens are 20-25 having usually distinctly 2-celled filaments. The pistil is club shaped, oblong and pubescent with short style and 6-7 parted ovary. The fruit surface is rough, bumpy and covered with pubescence. The fruit stalk/stem is short and stout. The fruit hangs well if not picked. The pulp is fine, gummy and acidic. Seeds are many, plump, ovate and brown in colour.

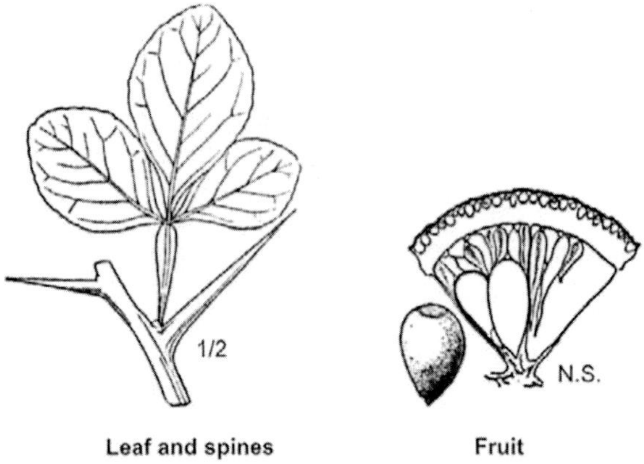

Fig. 29.11. *Poncirus trifoliata* (Source: Hume, 1957).

It hybridizes freely with the other genera, e.g. sweet orange x trifoliate orange producing a citrange. The botanical variety is *P. trifoliata* var. *monstrosa* (T. Ito) Swing.

Genus *Clymenia* Swingle

It originated in Oceania and contains *C. polyandra* (Tan.) Swing.

Genus *Microcitrus* Swingle (Australian wild limes)

The genus *Microcitrus* was separated from *Citrus* by Swingle (1915) based on the minute size of its flowers and juvenile leaves, slender twigs and small fruits containing 4-8 locules, besides considering the dimorphic foliage, parallel venation, short wingless petioles, short pistils and numerous free stamens. Seven species have been described and of these *M. australasica* (F. Muell.) Swing., *M. australasica* var. *sanguinea* (F.M. Bail.) Swing., *M. australis* (Planch.) Swing., *M. garrowayi* (F.M. Bail.) Swing., *M. inodora* (F.M. Bail.) Swing., *M. maideniana* (Domin) Swing. are native of Australia. The other two, *M. warburgiana* and *M. papuana* are native of New Guinea. The Australian species occur in rainforest extending from northern New South Wales to the Cape York Peninsula on the coastal side of the Great Dividing Range. One species, probably *M. australasica* also occurs in the Northern Territory in the vicinity of Darwin.

Mexican orange
(*Microcitrus australasica* (F. Muell.) Swing.; finger lime)

It occurs in rainforest growing into a tall narrow tree up to 10 m high. It has dimorphic foliage. Young seedlings develop only horizontal branches which have thorns about twice as long as the small juvenile leaves. After the initial horizontal branches with very short internodes, one or more upright shoots grow very rapidly and the size of the new leaves increases from the narrow cataphylls at the base to mature form (1.5-4 x 1.2-2.5 cm) with a blunt apex. The flowers occur singly in the axils of the leaves. They are round or pyriform with 3-5 petals and 12-20 free stamens. The fruits are long and narrow (6.5-10 x 1.5-15 cm) with thin, green, yellow, purple or black skin, blunt points at both ends and 5-7 locules. The juice vesicles are compressed so that when the skin is cut they tend to burst out. They contain an acidic juice with very little sugar. Seeds are plump and are like those of a lemon but are much smaller. In the mature form slender single spines arise alongside the petiole at each node.

Red pulp finger lime
(***Microcitrus australasica*** var. ***sanguinea*** (F. Muell.) Swing.)

It is a very small to large-sized tree. The leaves are tiny, about 4 cm long and the branches become very dense and spiny having one thorn in every leaf axil. New growth is purple in colour and 3 cm wide flowers are white and pink during the main flowering season from February to April. The fruits are about 7.5 cm long, but on juvenile trees these can be less than 2.5 cm long. The fruit skin is usually greenish black to very dark purple and thin. Once the fruit is cut the tiny round juice vesicles slowly seep out of the fruit without squeezing and resemble "citrus caviar". The vesicles are usually clear green but can be light pink. The juice is very tart, much like that of Mexican lime. In California, although the tree produces fruits year round, the main fruiting season is November-December. The fruits are technically edible. The most common use is to garnish or flavour the culinary preparations.

Australian round lime
(***Microcitrus australis*** (Planch.) Swing.; dooja)

It thrives in drier conditions more than *M. australasica*. It occurs as a tall slender and upright tree (9–18 m high) on the edges of rainforest in the vicinity of Brisbane and further north. It has dimorphic foliage with narrow, linear, 2.5-5 cm long, juvenile and mature leaves that are widest towards the middle. Fruits are round and rough skinned, of 2.5-8 cm diameter and have 5 or 6 locules. The skin is green turning lemon coloured at maturity. The flesh is similar to that of *M. australasica*.

Russell river lime
(***Microcitrus inodora*** (F M Bail.) Swing.)

It is known to occur only in lowland rainforest in thc high rainfall area. It is so named because of the absence of perfume in the flowers. It is a 2-4 m high shrub or small tree with large leaves and short wingless petioles that are not articulated with the blade and fruits. The fruits are oval or oblong, ribbed, up to 6.5 x 3.2 cm having 8 locules and like other *Microcitrus* species, have pulp with a very acid juice. This species differs from the others except *M. maideniana* in having two spines in each leaf axil.

Maiden's Australian wild lime
(***Microcitrus maideniana*** (Domin.) Swing.)

This species has the same origin as *M. inodora* but resembles it in having ribbed fruits and twin spines but differs in having a distinctly sunken apex.

Mount white lime
(*Microcitrus garrowayi* (F.M. Bail.) Swing.)

It occurs in the far north of Queensland and resembles *M. australasica* but has broader leaves and shorter and thicker fruits with fewer locules (4-5). The external appearance of the skin is more rough than in the finger lime owing to the large convex oil cells.

New Guinea wild lime
(*Microcitrus warburgiana* (F.M. Bail.) Swing.)

It occurs in the New Guinea.

Genus *Citrus* L.

The plants are native of the Old World tropical region. The trees or shrubs are evergreen. They are normally spiny. The solitary spines are located on one side of the vegetative bud. The new shoots are characteristically angular but eventually become rounded. The flowers are white and generally fragrant. The characteristics of *Citrus* species are asexual reproduction, high mutation frequency and cross compatibility between the species. However, that gave rise to the great morphological and ecological diversity among the species. Engler (1931) distinguished 33 genera and 210 species by their syncarpous ovary with one or two, sometimes several, ovules per carpel, their indehiscent fleshy fruits with a soft, parenchymatous pericarp bearing numerous schizolysigenous oil glands (sometimes with a hard exocarp as in *Aegle* and *Swinglea*), often with pulp vesicles in the locules, absence of endosperm and seeds with two or more embryos (Bayer *et al.*, 2009) (Fig. 29.12). Calyx is cup shaped and usually 4-5 lobed. Stamens are numerous, varying from 15 to 60. The filaments are united in bundles being basally connate. Styles are usually deciduous. The fruit is a hesperidium. The cotyledons are white to green.

The classification of citrus fruits has been complicated. Some of its peculiar characteristics make the classification difficult. First, it is difficult to ascertain centres of origin and diversity of *Citrus* types and its related genera, although it is believed that it lies in South East Asia, especially eastern India, northern Burma and southwest China, but possibly ranging from North East India eastwards through the Malay archipelago, northwards into China and Japan and south to Australia. A second factor that makes taxonomic classification difficult is the reproductive biology of the genus. Many citrus types are polyembryonic and reproduce asexually by seed through necellar embryony. Besides, There is a wide sexual compatibility between *Citrus* and related genera. All the types are interfertile although the widespread facultative apomixis prevents exchange of genes leading to some reproductive isolation. Thus one opinion has been that all of these belong to a single large species. There is also an extremely high rate of bud and limb mutation. The beneficial mutants, apomicts and hybrids have been vegetatively perpetuated. Being stable many of these lines have been given species status. Further their wide dispersal, the long history of cultivation and the paucity of remaining wild *Citrus* stands makes the task of taxonomists difficult and has led to major disagreements on citrus classification. The number of species recognized by them has, therefore, widely differed.

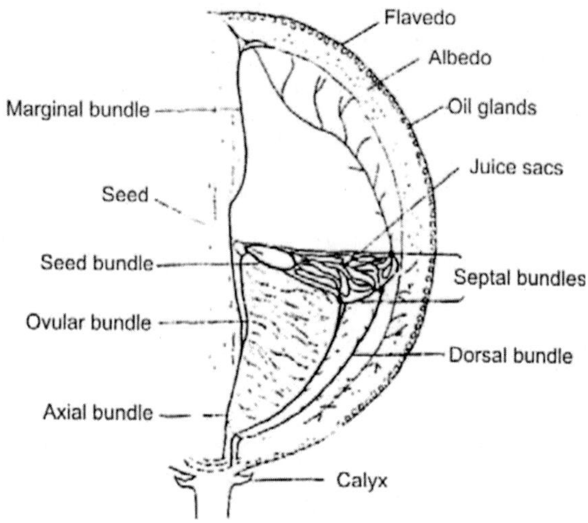

Fig. 29.12. Vascular arrangement (schematic) in citrus fruit (*Source:* Spiegel-Roy and Goldschmidt, 1996).

Earlier workers (Linnaeus, 1753; Hooker, 1875) had recognized only 3 or 4 species including also the kumquats.[1] Swingle (1948) published his work in a book, "*Citrus Industry*". He recognized 2 subgenera, *Eucitrus* and *Papeda*, and only 16 species and 8 botanical varieties.

[2] Tanaka (1954; 1961) classified the genus *Citrus* into 2 subgenera, Archicitrus (5 sections; Papeda, Limonellus, Ctrophorum, Cephalocitrus and Aurantium) and

---

[1] **Subgenus *Eucitrus*:** *C. aurantifolia* (Christm.) Swing., *C. aurantium* L., *C. aurantium* var. *myrtifolia* Ker-Gawl., *C. grandis* (L.) Osbeck, *C. grandis* var. *banokan* Tan., *C. indica* Tan., *C. limon* (L.) Burm. f., *C. medica* L., *C. medica* var. *ethrog* Engl., *C. medica* var. *sarcodactylis* (Noot.) Swing., *C. reticulata* Blanco, *C. reticulata* var. *austera* Swing., *C. sinensis*. (L.) Osbeck, *C. tachibana* (Mak.) Tan.; **Subgenus *Papeda*:** *C. celebica* Koord., *C. celebica* var. *southwickii* (Wester) Swing., *C. hystrix* DC., *C. ichangensis* Swing., *C. latipes* (Swing.) Tan., *C. macroptera* Montr., *C. macroptera* var. *kerrii* Swing., *C. macroptera* var. *annamensis* Tan., *C. micrantha* Wester, *C. micrantha* var. *microcarpa* Wester

[2] **Subgenus Archicitrus (section Papeda**: *C. balincolong* Tanaka, *C. boholensis* Tanaka, *C. macroptera* Mont., *C. micrantha* Wester, *C. southwickii* Wester, *C. westerii* Tanaka, *C. vitiensis* Tanaka, *C. celebica* Koord, *C. combara* Raf., *C. hystrix* DC., *C. kerrii* Tanaka, *C. latipes* Tanaka; **section Limonellus-Subsection Eulimonellu:** *C. aurantifolia* (Christm.) Swing., *C. limettoides* Tanaka, *C. latifolia* Tanaka; **subsection Megacarpa:** *C. webberi* Wester, *C. bergamia* Risso & Poit., *C. javanica* Blume, *C. longispina* Wester, *C. macrophylla* Wester, *C. montana* Tanaka, *C. obversa* Hassk, *C. ovata* Hassk., *C. papaya* Hassk., *C. pennivesiculata* Tanaka, *C. pseudolimonum* Wester, *C. davaoensis* Tanaka, *C. excelsa* Wester, *C. hyalopulpa* Tanaka; **section Citrophorum-subsection Citrioides:** *C. alata* Tanaka, *C. limonimedica* Lush., *C. medica* L., *C. nana* Tanaka, *C. odorata* Wester; **subsection Limonoides:** *C. limon* (L.) Burm. f., *C. limonia* Osbeck, *C. limetta* Risso, *C. longelimon* Tanaka, *C. meyerii* Y. Tanaka, *C. jambhiri* Lush., *C. pseudolimon* Tanaka; **subsection Decumanoides:** *C. assamensis* Dutta & Bhatt, *C. aurata* Risso, *C. balotina* Poit. & Turp., *C. lumia* Risso & Poit, *C. macrolimon* Tanaka, *C. megaloxycarpa* Lush., *C. mellarosa* Risso., *C. peretta* Risso., *C. pyriformis* Hassk., *C. rissoi* Risso, *C. sarbati Tanaka*, *C. duttae* Tanaka, *C. karna* Raf.; **section Cephalocitrus-subsection Decumana:** *C. grandis* (L.) Osbeck, *C. panuban* Tanaka, *C. truncata* Hort. ex Tanaka, *C. pseudograndis* Hort, ex Tanaka, *C. pseudogulgul* Hort. ex Shirai, *C. suizabon* Hort. ex Tanaka; **subsection Intermedia-Group Flavicarpa**: *C. aurantiaca* Hort. ex Tanaka, *C. leiocarpa* Hort., *C. mitsuharu* Hort. ex Y. Tanaka, *C. omikanto* Hort. ex Y. Tanaka, *C. paradisi* Macf., *C. pseudoparadisi* Hort. ex Y. Tanaka, *C. flavicarpa* Hort. ex Tanaka, *C. glaberrima* Hort. ex Tanaka, *C. hiroshimana* Hort. ex Y. Tanaka, *C. tosa-asahi* Hort. ex. Y. Tanaka; **Group Aureocarpa:** *C. asahikan* Hort. ex Tanaka, *C. intermedia* Hort. ex Tanaka, *C. kotokan* Hayata, *C. hassaku* Hort. ex Tanaka, *C. iwaikan* Hort. ex Y. Tanaka, *C. tengu* Hort. ex Tanaka; **section Aurantium-Subsection Medioglobosa:** *C. ampullacea* Hort ex. Tanaka, *C. anonyma* Hort. ex Y. Tanaka, *C. miaray* Wester, *C. natsudaidai* Hay., *C. obovoidea* Hort. ex Takahashi *C. otachibana* Hort. ex Y. Tanaka, *C. papillaris* Blanco, *C. rugulosa* Hort. ex Tanaka, *C. sulcata* Hort. ex Takahashi *C. taiwanica* Tanaka & Shim., *C. yamabuki* Hort. ex Tanaka, *C. yuge-hyokan* Hort. ex Tanaka, *C. himekitsu* Hort. ex

Metacitrus (3 sections; Osmocitrus, Acrumen and Pseudofortunella) and 144 species giving species rank to numerous cultivars and presumably hybrids.

In India, the information on "cultivated limes and lemons of India and Ceylon" by Bonavia (1888-90) and on 20 species and 24 botanical varieties by Lushington (1910) was published. After the publications of Tanaka, [3]Hodgson (1961) gave his opinion, which was between those of Swingle and Tanaka. He accepted *C. latifolia, C. myrtifolia, C. natsudaidai, C. unshiu, C. deliciosa, C. tangerina* and *C. reshni,* as valid species and justified giving species status to hybrid forms as these are capable to perpetuate true to form retaining their identity under natural reproduction through seed. He made a simple classification of citrus fruits into four groups (acid members, oranges, mandarins and pummelos) listing 31 species.

Tanaka, *C. iriomotensis* Tanaka, *C. pseudopapillaris* Tanaka, *C. medioglobosa* Hort. ex Tanaka; **subsection Aurantioides-Group Racemosa**: *C. aurantium* L., *C. maderaspatana* Hort. ex Y. Tanaka, *C. rokugatsu* Hort. ex Y. Tanaka, *C. myrtifolia* Raf., *C. yanbaruensis* Tanaka; **Group Contracta**: *C. canaliculata* Hort. ex Y. Tanaka; **subsection Sinensioides:** *C. sinensis*. (L.) Osbeek, *C. sinograndis* Hort. ex Tanaka, *C. temple* Hort. ex Y. Tanaka, *C. tankan* Tanaka, *C. oblongo* Hort. ex Y. Tanaka, *C. funadoko* Hort. ex Y. Tanaka, *C. iyo* Hort. ex Tanaka; **subsection Osmocitrioides- Group Tenuicarpa**: *C. luteo-turgida* Tanaka, *C. ujukitsu* Hort. ex Tanaka, *C. tamurana* Hort. ex Tanaka; **Group Compacta**: *C. aurea* Hort. ex Tanaka; **Group Paranobilis:** *C. shunkokan* Hort. ex Tanaka; **subgenus-Metacitrus-Section Osmocitrus-subsection Euosmocitrus:** *C. pseudo-aurantium* Hort. ex Y. Tanaka, *C. sudachi* Hort. ex Shirai, *C. takuma-sudachi* Hort. ex Tanaka, *C. wilsonii* Tanaka, *C. yuko* Hort. ex Tanaka, *C. hanaju* Hort. ex Shirai, *C. ichangensis* Swing., *C. inflata* Hort. ex Tanaka, *C. junos* Sieb. ex Tanaka, *C. kizu* Hort. ex Y. Tanaka; **subsection Pseudoacrumen:** *C. nippokoreana* Tanaka; **section Acrumen-subsection Euoacrumen:** *C. nobilis* Lour., *C. unshiu* Marc., *C. yatsushiro* Hort. ex Tanaka; **subsection Microacrumen-Group Anisodora**: *C. keraji* Hort. ex Tanaka, *C. oto* Hort. ex Y. Tanaka, *C. tarogayo* Hort. ex Y. Tanaka, *C. inflato-rugosa* Hort. ex. Tanaka; **Group Citriodora-subgroup Megacarpa**: *C. paratangerina* Hort. ex Tanaka, *C. benikoji Hort.* ex Tanaka, *C. platymamma* Hort. ex Tanaka, *C. suavissima* Hort. ex Tanaka, *C. succosa* Hort. ex Tanaka, *C. suhuiensis* Hort. ex Tanaka, *C. tardiferax* Hort. ex Tanaka, *C. clementina* Hort. ex Tanaka, *C. crenatifolia* Lush., *C. deliciosa* Ten., *C. genshokan* Hort. ex Tanaka, *C. reticulata* Blanco, *C. tangerina* Hort. ex Tanaka; **subgroup Microcarpa-Microgroup Angustifolia:** *C. tachibana* (Mak.) Tanaka, *C. erythrosa* Hort. ex Tanaka, *C. tardiva* Hort. ex Shirai, *C. indica* Tanaka, *C. kinokuni* Hort. ex Tanaka, *C. pseudo-sunki* Hort. ex Tanaka, *C. ponki* Hort. ex Tanaka, *C. reshni* Hort. ex Tanaka, *C. oleocarpa* Hort. ex. Tanaka, *C. sunki* Hort. ex Tanaka; **Microgroup Latifolia:** *C. depressa* Hay., *C. amblycarpa* Ochse, *C. lycopersicaeformis* Hort. ex Tanaka, *C. hainanensis* Tanaka, *C. telocarpa* Hort ex Tanaka, *C. tumida* Hort. ex Tanaka; **section Pseudofortunella**: *C. madurensis* Lour.

[3] **Acid members:** (*C. medica* (citron), *C. limon* (lemon), *C. aurantifolia* (lime), *C. jambhiri* Lush. (rough lemon), *C. limettioides* (sweet lime), *C. limetta* (lemon X lime); **Oranges:** *C. aurantium* (sour orange)., *C. sinensis* (sweet orange); **Mandarins:** *C. reticulata* (mandarin orange), *C. unshlu* (Satsuma mandarin), *C. nobilis (C. reticulata x C. sinensis);* **Pummelos** (Pamplemouse of Europe): *C. paradisi* (grapefruit), C. *maxima (C. grandis)* shaddock or pummelo; besides other species of somewhat doubtful classification: *C. limonia* Osbeck-Rangpur lime. Canton lemon, Cravo lemon, Mandarin lemon (*C. reticulata* var. *austera* or *C. aurantifolia x reticulate, C. bargamia* Risso-Bergamot (*C. aurantium* var. *bergamia, C. limon x C. aurantium* hybrid), *Citrus myrtifolia* Raff. chinnoto of Italy (*C. aurantium* var. *myritifolia), C. madurensis* Lour. or *C. mitis Blanco; Calamondin* (*C. reticulata* var. *austera x Fortunella* Sp.).

[4]Singh and Nath (1969) proposed 31 species and 25 botanical varieties and pointed out that there are some basic species of *Citrus* under *Eucitrus* (*C. medica, C. limon, C. grandis, C. aurantium, C. aurantifolia, C. reticulata* and possibly *C. limettioides, C. sinensis* and *C. tachibana*) and the others originated through mutations and hybridization among these species or with *Fortunella* and members of the Papeda group and accepted the subgenus *Papeda* of Swingle and divided the subgenus *Eucitrus* into 8 groups, which correspond closely to the 7 sections of Tanaka.

Until the mid 1970s, Citrus taxonomists based their conclusions solely on morphological and geographical data. Modern techniques have now become instrumental in deciphering the taxonomic situation in *Citrus*. Studies using chemotaxonomy and plant morphology suggest that there are only three major affinity groups within *Citrus*; first medica group consisting of *C. medica* (citron), *C. aurantifolia* (common lime), second reticulata group consisting of *C. reticulata* (mandarin), *C. sinensis* (orange), *C. paradisi* (grapefruit), *C. aurantium* (sour orange) and *C. jambhiri* (rough orange) and the third maxima group consisting of *C. maxima* or *C. grandis* (pummelo). Barret and Rhodes (1976) made phylogenetic evaluation of 146 morhological, biochemical, tree, leaf, flower and fruit characters and suggested that only three *Citrus* types, citron (*C. medica*), mandarin (*C. reticulata*) and pummelo (*C. maxima*) constitute valid species while the others appear to be hybrids -

---

[4] **subgenus *Eucitrus*: Decumana Group** (*C. pennivesiculata* Tanaka (gajanimma attara or baduapulli; similar to satkara, *C. macroptera* var. *annamensis*, of Assam), *C. pennivesiculata* var. *assamensis*. (adajamir of Assam, similar to *C. assamensis* of Bhattacharya and Dutta, 1956), *C. semperflorens* Lush. (sadaphal or nardaba of India; resemble pummelo, citron and *C. pennivesiculata*), *C. paradisi* Macf. (grapefruit), *C. grandis* Osbeck (shaddock, pummelo). *C. grandis* var. *megaloxycarpa* (amilbed or bhogatey of lower Himalayas; *C. megaloxycarpa* of Lushington, 1910), *C. grandis* var. *keem* (keem of India; *C. megaloxycarpa* of Lushington, 1910), *C. grandis* var. *rugosa* (mahalung or lemon thumb of India), *C. magaloxycarpa, C. magaloxycarpa* var. *keem,* keem proper, mahalung, holong-tenga, lemon thumb); **Medica Group** (*C. medica* Linn. (citron), *C. medica* var. *ethrog* Engl. (small fruited citron), *C. medica* var. *sacrodactylus* (Noor) Swing. (finger citron), *C. medica* var. *limonum* (lemons found in India; correspond to *C. limonum* Lush.), *C. limonimedica);* **Limonoid Group** (*C. limon* Burm (lemon; not found in India), *C. medica* var. *sphaerocarpa* (Kagzi Kalan or Nepali Kagzi, of India; *C. bergamia* Risso et Poit., *C. macrophylla* Wester, *C. limon* var. *sphaerocarpa* of Lushington, *C. limon* var. *decumana* (hill lemon or gulgul of India; *C. limonum* var. *decumana* of Lushington), Acidless Pani Jamir of Assam, *C. karna* Raf. (Karna Khatta), *C. pseudolimon, C. limonia* Osbeck. (Rangpur lemon or otaheit orange or cravo lemon or canton lemon), *C. limonia* var. *kusale* (Kusaie lime), *C. limonia* var. *kusai,* soh-jhalia, acidless rough lemon (mitha-tulia), *C. jambhiri* Lush. (Florida Rough), *C. jambhiri*-Intermediate type (kata jamir), *C. jambhiri*-katajamir type), *C. jambhiri* var. *chrysocarpa* Lush. (soh-jhalia of Assam), *C. jambhiri* var. *dulcis* (mitha-tulia or mithi of India), *C. jambhiri* var. *laevis* (kata jamir of Assam and Italian 76); **Aurantium Group** (*C. sinensis* Osbeck (sweet orange), *C. aurantium* (bitter sweet group), *C. aurantium* Linn. (i.e. sour orange), *C. aurantium* var. *myrtifolia* Ker-gawl (myrtle leaf orange or chinnoto or chinois of Europe), *C. aurantium* (natsudaidai), *Citrus natsudaidai* Hayata); **Aurantoid Group** (*C. rugulosa* Hort. ex Tanaka (attani of India; possess characters of pummelo and mandarin; can include tangelos, Indian grapefruit), jeneru tenga, *C. maderaspatana* Hort. Ex Y. Tanaka (kichili of south India; resemble sour orange), *C. junos* Sieb. ex Tanaka (yuzu of Japan), *C. junos* var. *mandarina* (jeneru tenga of North-East India; similar to *C. nobilis* of Bhattacharya and Dutta (1956) and Tanaka (1961) and to *C. pseudoaurantium* of Tanaka); **Acrumen Group** (*C. nobilis* Lour. (king mandarin including tangors like temple or kunembo; considered a natural tangor), *C. reticulata* var. *unshiu* (*C. unshiu* of Tanaka, 1954 and Hodgson, 1965, Satsuma mandarin), *C. deliciosa, C. reticulata* var. *deliciosa* Mediterranean or willow leaf mandarin; *C. deliciosa* of Tennore), *C. reticulata* var. *crenatifolia* (kaula mandarin of India; *C. crenatifolia* of Lushington), *C. reticulata* Blanco. (Nagpuri santra, ponkan), *C. reticulata* var. *austera* Swing. (smooth skinned mandarins like Soh-siem of Assam and Sunki of China), *C. lycopersicaeformis* Lush. (billi kichili, karpura tenga of Assam, kodai-kithuli of south India, Naga santra, heen naran of Ceylon, Reshmi, Konkni of north India, and the narrow leaved smooth fruited Cleopatra)); **Limonellus Group** (mithotra of Punjab, *C. limettioides* Tanaka (Indian or Palestine sweet lime), *C. limetta* Risso, *C. limettioides* var. *latifolia* (Tahiti lime, Persian lime, soh-synteng of Assam, Meyer lemon), *C. aurantifolia* Swing. (acid lime)); **Pseudofortunella Group** (*C. indica* Tanaka (Indian wild orange), *C. tachibana* Tanaka, *C. madurensis* Lour. (calamondin or hazara, shikikitsu of Formosa and Okinawa). **subgenus Papeda :** C. *ichangensis* Swing., *C. latipes* Tanaka, *C. micrantha* Wester, *C. micrantha* var. *microcarpa* Wester, *C. hystrix* DC., *C. macroptera* Mont, *C. macroptera* var. *annamensis* Tanaka, *C. macroptera* var. *Kerri* Swing., *C. celebica* Koord, *C. celebica* var. *southwickii* *Wester, C. combara.*

sweet oranges of mandarins and pummelo; grapefruit of sweet orange and pummelo; lemon of citron and rough lemon or pummelo; and lime possibly of citron, pummelo and *Microcitrus*. Similar conclusions have been communicated by Scora (1975). Citrus biotype interrelationships have been reviewed by Scora (1988).

Molecular data have provided information about the origin of many citrus taxa (Nicolosi *et al.*, 2000; Barkley *et al.*, 2006; Bayer *et al.*, 2009; Li *et al.*, 2010). Early types of pummelo and mandarian probably hybridized to produce the sweet and sour oranges and pummelo provided the cytoplasm. Lemon derived from citron and sour orange and the lime as hybrid of citron and one or more other taxa. Mandarian was the donor of at least some lime cytoplasm. The grapefruit was likely product of pummelo and sweet orange. A number of different karyotypes are found in grapefruit, probably resulting from independent crosses between pummelos of variant karyotype constitution and sweet oranges (de Moraes *et al.*, 2007). The studies have provided some insight into *Citrus* phylogeny and three species concept (*C. reticulata, C. maxima* and *C. medica* separated into three distinct clusters) has been generally supported (Pang *et al.*, 2007). Molecular evidence supported the hybrid origin of many so-called species, for example sweet orange, grapefruit, lemon, etc. and their putative parental species have been identified.

DNA markers, with their phenotypic neutrality, abundance and imperviousness to environmental conditions, are useful to sort out relationships between the various *Citrus* types. Results of AFLP analysis demonstrated that *Poncirus, Microcitrus* and *Eremocitrus* are distant from *Citrus* (Pang *et al.*, 2007). Further evidence indicated that *P. polyandra* should not be the hybrid between trifoliate orange and *Citrus* genotypes and it deserves species status. A strong affiliation exists between *C. halimii* and *Fortunella* and the results did not support *C. halimii* as the fourth basic species. Both the separation of subgenus *Papeda* from *Citrus* and of subgenus *Archicitrus* from *Metacitrus* were not clearly resolved in the dendograms. *C. ichangensis* is a distinct species very different from other *Citrus* genotypes and that it was improper to classify it into subgenus *Metacitrus* (Pang *et al.*, 2007). However, the classification given by Swingle (1948) is still widely followed.

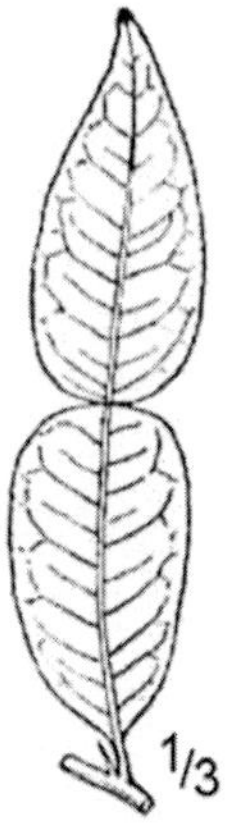

Fig.29.13. *Citrus ichangensis* leaf (Source: Hume, 1957).

Citron
(*Citrus medica* L.)

Most likely the citron originated in northeastern India and the nearby area. This was the first citrus fruit that was known to the Romans. The Romans and Greeks liked it in spite of the fact that its fruit is not edible and the tree has little ornamental value possibly for the penetrating and lasting fragrance and softness of fruits. Citrons are grown mostly for their peel, which is candied.

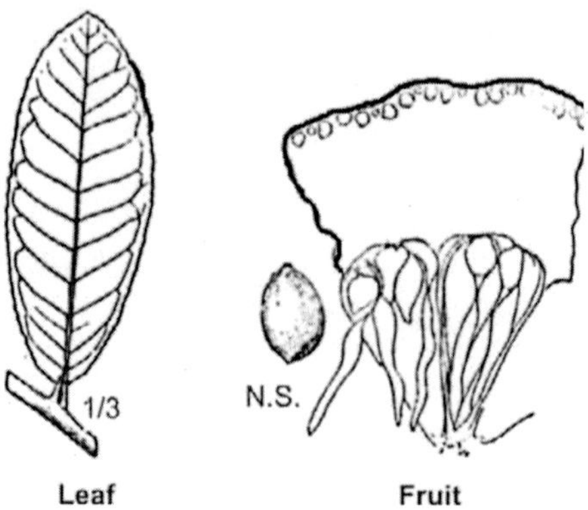

Fig. 29.14. *Citrus medica* (Source: Hume, 1957).

Citron is the second type Barret and Rhodes advanced as a true species. It is monoembryonic and relatively homozygous. Molecular marker data support its species status. Its unique chloroplast hybridization pattern also supports species status and indicates that citron always acted as the male parent.

The plants are shrubs or small trees growing up to 2.4-4.4 m in height. The wood is somewhat soft. The current growth is angular, which later becomes terete (circular). Young growth is purple tinged. The angled twigs have short or long and stout spines in the axils of leaves. The leaves are alternate, shiny, ovate lanceolate to elliptic ovate, medium large in size (6.25-18 cm) with toothed margin, pointed or rounded apex, rounded base and the short petiole almost wingless and not articulated with lamina. Flowers are axillary emerging over an extended period in few flowered racemes. These are large, 3-4 cm across, hermaphrodite and staminate flowers, large proportion being staminate having 5 lobed sepals, 5 petals that are pink tinged outside, 30-40 stamens, single style and thick ovary usually large having 10-13 locules. The fruit is oblong, obovoid or oval, 9- 30 cm in size having small segments and pale yellow and very thick rind having rough and very bumpy surface. The pulp or vesicles are greenish and sour in taste. Seeds are white, numerous, small and ovoid.

Cultivars

The cultivars are mainly of two types:

a. With pinkish new growth, purple flower buds and purple tinted petals, acid pulp, dark inner seed coat and chalazal spot (acidic cultivars). The well known cultivars are Diamante, commercially grown in Italy and Etrog, the main cultivar of Israel.

Diamante (Cedro Liscio): It is an important commercial cultivar of Italy. Its origin is not known. The trees are small in size, spreading and thorny with some large stout spines having open crown. The new growth including buds and flowers are distinctly purple pigmented. The fruits are usually large, long oval to ellipsoid and lemon yellow having furrowed basal cavity and pointed nipple like apex and many seeds. The rind is very thick and fleshy having lobed or ribbed and smooth surface. The flesh is crisp, dry and acidic.

Etrog (Atrog, Ethrog): New growth, flowers and flower buds are all purple pigmented. The tree is small, productive and has moderate growth. Leaf apex is obtuse. The fruit is lemon yellow, small to medium, ellipsoid with distinct neck, sharp nipple at apex and persistent style. The rind is fleshy and thick with bumpy and ribbed surface. The flesh is firm and crisp, acidic and has low juice content.

b. With no pink or purple tint in the new growth, shoots, floral buds and flowers, colourless inner seed coat and pale yellow chalazal spot and acid less pulp (non acidic cultivars). Corsican is a well established cultivar in Corsica (France) and California.

Fingered citron
(*Citrus medica* var. *sarcoclactylis* (Hoota and Nooten) Swingle)

It is well known in Japan, China and India. It is cultivated on a large scale in China. The fruit is split into a number of finger-like segments and has very little pulp and valued for its rich fragrance. It has dwarf growth habit and is planted as an ornamental plant.

Grapefruit
(*Citrus paradisi* Macf.)

The grapefruit originated in West Indies (Scora, 1988) and is a natural hybrid of pummelo and sweet orange. RAPD and SCAR marker data indicate that it was derived from a backcross between sweet orange and pummelo. The biotype might be homozygous. The plants are large (10-15 m tall), spreading, evergreen and round topped trees having dense foliage, angular and sparsely pubescent young twigs. The angular twigs have short and supple thorns. The green leaf is ovate, smaller than in pummelo and larger than in sweet orange, 7.5-15 x 4.5-7.5 cm in size with minute and rounded teeth on the margins and broad, oblanceolate or obovate petiolar wings (less than in pummelo). The flowers are borne axillary, singly or in clusters. These are about 4-5 cm across, pentamerous having 5-lobed calyx, 5 white and fragrant corolla, 20-25 stamens, single style and 12-14-loculed ovary. The fruit is nearly round or oblate to slightly pear shaped, borne in clusters, 8-15 cm in diameter, greenish yellow to pale lemon coloured with occasional pink to red blush having thinner rind and smaller vesicles (compared to pummelo) adhering to the rind. The pulp is pale yellow or nearly whitish or pink to deep red. The seed is white, smooth and elliptical and pointed having white cotyledons and is highly polyembryonic.

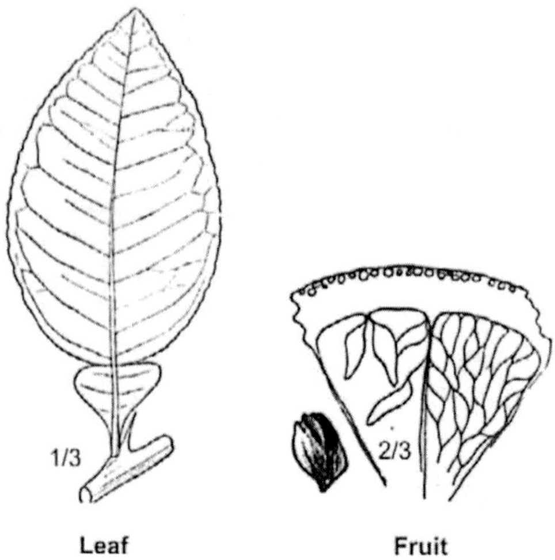

Fig. 29.15. *Citrus paradisi* (Source: Hume, 1957).

Common cultivars

Duncan: The trees are large, very vigorous and cold hardy. The fruits ripen in early to mid season. These are seedy, large, broad ovate to oblate globose, pale to light yellow in colour having light, small and spreading areolar ring and furrows at the base. The surface of the rind is even, smooth, glossy and medium in thickness. The flesh is light yellow buff, soft and very juicy having very characteristic strong and pleasant flavour.

Marsh (syn. **White Marsh, Marsh Seedless**)**:** The cultivar is highly valued for the seedless fruits and for its late ripening habit adapted to very hot climates. It is very productive. The tree is large and fast growing having spreading branches. The fruit is medium in size, spherical or oblate in shape and of light yellow colour having areole ring almost absent and few to no seeds. The rind is semi thick having even and smooth surface. The flesh is light yellow, soft and very juicy with pleasant flavour, which is less than in the seedy cultivars. The fruits have very good storage ability and can remain hanging on the tree for long time.

Pigmented cultivars

Foster (Foster Pink): The cultivar originated as a limb sport of cv. Walters and is the first pigmented grapefruit cultivar that ripens in mid season. The trees are large, very productive and vigorous. The fruit is very seedy, flat to spherical and medium large in size having inconspicuous areole, small and diverging furrows at the base and light yellow colour overlain by pink blush, which is also seen in the albedo (mesocarp). The rind is smooth and medium thick. The flesh is soft and light yellow but develop pink colour and good flavour under favourable conditions.

Foster Pink: The cultivar is a limb sport of Foster having better rind colouration and seedlessness.

Redblush (syn. **Ruby, Red Marsh, Red Seedless):** The cultivar originated as a limb sport of Thompson. The tree characters are between those of Thompson and Marsh. The fruit resembles that of Thompson in most of the characters except intense pigmentation of the flesh, albedo (mesocarp) and the pink fruit surface with bright red blush. The fruits hang well on the tree as in the parent cultivar.

Thompson (Pink Marsh): The cultivar originated as a limb sport of cv. Marsh. Being seedless it became very popular within a short period. The trees are large having vigorous growth and are very productive. The fruit is almost seedless or has few seeds. These are medium sized and spherical oblate having indistinct areole and light yellow colour. The rind is relatively thick, smooth and tough. The flesh is soft, juicy and dark yellow and develops pink colour under desirable conditions. The albedo is not coloured but has good flavour. The fruits have excellent storage and shipping quality and hang well on the tree for long period.

Melogold: It is an early maturing triploid originated from a cross between an acidless or low acid pummelo and the tetraploid white seedy grapefruit at the University of California, Riverside, USA. Although the fruit shape is not dissimilar to Marsh except for a slightly tapering neck (sheep nose), a feature often taken as an indication of poorer internal quality is its very large size, often averaging 700 g compared with 520 g and 450 g for Oroblanco and Marsh. Unlike the pummelo, the segments of Melogold cannot be peeled and separated from the wall without making it messy. The rind is smooth but develops colour slowly like Oroblanco, achieving full yellow colour several weeks after these are internally mature. Fruits are virtually seedless.

Oroblanco (Sweetie): It originated from the same cross that produced Melogold at the University of California, Riverside, USA. The trees are vigorous and attain the size similar to those of Marsh grapefruit but have broader leaves. Fruit shape is also similar to that of Marsh but is slightly flattened and somewhat larger in size. The rind remains green longer and is slightly thicker. Other internal physical characteristics are similar to Marsh, but the flesh is somewhat paler and has a wider open core. The flesh is tender and very juicy but is slightly less than in Marsh owing to a thicker rind. Fruit is completely seedless when grown in isolation.

The fruit is noticeably sweeter than that of Marsh as the sugar content is 2% higher and acid level is 0.6% lower. The higher sugar/acid ratio without any bitterness characterizes the Marsh fruits produced in cooler areas.

Lemon
(***Citrus limon*** (L.) Burm. f.)

The tree is small to medium and grows up to 3-6 m in height having slender and spreading twigs and sharp, stiff and stout thorns. The leaves are simple, light green, alternate and oblong elliptic or long ovate in shape and of medium size (10 x 6 cm) having finely toothed margin, short and narrowly winged petiole with distinct articulation. The flowers are borne singly or in clusters of 2 or more in leaf axils produced in all the seasons. The flower is 3-5 cm in diameter having 4-5 lobed calyx, 4-5 petals that are pink in bud and white above and purplish below, 20-40 stamens, single style and 8-10-loculed ovary. The characteristic features of the fruit are oval oblong shape with a distinct nipple like protuberance at the stylar end, 5-12 cm size having light yellow, thick and fragrant rind densely dotted with glands and adhering tightly to the highly acidic and tasty pale yellow pulp vesicles. The seed is plump, ovoid elliptic or ovate, pointed and smooth and long with inner white colour having seeds with greenish cotyledons and is polyembryonic.

Fig. 29.16. *Citrus limon* (Source: Hume, 1957).

Barret and Rhodes (1976) speculated that lemons are a complex hybrid similar to limes carrying a greater proportion of citron genes. Molecular marker data indicate that lemon originated from citron and sour orange, with sour orange as maternal parent. ISSR marker suggested its possible polymorphic origin. Owing to nucellar embryony, uniform rootstock material can be raised from the seeds. Some of the important cultivars are Lisbon, Eureka and Villafranca.

Eureka: The trees are medium in growth, almost thornless and spreading having open crown and sparse foliage. It is a very precocious and productive cultivar fruiting round the year with main flush in winter, spring and early summer. It is susceptible to cold. The fruits are small to medium, oblong elliptical and yellow in colour with short neck, small collar and rounded base but usually with a nipple having about 10 segments adhering tightly and small and solid central axis and few to no seeds. The rind is semi-thick, slightly rugose and pitted surface and sunken oil glands. The flesh is greenish yellow, tender, very juicy and finely grained having abundant juice which is very acidic but has excellent flavour.

Lisbon: The cultivar originated in Portugal. The tree is upright, large and thorny with vigorous growth and dense foliage and is very productive. Fruiting is more in winter and early spring. The fruit is yellow, medium, elliptical oblong having tapering base, indistinct neck and sharp pointed apex, prominent nipple and irregular areolar furrow, small and solid core and about 10 segments and few to no seeds. The rind is quite thick adhering tightly having pitted, smooth and less ribbed surface than in Eureka. The flesh is greenish yellow, tender, juicy, finely grained and very acidic in taste.

Villafranca: The cultivar originated in Sicily. It is not very important commercially compared to Eureka and Lisbon. The tree characters resemble those of Lisbon but the plants are more erect and upright in growth and have less foliage and few thorns. Fruit characters are similar to those of Eureka but fruiting is more during the winter as in Lisbon and ripens during December-January.

Assam lemon or Nepali oblong or Pat Nebu: It is originally a cross between citron and lemon. It originated in Assam. It is an ever bearing cultivar. The tree is vigorous having wide spreading and almost spineless branches and open and flat crown. The foliage resembles that of the citron. The fruit is medium large, oblong ovate to long elliptic having broad nipple and round base and none to few seeds. The rind is relatively thick, greenish yellow, glossy and smooth having 11-13 segments and usually hollow axis. The flesh is fine grained, greenish yellow, juicy and not too acidic. It is preferred for its aroma. An ecotype which is having high aroma is called Gandharaj.

Meyer: The tree is medium sized, spreading, hardy, almost thornless and productive. The new growth and flowers have purple pigmentation. It flowers throughout the year but more during the spring so that fruiting is more in winter. The fruit is medium, short elliptical or oblong and slightly ribbed having very small neck, round base and apex lightly furrowed and broad nipple, 10 segments and few seeds. The rind is thin and adheres tightly having smooth yellowish orange to orange surface. The flesh is light orange yellow, soft and very juicy having the typical lemon flavour.

Femminello Ovale: It is an old Italian cultivar. Tree is fairly vigorous and almost thornless. Fruit is round elliptic having yellow skin, juicy pulp and few seeds.

Kagzi Kalan: It is a selection. Plants are semi-vigorous, low spreading, prolific bearer and starts bearing three years after planting. The fruits are medium to large in size, extremely juicy and have few or no seeds. It is suitable for processing.

Pant Lemon-I: It is a selection from Kagzi Kalan. Fruits are medium in size, round and juicy having smooth and thin rind. It is tolerant to canker, tristeza and dieback. It can be the best replacement for Kagzi Kalan.

Baramasia or Seedless: Its trees are small, straggling, compact and somewhat spinous. Fruits are medium sized, ovoid globose to subglobose or ellipsoid oblong having medium thick rind, pyrethrum yellow or lemon chrome colour, pitted surface with inconspicuous ribs, slightly protruded to even base, rounded to truncate and at times slightly nippled apex, solid to semi-hollow axis with 13 to 15 segments in round fruited and 10 to 15 in long fruited type and none to a few seeds in both the strains. The pulp is medium firm and fine, cream coloured having medium to large and conical vesicles, abundant acidic juice of very good quality and flavour with 6°B TSS and 2. 17% acidity.

Galgal or Hill Lemon: It is an indigenous cultivar growing in a semi-wild and wild state in the submontane tracts of northern India. It is a heavy cropper and is most tolerant to heat and cold. Its trees are large, spreading, compact and rather spinous. Fruit is medium to large, of variable shapes from oblong ovoid to oblong globose and at times ovoid, thick to medium thick and leathery rind, apricot yellow or sunflower colour, rugose to wrinkled surface, short-necked or broadly necked base, nippled apex, medium to large and semi-hollow to hollow axis and 9 to 15 segments. The pulp is fine or coarse and tender or firm, straw coloured, large, slender to medium and irregular vesicles, very juicy and acidic, few to many (3 to 27) seeds, 10°B TSS and 4.8% acidity and of good quality.

Limes

Like the citron and lemon, the limes in all likelihood originated in northeastern India. Limes are one of the citrus biotypes believed to be apomictically perpetuated. The limes include both acid and sweet cultivars. The sour limes consist of two kinds, the small-fruited Mexican type (West Indian, Key) or Kagzi Nimbu (*Citrus aurantifolia*) and the large fruited acid lime, e.g. Tahiti or Persian lime (*C. latifolia*) which is triploid and therefore seedless. In the natural habitat, a number of forms that have been recognized differ markedly in characters such as size, form, shape, spine, seediness, etc. Barret and Rhodes (1976) suggested that the lime probably arose from a trihybrid intergeneric cross involving *C. medica* (citron), *C. grandis* (pummelo) and *Microcitrus* species. All RAPD and SCAR markers present in Mexican lime are also present in citron or in *C. micrantha* (a small fruited papeda) suggesting that these two species were the parents. RFLP data also support that citron is a parent.

Sour lime

(***Citrus aurantifolia*** (Christm.) Swingle; small fruited acid lime; West Indian lime; Mexican lime; Kagzi Nimbu)

The plants are more tender than those of the other citrus species. Plants are much branched and small trees (5 m tall), irregular in outline having slender and spreading twigs heavily marked with short and very sharp axillary spines. Leaves are simple, alternate, small (4-8 x 2-5 cm), ovate elliptic to sometimes oblong ovate with rounded base and minute rounded teeth (crenulate) on the margin and narrowly winged petiole. Inflorescence is short and axillary, usually 1-7 flowered, sometimes 10-flowered raceme. Flowers are white, small and emerge over a long period having cup shaped and 4-6 lobed calyx, white, 4-5 (12 x, 4 mm) petals, 20-25 stamens, 9-12-loculed ovary and a distinct style. Fruit is a small berry, about 3-6 cm in diameter, greenish yellow, oval, ovoid or globose in shape with distinct apical papillae at calyx end. The thin rind or peel having numerous glands adheres tightly to the vesicles. The pulp is very acidic, yellow green, juicy and fragrant. The seed is plump, small, oval and highly polyembryonic having green cotyledons. Sour lime hybridizes easily with other citrus species.

Cultivars

West Indian (Mexican sour lime, Key): The plants are very susceptible to cold. The trees are medium sized, spreading and bushy having many slender and short branches densely marked with small slender spines. New growth, flower buds and flowers are purple tinged. The leaves are broad lanceolate having winged petioles and light green and dense foliage. The tree flowers throughout the year but more during spring and late summer. The fruit is small, obovate, round or short elliptical having round base but with small furrowed nipple and 0-12 segments and is moderately seedy having few polyembryonic seeds. The rind is thin and greenish yellow with fine and smooth surface and adheres tightly. The flesh is greenish yellow, fine, soft, juicy and very acidic having the characteristic aroma.

Fig. 29.17. *Citrus aurantifolia* (Source: Hume, 1957).

Jai Devi (PKM 1): It is a selection from Kadayam type from Thirunelveli district of Tamil Nadu. Plants are vigorous and produce fruits round the year. Fruits are large (52 g), round in shape and juicy (52.3%) with attractive yellow colour having high ascorbic acid (34 mg/100 g juice) and 8ºB TSS. It is a high yielding (37 kg per tree) cultivar and is less susceptible to leaf miner and citrus butterfly.

Rasraj (*Citrus aurantifolia x C. limon;* Kagzi Lime and Nepali Round Lemon): It is an interspecific polyembryonic hybrid resistant to bacterial canker. The hybrid derives genes for canker resistance from Nepali Round lemon and other characters from acid lime. Trees are vigorous. The fruit is yellow having thicker skin than that of lime having 55 g weight, 6% acidity and 8ºB TSS. It contains 70% juice and 12 seeds.

Sai Sharbati: It is a local selection from western Maharashtra. Fruits are smooth having uniform thin skin and high juice and acidity. It is a high yielding cultivar (1205 fruits/tree in seventh year without bahar treatment) with bigger fruit size. It is tolerant to canker and tristeza.

Phule Sharbati: It is superior to Sai Sharbati having better plant canopy and more fruits (52.19 t/ha, 11.28% more than Sai Sharbati).

Vikram: It is a selection from local Kagzi lime. It is a high yielding cultivar bearing fruits in clusters of 5-10 during June-July and November-December and some off season fruits in September and May-June. The fruit weighs 46 g. It gives 30-32% higher fruit yield.

Pramalini: It is a selection from local Kagzi lime. It is a high yielding cultivar bearing fruits in clusters of 3-7 during December and off season fruits in September and May-June. The fruit has thin peel, 58% juice (compared to 53% in Vikram and 52% in normal lime) and weighs 47 g. It gives 30% higher fruit yield.

Kagzi Lime (Kagzi Nimboo): Fruit has 40 g weight and smooth surface texture. Two main fruiting seasons are observed; May-June and November-December.

Phule Sharbati Balaji (Tenali Selection): Fruits are medium to large in size. The trees are high yielding and resistant to citrus canker and bark eruption.

Seedless: It is a selection. Fruits are oblong in shape, primrose in colour and have thin skin. The trees are prolific bearer giving double the yield of that from the common lime but are late maturing.

Tahiti or Persian lime
(*Citrus latifolia* L.)

It is a triploid (2n = 3x = 27), hybrid of lime and citron. This is more hardy and more adapted to subtropical conditions. It is cold resistant similar to the lemons. Flowering occurs throughout the year but more during the spring. Purple pigmentation is present on shoots and flowers. The tree is vigorous, very spreading and almost thornless having dense green foliage. The leaves are broad lanceolate with winged petioles. The fruit is almost seedless, small to medium, variable in shape from oval, oblong, obovate to short elliptical having usually 10 segments, small and solid axis, round base, short neck, faint furrows and round apex and areolar area forming a small nipple. The rind is thin, lemon yellow and adheres tightly. The flesh is greenish yellow, tender, juicy and more acidic. The fruits are almost seedless or may have few seeds and are less fragrant than the common lime. It is monoembryonic.

Bearss (Bearss Seedless, Persian): Both tree and fruit characters correspond closely with the Tahiti lime. The flowers are devoid of viable pollen and the fruits are seedless.

Sweet lime
(*Citrus limettioides* Tan.)

The sweet lime is a hybrid of sweet lemon or sweet citron. It originated in northeastern India. The Indian sweet lime is called Mitha Nimbu. Sweet lime is of value in India, Near East, Egypt and Latin America and is said to have medicinal properties especially in prevention and treatment of fevers and liver ailments. It is said that Soh Syntcng of Assam is the acid form of this fruit. As in acid limes, a number of its forms have been recognized that differ in characters such as fruitfulness, fruit shape and size, presence of nipple, etc.

The tree is medium to large and spreading with fairly long and slender thorns. The leaves are medium in size, pale green and characteristically somewhat cupped or rolled having wing-margined petioles. The flowers are large and not coloured. The fruits are medium in size, globose to ellipsoid having flat papilla, thin rind, pale yellow flesh with distinct aroma and abundant juice. Seeds are medium in size and cotyledons are white and polyembryonic.

Cultivars

Indian (Palestine): The tree is large and spreading having irregular growth habit, thick and thorny branches and dense foliage. New growth is white (free of colouration). The leaves are rolled, cupped, long oval having winged petioles. The fruit shape varies from oblong to short elliptic to subglobose having round base and apex having flat nipple, circular furrow, semi-hollow axis and 10 segments and few highly polyembryonic seeds. The rind is very thin having smooth surface, distinct oil glands and greenish to orange yellow colour. The flesh is light yellow, juicy, free of acidity but with slight bitterness.

Dweet: The cultivar evolved from a cross between Mediterranean sweet orange and Dancy tangerine. The tree characters are intermediate of those of the two parents. The fruits mature late in the season and do not hang well on the tree. These are medium to large, globose oblate and reddish orange with distinct neck and pebbled surface and have numerous seeds. The rind adheres tightly, also puffs and peels very poorly. The flesh is orange, firm and very juicy having rich flavour.

Mency: The cultivar originated from a reciprocal cross with Dweet. The fruits ripen early and do not hang well on the tree. These are susceptible to sunburn. The fruits are small, somewhat oblate and necked and reddish orange having pebbled surface due to oil glands. Rind adherence is not tight and peels easily. Flesh is orange, juicy with acidic flavour. Seeds are many.

Mandarin
(*Citrus reticulata* Blanco)

The plants are relatively small (2-8 m tall) having spiny and slender twigs. The leaves are small, dark green, shining, green above and pale below, narrow at both apices, narrowly or broadly lanceolate or elliptic with acute base and tip, crenate margin and usually narrowly winged petiole. The flowers are borne singly or a few together in the leaf axils. The flowers are small (1.5-2.5 cm diameter), white, fragrant and pentamerous having 5 calyx, 5 white corolla, 20 stamens, single style and 10-15-locular ovary. Fruit is a depressed globose or subglobose berry, 5-8 cm in diameter, yellow or orange in colour having bright orange or red orange and deep orange pulp. The rind is thin and loose separating easily from the segments. The segments are sweet and juicy. The seed is small, pointed at one end and green or white from inside and are polyembryonic.

This is a very large group of varied and distinct types of highly acclaimed citrus fruits important in the Orient. These are of value for fruit, ornamental purpose and as rootstock. Sweet mandarin types are used as dessert fruit or to extract juice or segments for canning. Essential oils and pectin are derived from the rind. The mandarins closely resemble oranges and are usually referred as loose skinned oranges, a confusing usage. Mandarins are the most phenotypically heterogeneous group of *Citrus*. Both monoembryonic and polyembryonic clones exist as do self fertile and self incompatible types. These exhibit wide variation in respect of fruit size, colour, flavour, maturity time, adherence of rind, etc. Molecular marker data reveal great heterogeniety within the group. *C. reticulata* has a unique chloroplast banding pattern. Therefore, this was the third *Citrus* type assigned species status by Barret and Rhodes. Most important cultivars like Dancy, Satsumas, Nagpur and Clementine have small to medium fruits while large fruited mandarins are King and Ortanique. Cleopatra is a very small fruited mandarin. Mandarins have been divided into groups such as Satsuma (*C. unshiu* Marcovitch), King (*C. nobilis* Loureiro), Mediterranean (*C. deliciosa* Tenore) and Common (*C. reticulata* Blanco).

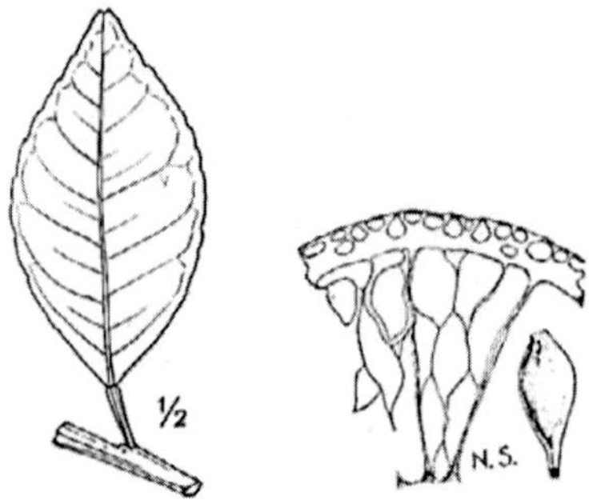

Fig. 29.18. *Citrus reticulata* (Source: Hume, 1957).

In some countries, the names mandarins and tangerine are used indiscriminately. However, mandarins are used for yellow fruited cultivars and tangerine for deep orange types.

Satsuma mandarin
(***C. unshiu*** Marcovitch)

This group includes very popular cultivars extensively cultivated in Japan called Unshiu Mikan. It is considered to be the most cold tolerant among the citrus fruits. The trees are slow growing having moderate growth and spreading and drooping branches that are almost thornless. The leaves are long lanceolate and dark green having tapering base and apex and long and thin winged petiole. The heat requirement being very low its fruits ripen earlier than those of other oranges. The fruit ripening is very early to mid season. The fruits have to be picked quickly when mature otherwise their quality deteriorates. The fruits are small to medium, oblate sub-globse and orange coloured having small and faint areole and pressed navel. The rind is smooth, leathery and thin having prominent glands and is easy to separate. The flesh is orange coloured, soft, melting and subacid in taste and has rich flavour. The segments are 10-12 with hollow axis and loose carpellary membranes. The fruits have good storage ability.

Cultivars

The following 5 groups of Unshiu Mikan have been recognized:

1. Wase (Early): All the clones of this group are early ripening.
2. Zairai (native or indigenous): This group includes old clones whose origin and parentage is not known.
3. Owari: This group includes cultivars that became popular in Owari Province.
4. Ikeda: The fruits of this group mature late and are small subglobose and seedless having moderate quality.

5. Ikiriki: The cultivars of this group are not productive but bear seedless and excellent quality fruits.

Owari (Satsuma): It is an old Japanese cultivar of unknown origin. The tree is slow growing having spreading and drooping branches and is very productive. The fruits are early to mid season in maturity. The fruit is oblate subglobose and seedless with small neck having 10-12 segments and hollow axis and is seedless. The rind is thin, smooth and leathery but both roughness and looseness increases at maturity. The peel is separable. The flesh is orange, tender, melting and subacid in taste.

Silverhill (Silverhill Owari): It is a cold resistant and productive cultivar. The tree is upright and very vigorous in growth. The fruits are more flat than of any other Satsuma mandarin and are of medium size and of excellent quality and mature early' in the season. The rind is quite thin, smooth, orange in colour and leathery. The flesh is orange coloured, soft, juicy, low in acidity and sweet. The storage quality of fruits is good.

King mandarin
(*C. nobilis* Loureiro)

The name of this group has been derived from that of the American cultivar King. Its characters are intermediate between those of orange and mandarin. The cultivar King, a representative of this group, is a natural tangor.

King (King of Siam): The tree is upright and moderate in vigour having open crown, few thornless to moderately thorny branches and dark green foliage with narrowly winged petioles. It is more cold hardy than many other mandarins and is a very productive cultivar. The fruit maturity is late to very late. The fruit is oblate or depressed globose with short neck, furrowed and depressed base, hollow and large axis, faint areole, 12-14 segments and few to many seeds. The rind is thicker than in the other mandarins, have medium adherence, peels easily and is deep yellowish orange to orange in colour having smooth to rough surface. The flesh is deep orange, soft and juicy with good flavour. The fruits hang well on the tree.

Mediterranean mandarins
(*Citrus deliciosa* Tenore)

The typical features of this group of the famous mandarins from the Mediterranean basin are derived from the species, *C. deliciosa* Tenore. It has plump and spherical seeds, small, narrow and deep green, sharp pointed leaves having aromatic glands and very aromatically flavoured juice. The important cultivar is Mediterranean (Willowleaf, Mediterranean Common).

Willow Leaf (Mediterranean Common): The trees are cold hardy, small and slow growing having broad crown, drooping, thin and thornless branches and narrow lanceolate leaves. It is an early to mid season cultivar and bear irregularly. The fruit is oblate, medium in size with few lobes, distinctly necked and furrowed, without areole but with small navel at the apex and having 10-12 segments, loose and hollow axis and many small and plump seeds. The rind is thin, loosely attached, yellowish orange, shiny and smooth. It separates rapidly in ripe fruits. The flesh is light orange, soft and juicy having sweet and aromatic flavour. The cotyledons are light green. The fruits lack storage ability.

Common mandarins
(*Citrus reticulata* Blanco)

This group contains a large number of cultivars, more than in any other mandarin group, having wide variation in plant and fruit characters.

Clementine (Algerian tangerine): It originated as an accidental hybrid in North Africa. The tree is cold resistant, medium in size having round crown, thin and almost thornless branchlets, dense foliage and narrow leaves, dark green above and lighter green below and lanceolate in shape. Fruit bearing is regular and consistent. The fruit is variable in shape from oblate to oblong or pyriform having necked or collared base and the apex is depressed or has small navel and 8-12 segments. The rind is firm and adhering, peels easily, deep orange to reddish in colour and smooth and glossy and pebbled with prominent oil glands. The flesh is deep orange, soft, melting, juicy and sweet and subacid in taste having aromatic flavour. The fruit is monoembryonic having few seeds with green cotyledons.

Dancy: The tree characters nearly resemble Clementine but the venation is not so prominent. It has alternate bearing tendency. The fruit is oblate-broad, obovoid or pyriform having broadly depressed apex and sometimes distinct neck, large and hollow axis and about 12 segments that separate easily. The rind is thin, leathery and loose and peels off readily. The surface is smooth, shiny and bumpy because of the oil glands. The flesh is deep orange, melting, tender, slightly more acidic and juicy having pleasant flavour. The fruits store quite well. The seed characters are like those of Clementine. The fruit has a few to some seeds that are polyembryonic having green cotyledons.

Kinnow: It was developed by H.B. Frost at California Citrus Experiment Station in 1915 from the cross between King and Willow Leaf. It has strong alternate bearing tendency. The tree is cold resistant. The trees are vigorous, large and symmetrical having tall and upright growth, slender and almost thornless branchlets and dense foliage. The leaves are broad lanceolate. The fruit is medium in size, oblate with both base and apex depressed and having solid to semi-hollow axis and 9-10 separable segments. The rind is thin, adhering but peelable, yellowish orange, smooth, shiny, leathery, tough and very smoothly pitted. The flesh is deep yellowish orange, very juicy, highly flavoured and aromatic and having high sugar, moderate acidity and 25-28 seeds. The fruit contains numerous pale greenish yellow and polyembryonic seeds. The fruits hang well on the tree. It is a late maturing cultivar.

Nagpur (Ponkan, Warnucro): The cultivar, the famous Pokan of South China, is of most commercial significance in India. The trees are upright in growth and are quite vigorous having moderate cold resistance. It is an early to mid season cultivar and is alternate bearer. The fruit is globose or oblate having large base (for mandarin) with small neck, distinct furrows and depressed apex sometimes with navel, large and hollow axis and 10 segments that separate easily. The rind is quite thick, orange coloured and has smooth pebbled surface due to sunken oil glands and adheres loosely. The flesh is orange, tender, juicy and melting having pleasant aromatic flavour. The fruit contains few small and plump seeds having green cotyledons.

Wilking: Also resulting from the cross between King and Willow Leaf, this cultivar is a sister of Kinnow. The tree characters are almost similar to Kinnow. It is a mid season cultivar and produces small fruits during the 'on' year and none in the 'off' year. The fruits are oblate, small to medium in size having semi-hollow axis, fattened base and apex and 9-12 segments that separate easily. Puffing is observed in the fruits. The rind is orange coloured, shiny, pebbled and medium peelable. The flesh is deep orange, very juicy, firm and rich in flavour. The fruit contains many seeds having greenish yellow to yellow cotyledons.

Page: It is an early ripening cultivar obtained from a cross between Minneola tangelo and Clementine mandarin. The tree has moderate vigour and almost thornless branches. The fruit matures early and is oblate subglobose having rounded apex and base, somewhat open and solid axis and 10 segments. The rind is somewhat thin and leathery that separates easily and has reddish orange, smooth and pebbled surface. The flesh is deep orange, juicy, tender and sweet and rich in flavour. The fruit contains many seeds having nearly white coyledons.

Ortanique: The tree is moderate in growth having spreading, drooping, thornless and slender branches and dense foliage with winged petioles. It is a mid season to late maturity cultivar. The fruit is variable in shape from broad obovoid oblate to subglobose having depressed apex or with a navel, round or tapering base, short or collared neck, distinct areole, partly open and solid axis and 10-12 segments. The rind is thin, tightly adhering but peelable, yellowish orange, glossy and finely pitted. The flesh is orange and juicy having characteristic rich flavour. The fruit contains 10 moderately plump and polyembryonic seeds.

Fallglo: It is a hybrid of Temple and Bower. The cultivar matures early and harvested between late October and November in the northern hemisphere. Growth and leaf characteristics are similar to Temple orange, being densely foliated, moderately vigorous having upright lanceolete leaves with reduced petioles. Fallglo trees are less freeze hardy than those of Orlando or Sunburst and thus may be less suitable for planting in areas where freezes are likely to occur. The fruit has oblate shape with a distinctive navel. The peel is smooth and contains prominent oil glands and can be easily separated. The peel and juice are deep orange in colour. The fruit quality is comparable to that of Temple. It does not seem to require cross pollination to achieve optimum production. The fruit contains 30-40 seeds depending on the extent of cross pollination. Seeds of Fallglo produce only zygotic progeny.

Ambersweet: It is an early ripening hybrid. Tree growth is moderately vigorous and somewhat upright. Trees have intermediate freeze hardiness. Fruits are moderately large, convex shaped and tapered at the stem end. Occasionally fruit produces a small navel. Juice colour is dark orange. Fruits may be seedless to seedy (one or more) depending on degree of cross pollination. Seeds produce zygotic progeny. Most of these have white cotyledons, although some green ones may be seen.

Sunburst: The cultivar resulted from a cross between Robinson and Osceols. Fruits are oblate, highly coloured having thin peel and good keeping quality. It requires pollination with cross compatible cultivars like Temple. The seeds are polyembryonic.

Robinson: Gardner and Fellows crossed Orlando tangelo with Clementine mandarin at the USDA Research laboratory in Orlando, Florida in 1942. Four selections made from the resulting zygotic progeny were named Robinson, Lee, Osceola and Nova. The trees of Robinson are moderately vigorous. Fruits are oblate in shape and of deep orange colour having excellent quality. It is polyembryonic.

Ellendale: It is a late cultivar. The fruits are large and of high quality. The cultivar has been commercially grown in Australia as a dooryard tree.

Daisy: The cultivar was produced at Citrus Research Station, Indio, California, USA from a cross, Fortune x Fremont. Fruit is medium in size, 135 g weight and juicy (47%) and has excellent quality. The fruit has one to three seeds per section. It has tendency of alternate bearing and granulation.

Khasi: It is an important commercial cultivar of Assam. Fruit is depressed and globose to oblate having 142-227 g weight, orange yellow to bright orange colour, usually obtuse base (stalk end), occasionally short-necked and slightly ribbed having thin rind with smooth and glossy surface and very slight or no adherence, 8 to 13 (usually 10) segments, little to medium rag, uniformly orange coloured pulp vesicles with somewhat coarse and very melting, firm and soft to very soft flesh texture, abundant orange coloured juice of agreeable flavour and well blended with 11.2°B TSS and 0.34% acidity, 8 to 25 (usually 10 to 15) medium sized seeds that are green when cut.

Coorg (Coorg mandarin): It is the most important commercial cultivar in South India grown extensively in Coorg and Wynad. Fruit is oblate to globose and finely papillate, of uniform bright golden yellow colour having delicately wrinkled and glossy surface, necked base (stalk end) or depressed with glandular ribs extending through the collar, thin to medium rind with slight adherence, 9 to 11 segments, little rag, uniformly golden yellow pulp vesicles having fine and tender texture, good flavour, abundant juice of deep chrome colour and when properly ripe, well blended with 10°B TSS and 0.57% acidity. The 14 to 30 seeds are medium sized and are light green when cut.

The tree of a promising clone identified at IIHR, Bengaluru is vigorous and upright having dense canopy with or without thorns giving average yield of 800-1000 fruits at ten year age. Fruits are subglobose, depressed at distal end and frilled at the stock end. The ripe fruit is deep orange, orange yellow or yellow, juicy (40%), weighs 90 to 105 g with 10-12 segments. The juice has 10°B TSS, 0.4-0.6% acidity and 41.6 mg/100 ml ascorbic acid. The fruits have excellent shelf life.

Butwal: This cultivar is indigenous to Butwal district on the border of Nepal where it grows in a semi-wild state. Fruit is oblate with dull orange, pitted, rough and glossy surface, low collared base (stalk end) with wrinkled furrows extending through the collar, thin, firm and soft rind with slight adherence, 9 to 12 segments, little rag, uniformly pumpkin coloured and quite juicy pulp vesicles having fine and tender texture, good flavour and well blended with 8°B TSS and 0.96% acidity, 9 to 22 medium sized seeds that are green when cut.

Emperor: This cultivar is from Australia. The fruit is depressed, globose with a blush of red, coarsely pitted, glossy and finely wrinkled surface, short to long necked base (stalk end) with glandular furrows extending through the collar, medium thick, firm and very soft rind with slight adherence, 10 to 12 segments, little rag, uniform saffron yellow pulp vesicles having fine and tender texture and fair flavour, deep golden yellow juice with about 8°B TSS and 0.81% acidity, 6 to 33 medium sized seeds that are light green when cut.

Darjeeling or Sikkim: This is a popular cultivar similar to Nepal, Assam and Sylhet oranges. The cultivar Laddu is grown in small areas in Uttar Pradesh and Deccan region. The trees are fairly vigorous, dense, round and dome shaped. The bearing is profuse. The fruits have flattened shape, almost seedless vesicles with fairly sweet pulp and good flavour and are smaller than santra.

Mandarin-like citrus fruits

Besides the hybrids, tangors and tangelos, there are some fruits in the Orient similar to the mandarins such as Calamondin, Rangpur and Iyo.

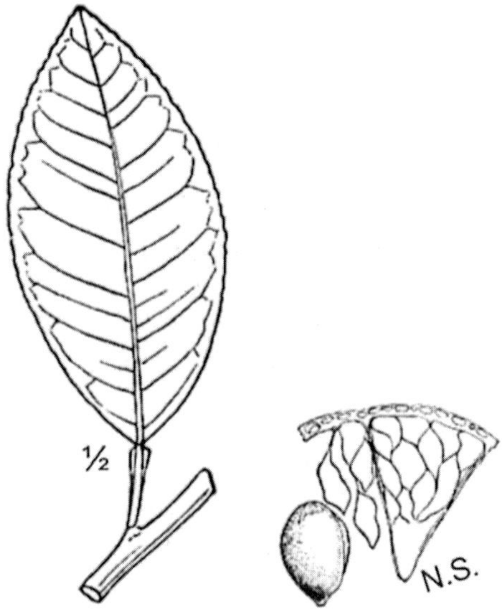

Fig. 29.19. *Citrus mitis* (Source: Hume, 1957)

Calamondin

(x ***Citrofortunella mitis*** (Blanco) J. Ingram syn. ***C. madurensis*** Loureiro, ***C. mitis***)

It has less value as a fruit but is widely planted as an ornamental tree. The trees are very cold hardy and are almost thornless having upright growth and small, broad and oval leaves. The fruits are produced profusely. These are very small and oblate to spherical with flat apex, small and hollow axis and 9 segments. The rind is thin, orange to orange red, smooth and pitted and peels easily. The flesh is orange in colour, soft, tender, juicy and acidic in taste. Seeds are few, plump, small and polyembryonic having green cotyledons.

Rangpur or Canton lemon
(***Citrus limonia*** Osbeck)

The tree is very cold hardy and vigorous in growth having spreading and drooping branches armed with small thorns. New growth, flower buds and petals are deep purple tinged. Leaves are elliptic to slightly ovate oblong. Flowers are mostly singly borne and purple tinged. Fruits are produced profusely. These are deep chrome, small to medium of variable shape from globose, round to broad obovate with short nipple at the apex, solid axis, collared base or with small neck and hang well on the tree. The fruit has 9-12 segments. The rind is yellowish to reddish orange, loose, thin, smooth and faintly pitted. The flesh is orange in colour, soft juicy and very acidic in taste. Seeds are many, small and polyembryonic with green cotyledons.

Ugli
(***Citrus reticulata*** Blanco x ***Citrus maxima*** Merril. or ***C. x paradisi*** Macf.)

It is a Jamaica tangelo and thought to be chance hybrid of mandarin and grapefruit. It was discovered growing wild in Jamaica where it is now grown possibly having developed in the same way as the grapefruit. However, it is usually taken to be a lemon-tangerine hybrid. Its name has been derived from the unsightly appearance of its rough, wrinkled, greenish yellow rind loosely enveloping the orange coloured pulp. The light green blemishes on the rind surface turn orange when the fruit is fully ripe. The fruit is slightly larger than a grapefruit and has fewer seeds. The flesh is very juicy and tends to be sweet as that of the tangerine rather than bitter as that of grapefruit. The rind is fragrant. The taste is often more sour than that of an orange and less bitter than that of a tangerine. Fruiting occurs from April to December. The fruit is marketed under the trade names "UgliFruit" and "UniqFruit".

Pummelo (Pomelo)
(***Citrus maxima*** Merril syn. ***C. grandis*** (L.) Osbeck) (2n=18)

Trees are medium tall (5-15 m), spreading, spiny and low branching having angular and often densely pubescent (may persist for a year or more) young twigs and prominent, large and long (5 cm) spines on branchlets, old limbs and trunk. Some forms are dwarf. Leaves are simple, alternate, dark green, leathery, shiny, large (5-20 x 2-12 cm), ovate to broad, elliptic having rounded to sub-cordate base, entire to slightly shallowly crenate margin, obtusely acute and sometimes slightly notched apex with undersurface pubescent along the veins and with petioles, broadly (7 cm) winged (broadest amongst all citrus species). Flowers are borne axillary, solitary or in clusters of 2 to 15 in 10-30 cm long terminal racemes. Flower is white, fragrant, large, measuring 3-7 cm in diameter, pentamerous having 5 lobed sepals, 5 cream coloured petals, 20-25 stamens, one style and 11-16-loculed ovary. The fruit is very large (largest of all citrus species), globose, subglobose or pyriform in shape, 10-30 cm in diameter, yellowish when ripe having very thick rind (1.25-2 cm) densely dotted with glands and clinging or more or less easily removed. The pulp vesicles are also very large, pale yellow or pink, well filled with sweet juice and divided into 11 to 18 segments. The vesicles tend to be dry in inferior types. Sometimes the rind is pink blushed and the pulp is greenish yellow or pale yellow to pink or red. The flavour is mildly sweet and bland to subacid or rather acid. Seeds are ridged, large, monoembryonic and yellowish white and of white colour from inside.

Pummelos are one of the three *Citrus* types Barret and Rhodes proposed as true species as these are monoembryonic, rather than being facultative apomicts and have a well developed reproductive isolation. Although pummelo is primarily cross pollinated, some selfed clones produce relatively vigorous progeny. Thus the species is relatively homozygous. *C. grandis* could be separated from the other *Citrus* types by its unique ISSR banding pattern. The pummelos fall into two major groups, pigmented and non-pigmented.

In Malaysia, pomelo is divided into the Thai, Chinese and Indonesian groups. The Chinese group consists of white flesh types with pyriform fruit and thick rind. The main cultivars are Shatian (J8) grown mainly in the southern State of Johor and Negeri Sembilan and Melaka of the Malayan Peninsula. The Thai group has smaller fruit (Saunt, 1990), grown in the northern State of Kedah. The Indonesian group has white-fleshed and acidless fruit with thick rind as in the cultivars Tambun White and Tambun Pink which are the most popular because of juiciness and low acid content (Ibrahim, 2000).

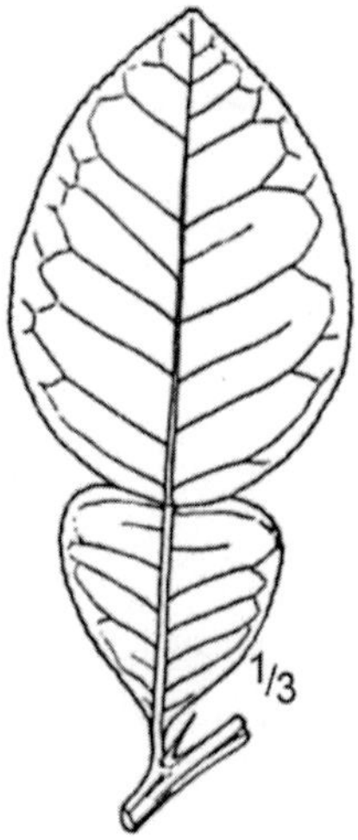

Fig. 29.20. *Citrus grandis* (Source: Hume, 1957).

Non-pigmented group

The common non-pigmented pummelos are mostly seedy and have moderate to high acid content and variable characters.

Banpeiyu: The cultivar is believed to have originated in the Malayan region and is very popular in Japan and Taiwan. The tree is large and spreading and grows very rapidly having pubescence on new growth. The leaves are large with broad petiolar wing. The fruit size is one of the largest among citrus fruits, spherical to subglobose in shape and light yellow in colour. The rind is thick tightly adhering the vesicles and with smooth surface. The flesh is pale yellow, soft and juicy having 15-18 segments, solid and large central axis, thin and tough carpellary membranes, excellent flavour and pleasant taste with good sugar-acid blend. It is a mid season cultivar. The fruits are seedy and have very good storage ability.

Hirado (Hirado Buntan): The cultivar originated as a chance seedling in Nagasaki Prefecture, Japan and is more resistant to cold. It is early to medium in ripening. The trees are medium to large and vigorous in growth having leaves with broad winged petioles. The fruit is seedy, large, oblate, bright yellow and glossy having both ends depressed, many segments and thin and tough carpellary membranes. The rind is medium thick having smooth surface. The flesh is light green to yellow and semi dry having pleasant flavour and balanced acid and sugar blend with bitterness. The fruits store well and are rated very highly.

Kao Phuang: This is one of the most famous cultivars grown in Thailand. The tree is upright and very vigorous and has early to mid season fruit ripening. The fruits are large, lemon yellow in colour, seedy, broad pyriform in shape with small and compact axis, semi-thick carpellary membranes and depressed apex and distinct neck at the stalk end. The rind is semi-thick having medium adherence with the vesicles which are firm and large and separate easily and have shiny and smooth surface. The flesh is juicy having good flavour. The fruits hang well on the tree for long.

Tahitian (Moanalua; often called Tahitian grapefruit): Originally grown from seed it was brought from Borneo to Tahiti and later introduced into Hawaii. It is a typical pomelo but with a thin peel. The peel is greenish yellow. The flesh also has a greenish shade but may become amber on full maturity. The fruit is juicy and sweet having a flavour of melon or lime.

Siamese Sweet: It is an acidless, sweet pomelo from Thailand. The cultivar is one of the parents of Oroblanco and Melogold pomelo-grapefruit hybrids, Chandler pomelo and Cocktail pomelo-mandarin hybrid. The tree is dwarf with drooping branches and hairy new growth. The fruit is oblate to broad ovoid having white, large and crisp pulp with easily separating, mild-flavoured but faintly bitter and non-juicy sacs.

Sweetie: It is a pomelo hybrid. Its parents are Siamese Sweet pomelo (a variant of Marsh grapefruit). The flesh is pale yellow and the taste is mild and sweet. This is one of the pomelos that retains some green on its rind even when ripe. The cultivar is also known as Oroblanco grapefruit. A similar cross is named Sweetie in Israel. The Israeli Jaffa Sweetie is sold as a cultivar of grapefruit and can remain dark green even when fully ripe and sweet.

Thong Dee (Khao Thongdi, Golden): It is from Thailand. Both tree and fruit are very big. The fruit is usually seedless but may have seeds if there are citrus trees with seeds in the vicinity. The fruit is oblate and large (15 cm wide) having 1 cm thick peel, large pulp sacs separating easily from the segment walls and white pulp with light brown streaks. The flesh is juicy, seedy and has good flavour. The cultivar has several strains, some produce seedless fruits with pinkish flesh.

Cuban Shaddock: It is from Cuba and is of unknown parentage. The tree is vigorous, upright, large, thorny and spreading with dense foliage and is productive but sensitive to cold. Leaves are large, oblong elliptic and blunt-pointed. Flowers and new growth are not purple tinted. The fruit is large to very large, globose to obovate and depressed having characteristics of citron and lemon. While the fruit sometimes resembles a big Ponderosa lemon (citron x lemon hybrid), the leaf resembles that of pomelo and not of lemon. Seeds are numerous and dark yellow at maturity. Rind is rough, bumpy and commonly somewhat furrowed, thick and spongy. Flesh is yellowish green, coarse-textured and juicy having acid flavour.

Shatian (PO51): It is common in Malaysia. The fruit is of medium size and weighs 1.5-2.0 kg and is almost round with flattened base. The skin is light green with yellowish tinge. The pulp is white, very juicy and sweet with 9-10°B TSS.

Tambun White (PO52): It is common in Malaysia. It is sometimes called Pomelo Manis originally found in the Tambun area near Ipoh, Malaysia. The fruits are medium sized (1.4-1.7 kg), almost round with flattened base and are seedless having high juice content. The skin is light green with yellowish tinge. The delightful citronella fragrance of Tambun makes it clearly one of the best pomelo in the world.

Mato Buntan: It is said to have been taken from South China to Taiwan and later to Japan. Its fruit is highly priced among the pomelo cultivars in Chinese Taipei and in Japan. The tree is dwarf and small, round-topped and drooping having short and thick twigs and shoots. Leaves are large and thick. The fruit is medium large, broadly obovoid to pyriform, seedy and light yellow at maturity. Rind is medium thick, pebbled from protuberant oil glands and tightly adherent. Segments are numerous (12-16) and membranes are thin but tough. Flesh is light greenish yellow, crisp, somewhat tough and lacking in juice. Flavour is sweet, mildly acid and sometimes with a trace of bitterness. It is early in maturity.

Reinking: It is a seedling selection from a cross of Kao Phuang and Shamouti orange made at Indio, California. The tree grows to a large size with big leaves and drooping branches. The fruit is large and pear-shaped and seedy with a slightly flattened bottom having slightly pebbled, thick and yellow rind. The flesh is light yellow and juicy having good texture and flavour. The fruits mature early and hold fairly well on the tree.

Honey: This is a local Chinese cultivar grown for more than 500 years in Pinghe area. It is said to be seedless and sweet with semi-transparent flesh. The colour is light green to lemon yellow. The fruit weighs about 2 kg. The main harvest is from mid September to mid October, but the fruit is available until February.

Sarawak: It is sometimes referred to as Tahitian pomelo. The tree is large and vigorous. The fruit is round with a flattened bottom and has a greenish yellow rind that is thinner than the typically thick pomelo rind. The flesh is greenish, juicy and sweet with a melon like or lime like flavour. It is early to mid season in maturity and the fruit holds well on the tree.

Nam Roi: The cultivar is from Ninh Minh, Vietnem. The fruit is pyriform with greenish yellow skin weighing 1.0-1.2 kg and is seedless but sometimes with more than 10 seeds. The flesh is juicy and light yellow separated very easily from mesocarp when ripe. The juice tastes good, sweet (8.33°B Brix) but is slightly sour and has strong flavour (SOFRI, 2002).

Pigmented group

A number of deep red and pink fleshed types originated in the Orient. The attractive pigmentation is owing to the carotenoid lycopene.

Siamese Pink (Siam): It is considered to be one of the finest cultivars owing to its strong flavour. The tree is very large, fast growing having spreading branches and slightly pubescent shoots, twigs and petioles. The fruits mature late and are almost seedless. The fruit is large, broadly obovate short pyriform, depressed at apex and light yellow with pink colouration having many segments and quite thick and tough carpellary membranes but open at axis when mature. The rind is medium thick, smooth and glossy and adheres tightly. The flesh is pinkish, very juicy and coarsely grained having subacid flavour with some bitterness resembling that of grapefruit.

Siamese Sweet: The cultivar is commonly called sweet pummelo as it is practically acidless. The tree is dwarf having drooping branches and distinctly round and pointed leaves. The new growth is markedly and densely pubescent. The fruits are broad oblate to obovoid. The vesicles separate easily and are large, crisp and sweet in taste but lack in juice content.

Chandler: It is a hybrid cultivar obtained from a cross between Siamese Sweet (white, acidless) and Siamese Pink (acid). The fruit has pink flesh and the other characters are intermediate between those of the two parents. Fruits mature early in the season and have very good storage potential. The fruits are medium to very big sized, round, almost perfect globe, and light yellow having slight pubescence and many seeds. The rind is smooth and medium thick. Segment walls are thin. The vesicles have firm flesh that later turn soft and moderate in juice with pleasant flavour superior to that of either parents with about 12% sugar and subacid taste. The colour of peel and flesh varies depending on climate and soil conditions.

Jaffa Red Pomelo: The name is in Israel for the Chandler pomelo, a cross between the Siamese Sweet and Siamese Pink. The fruits are often more pear-shaped than Chandler's. The colour depends on climate and soil conditions. The skin remains dark green and the flesh is often strong dark red. The flesh is sweet and subacid.

Ogami (syn. **Egami):** This cultivar has performed very well in Florida. Fruit is seedy, quite large, broadly round and flat (oblate). The rind is smooth, shiny, moderate in thickness and pink to deep pink in colour. The flesh is deep pink and the pigmentation extends far into the albedo (mesocarp).

Melomas: It has small to medium sized fruit weighing 1.0-1.2 kg. The fruit is oblate to round with clean and thin peel with fine hairs on the surface and yellowish green at maturity. The pulp is juicy and aromatic and has slightly pinkish tint, good flavour and high sugar content 11-12°B TSS, low acid and no bitter taste.

Pomelit: It is a selection from Thong Dee seedlings in Orlando, USA and is similar to Djeroek Delima Kopjor of Indonesia. The tree grows vigorously, but is reported to be susceptible to branch breakage under heavy fruit load. The fruit is seedy and round with a slightly flattened base. The rind is thin, greenish yellow to yellow and smooth. The flesh is pink, tender and juicy with good flavour and fine texture atypical of pomelo. The fruit matures early and holds well on the tree.

Cocktail: It is a cross between Siamese Sweet pomelo and Frua mandarin made in Riverside, California. The trees are large and vigorous. Fruit matures in early winter and holds well on the tree. The fruit varies in size between that of orange and grapefruit and is seedy and very juicy having mandarin coloured peel thinner than found in most pomelos. The flesh is juicy and subacid with pleasant flavour.

Valentine: It is a cross between Dancy mandarin and Ruby blood orange made at California, USA, a hybrid with a Siamese Sweet pomelo. The fruits are round to somewhat pyriform, usually with a slight neck at the base. Rind colour is medium to dark yellow. It combines large size and low acidity of Siamese Sweet pomelo, complex floral taste of Dancy mandarin and juicy red pulp of Ruby blood orange.

Da Xanh: It is green-skinned and round pomelo from Vietnam. The tree is vigorous in growth. The fruit weighs 1.0-1.5 kg with 12 or more seeds. The flesh is juicy, pinkish red when ripe having 11-12°B Brix sweetness, excellent in taste and strong flavour.

CHESP-8: This is pink flesh seedling selection. The tree is medium, 4-5 m tall, spreading and prolific bearer. Leaf is 20-22 cm long and 8-10 cm wide. It flowers in June-July and the fruits mature in February–March. The fruit is spherical, yellowish and weighs 1.35 kg having dark pink flesh giving 55-60% pulp recovery. The juice has 10°Brix TSS and 0.38 per cent acidity. Fruit yield is 250 to 350 kg per tree and each fruit contains 75-80 seeds.

Sweet orange
(***C. sinensis*** (L.) Osbeck.)

The medium to large evergreen tree grows upright up to 6-15 m in height with usually rounded crown, twisted and angled twigs often with stout spines (on young seedlings). When young, the twigs have slender semi-flexible and bluntish spines in the axils of leaves. The leaves are 6.5-15 cm long and 2.5-9.5 cm wide, rounded at base with short pointed apex, sometimes slightly serrated crenate margins and have small to medium (1-2.5 cm) narrowly winged petioles. The flowers are axillary, borne singly or in clusters (racemes) of 2 to 6. The flower is 2-3 cm in diameter, pentamerous, bisexual and fragrant having 5-lobed calyx, usually white corolla, 20-25 stamens, single slender style with globose stigma and 10-14 locular ovary. The fruit is globose, subglobose, oblate or somewhat oval, 4-12 cm in diameter having solid central axis and orange or yellow and about 0.5 cm thick rind adhering tightly to the juicy, subacidic vesicles. The pulp is yellow to orange red in colour, which often remains green in the tropics. The fruit may be seedless to many-seeded. The seed is irregular in shape, usually obovoid and is white externally as well as internally. Seeds are highly polyembryonic.

The oranges fall into four groups; viz. the most important Common or Blond oranges, the acidless oranges, the blood oranges having red pigmentation in the flesh due to accumulation of anthocyanins (light blood orange and deep blood orange), and the Naval oranges each having a number of commercially cultivated popular cultivars. Like sour oranges, sweet oranges are thought to be predominantly of *C. reticulata* (mandarin) genotype introgressed with *C. grandis*. Microsatellite probe assays point to a monophyletic origin for sweet orange followed by somatic mutation and selection of desirable clones.

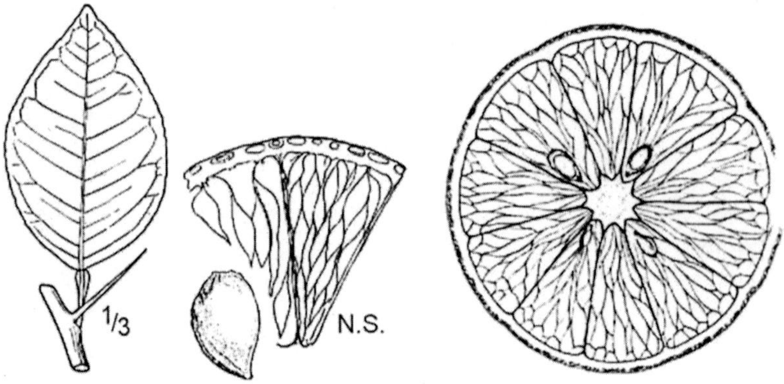

Fig. 29.21. *Citrus sinensis*: leaf and fruit sections (Source: Hume, 1957).

Common sweet oranges

Jaffa (Florida Jaffa): It is a clone of Palestine beledi seedling group. The tree is upright, medium large, quick growing having dense foliage and moderate in production but has alternate bearing. It is a mid season ripening cultivar. The fruit is obovate globose or ellipsoid and medium sized having raised basal end with radiating furrow, orange coloured pitted and pebbled rind of moderate thickness. The flesh is pale orange, tender and juicy having pleasant flavour and few seeds. The fruits do not keep well on the tree but have good shipping quality.

Joppa: The cultivar resembles Jaffa but has stiff and thornless branches and stout branchlets and starts bearing earlier and prolifically. It is a mid season cultivar. The fruit is medium in size and of globose to oblong in shape having orange coloured, medium thin and faintly pebbled rind. The flesh is light orange, soft and juicy having rich aroma and few seeds.

Foster: The tree is semi-spreading and medium in growth having dense foliage. It is a mid to late season cultivar. The fruit is medium large, roundish oblong in form, orange in colour having depressed apex and slightly collared base. The rind is thin and pitted having large oil glands and the pith is rather open. The fruit is of excellent quality, orange coloured flesh, coarse in texture having large vesicles or sacs. The flesh is very juicy with a fine blend of acidity and sweetness and rich flavour. The seeds are large and oval in shape.

Hamlin (Norris): The cultivar originated as a chance seedling and was named after the owner of the orchard, Mr. A.G. Hamlin. It tolerates cold better than most other oranges and is productive. The tree is medium large with moderate vigour. Fruits mature very early in the season. The fruit is small to medium, globose oblate and orange in colour at maturity having radially furrowed collar and a very faint areolar ring. The rind is thin with smooth finely pitted surface. The flesh is juicy, tender and orange coloured having pleasant sweet flavour and low acid. The seeds are few to none. It is an early cultivar.

Pineapple: The cultivar is a chance seedling. It is very productive. The fruits are considered excellent for processing. The tree is very sensitive to frost and is medium large with moderate vigour. It is a mid season ripening cultivar. The fruit is spherical obovate and medium sized having somewhat depressed and radially furrowed base and faint areolar ring. The rind is medium thick, bright orange and finely pitted and pebbled. The flesh is light orange, soft, juicy, sweet flavoured and moderately seedy. The two limb sports of this cultivar that are seedless are Seedless Pineapple and Blaquenmines.

Parson (Parson Brown): The cultivar originated as a chance seedling and is known for its early maturity. The trees are large, vigorous and productive. The fruits are globose, medium large and orange in colour with radial furrows at the base and very faint areole but are relatively more seedy. The rind is moderately pitted and pebbled. The flesh is orange, firm, juicy and highly flavoured.

Valencia (Valencia Late, Hart's Tardiff and Hart Late): The cultivar has wide adaptation and has tendency for heavy cropping but is alternate bearing. The tree is large, upright and vigorous. It is a late season cultivar. The fruit is oblong spherical, medium large and orange in colour. The rind is quite thick, leathery, tough and smooth and pebbled having a faint areolar ring. The flesh is orange, very juicy and sweet but with some acidic taste and good flavour and few to no seeds. The fruits hang well on the tree and have very good storage and transportation ability and excellent processing quality.

Pigmented or blood oranges

This group of oranges differs from the common sweet orange in that the fruits generally have pink or red colouration on the rind, in the flesh and juice and has a distinct flavour. Most cultivars of this group originated in Malta and Sicily of the Mediterranean basin.

Ruby (Ruby Blood): This cultivar was introduced from the Mediterranean region. The trees are not very large, compact and moderate in vigour and production. It is a mid season cultivar. The fruit is globose oblong and medium-sized having faint areolar spot, circular furrow and intense red or reddish blush and few seeds. The rind is not very thick, pitted and pebbled. The flesh is soft and juicy having rich flavour and orange colour that turns red under favourable conditions.

Spanish Sangunelli (syn. Sanguine Sanguinella Negra): The cultivar originated as a limb sport of Dobefna and has high productivity and late to mid season maturity. The tree is small to medium in size, spineless and has light green foliage. The fruit is large and very attractive and somewhat asymmetrical. The rind and flesh have intense red colour. No cultivar matches it in colouration. The surface of the rind is smooth. The flesh is sweet, firm and juicy having excellent flavour and fine texture. The fruits are seeded and have very good shipping and storage ability.

Torocco (Tarocco dal Muso, Taracco di Francofonte): It is an Italian cultivar. The tree is medium in size and irregular in bearing having moderate productivity. The cultivar has mid season maturity. The fruits are quite large, variable in shape from globose to broad to obovate, prudently furrowed and pebbled moderately and has yellowish orange collar blushed with red. The flesh is orange to red pigmented, somewhat thick, firm and juicy and adheres tightly and is rich in flavour. The seeds are few to none. The fruits do not retain quality if left for long on the tree but have good shipping and storage ability.

Dobkfina (Oval Sangre, Blood Oval): The small trees have open crown, spreading branches, poor growth and light green foliage. The trees are precocious and heavy bearer. The fruits are small to medium, oval to oblong, yellowish orange blushed with attractive rose colour and almost seedless. The rind is medium thick, leathery, firm, smooth and adheres tightly and is difficult to peel. The flesh is firm, pink tinged and juicy having pleasant taste and not very strong flavour. The fruits do not hang well on the tree but their shipping and storage quality are good.

Blood Red (Blood Red Malta): Origin of this cultivar is not known but it probably came from the Mediterranean basin. It is widely grown commercially in India and Pakistan. The tree has compact growth habit. Fruit skin is cadmium yellow and the pulp is blood red in colour with abundant juice having characteristic flavour. Seeds are numerous.

Maltaise Sanguine (Portugaise): Its origin is uncertain but it is said to be the Egyptian cultivar Baladi Blood introduced from Malta. Maltase Blood and Blood Red Malta seem to be clones of this cultivar. The trees are medium large, moderate in vigour and very productive. The fruit is oblong globose, medium in size and has well developed blood colour both on rind and in the flesh. The rind is moderately pebbled, soft and easy to peel. The flesh is soft and tender having excellent taste and flavour. The fruits have few seeds and good storage and shipping quality.

Malta (Common): The fruits have orange yellow colour, ellipsoid to spherical shape, broad and rounded base (stalk end) and apex (stigma end), medium to large size, 227 g or more in weight having smooth, finely pitted surface, indistinct stigmatic point situated in a very small circular depression, medium thick and leathery rind, 10 well defined segments, thick membranes, solid core (centre), 25 to 30 seeds, light orange and fine grained pulp, long and plump juice sacs with abundant juice of good flavour and good blend of sweetness and acidity.

Batavian orange: The cultivar is similar to Sathgudi and can hardly be distinguished from it except that rind colour is light yellowish green having pale yellow patches. It is commercially grown in South India in Andhra Pradesh and Orissa where it is rated next to Sathgudi orange. The flavour of pulp is inferior to that of Sathgudi.

Navel oranges

These are established types in Mediterranean region since centuries. Regular development of navel and seedlessness of fruits owing either to less or no viable pollen are the distinct features of Navel oranges.

Washington Navel (Riverside, Bahia, Baia or Baiana): It is considered to be a limb sport of the cultivar 'Selecta' of Bahia, Brazil. The tree is medium in size and vigour with round topped crown. The anthers have no pollen and the fruits are seedless and mature early. The fruit is large, spherical obovoid ellipsoid having projected apex with broad nipple and collared base and medium large navel. The rind is medium thick having smooth and pebbled deep orange surface. The flesh is deep orange, firm, less juicy and rich in flavour and taste. The fruits ship and store well but have poor processing quality.

Navelencia: The trees are less vigorous than those of Washington. The fruits lack colour similar to those of Washington. The fruits have paler and smoother rind having the prominent navel protuberance, mostly closed, and are almost seedless. The flesh has lighter colour (as of rind) and is more firm and juicy than that of Thompson and has good flavour. The fruits mature earlier than those of Washington but later than those of Thompson and hang well on the tree.

Thompson: It is a limb sport of Washington. The tree is compact, semi-dwarf and less vigorous than that of Washington. The fruit is seedless, medium large, globose to ellipsoid obovate in shape with open navel. The rind and flesh are less coloured. The rind is smooth, glossy and finely pitted. The flesh is more firm and juicy with good taste and flavour. The fruits ripen early.

Acidless or sugar oranges

This small group comprises very low acid cultivars resembling sweet lemons, limes or limettas. These have low acidity, sweet flavour and insipid taste and cream coloured chalazal spot on the seed, which is usually darker in acid types and more so in sweet oranges. But the Indian cultivar Mosambi, said to belong to this group, has dark brown chalazal spot. The typical acidless orange cultivars are very few. The cultivars such as Shamouti Meski and Maltaise Meski developed respectively from the common oranges Shamouti and Maltaise Blonde.

Maltaise Meski: This is an acidless cultivar originated in Tunisia. Its tree, foliage, flower and fruit characters are similar to those of the parent Maltaise Blonde.

Shamouti Meski or Shamouti Moghrabi: It is a Labonese cultivar. It is indistinguishable from Shamouti (Palestine Jaffa) in all characters except that it is acidless (insipid) and has more seedy fruits.

Mosambi (Mosambique): It is a very popular cultivar grown commercially in India. It is an early cultivar and has tree characters as those of blonde oranges (e.g. Jaffa and Shamouti). The fruit is medium large, oblate globose to obovoid and light yellow to pale orange in colour having shallow areolar ring. The rind is quite thick having faint stripes with longitudinal ridges and grooves. The flesh is light yellow and juicy with low acidity having insipid taste and few seeds.

Phule Mosambi: It has high yield potential. Fruit has thinner rind, better quality (47.37% juice) and less incidence of pests and diseases.

Sathgudi: The origin of this cultivar is not known. It is a popular cultivar grown extensively in India. The tree is vigorous and produces moderately. The fruit is medium large and globose having solid and thick axis and medium thick, tough, leathery, pitted and pebbled rind that peels easily and is moderately seedy. The flesh is orange in colour and juicy having low acidity, sweet taste and fairly good flavour. It is a mid season cultivar.

Shamouti (Palestine Jaffa, Jaffa and Chamouti): It is a mid season cultivar originated near Jaffa in Israel. The tree is upright and moderate in growth and vigour having thick and thornless branches and narrow winged petioles. It is a mid season cultivar. The fruit is medium large, oval ellipsoid with flat base, narrow collar, round apex and slightly orange coloured and is seedless or nearly so. The rind is thick, leathery, smooth and pitted and peels easily. The flesh is light orange, firm, juicy, fragrant and sweet in taste. The fruits have very good shipping and storage quality.

Marrs (Marrs Early): The cultivar is precocious, very early maturing and heavy bearer. The trees are moderate in vigour. The fruits are borne in clusters and are round to oblate and medium orange in colour having seeds moderate in number. The fruits get well coloured under favourable conditions. The rind is pitted. The flesh is orange in colour, juicy and sweet having low acidity. The fruits hang well on the tree retaining the quality. Picking should be delayed for good quality.

Sour orange
(*C. aurantium* L.)

Plants are relatively tall (10 m) having small, thin and slender thorns and twigs that are angular and with less sharp thorns when young. The leaves are alternate, simple, dark green, shiny, ovate to elliptic, large, 6.5-13.75 cm in length and 3.75-10 cm in width, broadest in the middle with faintly toothed (crenulate) margin, round to short pointed apex, 2-3 cm long petiole and broad petiolar wings. The large, white, very fragrant flowers are borne axillary, singly or in small clusters. About 5-10% flowers are staminate. The stamens are 20-25 and the ovary is 10-12-loculed. The fruit is round, oblate or oblong oval, 1.6-4 cm in size and very aromatic having bright reddish orange and thick rind with rough and bumpy surface and usually hollow core. The pulp is orange in colour and very sour to bitter in taste. The seeds are few to numerous and polyembryonic.

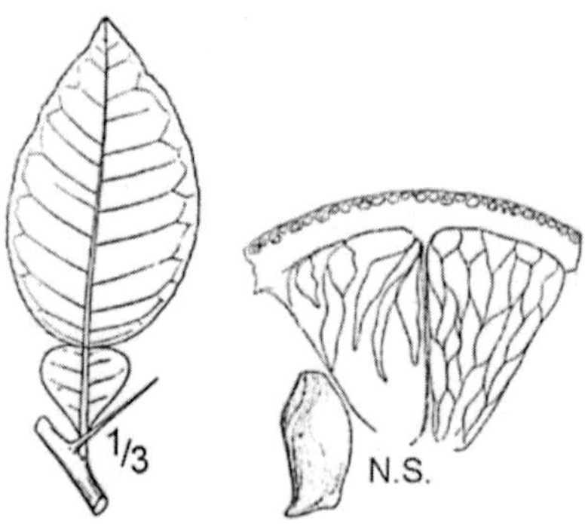

Fig. 29.22. *Citrus aurantium* (Source: Hume, 1957).

These are of three types: the common type used as rootstock and the well-known biotype Seville orange used for the preparation of marmalade; the bitter sweet orange, similar to the common type but less acidic; and the ornamental type grown for flowers from which neroli oil is extracted. The sour oranges are highly apomictic Barret and Rhodes believed sour oranges to be of predominantly *C reticulata* (mandarin) genotype introgressed with genes from *C. grandis* (pummelo). The cultivars are Common Bitter Orange and Bitter Sweet.

The bergamot orange is a hybrid grown for its distinctive rind oil used in Earl gray tea and in perfume. RAPD and SCAR markers show that bergamot is a hybrid of citron and sour orange.

Rough lemon or Jambhiri
(***Citrus jambhiri*** Lush.)

The tree is tall, erect and vigorous having robust trunk, columnar crown and upright branches. Young shoots are glabrous, light purple tinged and moderately thorny. The foliage is dense. The leaves are medium large, ovate elliptic having dull green upper surface, less green lower surface, crenate margin, notched apex, cuneate base, rather distinct venation and feeble aroma. The leaf petioles are long, usually wingless and rounded, rarely with rudimentary wings distinctly articulated with leaf blade. Flowering usually takes place only once in a year. Flowers are small, usually solitary or 2 to 3 arising from the axil of the leaf and light purple tinged. The fruits are subglobose or rather spherical having warty, wrinkled and rough surface, smooth when the fruit is fully ripe. The fruit is of medium size (about 7 cm.) with almost rounded or slightly necked base, necked apex, very prominent areole and a circular furrow with a blunt nipple sunken in the areolar area, white mesocarp, hollow, circular or irregular axis, conspicuous vascular bundles and 8 to 11 segments with little adherence and stylar scar depressed in small apical pit. The rind is medium thick, leathery and puffing with slight adherence and large oil glands. The pulp is yellow in fine, tender, slender, fusiform and melting vesicles having abundant juice with 3.74% acidity and 7.9°B TSS. The fruit contains 13 to 31 more or less ovate and highly polyembryonic seeds having dark purple chalazal spot and greenish white cotyledons. The inner seed coat is brown. Several forms of jambhiri are found in India.

Karna Khatta
(***C. karna*** Raf.)

The trees are medium tall and thorny having moderately dense foliage with open centre and drooping branches particularly when loaded with fruits. The young shoots are tinged. Leaves are medium-sized, ovate oblong, medium leathery, dark green, articulated with wingless or very narrowly winged petioles, usually serrate, very rarely crenate margins and slender and obtuse apex. The bisexual and staminate flowers are purple tinged. Fruit shape is variable, usually ovate oblong, occasionally almost spherical having smooth to slightly undulating and finely pitted surface, golden yellow to light orange colour, medium size, basal area slightly depressed and ribbed with usually very broad mamillate and occasionally obtuse apex and solid axis. The rind is thick, spongy with medium to strong adherence and medium large oil glands. It has 9 to 13 segments with medium adherence, orange coloured pulp vesicles having coarse texture, plenty of light orange juice, soft and medium rag and 20 to 30 seeds with purple chalazal cap and white cotyledons.

Subtribe *Balsamocitrinae* (7 genera, 13 species; hard shelled citroid fruit trees)

A. Tabog group: The group has one genus.

Genus *Swinglea* Merrill

It originated in Philippines having one species, *S. glutinosa* (Blanco) Merr.

B. Bael fruit group: 4 genera.

Genus *Aegle* Correa

It originated in India having one species: *Aegle marmelos* (L.) Corr.

Genus *Afraegle* (Swing.) Engler

It originated in West Africa having 4 species: *A. asso* Engl., *A. gabonensis* (Swing.) Engl., *A. milbraedii* Engl. and *A. paniculata* (Schum) Engl.

Genus *Aeglopsis* Swingle

It originated in West Africa having 2 species: *A. chevalieri* Swing. and *A. eggelingii* M.R.F. Taylor.

Genus *Balsamocitrus* Stapf.

It originated in Uganda having one species *B. dawei* Stapf.

Bael
(*Aegle marmelos* (L.) Corr.)

The bael (*Aegle marmelos* (L.) Correa) is indigenous to India. It is also known as Bengal quince. In India it is known from prehistoric times and has a great mythological significance. It is found growing wild in dry forests on hills and plains of central and southern India and Pakistan, Sri Lanka, Myanmar, Thailand, Indonesia, the Philippines, Vietnam and Bangladesh (Roy, 1990a; Verheij and Coronel, 1991). It is common in the Indian States of Uttar Pradesh, Orissa, Bihar, West Bengal and Madhya Pradesh and is cultivated on a limited scale in North India.

The tree is deciduous. It grows 6-8 m in height. The bark is shallowly furrowed and corked. The leaves are aromatic, alternate and trifoliate composed of 3 oval, pointed, shallowly toothed leaflets, 4-10 x 2-5 cm in size, terminal one with long petiole and sharp axillary thorns. The bisexual flowers are borne in clusters and are nearly 2 cm wide, sweet scented and greenish white. The calyx is shallow with 5 short and broad teeth that are pubescent outside. Five petals are thick, pale greenish white and are dotted with glands. Stamens are numerous and coherent in bundles. The ovary is oblong, ovoid and slightly tapering. The fruit is usually globose with about 3 mm thick, nearly smooth, greenish yellow, and hard pericarp filled with soft, yellow, orange, very fragrant and pleasantly flavoured pulp. The number of cells (seed cavities) in the fruit, arranged in a circle, is equal to the number of cells in the ovary. Seeds are numerous, woolly pubescent, compressed and arranged in closely packed tiers in the cell surrounded by transparent mucilage (Morton, 1987; Verheij and Coronel, 1991).

Cultivars

The important cultivars in India are Mirzapuri, Darogaji, Ojha, Rampuri, Gorakhpur, Azamati, Khamaria, Varanasi, Kaghji Gonda, Kaghji Etawah, Sewer Large and Deoria Large. Lot of variability exists with regard to fruit shape, size, shell thickness and quality in Uttar Pradesh, Bihar and West Bengal States of India. Some superior selections such as NB4, NB5, NB7 and NB9, which produce good quality fruits of medium size, having low fibre and mucilage, have been made. Three promising lines, Basti Collection-1, Basti Collection-2 and Basti Collection-4, superior in physico-chemical characters, have been identified (Srivastava *et al.*, 1998).

Narendra Bael-4 (NB-4): Trees are spreading and heavy bearer. Fruit is oblong in shape and of excellent quality.

Narendra Bael-5 (NB-5): It is a prolific cultivar. Fruits are medium in size, round with thin skull having low fibre and seed content.

Narendra Bael-7 (NB-7): Fruits are very large in size, flattened round and yellowish green in colour.

Narendra Bael-9 (NB-9): It is a heavy bearing cultivar. Fruits are medium to large in size with oblong shape having low fibre and seed content.

Narendra Bael-16 (NB-16): Fruits are elliptical round, of average weight (1.3 kg) and yellow pulp having 31°B TSS and low fibre content.

Narendra Bael-17 (NB-17): The fruits are attractive, of average weight (1.75 kg) and low fibre content.

CISH B-1: It is a mid season cultivar and the fruit matures during April-May. The trees are tall, vigorous with dense canopy, erect growth and precocious and heavy bearer. Fruit shape is oval to oblong of 1 kg average weight suitable for canning and preparation of slices.

CISH B-2: Trees are dwarf with medium spreading habit and precocious having moderate bearing. Foliage is sparse and almost thornless. Fruits are oblong to round in shape having pleasantly aromatic pulp. These are suitable for processing.

Goma Yashi: It produces good quality fruits of large size, ovate shape and greenish yellow colour. Flesh colour is straw.

Pant Aparna: The trees are dwarf with drooping foliage and almost thornless, precocious and heavy bearer. The leaves are large, dark green and pear shaped. Fruits are globose in shape with 1 kg average weight.

Pant Shivani: It is an early to mid season cultivar. Trees are tall having vigorous, dense and upright growth. It is precocious and heavy bearer. Fruit weight ranges from 2 to 2.5 kg.

Pant Sujata: The trees are medium dwarf with drooping and spreading canopy and dense foliage. It is precocious and heavy bearer. Fruit weight varies from 1 to 1.5 kg.

Pusa Urvashi: It is a mid season cultivar. The trees are tall and vigorous with dense upright growth. It is precocious and heavy bearer. Fruit is ovoid oblong having 1.5 to 2.5 kg weight.

C. Wood apple group (2 genera)

Genus *Limonia* L. (x=9)

It originated in South and South East Asia. It has one species *L. acidissima* L. found growing in India, Sri Lanka and Java. The trees are polygamous deciduous or semi-evergreen having solitary and straight spines, imparipinnate leaves, opposite leaflets and winged petiole and rachis. Infloresences are compound racemes. Flower is bisexual or male, (4)-5-(6)-merous, diplostemonous having caducous calyx, anthers longer than filaments, short disk, ovary at first incompletely 4-6-loculate and then becoming 1-loculate with numerous ovules on 4-6 parietal placentae, thick and attenuate style and oblong and fusiform stigma. Fruit is large, globose with hard and woody shell. The pulp vesicles are absent and the single cavity has numerous densely packed seeds embedded in the gum-like pulp. Seeds are oblong and compressed and the testa is minutely pilose (Kubitzki, 2011).

Wood apple
(***Limonia acidissima*** L. syn. ***Feronia limonia*** (L.) Swingle,
F. elephantum Correa; elephant apple)

The plant is native of India, Indochina and Sri Lanka and is found throughout the plains of India. It also grows wild in drier regions of Burma and Indochina. The tree small, deciduous with short, erect and cylindrical stem, 9-12 m in height and 60-120 m in girth and very thorny branches. The leaf is up to 12 cm long, imparipinnate with narrowly winged rachis and petiole. Leaflets are 5-7, opposite, odd-pinnate, usually obovate in shape, the terminal one up to 4 cm long, blunt apex, entire margin, generally small petioles, axillary and long and straight spines. The lamina is dotted with oil glands and is faintly aromatic when crushed. The flowers are polygamous (staminate and perfect), pentamerous, white, green or reddish purplish, usually borne together in lax terminal or axillary inflorescences having 10-12 stamens and cylindrical stigma. Fruit is a hard-shelled and globose berry, up to 10 cm in diameter having whitish, scurry and woody surface and filled with pinkish, acidic and aromatic pulp containing numerous slimy seeds. The seed is 5-6 mm long and hairy having thick green cotyledons (Morton, 1987; Verheij and Coronel, 1991).

Cultivars

Two types are recognized, one with small acidic fruits and the other with large sweet fruits (Morton, 1987). The selection HB-19, made at Parabhani, Maharashtra, has large fruits with higher pulp content. Considerable variability is available from which more promising selections can be made.

Genus *Feroniella* Swingle

It originated in South East Asia 3 species; *F. lucida* (Scheff.) Swing., *F. oblata* Swing. and *F. pubescens* Tan.

Genus *Casimiroa* La Llave & Lex.

It has 2 species, *C. edulis* and *C. tetrameria* Millsp. found in the highlands of South America. While the fruits of *C. edulis* are free of the unpleasant turpentine flavour, those of *C. tetrameria* (matasano), although larger and more attractive have the lingering disagreeable flavour. It is a native of the central American and Mexican highlands.

Grapefruit (Dr. Suneel Sharma)

Marsh Seedless

Thompson

Ruby

Sweet Orange (Dr. Suneel Sharma)

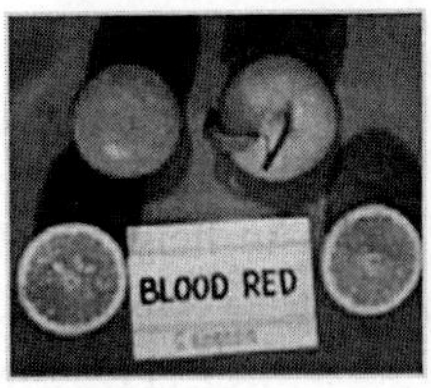

Blood Red

Jaffa

Valencia

Satgudi

Pomelo

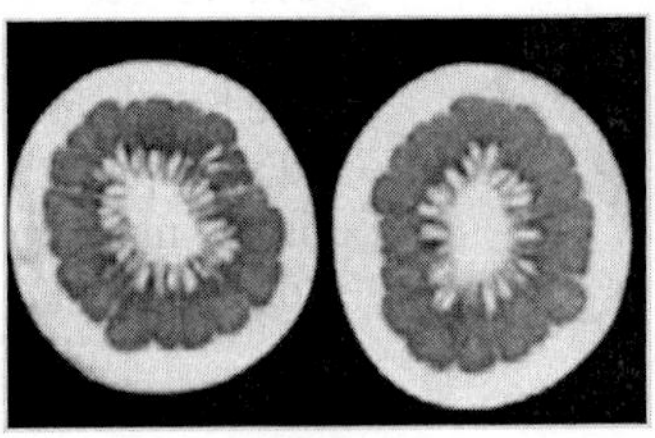

CHESP-8

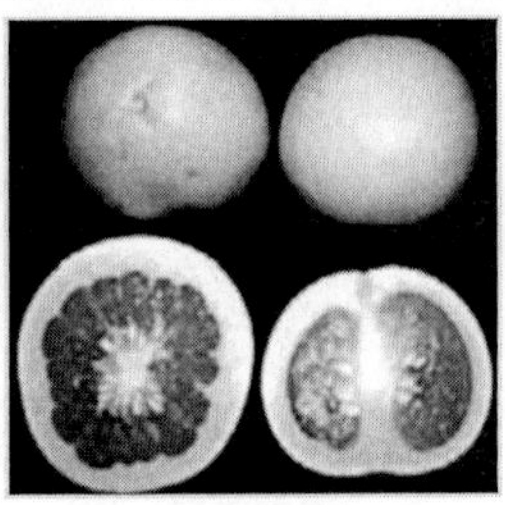

Chandler

Siamese Sweet

Mandarin

Nagpuri mandarin

Wilking

Coorg mandarin

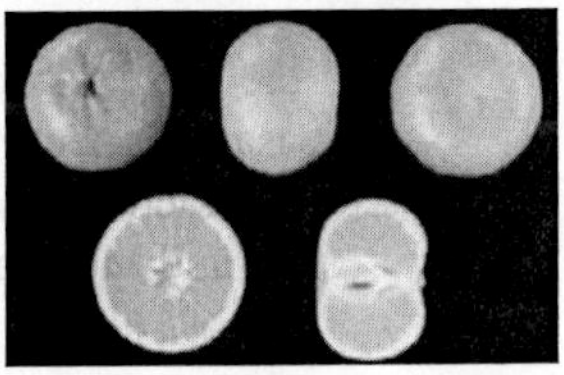

Clementine

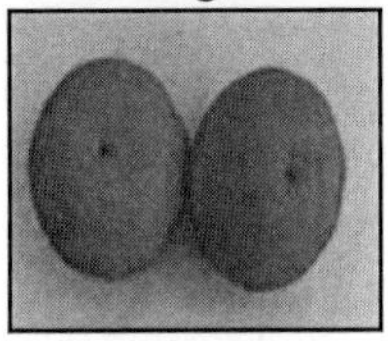

Willow Leaf

Kinnow

Lemon (Singh *et al.*, 2008; Dr. Suneel Sharma)

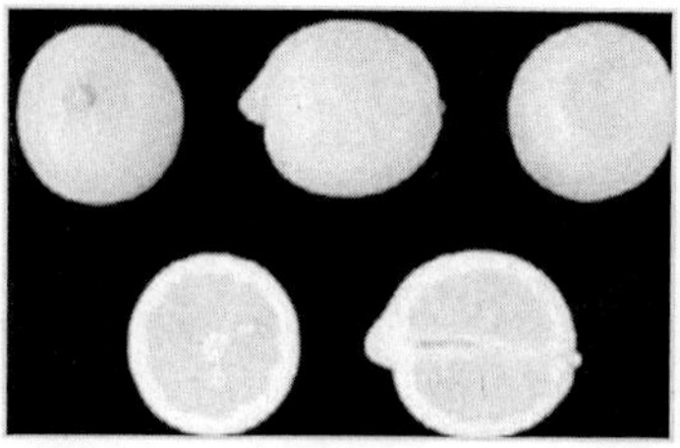

Eureka Lemon

Assam Lemon

Kagzi Lemon

Karna Khatta

Jatti Khatti

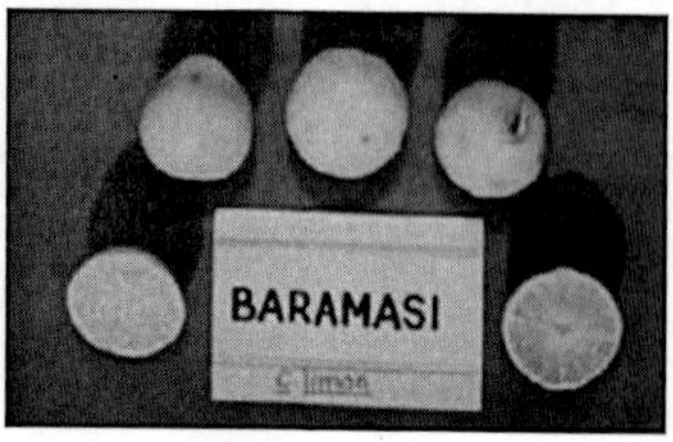

Baramasi

Rasraj

Chakradhar

Vikram

Sai Sharbati

Tahiti lime

Sweet lime

PDKV lime

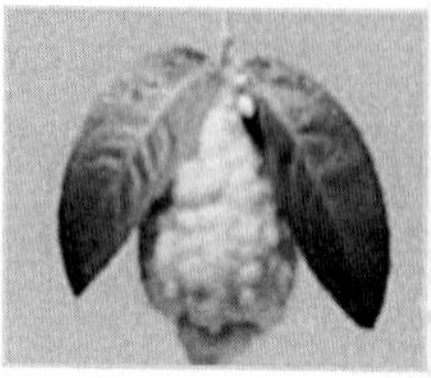

Citrus medica

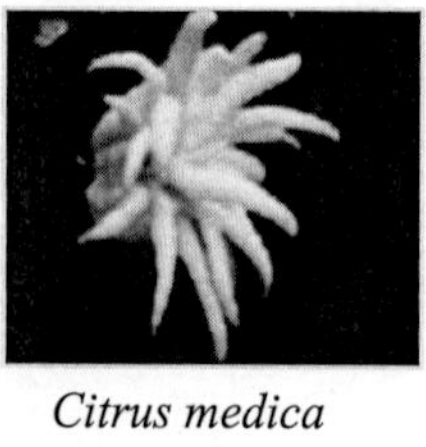

Citrus medica

Citrus macroptera

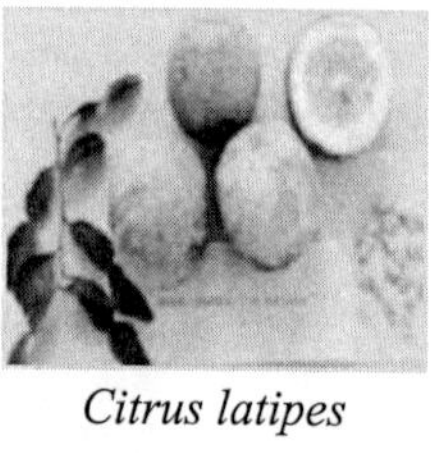

Citrus latipes

White sapote

(***Casimiroa edulis*** Llave and Lex.; casimiroa; cochil sapote; matasano)

The evergreen tree is large and attains more than 10 m height. The bark is bright green in young and ash gray in mature branches. Lenticels are prominent on young and old wood. The leaves are palmately compound consisting of 3-7 leaflets that are elliptical to lanceolate, 7-13 x 2.5-5 cm in size and have sharply or bluntly pointed apex and entire to crenulate margins. The lower leaf surface is glabrous, coppery when young and later turns glossy bright green. The flower is small, 1 cm in diameter, pale green to cream, borne in terminal clusters in the axils of mature leaves or on the bases of one year or older shoots. A panicle has 15-100 or more flowers, which are bisexual but sometimes functionally unisexual. Calyx is acute with 5, short and pubescent lobes. Petals are 5, small and oblong. Stamens are 5 and stigma is 3-5 lobed. The fruit is a greenish yellow drupe, 7-10 cm in diameter and spherical to oblate in shape. The yellowish or white and melting flesh is sweet and pleasant but sometimes have a slightly resinous flavour. Skin is thin, yellow to light green and fragile. Seeds are 1-5, oblong elliptic, 2 x 1 cm in size, compressed, normally 2 or 4, large and enclosed in a stony endocarp (Morton, 1987; Verheij and Coronel, 1991).

Cultivars

A number of cultivars with widely differing characteristics have been selected mainly in Florida and California. Important cultivars of white sapote developed in California and Florida are Blumenthol, Colemen Dade, Harvey, Pike, Suebella, Wilson, Gillespie, Golden, Maechtlen, Malthy, Parroquia, Pike and Yellow (Morton, 1987). Most of the cultivars produce only functionally female flowers due to the absence of pollen in the stamens and therefore, require pollination, e.g. Reinikie, Commercial, Golden Globe, Blumenthol, Yellow and Dade. The cultivars Ortego and Vernon that produce abundant pollen and flowers over extended period of time may prove useful as pollinizers. Cultivars are also classified on the basis of time of fruiting. Dade and Pike bloom in spring and mature fruits in autumn. Wilson, Suebelle, Lemon Gold are evergreen types and continuously bloom and mature fruits.

Intergeneric hybrids

I. Hybrids of *Poncirus*

Citranges: These are named by using parts of the words 'citrus' and 'orange'. Hybrids between *Poncirus trifoliata* and *Citrus sinensis* were obtained by USDA in 1904 in an effort to produce cold resistant stocks. In general, these hybrids show some degree of intermediacy between the two parents. Some are nearly deciduous like the trifoliate orange, whereas the others are almost evergreen. The leaves are mainly trifoliate but larger than of the trifoliate orange. Some hybrids produce unifoliate leaves that are wholly evergreen. The cultivars have different intensities of thorniness. Their degree of cold resistance varies considerably, but all are hardier than sweet orange. Those with trifoliate orange characters are hardier, but none is as hardy as trifoliate orange. The fruits range in colour from yellow to deep orange, the surface may be hairy or glabrous, rough, wrinkled, ribbed or smooth and the size ranges from as small as that of trifoliate orange to as large as that of sweet orange. The rind is thin and the pulp is juicy and highly flavoured but is sour. Most of the cultivars reproduce true-to-type owing to a very high degree of nucellar embryony. In general, citranges are too acidic and bitter to be eaten fresh, but when fully ripe may be relished with sugar. Most of the citranges are very vigorous and disease resistant and more cold hardy than any of the commonly used stocks other than trifoliate orange. Dwarfing influence on grapefruits, Satsuma, sweet orange and lemon was exhibited by citrange cultivars like Coleman, Cunnigham, Etonia, Morton, Rusk, Rustid, Sanford, Savage and Troyer produced by Swingle during 1897-1910. Cunningham is also hardy and cold resistant, which is evidently a valuable rootstock for Satsuma and other mandarins. Rusk citrange has been widely used as a rootstock for Satsuma orange.

Citrangequats: As the name suggests, these are the hybrids of the citrange with the kumquat (*Fortunella* species). Thus they are trigeneric hybrids possessing combined heritage of three species from three different genera (*Poncirus*, *Citrus* and *Fortunella*). Citranges are subject to spring frost injury to the tender young growth, while kumquats resist it because of their winter dormancy. Therefore, Swingle (1909) crossed the citranges with kumquats to incorporate this character, mainly of Nagami kumquats. However, the hybrids did not carry this character but proved promising in other respects. The Thomas-ville, the hybrid between oval kumquat (*F. margarita*) and citrange Willits proved suitable for home gardens as it makes good marmalade. Owing to immunity to citrus canker, like its parent kumquat, it was found to be a good rootstock for citrus. Telfair is also a hybrid of the above parentage while Sinton is a hybrid between the oval kumquat and Rusk citrange.

Citrangedins: These are trispecific hybrids, i.e. hybrids between citranges and Calamondin (*Citrus mitis*). The only cultivar that proved promising is Glen citrangedin (Swingle, 1909) originated as a hybrid between Calamondin (♀) and Willits citrange (♂). The fruit is valuable for making limeade and for culinary purposes. The tree is hardy.

Citrangors: It is a back cross hybrid between citrange and *C. sinensis.*

Cicitranges: It is a back cross hybrid between citrange and *Poncirus trifoliata.* These hybrids often look much like trifoliate orange in both leaf and fruit characters.

Citrumellos: These are hybrids between *Poncirus trifoliata* and *Citrus paradisi.* These are very similar to citranges. At least one of them, the Sacaton has high intensity of nucellar embryony like Troyer citrange and is a promising rootstock.

Citrandarins *(Poncirus trifoliata x C. reticulata)*: These hybrids are much like citranges except that owing to one parent being mandarin orange the amount of bitter oil in the pulp vesicles is greatly reduced.

Citremons *(P. trifoliata x C. limon)*: Highgrove hybrid is very fast growing having fruits like lemon but are rougher and more seedy. It may be used as a rootstock.

Citradias (P. trifoliata x C. aurantium): These hybrids greatly resemble citranges, but seem to be more cold resistant. The hybrid Brounell citradia makes good stock for citrus.

Citrumquats *(P. trifoliata x F. japonica* or *F. margarita)*: The citrumquat proved an exceedingly difficult hybrid to breed. This hybrid possesses the highest degree of cold resistance coming from both the parents.

II. Hybrids of *Fortunella*

Procimequat [(*Fortunella japonica x C. aurantifolia* cv. Mexican) *x F. hindsii*]: It is a complex hybrid having triploid chromosome number as a result of cross pollination of Eustics limequat with the tetraploid species *F. hindsii.*

Limequats *(C. aurantifolia x F. japonica)*: These are hybrids between Mexican lime and oval kumquat (Swingle, 1909) having fruits very similar to those of lime but are hardier than limes. Among the cultivars, Lakeland and Tavares are good while Eustis is better.

Orangequats (*C. reticulata* cv. Satsuma x *F. japonica* x *F. margarita* cv. Meiwa)

III. Hybrids of *Eremocitrus*

Eremolemons (*E. glauca* x *C.* limon cv. Meyer): These hybrids have unifoliate leaves. The Coachella eromolemon has high tolerance for boron and a greater tolerance for soil and water salinity than any other *Citrus* species.

Eremoranges (*Eremocitrus glauca* x *Citrus sinensis*)

Eremoradias (*E. glauca* x C. *aurantium*)

Citrangeremos (*E. glauca* x Citrange)

Intrageneric hybrids

Lemonange (*C. limon x C. sinensis*)

Lemonimes (*C. limon x C. aurantifolia*): Cultivar Perrine lemon was developed by Swingle.

Lemandarins (C. *limon* x *C. reticulata)*

Tangors
(*C. sinensis* x *C. reticulata*)

These were developed by safeguarded cross pollination by Swingle. These have also originated as chance seedlings in China, Japan and other eastern Asiatic countries. The King orange of Indochina is a tangor. The Temple orange of Florida is probably another tangor.

Tangelos
(C. *reticulata* Blanco x *C. paradisi* Macf.)

These hybrids look very much like oranges and are often beautifully coloured and have delicious flavour (Webber and Swingle, 1905). Similar hybrids developed in China, Indochina, Japan and other Far Eastern countries through insect pollination of mandarin oranges and pummelos. Swingle claimed that Natsudaidai of Japan is almost certainly such accidental hybrid and thus the name *C. natsudaidai* Hayata is not justified.

Of the several good flavoured tangelos, only Pearl tangelo (Imperial x Willow Leaf) proved outstanding. The tangelos Sampson, Thornton and Williams were developed by crosses between Dancy tangerine and grapefruit. Sampson has vigorous growth and therefore has high resistance to gummosis. Owing to uniform nucellar seed reproduction, it has shown great promise as rootstock in California. Orlando, Seminole and Minneola are other outstanding tangelos.

Clement: Ii is a hybrid between Duncan grapefruit and Clementine mandrarin. The tree and foliage characters are intermediate of those of the parents. The cultivar is productive and is early to medium in ripening. The fruit is medium large, subglobose to oblong and light orange to yellow in colour. The rind is quite thick, pebbled and peels easily. The flesh is dull yellow, soft and moderately sweet.

Minneola: It is a hybrid of Duncan grapefruit and Dancy tangerine. It is less cold hardy than Orlando and requires cross pollination for regular high yields. The trees are vigorous and productive. It is a mid season cultivar. The fruit is large and oblate obovate having distinct neck. The rind is medium thick with moderate adherence, deep reddish orange colour having smooth and pitted surface. The flesh is orange, juicy and tender having rich, aromatic and somewhat tart flavour.

Orlando: It is a hybrid of Duncan grapefruit and Dancy tangerine. It is early maturing like Minneola. The tree is less vigorous and more cold hardly. The leaves are cup-shaped. The fruit is medium large and broad oblate to subglobose without neck. The rind is orange in colour and thin having pebbled surface, quite tight adherence and does not peel easily. The flesh is orange, very juicy, tender and somewhat sweet and seedy.

Seminole: Its parents are same as of Orlando and Minneola. The tree is more fruitful, productive and medium large having small to medium, rounded and cupped leaves and late in fruit maturity. The fruit is broad oblate, deep reddish and usually having hollow axis. The rind is thin, pebbled, adhere moderately and is peelable. The flesh is dark orange, juicy, soft, somewhat acidic in taste and seedy.

Family Meliaceae Juss.

It contains about 50 genera and 550 species (Heywood *et al*., 2007). They are pantropical with an extension into northern and western China (*Toona*), particularly common as understorey trees in rainforests but also growing in mangroves and even in arid regions.

The plants are small to large trees and occasionally shrubs, sometimes "tuft trees," rarely herbaceous with a woody stock (*Munronia*). The bark is often aromatic, smelling sweet when slashed, produces bitter triterpenoid compounds, usually with scattered secretory cells and often reddish inner bark. The leaves are alternate, spiral, usually pinnate (rarely 2-pinnate), 3-foliate or 1-foliate and exstipulate. In the Malaysian genus *Chisocheton*, the leaf tip buds continue to produce new pinnae periodically for many years. Similar indeterminate growth is found in the American and African genus *Guarea*. The flowers are actinomorphic, functionally unisexual but appearing bisexual (plants monoecious, dioecious or polygamous) borne in cymose panicles on the trunk or branches or in the axils of the undeveloped leaves or terminal or rarely epiphyllous or cauliflorous. The inflorescences show considerable variation. The flowers have distinct sepals and petals. The sepals are (3(4-5(-8)), free or partly united. The petals are 3-5(-7), distinct or partly united. The stamens are 5-10 having free filaments (*Cedrela, Toona*) or partially to completely united into a tube. The ovary is superior, syncarpous with 2-13 carpels and 2-13 locules with axile placentation and 1-2 or more ovules per locule. The stigma is usually capitate or discoid. The fruit is a capsule, berry or drupe, rarely a nut. The seeds are one to many, winged or fleshy or unwinged and with a fleshy arillode or testa (sarcotesta).

Genus *Lansium* Correa (2n=144)

The genus is possibly restricted to Malesia and contains 3 species. The trees have simple hairs. Leaves are pinnate having sub-opposite to alternate leaflets, the most apical one appearing on side and the petiolules are pulvinate at the base. Inflorescences are spikes, racemes or more rarely basally branched panicles with spicate or racemose branches borne on twigs, branches or bole. Flowers are unisexual (tree dioecious) and bisexual. The bisexual flowers are larger than the male having deeply 5-lobed imbricate calyx. The 5 petals are free from each other but are united with the staminal tube in proximal third to half, imbricate, globose to cyathiform and ±undulate margin. Anthers are (8)10 in one whorl inside the throat of the tube, their tips slightly or not exerted, without appendages. The ovary is 3-5-locular with one ovule in each locule, long and broad columnar style and its flanks ribbed with the impressions of the surrounding anthers. Fruit is 1-5-seeded berry. Seed is usually arillate and the aril completely envelops the seed. The embryo has thick plano-convex and superposed free cotyledons.

***Lansium domesticum* Correa;** langsat; lanson; lanzon; duku, longkong (2n=44).

It is a native to South East Asia, may be southern Thailand. It is a polymorphic species. The tree is slender, tall and up to 30 m in height with hollow trunk and green bark. The leaves are pinnately compound, 30-50 cm long with 5 to 7 alternate leaflets, obovate or elliptic oblong, dark green and glossy on the upper surface and paler and yellow beneath. The flowers are produced in clusters borne in hairy spikes arising from the trunk and large branches. The flowers are sweet smelling and attractive, solitary, yellow, bisexual and subsessile. Sepals are five and 5 petals are joined to a tube of ten fused stamens and the single style carries a 5-lobed stigma. The fruits resemble loquat to some extent except that these have a dull straw or brownish yellow-coloured pubescent leathery skin. The flavour of the fruit is very pleasant. The fruits are oval, 2.5-5 cm in diameter, densely pubescent and develop in clusters of up to 25 fruits. They have yellow leathery skin that does not adhere to the flesh and contain milky white latex. The flesh consists of white, translucent, sweet and juicy aril that separates into 5 segments and may contain a green coated seed firmly attached to the flesh (IBPGR, 1986b; Morton, 1987). The reproductive and vegetative characters are very similar but few distinct types are known (Rao and Ramanatha Rao, 1998).

Duku (*L. domesticum* var. *domosticum*): It is from Thailand. It is adapted to humid tropics. The trees are spreading with dome shaped canopy, knobs on the trunk, bright green leaves and short spikes. The leaf is long, tapered and thin having oval tip and shallow veins and smooth underneath, very similar to coffee leaf. The fruits are borne in clusters of 8-12 per raceme and are larger than those of longkong and langsat. The fruit is round (6-10 cm dia.) and sweet and has less flesh, strong aroma and few seeds without bitterness and less latex at maturity. The peel is thick (6 mm), easily peeled and light yellow. The skin is rough having short hairs. The seed is long and produces only one seedling on germination. Cultivars of this group have been named after their place of origin, e.g. Du-1, Menteng, Depok, Rejoso, Long Khong and Karangkajen. The important cultivar in Malaysia is DU-1 (Nakasone and Paull, 1998).

Langsat (*L. domesticum* var. *pubescens*): It is more adaptable. The tree is slender having white striped trunk with knobs, upright branches and sparse dark green foliage. The leaf is long, tapered and thin having shallow veins, hairy and smooth underneath. The spikes are 10-30 cm long and 15-25 fruits are borne in each spike. The fruit is ovoid, smaller than those of longkong and duku and has white translucent flesh having subacid taste and contains 1-5 seeds and latex until fully ripe. The seed is small, green and germinates giving only one seedling. The peel is thin, very smooth and having soft skin with very short hairs (fuzzy) and develops pale yellow colour at maturity. The flesh is translucent and has less sweet aril and larger seeds. The flesh sticks to the seeds. The seed is small and green and germinates giving only one seedling. Cultivars of this group are Paete, Mindango and Uttaradit. The cultivar Uttarodit and Paete are from Thailand (Nakasone and Paull, 1998).

Longkong: The trunk is round and upright having little roughness and without white stripes. The leaf is large, thick, dark green and glossy having oval and wavy tip and deep veins and is smooth underneath. The fruit is large and oval and borne in clusters of 10-40 on compact raceme. The peel is thick, hard, less rough and hairless and develops light yellow colour and has no latex when ripe. The flesh is translucent, hard and aromatic having few or no seeds. It has few large, oval, yellow greenish and bitter seeds which are fertile and produce many seedlings on germination.

The smaller seedless and sweetish types are edible (IBPGR 1986b; Morton, 1987; Verheij and Coronel, 1991). The cultivars collected at Sri Wijaya University, Palembang and Sumatra are intermediates or close relatives of the two major types, duku and langsat. These have local names such as Rasuan, 1105, 1089, 1092, Komering 3, 12, 17, Palembang 175, 176, 358 and 561, Lonijang-2, Bulat and Penyandigan. The cultivars in Java are Duku Condek and Duku Metesih. Many cultivars have local names like Duku Johore and Duku Trengganu. One seedless cultivar of Thailand is Longkong Norathiwat (Rao and Ramanatha Rao, 1998).

Genus *Sandoricum* Cav. (2n =16, 22, 44, 64)

Of the 5 species, all but *S. koetjape* are restricted to West Malesia. All the five are wild in Borneo (Kubitzki, 2011). The trees have simple hairs. Leaves are trifoliate. Flowers are borne in axillary thyrsus and have ±truncate to shallowly or 5 lobed calyx, (4)5, distinct, imbricate petals, cylindrical staminal tube ribbed distally, margin with 5 or 10 short lobes, 10 glabrous anthers, tubular nectary free disk with coarsely toothed margin, 4 or 5-locular ovary slightly sunk in the receptacle, each locule with 2 collateral ovules and style head with 4 or 5-lobed stigma. Fruit is a 1 to 5-locular drupe, 1 (or 2)-seeded pyrenes, rather dry fleshy or soft and fibrous, outer mesocarp fleshy or spongy, inner mesocarp fibrous and thin and cartilaginous endocarp. Seeds are exarillate, pachychalazal, kidney-shaped, laterally compressed having thin sarcotesta with no endosperm, thick, pano-convex, collateral cotyledons and apical radicle extending to the surface or slightly exserted.

Santol
(***Sandoricum indicum*** syn ***S. koetjape*** (Burm.f.) Merr.)

Santol is native of South East Asia. The cultivated forms may have arisen in West Malesia. It is found throughout Myanmar, Vietnam, Malaysia, Indonesia, Borneo and the Philippines. It is cultivated in Thailand and the Philippines. Santol trees are semi-deciduous, quick growing and 30 m in height and have a smooth, pinkish and straight trunk. Vegetatively propagated trees tend to be smaller and have more bushy growth. The dark green leaves are alternate, compound with 3 leaflets each up to 30 cm long. The lightly scented bisexual flowers are small, greenish and arise in loose panicles from axillary buds along the branches. The 5 sepals form a small cup, the 5 petals are reflexed and separate and 10 stamens join into a tube. Pistil is with a large stigma and 5-celled ovary. The fruits are round, slightly flattened, 50 to 100 mm across with a thick, firm and downy rind and lengthwise slight indentations. The rind colour varies from dull to golden yellow, sometimes with a tinge of pink. The aril is relatively thin (up to 15 mm), white, juicy and translucent and attached with 3 to 5 seeds, which are about 15 to 20 mm long (Morton, 1987; Verheij and Coronel, 1991). The Important cultivars grown in Thailand and Philippines are Tep Ros, Peai Fai, Tab Tim, Took Thong, Tong Ta Wee and Nim Nauan.

⑧ ⑧ ⑧

Chapter 30

MALVALES

The plants are trees, shrubs, woody lianas or herbs mostly with tufted or stellate hairs or peltate scales or less often with unicellular hairs, sometimes also with glands, more rarely lepidote, exceptionally with prickles or spines. Mucilage cells, cavities or canals are usually present in the parenchymatous tissues and sometimes in the wood and roots. Vessels are with simple perforations and lateral pitting is usually alternate. Fibres are with simple pits. Rays are heterogeneous to homogeneous. Axial parenchyma is apotracheal or paratracheal. Some cells of the rays and the axial parenchyma usually contain prismatic and mostly solitary crystals. Phloem in young stem is stratified tangentially into hard and soft layers and generally has wedge-shaped rays Sieve element plastids are of S-type. Nodes are trilacunar, pentalacunar, multilacunar or rarely unilacunar. Leaves are alternate or opposite, rarely verticillate, often palmately veined, simple to palmately compound, usually stipulate and petiolar vascular system usually complex. Stomata are anomocytic or paracytic, sometimes anisocytic or encyclocytic.

Flowers are borne in various types of basically cymose inflorescences or less often solitary, mostly bisexual, usually actinomorphic, mostly 5-merous and commonly with a double perianth, often provided with an epicalyx. Nectaries usually consist of tufts of glandular hairs and are mostly borne at the base of sepals. Sepals are free or basally connate, valvate or rarely imbricate (Monotaceae except *Pseudomonotes*). Petals are free or sometimes basally adnate to the filament tube, contorted, imbricate or valvate, sometimes reduced or wanting. Nectary disc, if present, is of glandular trichomes on ventral face of the sepals, more rarely on petals or androgynophore. Stamens are basically in two cycles, five to numerous, rarely 1-2 (*Pimelea*), developing in centrifugal sequence when numerous, free or often the filaments connate into a tube or 5-10(15) fascicles, members of the outer cycle often wanting or transformed into staminodia. Anthers are basifixed or dorsifixed, tetrasporangiate and 2-locular or disporangiate and 1-locular opening longitudinally or less often by apical pores. Tapetum is secretory or amoeboid. Microsporogenesis is simultaneous. Pollen grains are 2 or 3-celled and from 3-colpate or 3-colporate to pentaporate. Gynoecium is of two to many (often five) united carpels with free or more or less united stylodia. Ovary is superior or rarely inferior with as many locules as the carpels or seldom unilocular by failure of the intruded parietal placentas to meet in the centre. Ovules are 1 to many in each locule or on each placenta, ascending or pendulous, anatropous to campylotropous and hemitropous, rarely orthotropous, bitegmic and crassinucellate. Female gametophyte is of *Polygonum*-type. Endosperm is mostly nuclear. Fruits are usually dry, dehiscent or indehiscent, less often fleshy and rarely winged. Seed coat is generally formed by both the integuments and is usually exotegmic. Embryo is straight or less often curved. Endosperm is present or wanting. Fatty acids containing a cyclopropenyl ring are often produced. Both the anatomy of vegetative organs and the palynological data confirm its origin from Flacourtiaceae. It includes 10 families and 3560 species.

Family Muntingiaceae

It is small family having about 4 genera and 4 species (Heywood *et al.*, 2007). The plants are shrubs or small trees, lacking mucilaginous cavities and have stellate, simple and glandular hairs. The leaves are simple, alternate, asymmetric and cordate at the base, serrate and palmately nerved. Stipule-like structures, usually dimorphic, occur at each leaf and are filiform or foliaceous and peltate or vestigial. The inflorescences are supra-axillary. Flowers are solitary or in clusters and are actinomorphic and bisexual. The sepals are 5, valvate and fused at the base. The petals are 5, imbricate, crumpled in the bud and are caducous. The stamens are numerous with free, filiform filaments and dithecal, longitudinally dehiscent and usually with basally attached anthers. The ovary is inferior (superior but embedded in receptacle in *Muntingia*) with 5 carpels and locules (rarely more), with pendulous placentae and numerous ovules. The style is short and thick and the stigma is capitate to slightly lobed and grooved. The fruit is a berry having numerous seeds.

Genus *Muntingia* L.

The genus has one species, *Muntingia calabura* L., widely distributed and cultivated throughout the neotropics and escaped from cultivation in the eastern tropics (Kubitzki and Bayer, 2003). The plants have stipule-like unilateral appendages, restricted to the upper side of plagiotropic shoots. The anthers are rounded and longitudinally dehiscent. Ovary is surrounded by a disk-like elevation of the receptacle. Ripe fruits are red. Seeds are embedded in the pulp.

Jamaica cherry

(***Muntingia calabura*** L.; Panama berry; Japanese cherry; jam fruit; capulin)

It is a neotropical species which has become pantropical. It is cultivated all over the tropics. It grows on roadsides all over South East Asia where it was introduced in the 19th century.

It is a very fast growing evergreen slender tree reaching 7.5-12 m in height with spreading and nearly horizontal branches. The leaves are alternate, lanceolate or oblong, 5-12.5 cm long, dark green and minutely hairy on the upper surface, gray or brown hairy on the underside and irregularly toothed. It produces flowers continuously on fan like branches, which become erect after leaf fall. The horizontal branches are pendent towards the tip. Inflorescences emerge along with the growing shoot subtending the leaf. The flowers are borne in groups of 1-3 in supra-axillary fascicles. These are hermaphrodite, pentamerous with white petals, 10-25 stamens in the first emerging flower and more than 100 in the last fascicle. The fruit is a dull red berry, 15 mm in diameter with several thousand tiny seeds in the soft pulp. The fruit matures shortly before the leaf falls (Morton, 1987; Verheij and Coronel, 1991). Yellow and white fruited types are known. There is a lot of scope for selection of superior types from the natural variability.

Family Malvaceae (Cotton, Mallows and Hollyhocks)

The family Malvaceae, or the mallows, based on the molecular techniques, has been expanded to include all the commonly recognised families which were always considered closely allied to it to compose a monophyletic group. Adopting this circumscription, the family is estimated to contain 243 genera with 4225+ species incorporating 9 subfamilies (Byttnerioideae, Grewioideae, Sterculioideae, Tilioideae, Dombeyoideae, Brownlowioideae, Helicteroideae, Malvoideae and Bombacoideae). Representatives occur in all except the coldest parts of the world but are most numerous in the tropics.

The plants are herbs, shrubs and trees, both evergreen and deciduous. They have fibrous bark and produce mucilage when wounded. Some *Brachychiton* and *Hildegardia* have inflated trunks (bottle trees). Leaves are generally alternate, often palmately lobed or compound and palmately veined. The margin may be entire, but when dentate, a vein ends at the tip of each tooth (malvoid teeth). Stipules are present. The stems contain mucus canals and often also mucus cavities. Hairs are common, and are most typically stellate. The flowers are commonly borne in definite or indefinite axillary inflorescences, which are often reduced to a single flower, but may also be cauliflorous, oppositifolious or terminal. They often bear supernumerary bracts. They can be unisexual or bisexual, and are generally actinomorphic, often associated with conspicuous bracts, forming an epicalyx. They generally have five valvate sepals, most frequently basally connate, with five imbricate petals. The stamens are five to numerous, and connate at least at their bases, but often forming a tube around the pistils. The pistils are composed of two to many connate carpels. The ovary is superior, with axial placentation and capitate or lobed stigma. The flowers have nectaries made of many tightly packed glandular hairs, usually positioned on the sepals. The fruits are most often loculicidal capsules, schizocarps or nuts. Pollination is mainly by bees. Seeds are sometimes have hairs or are arillate and occasionally winged. Embryo is straight or curved. Endosperm is present often with cyclopropenoid fatty acids.

Genus *Grewia* L. (n = 9, 18)

It has about 280-300 species in Africa, Madagascar, Indo-Malesia, warm temperate China, Australia (Queensland) and the western Pacific (Marianas, Fiji, Samoa, New Caledonia), preferentially in semi-arid forest and woodland, often in secondary and rarely in the temperate zones (Kubitzki and Bayer, 2003).

The plants are trees, shrubs or climbers. Leaves are dentate. Inflorescences are solitary or fasciculate flower triads with entire involucral bracts. Flowers are 5(4)-merous and hermaphroditic or rarely polygamous or dioecious in plants having petal limb usually longer than gland. The gland is barbellate at the margin. The sepals are distinct and glandular at the base. Androgynophore is present. The stamens arise from the apex of the raised torus. Stamens are numerous and free. Ovary is (1)2-3(4)-locular, with 2-12(-40) ovules in each cell. Style is cylindric or dilated at apex with dilated or lobate stigma. Fruit is dupaceous, entire or 2-4-lobed, ±fleshy with 3-4 (-16) pyrenes each containing 1(2) seeds.

Phalsa
(***Grewia asiatica*** L. syn. ***subinaequalis*** DC)

It is a deciduous bush but can grow into a tree as high as 10 m if left unpruned. The stem is hardy, woody and glabrous with rough bark. The leaves are 7.5 to 15 cm long, dull green on the upper surface and light green beneath, petiolate, stipulate, orbicular or broadly ovate having reticulate venation and are 5-nerved at the base. The young leaves are softly tomentose on both upper and lower surfaces becoming less hairy and rough with maturity. The stipules are subulate. The buds are elongated. The inflorescence has been described as cymose, but critical observation indicates it to be irregularly corymbose. The yellow flowers are borne in axillary clusters and have 2-7 peduncles each having 3-6 pedicellate flowers. The flower is 1.9 cm in diameter, bisexual, pentamerous, yellow, having 1 cm long pedicel, 5 oblong, 1-1.5 cm long sepals, 5 obovate, 6-8 m long and yellow petals, 70-80 free stamens, 4 to10-locular ovary and irregularly lobed stigma. Fruit is a globose drupe, 1.8-2.2 cm in diameter, indistinctly lobed, red or purple and finely warty with stellate hairs. The flesh is fibrous, greenish white, stained purple red having 1-2 hemi-spherical 5 mm wide seeds (FAO, 1982; Morton, 1987; Verheij and Coronel, 1991). There is no distinct cultivar of phalsa. Some selections have been named as Local, Sharbati, Tall and Dwarf. Dwarf type is more productive and juicy.

Durionaceae (now included in Malvaceae)

The plants are tropical trees of lowland evergreen forest in South East Asia. These are often tall with an indumentum mainly of lepidote scales, but stellate hairs also occur. The leaves are alternate, simple, usually oblong, pinnately nerved and entire, the lower surface densely covered in scales (sparsely in *Neesia* and also mixed with stellate hairs), pulvinate petioles and generally caducous and linear stipules. The inflorescence is usually axillary, of 1-2 flowers, but often fasciculate or sub-fasciculate, ramiflorous or cauliflorous. The flowers are actinomorphic and bisexual. An epicalyx is present, usually entirely enclosing the flower in the bud and splitting into 2-5 lobes. The calyx itself is shallowly to deeply divided into 5 valvate lobes. The corolla is calyptrate in *Neesia* and *Coelostegia* but otherwise is divided into 5 free linear to spathulate petals. An androgynophore is absent. The stamens are numerous, either free, united at the base or united into 5 phalanges that are either free or connate into a staminal tube, the filaments becoming free at various heights. The anthers are sometimes dithecal and longitudinally dehiscent, but are more usually monothecal and borne in clusters, dehiscing by terminal pores or longitudinally. The pollen is sub-oblate, 3-colporate, lacking spines but with low, large and smooth verrucae. The ovary is superior or partly embedded in the receptacle, sub-ovoid or slightly 5-angled, clothed in peltate scales (except *Neesia*), 5-locular with 2 to numerous axile anatropous ovules in 2 ranks in each locule. The style is usually well developed and the stigma is peltate or capitate. It is speculated that some species of durian are bird pollinated. The fruits are globose to ellipsoid, spiny to sub-spiny, loculicidally dehiscent capsules, each of 5 locules with 1 to several large raphe, funicular arillate seeds. Rarely the fruits are indehiscent or the seeds are exarillate. It contains 6 genera and 45 species (Heywood *et al.*, 2007).

Genus *Durio* Adalts (2n = 28, 56)

The trees are usually buttressed, occasionally with pneumatophores. Leaves are abaxially with peltate scales and stellate hairs. Inflorescence is axillary, sometimes ramiflorous or cauliflorous, epicalyx completely enclosing the floral bud, splitting into 2 lobes. Sepals are ±fused. Petals are (4)5(6), longer than calyx and sometimes clawed and deciduous. Stamens are numerous, distinct or in 5 phalanges, sometimes forming a staminal tube. Filaments are with numerous, longitudinally dehiscent locules. Ovary is (3-)5(6)-locular with 2 to several ovules per locule having usually pubescent style and small and globose stigma. Fruit is usually capsular, occasionally indehiscent with pericarp covered with slender or stout spines, glabrous within. The seeds are usually arillate with flat and thick cotyledons.

There are 30 known species of genus *Durio*, 19 originated from Kalimantan, 11 from West Malaysia and 7 from Sumatra. Out of these only 8 species produce edible fruits. These are *D. zibethinus* (durian), *D. dulcis* (durian maragang, lahong), *D. grandiflorus* (munjit), *D. graveolens* (durian merah, kuning, tabelak), *D. kutejensis* (durian kulu, pulu, lai), *D. lowianus* (chaarian), *D. oxleyanus* (durian sukang, keratogan) and *D. testudinarum* (kura-kura). All the species are restricted to Malay Peninsula, Borneo and Sumatra except one from the Philippines and the cultivated durian from Ceylon and India to New Guinea and the Philippines (Kubitzki and Bayer, 2003).

In Malaysia, 24 species have been identified and described based on leaf, flower and fruit characters (Salma, 2000). Two new species, viz. *D. johoricus* and *D. perakensis* have been named. *D. acutifolius* (durian antu, durian burong, durian enggang, durian loiang, durian lahung, lai hengang, lai kayo, lai koko, mehawan, tupaloh, tuwola pumpulu) has been merged with *D. griffithii* as *D. griffithii* var. *acutifolius*. Three new varieties have been recorded i.e. *D. singaporeansis* var. *jerangauensis*, *D. grandiflorus* var. *tomentosus* and *D. testudinarum* var. *crassifolius*. Inter and intra-specific variation of these species in fruit quality such as aril thickness and colour, taste, creaminess and flavour provide a vast array of genetic resources for improvement. Many of these have been observed to have resistance against pests and diseases besides some ecotypes have tolerance to ecological conditions. *D. lowianus* is resistant to *Phytopthora* stem canker has been recommended as a rootstock for commercial clones (Mohd. Shamsudin *et al.*, 2000). *D. carinatus* thrives on peat swamps and may be suited as a rootstock in wet areas while *D. lanceolatus* is adapted to sandy soil and has tolerance to drought. *D. kinabaluensis* is adapted to high altitude and hill slopes above 1000 m above sea level (Salma, 2000). However, among these species, *D. zibethinus* is still of economic importance and is grown commercially in Malaysia and many other countries of South East Asia.

Durian

(*Durio zibethinus* Murr. syn. ***D. foetida*** Thunb., ***D. acuminatissima*** Merr.) (2n=56)

Durian is highly valued and most desired seasonal fruit in South East Asia. It is native of Malaysia and Indonesia. It has spread to Australia, Cambodia, Dominica, India, Myanmar, Papua New Guinea, Philippines, Singapore, Sri Lanka, Thailand and Vietnam. Although it is planted on a small scale throughout the tropics, commercial production is mostly in Thailand, Malaysia, Indonesia, Philippines, Vietnam, India, Australia and Zanzibar. Durian is strictly tropical and thrives well in areas with warm and humid climate. A warm and moist condition is ideal. The optimum temperature is 25-30°C with 80% relative humidity and an evenly distributed annual rainfall of 1500-2000 mm limits losses from root rot. Although it cannot tolerate a prolonged dry period, a relatively dry spell stimulates and synchronizes flowering. It grows successfully near the equator, and up to 18° from the equator in Thailand and Australia. It is found mainly in lowland forests of South East Asia (mainly Malaysia). Durian production is the best from sea level to about 213 m elevation.

The evergreen tree has straight trunk and is rounded, large and buttressed and grows upright up to 45 m height in the forests but is rarely more than 20 m in orchards. It has irregular dense crown with dark gray bark. From 72 to 87% of the roots remain in the top 45 cm soil within 85% of the canopy radius. The leaves are unifoliate, elliptic obovate, alternate and 6-25 x 2-91 cm in size. They are dark green, glossy and coriaceous on the upper surface and silvery or bronze and tomentose with a layer of stellate hairs on the lower surface.

The plant has a long juvenility period of about 5-10 years. In Puerto Rico, flowering takes place in April-May and fruits ripen in August-September. The flowers are perfect, malodorous, whitish to golden brown, large, 5 to 7.5 cm in diameter surrounded by an involucre having a strong odour. These are borne in huge inflorescences in clusters of 3-30 directly on the main branches. The calyx is composed of 2 parts; the exterior 2-3 free sepals and the interior 3-5 broadly cut fused sepals. The corolla consists of 4-5 free petals that are white streaked with green. Some cultivars have red petals. The stamens are grouped in 5 lobes bearing up to 12 reniform anthers, single large orange or yellow stigma and 5-celled ovary. The flowers open at about 15.00 hrs and remain open until dark. The stigma remains receptive overnight and it is believed that small bats pollinate them. Durian trees tend to be self incompatible. Therefore, two or more cultivars have to be planted in close proximity to ensure cross pollination and good fruit set. The tree yields 10-40 fruits in 1st year reaching up to 200 fruits after 10th year. The average per tree yield is about 50 fruits per year, each fruit weighing from 1.5-4 kg.

The fruit looks like a jackfruit except that the spines are triangular. The fruits are large, 15-30 cm long, round to oblong in shape with 5 segments having olive green or yellow and woody and thick exocarp fully covered with hard, sharp and hexagonal spines on large and long peduncle making difficult to handle the fruit. The shell splits into five segments when ripe, exposing 1-7 light brown seeds enclosed in the sweet, aromatic and creamy whitish, yellowish or pinkish aril. The large brown edible seeds are 25 cm in size (FAO, 1982; Martin *et al*., 1987; Verheij and Coronel, 1991).

The very fragrant aril can be overpowering when indoors. Its taste is sweet and spicy like that of custard or vanilla ice cream with overtones of spices, onions and banana. It is best described as smelling like hell and tasting like heaven. The British naturalist Alfred Russel Wallace described its flesh as "rich custard highly flavoured with almonds". The durian fruit consists of only 26% of the edible portion that is aril. The ripe aril is usually eaten fresh (right after opening the fruit) or frozen, but it is also used to flavour ice cream, cookies and candies or preserved by canning, freezing, drying, boiling with sugar, fermenting or salting. It is also used to prepare sauces and relishes, jam, chips and soup. Young shoots and unripe fruit may be boiled and eaten as vegetable. The seeds are claimed to be edible when boiled or roasted although they contain a poisonous substance that causes shortness of breath. Several parts of the tree have medicinal properties.

Cultivars

There are more than 100 cultivars of durian. A good durian cultivar must have (a) good aril recovery of 30% or more; (b) yellow to deep yellow, firm, creamy aril; (c) small seed; (d) high and consistent yield of 70-100 fruits/tree; and (e) resistance to major pests and diseases. There is considerable genetic diversity in terms of fruit size and shape, skin colour and thickness, aril volume, colour, aroma, texture, flavour, seed number and size and peduncle length.

Malaysian cultivars are generally smaller than the Thai. Many have much thinner spikes and usually the flesh around the seeds is not as thick as in Thai cultivars. Though these usually have a more rich flavour than in most Thai types and may have highly variability in fruits of the same tree.

D-2 (Dato Nina): The fruit has thick, golden yellow pulp of excellent flavour. It is poor yielder.

D-7 (Repok B-2): The cultivar has yellow, creamy pulp with a pleasant flavour but is poor in keeping quality.

D-10 (Durian Hijau): The cultivar has characters similar to D-7.

D-16: The cultivar has light yellow creamy pulp and is a good yielder.

D-24: This is an important commercial cultivar in Malaysia. The tree bears around 150 fruits that are ovoid, sweet, each of 1-2 kg having thick, creamy golden yellow pulp and large seeds.

D 99 (Kob): It has round fruit of 1-1.4 kg with thick, soft, sweet yellowish pulp and is high yielding (around 130 fruits/tree/year) and tolerant to drought and *Phytophthora palmivora*.

D 123 (Chanee, Gibbon): It has ovoid fruit of 2-4 kg with thick, soft, sweet golden pulp. This is an important commercial cultivar in Thailand.

D 158 (Kan Yao): It has ovoid fruit of 2-4 kg with sweet, thick and golden pulp.

D 159 (Mon Thong, Golden Pillow): The fruit weighs from 2 to 5 kg and is slightly elongated with a pointed tip. The flesh is thick, golden yellow, sweet and has a mild aroma. The edible portion is 30%.

D 169: The fruit has elongated ovoid shape, 1-2 kg weight, yellow, thick, sweet and soft pulp.

The other important cultivars in Malaysia are D-98 (Katoi), D-114, D-116 (Batu), D-117 (Gom-bak) and Hew 3.

In Thailand, the popular cultivars are Geanyao, Chanee, Chompoorsi, Kan Yao (long stalk), Mon Thong, Kob, Golden Pillow, Chanee, Deception, Frog, Gibbon, Kradum Thong, Chompoosri and Luang (Page, 1984; Morton, 1987; Verheij and Coronel, 1991).

Deception: The cultivar is one of the most popular in South East Asia especially in Thailand. The fruit has lesser edible content than the impression provided from its very large size. The fruit quality is very good.

There are ten recommended cultivars grown in Philippines, two of which are from Thailand (Chanee and Mon Thong). The local cultivars include Alcon Fancy, Arancillo, Atabrine, Duyaya, GD 69, Lacson Uno, Mammer, and Puyat. Their weights range from 0.5 to 5 kg and the flesh has different shades of yellow.

Alcon Fancy: The fruit is round, weighing about 0.5 to 1.5 kg. The flesh is canary yellow, sweet and has slightly bitter taste and mild aroma. The edible portion is 34%.

Arancillo: The fruit is ovoid, weighing 0.65 to 2.0 kg. The flesh is mimosa yellow, firm, sweet and creamy and has a mild aroma. The edible portion is 38%.

Atabrine: It is a local selection. The cultivar has ellipsoidal fruit, weighing up to 3 kg.The flesh is maple yellow, sweet and very glutinous with slightly bitter taste. The edible portion is 25%.

Duyaya: This local cultivar has oblong fruit, weighing from 2 to 4 kg. The flesh is thick, bright yellow, sweet and has a mild aroma. The edible portion is 32%.

GD 69: The fruit is obovate and large, weighing 2-2.5 kg. The flesh is yellow orange, firm, thick and has mild aroma. The edible portion is 45%.

Lacson Uno: This local cultivar has obovate and large fruit, weighing 3-4 kg. The flesh is yellow, firm and sweet and has mild aroma. The edible portion is 38%.

Mamer: The fruit is ellipsoidal and weighs from 2 to 3 kg. The flesh is barium yellow, sweet and glutinous. The edible portion is 25%.

Puyat: The fruit is big and cylindrical, weighing 2-4 kg. The flesh is chrome yellow, firm, sweet and buttery and has a mild aroma. The edible portion is 40%.

The important cultivars of Indonesia are Sunan, Sukan, Hepe, Mas, Sitokong, Petruk, Sitokong, Bakul, Mas, Sitebel, Simanalagi and Simadat. Of these, Petruk, Sunan, Sitokong and Simas are popular.

Other cultivars

Mawn-Tawng: The fruits are of large size, somewhat more elongated with a pronounced beak at the base. The fruit colour is golden to yellowish brown on maturity and the flesh is creamy yellow, sweet and juicy. It has pleasant characteristic flavour of durian. Individual fruit may weigh 2-5 kg.

Gahn-Yao: It is rated among one of the best cultivars. Fruits are large (2-5 kg) and globular in shape having greenish brown colour at maturity. The flesh is creamy yellow, sweet and firm but becomes soft and mellow as the fruit ripens.

Golden Pillow: The cultivar is grown throughout Thailand. It is rated very high for its quality characters but is not very prolific bearer. The fruit is medium to large in size and elongated globular having greenish surface with light brownish tinge. The pulp is creamy yellow and is very sweet. It is considered among the sweetest cultivars.

Gra-Doom-Tawng: The cultivar produces globular and somewhat small fruit weighing about 1.5-3 kg. It has greenish brown surface at maturity. The flesh is bright dark yellow and firm. It softens later and is sweet and of very good quality.

Soft Flesh: The pulp is soft and very juicy with taste varying from insipid (flat) to sweet acidic to very sweet. It yields to the pressure of fingers when ripe.

Firm flesh: The flesh is firm, hard and crisp and its taste varies in sweetness.

Singapore or Ceylon Jack: It is an early fruiting cultivar. The fruit size is as that of the common jackfruit.

Rudrakshi: The fruits are round having smoother rind compared to that of jackfruit. The pulp is fleshy and of inferior quality.

***Quararibea* Aubl.** (2n = 88, 160)

More than 20 neotropical species are known, some of which have edible fruits. The trees are deciduous, often with swollen trunks. Trees have 4-5 verticillate branches, occasionally lepidote. Leaves are 3-11-digitate, simple and usually elliptic. Flowers are rarely zygomorphic, usually leaf-opposed, exceptionally cauliflorous having cupular to long tubular and occasionally ridged calyx, radially symmetric lobes of staminal tube, up to 4 mm long, each with (4-)6-8(-15) (sub-) sessile, monothecal anthers, superior to almost inferior, 2(-4)- locular and 2(3)-ovulate ovary and capitate lobed stigma. Fruits are drupaceous having fleshy fibrous 1-2-seeded mesocarp, scarce or no endosperm and fleshy cotyledons (Kubitzki and Bayer, 2003).

Matisia

(***Quararibea cordata*** (Humb. & Bonpl.) Vischer Garc. et Hern. syn. ***Matisia cordata*** Humb. & Bonpl; chupa-chupa; South American sapote; zapote chupa; sapote colombiano)

Matisia or chupa-chupa is a native of the hot lowland forests of the lower Andes in Colombia, Venezuela, Ecuador and Peru. It is found growing in western Amazonia, Cauca and Magdalena valleys and Colombia (Morton, 1987). This fruit of the Amazon has potential to be developed as food crop beyond its native area (Martin *et al.,* 1987; Leaky and Newton, 1994).

It is a fast growing erect tree of 20 m height with a relatively small and spreading crown and pagoda shaped branching. Bark is smooth and light brown gray in colour. Leaves are alternate, simple, cordate, 50x40 cm in size, coriaceous and glabrous. Flowers are bisexual and emerge from mature branches in groups of 3 to 10 and are 6 to 7 cm long. The calyx is large and green occupying about three-eighth of the total length of the flower, consisting of 5 sepals. The corolla, consisting of 5 creamy petals, is slightly larger than the calyx. Stamen filaments are fused to form an elongated column protruding beyond the petals and branches above into 5 spreading arms each bearing a series of paired anthers. The ovoid fruits are up to 12 cm long and 8 cm wide with persistent calyx. The fruit skin is hard, thick and green brown with woody surface and fibrous and yellowish flesh containing 5 big seeds that are connected with thin strings. Its flavour is sweet and pleasant and seems a composite of carrot, melon, mango and papaya. Seeds are angular bean shaped, 2-4 cm long and without endosperm (Page, 1984; FAO, 1986; Morton, 1987; Brücher, 1989).

Genus *Adansonia*

The genus *Adansonia* contains 8 species, 6 endemic to Madagascar, one to mainland Africa and one to northwestern Australia (Sidbe and Williams, 2002). Pollination is known to be carried out by bats, birds and moths.

Baobab

(***Adansonia digitata*** L.; monkey bread; baobab tree; cream of tartar tree)

It is native of tropical Africa and is distributed in Australia, Oceania including Papua New Guinea and eastern, southern and tropical West Africa (Martin *et al.*, 1987). In South East Asia, it is cultivated in Java and the Philippines. It has been introduced in India by seafaring traders.

The tree is erect, medium sized, thornless, long-lived and deciduous and grows up to 20 or 25 m in height and 2-7 m in girth with a relatively short and swollen trunk and spreading branches. Bark is smooth and grayish, often with purplish tinge. The shape of trunk can be cylindrical, bottle shaped, gnarled or tapering. Leaves are digitately 5-7-foliate. Leaflets are three in young plants and five or seven in older plants, which are 5x12.5 cm in size and obovate oblong or lanceolate in shape, dark green above and sparsely pubescent on the lower surface. Flowers are hermaphrodite, solitary, and 15 cm in diameter and are borne in long axillary peduncles. These are deeply lobed, leathery and silky hairy having 5 narrow petals, many white stamens with purple dot like anthers spreading 5 cm across from a 4 cm high tube. Pistil has a bristly stigma and 7-10-celled ovary with slender style. Fruit is a globose, ovoid or oblong cylindrical pod, 20-30 x 10 cm in size, woody, gray with soft yellowish felt outside and having farinaceous, whitish or yellowish and sometimes pink tinged pulp. Seeds are reniform, shining brown or blackish with thick testa (FAO, 1982; Duke, 1989; Sidibe and Williams, 2002).

Genus *Pachira* Aubl. (2n = 72, 88, 92)

The genus contains 144 neotropical and 150 palaeotropical species. About 50 species, mostly neotropical and 6 species occur in tropical Africa (Kubitzki and Bayer, 2003).

The trees, sometimes deciduous, are spiny and/or buttressed. Leaves are 3-11-digitate with leaflets having basal suture, sometimes petiolate, entire or serrate. Flowers, solitary or 2-3, are axillary, 7-35 cm long having cupular to tubular and truncate to lobed calyx, often with glands outside, persistent, sometimes accrescent, spathulate to linear petals, exceeding the calyx before anthesis. Staminal tube is with long and free filaments in distinct phalanges, 90-1000, monothecal stamens, 5-locular ovary with numerous ovules and lobed stigma. Capsules are often large, (sub-) lignose with copious kapok or velutinous endocarp and 25 to many seeds.

Malabar chestnut

(***Pachira aquatica*** Aubl.; Guiana chestnut; saba nut)

It is a native of tropical America. It appears to have been domesticated by the Indians in Central America. It is grown in Congo and is also found in Panama and Paraguay (Brücher, 1989). The related edible species are *Pachira macrocarpa* Walp. (cayenne nut), growing in Mexico and Costa Rica and *P. insignis* Sav. (maranno nut), known as shaving brush tree and found in Trinidad, Tobago and Venezuela.

The tree is evergreen growing up to 15 m in height and has buttressed trunk. Leaves are alternate and palmate (digitate) with 6-8 leaflets, elliptic lanceolate, strongly glabrous and 5-28.5x2.4-14.5 cm in size with acuminate to rounded apex. Fragrant bisexual flowers are 17.5 cm long having greenish white to pink petals and with many red stamens. The fruit is large, oblong ellipsoid, 30 cm long, 12.5 cm in diameter with brown skin and flesh having 10-25 seeds per capsule of 1 cm diameter.

Genus *Cola* Schott. & Endl.

It is a genus of trees native to the tropical rainforest of Africa. Species in this genus are sometimes referred to as Kola tree or Kola nut for the caffeine containing fruit produced by the trees that is often used as a flavouring ingredient in beverages. The genus is related to the South American genus *Theobroma*. The trees are evergreen growing up to 20 m tall with glossy ovoid leaves up to 30 cm long and star shaped fruit.

Kola nut

(***Cola acuminata*** (Beauv.) Schott. and Endl.)

Cola nut or kola nut is native of tropical Africa (Martin *et al.*, 1987). The nuts are source of the stimulating component in "Coca Cola". It has been widely introduced into the tropics. The related species having edible fruits are *C. nitida* A. Chev. (bitter cola) and *C. verticillata* (Thonn.) Stapf. ex A. Chev. (owe cola). It is cultivated in Nigeria, tropical America and Indonesia.

It is a long-lived and evergreen tree having smooth green bark growing up to 14-20 m tall. The leaves are alternate, 16-20 x 2.5-5 cm in size, leathery, obovate, acute and long acuminate with entire margin. Flowers are yellow, numerous, unisexual or bisexual, 15 or more and are borne in axillary or terminal panicles. Corolla is absent and the calyx is petaloid, 5-lobed, greenish yellow or white with purple markings on the edges. Male flowers have slender corolla, shorter than calyx, bearing a ring of 10 bilobed, 5-celled ovary and 5, linear, reflexed and superimposed styles. Ovules are anatropous attached in double row to the ventral surface of each carpel. The fruit is oblong obtuse, rostrate, warty and coriaceous to woody follicle, 5-17 x 5-7.5 cm in size with brown surface resembling the alligator skin. The thick and fibrous pericarp has cells filled with coloured, starchy and resinous matter. Each fruit has 5-12 seeds. The seed is 2.5-5 x 1.3 cm in size, soft, yellow and internally whitish, pinkish or purple becoming brown on drying. The cotyledons, often 3, are flatly ovate or auriculate (Duke, 1989).

Genus *Theobroma* L. (2n = 20)

The genus contains 6 sections with 22 species distributed in Central and South America in evergreen forests. The trees have pseudowhorled plagiotropic branches. Leaves are dimorphic spiral on orthotropic and distichous on plagiotropic shoots, simple or weakly lobed, sometimes without prominent basal veins. Inflorescences are axillary to cauliflorous. Calyx is sometimes splitting into 2-3 lobes. Petals are deeply cucullate with apical appendage (lacking in 2 spp.). Stamens are fused in groups of 2-3 forming a short tube with the prominent staminodes. Ovary is 5-locular with numerous ovules. Fruits are indehiscent, carnose to ±lignified and many-seeded. Seeds are embedded in sweet endocarp pulp. These have exalbuminous and corrugated cotyledons.

Cocoa

(***Theobroma cacao*** L.) (2n = 20)

It is one of the 20 species of the genus *Theobroma,* native to southern tropical America. Two subspecies have been identifed *T. cacao* subsp. *cacao* (Criolla populations) and *T. cacao* subsp. *sphaerocarpum* (Forastera and Trinitario populations). Cocao is a semi-deciduous perennial tropical plant reaching a height of 5-8 m with a dense foliage and rounded canopy. The main trunk is short, branching (jourquetting) at a height of 1.0 to 1.5 m from ground level in whorled and horizontal fashion. The branches are called jourquetts or fans. The terminal growth of the stem ends in jourquetting and further growth of the plant is continued by a sucker (chupon) which arises just below the jourquette and grows vertically for some time and again ends in a jourquette. This process is continuous. Leaves vary in colour according to the type from pale green or pink to dark red. The tree puts forth new shoots or leaves two or three times a year. Therefore, leaves of three different age groups can be seen at one time.

Cacao is cauliflorous bearing flowers and fruits on old wood of the trunk and main branches emerging from leaf axils (cushion). The inflorescence is a compressed cyme with a short and thick peduncle. The flowers are hermaphrodite. The leaves may fall at bloom time. Calyx consists of 5 pink or white and pointed sepals. Petals found at the base are narrow at the ends. The ovary is superior and is surrounded by a slim style bearing five small stigmas. Five long and pointed staminodes are present. All the cacao trees including the self incompatible types are fertilized by the pollens of other trees showing a high rate of cross pollination. A cacao tree may produce 6000 flowers but only one per cent set mature fruits. The fruit is a berry containing 25-40 seeds each surrounded by pulp developed from the inner integument of the ovule. Botanically the fruit (often called pod) is an indehiscent drupe. Pods are smooth or ribbed and are variable in size, shape (elliptic to ovoid) and colour (red, yellow, purple or brown). The seeds (20-50) are embedded in a white pinkish or brownish, acid to sweet and aromatic mucilage (pulp). Seeds (beans) are flat or round, white, brownish or purple, sweet or bitter and have two white or purple cotyledons. Beans are source of cocoa butter and cocoa mass.

Cultivars

Criollo: Pods are dark red or yellow, thin-walled, irregular round with pronounced point, soft with rough surface having 10 distinct longitudinal and prominent ridges, deep furrows and big warts (Fig. 30.1). Cotyledons are plumpy and white when fresh and turn cinnamon on fermentation (for 3 days) and are sweet and bland flavoured. The yield is comparatively low. The cultivar is susceptible to drought and diseases and is thus discouraged for cultivation. Being tender it is most suitable for use as intercrop in coconut and areca nut. Fruits have bland flavour. It is grown in America.

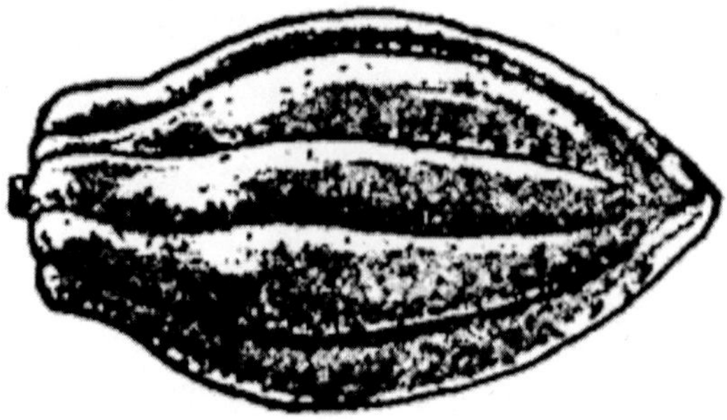

Fig.30.1. Criollo coco.

Forestero: Pods are yellow, thick and hard walled, melon shaped with rounded ends, smooth with shallow and inconspicuous ridges (Fig. 30.2). Cotyledons are flat and purple when fresh and turn dark chocolate brown on fermentation (for 6 days) and are bitter in taste and harsh flavoured. The yield is comparatively higher. Fruits have harsh flavour and bitter taste. The cultivar is a good yielder but the quality is not as good as that of Criollo. The cultivar is less susceptible to diseases and is thus recommended for cultivation. It is grown in Africa.

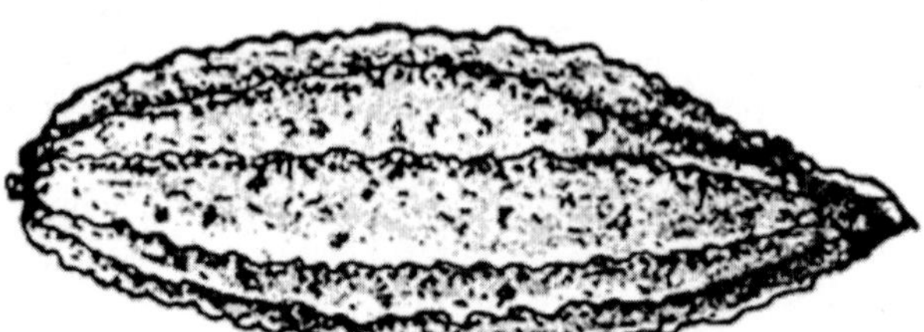

Fig. 30.2. Forestero coco.

Trinitario (Crillo x Forestero): The cultivar is from Trinidad and is hardy and productive, very heterogeneous exhibiting wide variation in shape and colour (green, red or purple) of pods. It yields cocoa of fairly good quality.

G VI 56: Pod yield is 78-95 per tree giving 7.25 beans per pod, each having 95 g fresh weight and 4 kg dry beans per tree.

M 16.9: Pod yield is 56-72 per tree giving 46 beans per pod, each having 111 g fresh weight and 3.3 kg dry beans per tree.

Amelonado: It is cultivated in West Africa. The pods are Forestero type and are melon shaped.

Amazon: It is a relatively new type collected from the Amazon forests and has high vigour and yield.

CPCRI, Kasaragod has recommended the high yielding selections 1-21, II-11, II-18, II-67, III-5 and III-101 introduced from Malaya for commercial cultivation. These have beans weighing more than one gram.

High yielding clones

I-14 (Jarangau Red Axil): Single bean weighs 1.17 g. Dry bean yield per tree is 1.18 kg and per hectare is 803 kg.

I-56 (PA& x Na 32): Single bean weighs 1.20 g. Dry bean yield per tree is 1.12 kg and per hectare is 761 kg.

III-105 (Amel x Pa 7): Single bean weighs 1.06 g. Dry bean yield per tree is 1.07 kg and per hectare is 728 kg.

NC42/94 (T86/2): Single bean weighs 1.06 g. Dry bean yield per tree is 1.05 kg and per hectare is 714 kg.

NC45/53 (P10 x Pt): Single bean weighs 1.05 g. Dry bean yield per tree is 1.149 kg and per hectare is 781 kg.

Hybrids

VTLCH 1 (Vittal Cocoa Hybrid) (II-67 x NC-42/94; Malaysian and Nigerian): It is an early and heavy bearing cultivar tolerant to water stress. It bears about 40 red to pink pods per tree per year, 43 beans per pod, single pod weighing 430 g and single dry bean weighing 1.01 g. It is suitable for both rainfed and irrigated areas.

CCRP-8 (CCRP-1 x CCRP-7): It is a heavy bearing cultivar tolerant to vascular streak dieback. It bears 53.9 pods per tree per year, 90.4 beans per pod, single pod weighing 389.3 g and single dry bean weighing 0.9 g giving 3.5 kg dry bean yield per tree per year.

CCRP-9 (CCRP-1 x CCRP-4): It is a heavy bearing cultivar, tolerant to vascular streak dieback. It bears 106.7 pods per tree per year, 36.7 beans per pod, single pod weighing 370.7 g and single dry bean weighing 0.8 g giving 3.1 kg dry bean yield per tree per year.

CCRP-10 (CCRP-3 x GVI-68): It is a heavy bearing cultivar, tolerant to vascular streak dieback. It bears 79.6 pods per tree per year, 41.6 beans per pod, single pod weighing 332.5 g and single dry bean weighing 1.1 g giving 3.5 kg dry bean yield per tree per year.

II-67 x NC42/94: Single bean weighs 0.99 g. Dry bean yield per tree is 1.245 kg and per hectare is 847 kg.

II67 x NC29/66: Single bean weighs 1.06 g. Dry bean yield per tree is 1.478 kg and per hectare is 1006 kg.

IC86 x 8ca6: Single bean weighs 0.95 g. Dry bean yield per tree is 1.145 kg and per hectare is 779 kg.

I56 x II-67: Single bean weighs 1.00 g. Dry bean yield per tree is 1.481 kg and per hectare is 1007 kg.

Cupuassu

(***Theobroma grandiflorum*** (Willd. ex Spreng.) K. Schum.; cupuazu; cupuacu; cupuaan; cupu-arju)

Cupuassu is native of Amazon basin of Brazil. It is cultivated in Amazonia. The related edible species are *T. bicolor* Humb and Bonpl. and *T. subinanum* Mart. It is an important fruit of the lower Amazonas (Para) of Brazil outside of which it is almost unknown (FAO, 1986; Martin *et al*., 1987; Brücher, 1989). It is also found in Colombia, Venezuela, Ecuador and Costa Rica. Cupuassu is locally important fruit and may have potential in other areas (Martin *et al*., 1987).

The tree is evergreen, small to medium in size and growing up to 20 m in height with a canopy spread of 7 m. The leaves are alternate, simple, coriaceous, short-petioled, elliptic oblong, 15-60 x 5-16 cm in size, dark green above and pale green to brown below. Each inflorescence has 3 to 4 flowers borne in the axils of leaves or along the bare branches. Each flower has a calyx composed of five triangular fused sepals, a corolla with 5 purple petals, 5 purple staminodes, 5 stamens and a 5-locular ovary. It is morphologically different from the other species because of absence of cauliflory and bearing flowers and fruits on the side branches. The fruit is ellipsoidal or oblong, 12 to 25 cm long, 10-12 cm in diameter and up to 1.5 kg in weight having about 50 seeds. It has a velvet brown skin and hangs on long pedicels. The woody pericarp easily breaks to expose the edible seeds, which are round, flat and have a certain similarity with the true cacao kernels, but have more fat with a consistency of cacao butter (FAO, 1986; Brücher, 1989).

⑧ ⑧ ⑧

Chapter 31

BRASSICALES (CAPPARALES)

The plants are trees, shrubs, often herbs containing myrosin and protein rich cells, dilated cisternae of the endoplasmic reticulum. Vessels have simple perforations and alternate lateral pitting. Fibres have simple or bordered pits. Rays are heterogeneous or homogeneous. Axial parenchyma is apotracheal or paratracheal, rarely diffused. Sieve element plastids are of S-type or sometimes of Pfcs-type. Nodes are unilacunar or trilacunar. Leaves are alternate or rarely opposite, simple or often variously compound with or without stipules. Stomata are anomocytic, anisocytic or paracytic. Flowers are mostly in terminal racemes, bisexual or sometimes unisexual, actinomorphic or zygomorphic with double perianth or sometimes apetalous having nectaries of receptacular origin. Perianth is 4-merous or less often 5-merous with free segments. Stamens are (2-)4 to many arising from a limited number (most often 4) of primordia and centrifugal when in more than one series. In some Capparaceae and in most of the Brassicaceae, two median primordia divide and therefore the number of stamens reaches six. In other cases all the four primordia branch once or more and as a result the androecium consists of 8 or more stamens. Anthers are tetrasporangiate opening longitudinally. Tapetum is secretory. Microsporogenesis is simultaneous. Pollen grains are 2 or 3-celled, mostly 3-colpate or 3-colporate. The extrastaminal nectaries are present. Gynoecium is paracarpous of two or less often 3-6 to 12 and rarely up to 60 united carpels with stylodia are united into a shorter elongated style having a simple or lobed stigma or the stigmas are separate and sessile. Ovary is superior, basically 1-locular or partitioned by replum (a thin, unvasculated frame-like partition connecting the two parietal placentas) with usually more or less numerous ovules, sometimes only with one ovule. Ovules are anatropous or more often campylotropous, bitegmic, crassinucellate or less often tenuinucellate. Female gametophyte is usually of *Polygonum*-type. Endosperm is nuclear. Fruits are of various types. Seeds have large, oily, more or less curved or folded embryo, mostly without endosperm or with scanty endosperm, rarely with well developed endosperm. Plants mostly produce glucosilonates (mustard oil glucosides, thioglucosides), sometimes alkaloidal amines and cyanogenic. Several genera are accumulators of heavy metals lacking ellagic acid. The order has 17-18 families.

Family Moringaceae (horse radish tree, drumstick tree, ben oil tree)

The family contains one genera and 13 species (Heywood *et al.*, 2007). The plants are trees (often with swollen trunks), shrubs, and subshrubs, sometimes with a tuberous rootstock, all with soft and usually brittle wood. The leaves are compound, 1-3-imparipinnate with a petiole, leaflets in opposite pairs and stipules represented by stipitate glands. The inflorescence is an axillary thyrse. The hermaphrodite flowers are 5-merous, red, yellow or white and range from actinomorphic (*M. ovalifolia*) to strongly zygomorphic (*M. borziana*) with a cup-like to tubular receptacle. The sepals and petals are 5, free and overlapping in bud, equal or unequal. The stamens are 5, alternating with the 3-5 staminodes and opposite the petals. The ovary is superior borne on a stalk of 3 fused carpels forming one locule with many ovules. The fruit is large elongated capsule with 3 valves and sometimes twisted.

Genus *Moringa* Adans

The genus has three sections; sect. *Donaidsonia* with regular hypogynous flowers, sect. *Moringa* with irregular perigynous flowers and short receptacles, and sect. *Dysmoringa* with irregular perigynous flowers and long and tubular receptacles (Kubitzki and Bayer, 2003).

Drumstick tree

(***Moringa oleifera*** Lam. syn. ***M. pterygosperma*** Gaertn, ***M. nuxben*** Perr.)

It is a native of India. It is distributed primarily in the Horn of Africa and extends through India along the southern coast of the Arabian Peninsula. It is a medium sized nitrogen fixing tree that is short, slender, deciduous and perennial having drooping branches and growing up to 10 m tall. The bark of the tree is thick (2.5 cm), soft, white, corky and has deep fissures. The leaves are feathery, opposite, pale green, compound, tripinnate, 30 to 60 cm long with many, small, of 1.3-2 cm x 0.3-0.6 cm leaflets. Flowers are fragrant, white or creamy white, bisexual, 2.5 cm in diameter, borne in pendulous axillary panicles having 5 sepals, petals and yellow stamens. Pods are pendulous, brown, triangular splitting lengthwise into 3 parts when dry, 30-120 x 1.8 cm in size containing about 20 seeds embedded in the pith. The pods taper at both ends and are 9-ribbed and contain seeds that are 1 to 2 cm wide and dark brown with 3 papery wings (Ramachandran *et al.*, 1980; Duke, 1989).

Cultivars

Several cultivars are grown. The cultivar Bombay having curly fruits is one of the best. Other cultivars have 3-angled or nearly round fruits. The length of fruits is 60 to 90 cm in cultivar Jaffna and 90 to 120 cm in cv. Chavakacheri Murunga. In Thirunelveli district of Tamil Nadu, cultivars Palmurungal (thick pulp, bitter taste) and Punamurungai are grown. In Trichy district of Tamil Nadu, cultivar Kodikalmurungai is grown which bears very short fruits. Cultivar PKM-1 has been released for cultivation (Ramchandran *et al.*, 1980).

Family Caricaceae (papaya or pawpaw family)

It is a small family having 6 genera (*Cylicomorpha, Carica*, *Jacaratia*, *Jarilla Horovitzia* and *Vasconcellea*) and 34 species (Heywood *et al.*, 2007), all of which except *Cylicomorpha* originated in the Americas. *Carica*, unlike the *Vasconcellea* is characterised by a unilocular ovary, a hollow stem, leaves with more than seven veins and lamelliform protuberances on the seed sclerotesta. Besides, it is incompatible to hybridise with *Vasconcellea* species. *Carica* and *Vasconcellea* both show a chromosome number of 2n = 18 (De Zerpa, 1980).

The plants are distinctly semi-herbaceous, small trees with a straight and rarely branched trunk having soft wood, milky latex in all parts in articulated laticifers. The latex may be pungent. The stems are sometimes spiny. In *Carica papaya,* the stem has little development of secondary xylem, the wood being formed largely from the phloem producing a soft and large pithed trunk. The leaves are alternate, crowded at branch tips, palmately 7-lobed forming a terminal crown of foliage, usually with long petioles and without stipules (stipules spine-like in *Vasconcellea stipulata*). The inflorescences are axillary, 1-flowered or a many flowered thyrse. The plants have varying sex forms but are usually dioecious (very rarely monoecious or polygamous). The flowers are unisexual, rarely bisexual (sometimes in *Carica*). The flowers are regular with the parts in fives. Staminate flowers are on open long racemes. Calyx is 5-lobed and much reduced. Petals are connate into an elongated 5-lobed tube. Stamens are 10 in two whorls attached at the throat of corolla and the anthers open inward. The pistillate flowers are sessile, usually solitary or in groups of 3. Calyx is reduced. Petals are short, twisted and fleshy and connate at the base. The stalks are short. The ovary is superior, of 5 fused carpels, with 1-5 locules having many anatropous ovules on parietal placentas. The style is short and crowned by 5 stigmas. The fruit is a berry. Each of the numerous seeds have gelatinous envelope, oily proteinaceous endosperm and a straight embryo.

Genus *Carica* L.

Carica is a monotypic genus and includes the economically important species, *C. papaya* L, cultivated since prehistoric times in the lowlands of East Mexico and Central America and now in all the tropical countries. The trunk is unbranched with hollow internodes and succulent with prominent leaf scars. The plant is rapidly growing and palm like. The leaves are very large, palmately lobed, 7-12-veined with long hollow petiole. As the plant grows, the leaves shed leaving a crown at the top only. The inflorescence is monoecious or dioecious. Flowers are 5-merous having stamens with submoniliform hairs, connectives hardly prolonged beyond the apex of anther, entirely 1-celled ovary, very short or no style, 5 usually anther like branched stylodia. Fruit is ovoid to 7 ellipsoid and pentagonal.

Papaya
(*Carica papaya* L.) (n=8, 2n=18)

The papaya is native of Mexico and may have originated from the lowlands of Central America, from Mexico to Panama (Nakasone and Paull 1998) as a result of natural crosses between some *Carica* species. Undomesticated papaya was a spindly plant with nearly inedible fruits. During domestication the species has undergone considerable changes in growth habit fruit size, flesh colour mating system (Manshardt and Moore 2003).

Table 31.1. Characteristics of the major flower types in papaya.

Character	Male (staminate)	Female (pistillate)	Hermaphrodite (elongata)	Hermaphrodite (intermedia)	Hermaphrodite (pentandria)
Inflorescence	Pedunculate pendulous panicle 25-100 cm long	Solitary or few flowered cyme, 3.5-5 cm long stalk	Short peduncled cluster, stalks less than 25 cm long	Short peduncled cluster, stalks less than 25 cm long	Short peduncled cluster, stalks less than 25 cm long
Corolla	Trumpet-shaped, 2.5 cm long, 5 lobes, light yellow cream petals	Almost free fleshy, 5 yellow/cream petals (fused at base)	Partially united 5 petals, fused from ¼ - ¾ of their total length	Undefined petals; may be fused up to 2/3 of their length or free	Almost free 5, fleshy, yellow/cream petals
Stigmas	None	5, fan-shaped on a short style	5, fan-shaped	5, fan-shaped	5, fan-shaped
Ovary	Rudimentary non-functi-onal ovary (or pistillode)	Ovoid oblong, 2-3 cm long, central cavity, numerous ovules	Elongate	Distorted, irregular shape	Ovoid oblong, 2-3 cm long, 5-furrowed
Stamens	10 in 2 whorls alternating with petal lobes	None	10 borne at the throat of the corolla in 2 clusters (5 long, 5 short)	2-10, some or all are distorted	5 attached by long filaments near base of ovary
Shape of resulting fruit	--	Spherical/ ovoid	Cylindrical to pear-shaped	Carpelodic (cat face)	Ovoid and lobed

The plant has hollow green or deep purple trunk growing straight and cylindrical up to 10 m height and can become 30-40 cm thick at the base, thinning to 5–7.5 cm at the crown. It generally branches only when injured. The trunk has prominent leaf scars. The leaves emerge directly from the upper part of the stem in a spiral on nearly horizontal 25–100 cm long petioles and form a loose open crown. The leaf blade is 25–75 cm in width and has prominent yellowish ribs and veins and is deeply divided into 5 to 9 main lobes. The life of a leaf is 2.5 to 8 months and 1.5–4 new leaves arise per week. All parts of the plant contain white latex. The flowers are borne on inflorescences which appear in the axils of the leaves. Papaya has a highly complex sexual expression. Some plants bear only short-stalked pistillate (female) or hermaprodite (perfect) flowers (having female and male organs) borne on short stalks, while others may bear only staminate (male) flowers clustered on long panicles. There are also monoecious plants having both male and female flowers. Pistillate trees tend to be stable while staminate and hermaphrodite trees may be 'ambivalent', going through seasonal sex reversals. The characteristics associated with the major flower types in papaya are given in Table 31.1. There is a range of possible variation and the proportion and type of flowers produced may vary even on the same tree. Hermaphrodite flowers are of two main types–elongata and pentandria

- with an intermediate type between these two. The fruits are technically classified as fleshy berries sometimes called pepo-like berries since they resemble melons by having a central seed cavity. These are borne axillary on the main stem, usually singly but sometimes in small clusters. The fruit may be globose, ovoid, obovoid, or pyriform, 7-35 cm long, and 0.25-10 kg in weight. It has a smooth exocarp (peel) and thick, fleshy mesocarp. Female plants produce medium to large round fruits with a large seed cavity having good quality. The hermaphrodite plants produce small to medium elongated fruit of good quality but have smaller seed cavity. The male plants with bisexual flowers may produce a few elongated poor quality fruits. The seeds are enclosed in aril. Being tropical the plants are susceptible to hot winds and frost but require heat for fruit maturity. The plant may survive for 25-30 years but the productive life is only 3 years.

Cultivars

The major papaya (Betik) cultivars in Malyasia are Foot Long, Batu Arang, Sitiawan and Subang 6. The cultivars used as dessert fruit are Eksotika, Eksotika II and Sekaki. The Eksotikas have better eating quality than the other three in terms of flavour and taste.

Eksotika: It is a self pollinated inbred cultivar having similar features as Sunrise Solo except for the larger fruit size. It is a very popular cultivar in Malaysia. Eksotika is a good bearer giving fruit yield of about 60 t /ha/year. The fruits are small to medium in size (400–800 g). The fruit has orange red flesh with a pleasant aroma and high TSS of 12-14ºBrix but does not keep well because of its soft texture. It is also quite susceptible to fruit freckles and malformed top disease.

Eksotika II (F1 hybrid): It was developed by hybridizing Line 19 with its sib Line 20 (Eksotika). The hybrid gives 14-33% higher yield due to the larger fruit weighing 600-800 g. The appearance of the fruit is more attractive having smooth skin and high tolerance to freckles. The flesh is firmer and the fruit stores longer and is thus preferred for export.

Sekaki (Hong Kong papaya): It is the second most popular cultivar in Malaysia after Eksotika. It is a cross pollinated cultivar and a prolific bearer (60-70 t/ha/year) with medium size fruits of 1.5-2 kg. The tree is dwarf and bears low near the ground. It is also easy to manage in the field because of its tolerance to malformed top disease. The fruit is attractive with smooth, even-coloured and freckle free skin. The flesh is red, firm but sweetness is not high (10ºBrix or less).

In Philippines, papaya (lechosa) is grown for local fresh fruit market, for canning and for papain production. The most popular strains of Solo papaya commercially propagated here are Kapoho and Sunrise. The other cultivars are Cavite Special (Pineras), Waimanolo, Sinta, Red Lady Papaya (F1 hybrid), Tainung No. 1, Tainung No. 2 and Tainung No. 3. Cavite and Sinta are hermaphrodite and red-fleshed. Cavite or Pineras has large and oblong fruit weighing 1.5 - 6 kg.

In Thailand, the main cultivars of papaya (*malakor, loko, ma kuai thet*) are Khaegdam, Kaegnuan, Koko and Sai-nampueng. Sai-nampueng and Khaegdam are red-fleshed hermaphrodite cultivars. Khaegdam is vigorous, bears fruit weighing about 1.2 kg with 10.6ºBrix TSS.

In Indonesia, the main cultivars are Semangka, Dampit, Jingga and Paris. The cultivars Semangka and Dampit are hermaphrodite and red-fleshed. Cibinong bears large red-fleshed fruit (2-3 kg) and is grown mainly for its high papain yield. In Singapore, the main cultivars are P 10 (Eksotika x Local cultivar), Ng Toh Seng, P 5, P 15 (Eksotika x Guam Island, P 2).

The main cultivars in Florida (USA) are Cariflora, Betty and Homestad. Hortus Gold, Sunnybank, Betty and Cariflora are dioecious. The fruits are yellow fleshed and typically round. Cariflora fruits are smaller and quite tolerant to the ring spot virus disease. The gynodioecious cultivars that are cross pollinated are Khaek Dam, Maradol, Cibinong and Cavite.

Solo is a self pollinated hermaphrodite type. It is so called because of its small fruit, about 350–500 g size, convenient as a single serving. It is the most popular cultivar in the world. Within this cultivar many lines such as Line 5, Line 8, Line 10, Kapoho (Puna Solo), Waimanalo (Line 77) and Sunrise have been developed. Kapoho is the major cultivar in Hawaii while Sunrise is the only pink-fleshed line widely grown in many parts of the world. The others are yellow-fleshed. The other cultivars are Sunset, Higgins (Line 17A), Wilder (Line 25), Dwarf Solo (Solo x Florida's Betty), Big Bluestem (Purplestem x Bluestem), Bluestem Solo and Blue Solo (Big Bluestem x Solo).

In the Caribbean region, the main cultivars are Wakefield, Graeme 5 and Graeme 7 from Barbados, the hermaphrodite and red-fleshed cultivar Maradol from Cuba and Santa Cruz Giant and Cedro from Trinidad are important. Maradol is short-statured cultivar and bears fruits very close to the ground. The fruit weighs 1-2 kg and is attractive having firm red flesh with 10-11°TSS. It has a characteristic musky flavour but is quite susceptible to anthracnose. In Mexico, the main cultivars are Verde, Gialla, Cera and Chincona. In South America, Betty, Bettina, 43-A-3 and Zapote in Brazil, Campo Grande, Tocaimera, Solo and Zapote in Colombia and in Venezuela Paraguanera and red-fleshed Roja are the popular cultivars.

The main South African cultivars are Hortus Gold, Honey Gold and Kaapmuiden. These are yellow-fleshed cultivars. Hortus Gold and Honey Gold are dioecious. In Trinidad and Tobago, the popular cultivar is Santa Cruz Grant. In Dominican Republic, the hermaphrodite yellow-fleshed cultivar Cartagena is popular.

The main cultivars in Australia are Bettina Improved Petersen, Petersen Hybrid No 5, Yarwun Yellow, Gold Cross, Goldy, Hong Kong, Guinea Gold, Golden Surprise, Sunny Bank, Sunnybank/S7 and Richter/Arline. Guinea Gold is yellow fleshed hermaphrodite cultivar. Improved Petersen, Sunnybank/S7 and Richter/Arline are dioecious and yellow-fleshed.

Papaya cultivars in India

Honey Dew (Madhu Bindu): This cultivar is extensively grown in North India. The stem and leaf stalks are purple. The fruit is oblong and purple on the stem end. The pulp is orange in colour having very few seeds, good taste and flavour.

Barwani: There are two types; one having deep orange pulp and the other having golden yellow or golden orange pulp.

Washington: The cultivar is grown all over the world. The stem, leaf petioles and basal part of the fruit close to the stem have purple pigmentation. The fruit is ovate oblong in shape and large in size of around one kg weight having good flavour, sweet taste and orange-coloured pulp.

Pusa Majesty: It is a selection from line Pusa 22-3. It is a gynodioecious cultivar tolerant to viral diseases. Fruit is round with short apex, medium size, smooth skin and firm, solid and yellowish flesh. The fruit has very good keeping quality and withstands long distance transportation

Pusa Dwarf: It is a selection from line Pusa 1-45 D. The plant is distinctly dwarf and hence suitable for high density planting and for kitchen garden. It is a dioecious cultivar. The plants start bearing even when about 30 cm tall. The fruit is oval and medium in size weighing 1-2 kg.

Pusa Nanha: It is a mutant and is extremely dwarf. It is a dioecious cultivar and is suitable for planting in kitchen gardens, pots and for roof top cultivation. It is ideal for high density orcharding.

Pusa Delicious: It is a selection from line Pusa 1-15. It is a gynodioecious cultivar having pistillate and hermaphrodite flowers on the same plant. The fruit is oval with a bulge near the distal end. The flesh is orange-coloured having very good eating quality and excellent flavour.

Pusa Giant: It is a selection from line 1-45 (V). It is a dioecious cultivar. Fruits are big-sized and attractive.

Pant 1: It is a seedling selection. Plants are dioecious, dwarf, high yielder and start bearing fruits at 40-45 cm height from ground level. Fruit is medium in size and 1 to 1.5 kg in weight. It is resistant to anthracnose and has been recommended for *tarai* (damp) area of Uttar Pradesh, India.

Pant 2: It is a seedling selection. Plants are vigorous and medium in size. It is tolerant to frost and wet feet conditions. Fruits are medium to large in size.

Pant 3: It is a seedling selection. Plants are medium in height and tolerant to frost and waterlogging. First flower emerges at a height of 115-130 cm from ground level. Fruits are small to medium in size (0.5-0.9 kg) and of excellent quality.

Coorg Honey (Coorg Honey Dew): It is considered to be a chance seedling of Honey Dew. The plants are gynodioecious, dwarf and prolific bearer. It has no male plants. Productivity of the cultivar is poor. The hermaphrodite and pistillate flowers are borne on the same plant. Each plant bears 40-60 fruits and an average yield of 200 tonnes per hectare could be obtained over a three year cropping period. The fruit is long to oval, weighting 2.0 to 3.5 kg and of excellent quality. The flesh is thick having more edible portion, sweet taste, pleasant flavour and good quality.

CO 1: It is a selection from cultivar Ranchi on selfing for nine generations. Plants are dioecious and dwarf. The fruit has orange yellow flesh, of fairly uniform quality, medium size and spherical shape.

CO 2: A pure line selection from local type. The plants are medium tall and dioecious. The fruit is large, obovate and initially bright green colour which later turns yellowish green. Flesh is soft, moderately juicy and orange yellow in colour. It is a dual purpose cultivar used for fruit, predominantly for papain production.

CO 3 (CO 2 x Sunrise Solo): Trees are gynodioecious and vigorous. The fruits are pyriform, medium in size having high sugar content and red flesh. It is suitable for table purpose.

CO 4 (CO 1 x Washington): It is a dioecious cultivar having medium to tall trees. Petiole, leaf and stem are purple. It is suitable for home gardening.

CO 5: It is a selection from Washington. It is a dioecious cultivar. Petiole is purple. It is cultivated mainly for papain production. It yields 1500-1600 kg of dried papain per hectare.

CO 6: It is a selection from the cultivar Pusa Giant and is dioecious having plants of dwarf stature. It has large fruits and is recommended for production of papain as well as fruits.

CO 7 (Pusa Delicious x CO 3): It is a gynodioecious cultivar. It has fruit size and quality superior to CO 3.

Pink Flesh Sweet: It is a selection. It is a gynodioecious cultivar. Fruits are medium in size having pink flesh suitable for dessert purpose.

Punjab Sweet: It is a dioecious cultivar. Plants start bearing at a height of 75 cm from ground level. Fruits are round to oval in shape and are yellow-fleshed.

Taiwan: This is gynodioecious cultivar having blood red flesh of good taste.

Arka Surya (Sunrise Solo x Pink Flesh Sweet): It is a gynodioecious cultivar. Fruits are medium in size weighing 600 to 800 g. The pulp is red in colour with 3.0 to 3.5 cm thickness and sweet taste having 13.5 to 15°B TSS.

Arka Prabhath (Surya x Local Dwarf): It is a gynodioecious cultivar having deep pink pulp about 3.0 to 3.5 cm in thickness. Fruits are sweet having 13.50 to 15°B TSS.

Sunrise Solo: This is a gynodioecious cultivar having pink flesh and good taste.

Solo: This cultivar is important in Hawaii.

HPSC-3 (Tripura Local x Honey Dew): This cultivar is resistant to papaya mosaic virus.

Red Lady (Taiwan 786): This is a gynodioecious cultivar having blood red coloured flesh with good taste.

Hybrids: Commercial F1 hybrids of papaya are rare. An important hybrid developed in Taiwan that has resistance to papaya ringspot virus disease is Tainung No. 5 which was developed from a cross between Florida (FL-77-5) and the Costa Rica Red.

Clonal cultivars: Honey Gold is a well known papaya clone. It is a dioecious cultivar selected in South Africa from plants propagated by leafy cuttings over generations. The clonal plants have greater uniformity especially in fruit shape which is a sex-linked characters. Clonally propagated Honey Gold clone gives fairly high annual yields of 25-30 tonnes/ha under subtropical conditions and is known to remain productive for 10 or more years. *Vasconcellea heilbornii* var. *pentagona* or babaco, which is grown to a small extent in Ecuador and New Zealand, is exclusively propagated by cuttings because it has parthenocarpic fruits.

Transgenic cultivars: The world's first transgenic papaya was SunUp, which was transformed with coat-protein mediated resistance to papaya ringspot virus disease. Rainbow is the first transgenic commercial cultivar developed in Hawaii from a cross between SunUp and the conventional cultivar Kapoho. Transgenic cultivars of Kamiya have also been developed by introduction of the coat-protein transgene from Rainbow through conventional hybridization and backcrossing. Transgenic papaya cultivars with delayed fruit ripening and resistance to papaya ringspot virus disease are being developed in Malaysia, Thailand, Indonesia, Philippines and Vietnam under the Papaya Biotechnology Network of South East Asia.

Genus *Vasconcellea* A. St. Hil.

Vasconcellea is the largest genus of Caricaceae. The genus has 21 (20 species and 1 hybrid, *Vasconcellea* x *heilbornii* considered as a separate species) of the 35 species of family Caricaceae (Badillo, 2001). Despite their wide distribution in tropical America, ranging from Mexico down to Argentina and Chile, the majority of the *Vasconcellea* species are confined to the northern Andean region (Colombia, Ecuador and Peru). This is also reflected in their temperature preferences, generally cooler climates with limited seasonality, as prevalent in the equatorial Andes. Ecuador holds 16 of the 21 described species (Badillo, 1999): *V. stipulata* (V. Badillo) V. Badillo, *V.* x *heilbornii* (V. Badillo) V. Badillo, var. *chrysopetala* (Heilborn), var. *fructifragrans* (Garcia. Barr. et Hern.), the largest mountain papaya Babaco, *V. weberbaueri* (Harms) V. Badillo, *V. parviflora* A. DC., *V. goudotiana* Tr. et Planch, *V. palandensis* (V. Badillo et al.) V. Badillo, *V. cundinamarcensis* V. Badillo, *V. longiflora* (V. Badillo) V. Badillo, *V. pulchra* (V. Badillo) V. Badillo, *V. monoica* (Desf.) A. DC., *V. omnilingua* (V. Badillo) V. Badillo, *V. microcarpa* (Jacq.) A. DC. (subsp. *baccata* (Heilborn) V. Badillo, subsp *heterophylla* (Poepp. et Endl.) V. Badillo, subsp. *microcarpa*), *V. cauliflora* (Jacq.) A. DC., *V. chilensis* (Planch. ex A. DC.) A. DC., *V. quercifolia* Saint-Hilaire, *V. candicans* (A. Gray) A. DC., *V. crassipetala* (V. Badillo) V. Badillo, *V. sphaerocarpa* (Garcı´a-Barr. et Hern.) V. Badillo, *V. glandulosa* A. DC. and *V. sprucei* (V. Badillo) V. Badillo). These occur up to 3500 m a.s.l and thus the region is considered as one of the 'hot spots' for these species. Kyndt *et al.* (2005) observed high intraspecific variation within *V. microcarpa* and *V.* x *heilbornii*. The high diversity also consists of hybrids with varying degrees of fertility. The fertile hybrids may cross with parental or nonparental species (Badillo, 1971). Two naturally occurring hybrids are *V.* x *heilbornii* and the hybrid between *V. monoica* and *V. cundinamarcensis*. Within *V.* x *heilbornii*, Badillo (1993) recognizes the cultivar Babaco and the varieties *chrysopetala* and *fructifragrans*. Five species; *V. horovitziana*, *V. omnilingua*, *V. palandensis*, *V. pulchra* and *V. spruce*; have been placed on the Red List of Threatened Species (IUCN, 2003).

Species of *Vasconcellea* are wild, semi-domesticated or domesticated plants with a shrub or tree-like and pachycaulous habit. Plants are usually dioecious but sometimes monoecious or polygamous. The medullar stem is mostly simple or scarcely branched. In some species it is covered with spiny stipules. The leaves are large to very large, from entire to compound in shape and are concentrated in a terminal crown. All organs produce white latex containing cysteine endopeptidases.

Flowers are pentamerous with white, green, yellow, orange or pink petals. The fruit is a berry with varying shape, dimension and colour (Badillo, 1993). Compared to the lowland *Carica papaya*, highland papaya fruits are generally smaller and have distinct texture, taste and aroma. In the Andes, these are consumed fresh, roasted, processed in juices, marmalades, preserves or dairy products or even prepared in sauces, pie fillings and pickles. A number of species show potential as raw material in the tropical fruit industry, fresh or in processed products and as a source of papain.

Vasconcellea species are often collectively called 'highland papayas' or 'mountain papayas' (NRC, 1989) because of their resemblance to *C. papaya* being the nearest relatives and for their typical ecological preferences for higher altitudes. The common highland papayas in Loja Province are *V. candicans* (A. Gray) A. DC., *V. cundinamarcensis* V.M. Badillo, *V. monoica* (Desf.) A. DC., *V. parviflora* A. DC., *V. stipulata* (V. M. Badillo) V.M. Badillo, *V. weberbaueri* (Harms) V. M. Badillo and *V.* × *heilbornii* (V. M. Badillo) V. M. Badillo, a natural hybrid. Of these, *V. cundinamarcensis*, *V. stipulata* and the natural hybrid between both species, *V.* × *heilbornii* are the most abundant highland papayas and consequently the most used ones (Jiménez *et al.*, 1998).

Vasconcellea spp. naturally possess a number of desirable traits including resistance to pathogens, cold tolerance and higher sugar content of fruit (Drew *et al.*, 1998) and thus show potential as genetic resources for improvement of the common papaya. They carry resistance genes, particularly for the most severe and widespread disease, the papaya ringspot virus (Drew *et al.*, 1998), cold tolerance and organoleptic characteristics (Manshardt and Wenslaff, 1989ab). However, inter-specific gene flow between *C. papaya* and *Vasconcellea* species faces considerable postzygotic barriers (Drew *et al.*, 1998).

Mountain pawpaw

(*Vasconcellea pubescens* A.DC. syn. ***V. cundinamarcensis*** V.M. Badillo, ***Carica candamarcensis*** Hook. f., ***C. cestriflora*** (A. DC.) Solms, ***C. chiriquensis*** Woodson, ***C. pubescens*** (A. DC.) Solms, ***C. pubescens*** Lenné & C. Koch, ***C. cundinamarcensis*** Linden, ***Papaya cundinamarcensis*** (Linden) Kuntze, ***P. pubescens*** (A. DC.) Kuntze, ***V. cestriflora*** A. DC.; chamburo, papaya de tierra fría, chihualcán, siglalón, chichuacacón, titi-ish, bonete, papaya de altura, papayuela, tapaculo)

Mountain papaya is abundant in temperate zones of the Andes (1500–3000 masl) ranging from Colombia and Venezuela to Bolivia. It is cultivated in Chile. The plants are polygamous, monoecious or dioecious, 3–10 m tall with stout, succulent, medullose and few-branched stems. Essentially pubescent in all parts, the leaves are clustered in a dense terminal crown. Leaf blade is palmately divided into 5 or 7 lobes. Male inflorescence is 1-15 cm long, many-flowered with pubescent axis and branches (Fig. 31.1). Female inflorescence is pubescent, short and few-flowered. Bisexual inflorescences similar to the male ones, sometimes only with male flowers, occur in certain seasons. Flowers are greenish with somewhat spreading, glabrous or pubescent calyx lobes and the corolla pubescent or subglabrous outside and glabrous or pilose within. Fruits are obovoid, slightly apiculate, yellow or orange, acid, fragrant, obtusely pentagonous or 5-sulcate, 6–15 cm long and 3–8 cm wide sometimes bigger. Seeds are numerous having smooth sarcotesta, reddish brown sclerotesta with wide low interrupted ribs. In some cases, the fruits produced on monoecious plants are rather oblique. The plant can withstand frost. It is a potential gene source for inducing cold tolerance and disease resistance in papaya.

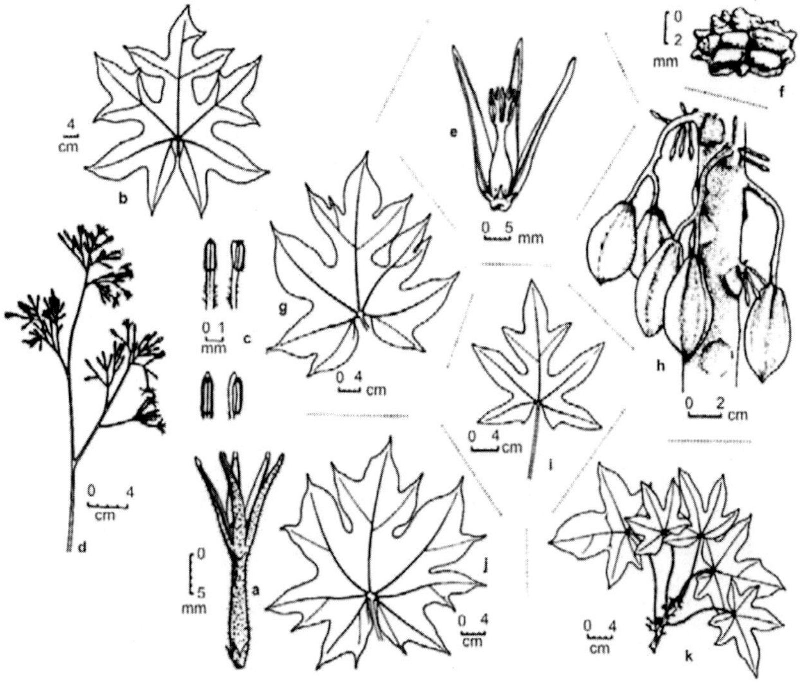

Fig. 31.1. a) male flower; b,g,i-k) leaf forms; c) upper and lower stamens; d) male or predominantly male inflorescence; e) female flower, two petals removed; f) seed; h) fruits from bisexual inflorescence (Badillo, 1993).

Toronche

(***Vasconcellea stipulata*** (V.M. Badillo) V.M. Badillo; toronche, siglalón silvestre)

This is known only in central-southern part of Ecuador and northern part of Peru between 1600 and 2450 m a.s.l. The plants are dioecious tall shrubs or arborescent, up to 10 m high, sometimes branched, often without leaves when flowering. Leaf scars are present on both sides with a conical, hard, pointed stipular and 4-7 mm long spine. Leaf lamina is glabrous, 5-nerved and 5-lobate with a 1-2 mm long stipular spine on each side of the petiole (Fig. 31.2). Male inflorescences emerge solitary or several from each trunk having 1.5–15 cm long peduncle. Female inflorescence is short and few-flowered. Flowers are orange red or orange or greenish. Calyx is opposed, rarely subopposite or alternate to the corolla lobes. Corolla is glabrous outside and pilose inside. Ovary is 5-celled and fusiform. Fruit is yellow, fragrant, 5-locular, ellipsoid, 8–9.5 cm long and 4–4.5 cm wide, truncate at base, narrowed in upper half with 10 longitudinal furrows, 5 of which are deeper. Seeds are numerous with 6-7 inconspicuous, longitudinal interrupted crests.

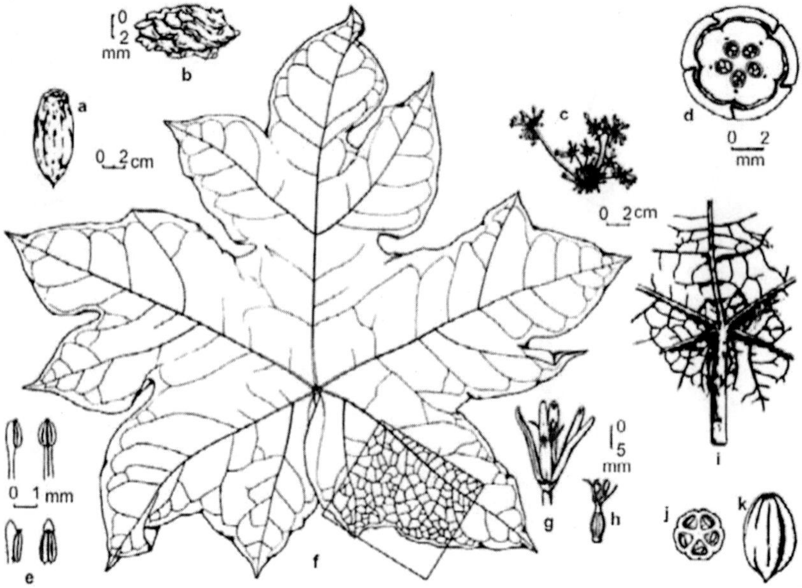

Fig. 31.2. *Vascncellea stipulata*: a) fruit; b) seed; c) male inflorescence; d) cross section of a female flower in flower bud showing corolla and ovary; e) upper and lower stamens; f) leaf; g) female flower; h) gynoecium of female flower; i) detail of abaxial leaf petiole connection; j) cross section of fruit; k) lateral view of fruit (Badillo, 1993)

Vasconcellea parviflora **A. DC.:** It is native to western South America (Ecuador, Peru) and is considered a bridge species for gene transfer in papaya.

Vasconcellea cauliflora **(Jacq.) A. DC. syn. *Carica cauliflora*:** It is a South American species. The leaves are 5-lobed. It is resistant to leaf mosaic virus. The fruit is small.

Vasconcellea goudotiana **Triana & Planch.:** It is a South American species.

Vasconcellea candicans **(A. Gray) A. DC.:** It is native to western South America (Ecuador, Peru). It has heart-shaped leaves (reflected in their four basal secondary veins and a small basal extension in length).

Vasconcellea quercifolia **A. St.-Hil. (alasacha):** It is native to South America. It is hardiest among the *Carica* species and has potential to induce disease resistance in papaya based on hybrids through embryo rescue technique. It is characterized by typical oblong leaves with many small lobes, usually 3-lobed. The fruit is of the size of a date but is of worthless quality.

Vasconcellea monoica **(Desf.) A. DC. (col de monte, peladera):** It is native of Brazil, western South America (Bolivia, Peru) from the region having 37.5 cm rainfall and more than 50°C temperature and thus can withstand heat and rain successfully. The plants are monoecious. The flower cluster contains both male and female flowers. The fruit becomes ready in 3-4 months. It has potential for inducing growth habit in papaya.

Babaco

(*Vasconcellea heilbornii* V. M. Badillo var. ***pentagona*** (Heilborn) V. M. Badillo syn. ***V.* × *heilbornii*** (V.M. Badillo) V.M. Badillo; ***Carica* × *heilbornii*** nm. ***pentagona*** (Heilborn) V.M. Badillo; ***Carica pentagona*** Heilborn); ***C. pentagonia*** Heilborn); ***Carica* × *heilbornii*** V.M. Badillo

The babaco is a natural sterile hybrid between *C. pubescens* Lenne and Koch syn. *C. candamarcensis* Linden. and *C. stipulata* Badillo (Horovitz and Jimenez, 1967). Its native home is highland region of Ecuador in northwestern South America between 1800 and 3000 m above sea level (Dawes and Pringle, 1983) and Colombia. It is cultivated intensively (although on a small scale) in Ecuador, Australia and New Zealand (Morton 1987). The plant is not known in the wild and is propagated by cuttings.

The tree is slender stemmed and herbaceous, about 2 m in height. The large palmate leaves are found near the apex of the trunk. These are somewhat pubescent to completely glabrous. Stipular spines are present or sometimes absent. The plants are dioecious, monoecious or polygamous resembling the papaya plant. Flowering occurs in leaf axils on the newly formed trunk. Flowers are orange, green or greenish yellow, calyx lobes alternating with corolla lobes. Solitary short-peduncled female flowers are green or greenish yellow inside. The pistillate plants are cross incompatible with those of the related species. As the plants are reproduced by clonal means the male flowers may not be seen in some cases. Different cultivars (previously nothomorphs) differ in stipules, female flowers and inflorescences. Badillo (1993) stated that several types can be found in Ecuador. A mutant with sweeter fruit has been identified in New Zealand. Fruit develops parthenocarpically and is aromatic, flavourful, 30 x 10 cm in size, five sided and may reach 30 cm in length, weighs over 1 kg and is usually seedless. The fruit size, however, varies depending on stem vigour and position of the fruit on the stem. The fruit is prismatic parthenocarpic, 1-locular, up to 50 cm long with no or few seeds. It is juicy having slight acidity and mild papaya flavour.

Chamburo

(*Vasconcellea* × *heilbornii* var. ***chrysopetala*** (Heilborn) V.M. Badillo syn. ***Carica* × *heilbornii*** nm. ***chrysopetala*** (Heilborn) V.M. Badillo; ***Carica chrysopetala*** Heilborn; higacho, toronchi)

The plants are tree-like, up to 7 m tall having stipular spines. The flowers are small, solitary and yellow orange especially in the upper part and the remaining are green. The fruits are ovoid, 5-locular, 11-20 cm long and 6–8 cm wide with obtuse apex having few to very few seeds.

Another very similar species is *Vasconcellea* × *heilbornii* var. *fructifragrans* (García-Barr & Hern. Cam.) V.M. Badillo syn. *Carica* × *heilbornii* nm. *fructifragrans* (García-Barr. & Hern. Cam.) V.M. Badillo; *C. fructifragrans* García-Barr. & Hern. Cam. (chamburo, chihualcán). However, it shows some notable differences, like well developed stipular spines, greenish yellow flowers, subpentagonous and almost cylindrical, 12-14 cm long or sometimes longer fruits.

Family Salvadoraceae

It is a small family of trees and shrubs of the Old World arid tropics. The family is native to arid, often saline areas in Africa, Madagascar, Arabia and India including New Guinea. It contains about 3 genera and 8-16 species (Heywood *et al.*, 2007).

The plants are shrubs to small trees with or without thorns and with opposite leaves. The leaves are leathery, entire and simple with minute stipules. *Azima* has 2 or more axillary thorns. The inflorescence is a terminal or axillary, simple or compound raceme with numerous small flowers. The flowers are actinomorphic, hermaphrodite (*Dobera*, *Salvadora*) or unisexual (polygamodioecious in *Dobera* or dioecious in *Azima*). The sepals are 2-4(5) joined in a tube with imbricate or valvate lobes. The petals are 4(5), free or fused at the base (*Salvadora*) with teeth or glands on the inner side. A disk in the form of separate glands may be present between the filaments. The stamens are 4(5) alternating with the petals. These are free in *Azima,* fused into a tube in *Dobera* or joined to the base of the petals in *Salvadora*. The ovary is superior, of 2 fused carpels and bearing a short to almost lacking style and 2-lobed stigma. The fruit is a berry or drupe and contains a single seed without endosperm and an embryo having thick cotyledons.

Genus *Salvadora* L.

Flowers are borne in racemes or compound racemes, nearly sessile and usually perfect having imbricate calyx, marcescent petals, stamens adnate at the base or middle of corolla tube alternating with small distinct nectariferous glands, 1-locular ovary and 1-ovulate and peltate or capitate stigma. Fruit is a drupe with crustaceous endocarp and a globose and oily seed. It contains about 4 species distributed from South and West Africa (Mauritania, Senegal) through Arabia and the Middle East to India and China (Kubitzki and Bayer, 2003). Of these, the genus *Salvadora* having *S. oleoides* Decne and *S. persica* L. are important. Both the species are distributed in Gujarat, Haryana and Punjab and to some extent in Andhra Pradesh and Karnataka States of India (Rao, 1996).

Indian mustard tree
(***Salvadora oleoides*** Decne; saltbush; tooth brush tree)

It is a native of Africa and is distributed in the northern Sahel into the central Sahara from the Atlantic to the Red Sea, through Arabia to India, East and South Africa and Namibia (Maydell, 1986). It is a much branched evergreen shrub or tree with dense and round crown growing 4-6 m in height with dull gray or gray white bark having deep cracks. Leaves are simple, 4-8 x 1-2 cm in size, opposite, thick to fleshy, glabrous, glaucous with very small stipules, elliptic ovate or ovate lanceolate in shape. The greenish white or greenish yellow flowers are borne in axillary or terminal lax panicles. Calyx is short and 4-lobed having petals joined at the base with 4 stamens. Fruits are round and smooth drupes, about 6 mm in diameter having one subglobose, smooth and brown seed, 4 mm in diameter (Maydell, 1986).

Family Capparaceae (Caper family)

The trees, shrubs and herbs are widespread in the dry areas of warm temperate and tropical regions usually having seasonal drought. It contains 17 genera and 470 species (Heywood *et al.*, 2007).

The plants are trees up to 30 m in height, shrubs, scramblers or lianas having alternate leaves, sometimes leafless with green stems. The leaves are simple in many genera or are (1-)3(-5) -foliate (only *Boscia*, *Cladostemon* and *Crataeva*) and without simple leaved or unifoliate species. Simple leaves are usually without stipules, otherwise the stipules are minute except in *Apophyllum*, where the leaves are ephemeral and the stipules are up to 7 mm long. The inflorescence is usually a terminal or axillary raceme. Single flowers are borne in leaf axils, rarely a corymb or fascicle. The flowers are usually hermaphrodite but are unisexual in *Apophyllum* (plants dioecious or polygamous) and *Capparis* (plants dioecious or monoecious), sometimes with male flowers by abortion of carpels (*Boscia, Maerua*). The flowers are 4-merous, actinomorphic to weakly or strongly zygomorphic (*Capparis*), the receptacle sometimes forming a pronounced disk, cone or tube, sometimes with 4 peripheral nectaries or other appendages. The sepals are usually 4, free, then unequal and sometimes petaloid (*Capparis*), often connate, then sometimes valvate or calyptrate or rupturing into 2-4 lobes. The petals are usually 4 but sometimes absent. When present these are pale and often white, cream or yellow in colour and subequal or unequal in size. The sramens and ovary are sometimes borne on a stalk (androgynophore). The stamens are (3-) 4-8 to (-many) but are reduced to 1-3 sterile ones in female flowers of *Apophyllum*. Staminodes are usually absent but are 2-7 when present. The ovary is usually borne on a stalk (gynophore) and is superior, usually 1-locular, sometimes becoming 2-locular by the development of a septum. The stigma is usually sessile but sometimes a short style is present. The fruit is usually many-seeded, rarely 1 to 2-seeded or 4- seeded (*Boscia, Cadaba*), but are 8-seeded in *Buchmannia*.

Genus *Capparis* L. (n=9, 10, 12, 18, 19, 20, 21, 22, Ca. 42, 49, 80)

About 250 species are pantropical with outliers in the subtropics and the Mediterranean region distributed from deserts to rainforests and from coastal to hilly regions up to 2500 m altitude, but are mostly confined to the region having seasonal droughts especially in Africa and tropical America (Kubitzki and Bayer, 2003).

The plants are shrubs or trees up to 16 m tall, evergreen or deciduous, glabrous or hairy. The hairs are simple and branched to stellate or scaly. Leaves are simple, stipulate or exstipulate. Stipules are often thorny. Flowers are solitary or fascicled in leaf axils or in terminal or lateral racemes or fascicled on strongly abbreviated axillary spurs or borne in a serial, supra-axillary row and are bracteate or ebracteate. Flowers are hermaphroditic or the plants are monoecious or dioecious and zygomorphic. Receptacle is often with a dorsally inserted nectary. Disk is usually present and is annular. Sepals are 4 or very rarely 2, free or more or less connate; when 2-seriate the pairs are often unequal, the inner pair being occasionally petaloid. Petals are 4, dorsal pair usually asymmetrical and with connivent bases, usually white, cream or pale yellow. Androgynophore is absent or is spurious. Stamens are 6-

50(-200) and all are fertile. Gynophore is usually elongated. Carpels are 2-6 and the ovary is single or up to 6. Fruit is baccata or pepo, globose or cylindrical, rarely torulose and few to many-seeded. Seeds are usually embedded in the pulp. Embryo is spirally incurved (cochleate) or almost straight. Cotyledons are cumbent, unfolding so that back of one cotyledon lies against the radicle.

Ker

(***Capparis decidua*** (Forsk.) Edgew. syn. ***C. aphylla*** Roth.; khari; teet; dela; kareel)

It is one of the prominent wild flora of the Thar Desert. The related edible species are *C. spinosa* L. (mediterranean caper) and *C. divaricata* Lam. The plant is a much branched, straggling and spinous shrub with grayish bark growing up to 4 m high with 0.5 m spread. Its branches are green, often zigzag, leafless and smooth. Leaves are found only on young shoots. The leaves are small (2x12 mm), caducous and succulent and appear for a brief period on tender shoots. Brick red flowers, 2.5 cm in diameter, emerge in groups of 2-8. The fruits are 0.7-1.5 cm in diameter and have many seeds. Immature fruits are dark green and acrid in taste. On ripening, they turn scarlet red and become sweet in taste and these are known as *pinju*.

The types having pink to whitish fruits occur (Arora and Pandey, 1996). No standard cultivar is known but great variability exists with respect to fruit size, pulp content and density of thorns from which selection of types having large fruit size, small seed, low acidity and high protein content can be done.

⑧ ⑧ ⑧

Chapter 32

SANTALALES

The plants are trees, shrubs, woody lianas or perennial herbs, mostly hemiparasitic and attached to the roots or stems of other plants (parasites or semiparasites without mycorrhiza). Vessels are with simple or sometimes scalariform perforations and scalariform to alternate lateral pitting. Fibres are with bordered or simple pits. Rays are heterogeneous or homogeneous. Axial parenchyma is mostly apotracheal. Sieve element plastids are of Ss or So-type. Nodes are trilacunar or unilacunar, rarely pentalacunar. Leaves are alternate or opposite, simple, entire, mostly exstipulate. Stomata are of various types. Inflorescences are axillary or terminal. Flowers are mostly small, bisexual or unisexual, usually actinomorphic, often apetalous or in the most reduced groups even naked. Calyx is often much reduced or absent. Petals are free or basally connate, mostly valvate, seldom absent. Stamens are usually isomerous and opposite the petals. The filaments are free or sometimes adnate to the petals (very rarely to the sepals) or connate into a column. Anthers are basifixed, tetrasporangiate, disporangiate or monosporangiate opening longitudinally or by apical pore or transverse slit. Tapetum is secretory. Microsporogenesis is simultaneous. Pollen grains are 2-celled or sometimes 3-celled with various types of apertures (from 3-colpate to pentaporate). Gynoecium is of 2-3(-4) united carpels, with short or rarely long style. Ovary is superior to inferior, mostly 1-locular or partitioned only at the base having the ovules pendulous from a columnar placenta extending upward beyond the basal partitions. Ovules are anatropous, epitropous, hemitropous or orthotropous, bitegmic to unitegmic and usually tenuinucellate, but are more often ategmic, sometimes not differentiated from the placenta and mostly characterized by clearly expressed reduction of nucellus. Female gametophyte is of *Polygonum* or sometimes of *Allium*-type, rarely of *Adoxa*-type. Endosperm is cellular or less often helobial. Fruits are of various types. Seeds are mostly without seed coat, usually with well developed endosperm and the embryo is terete and lug-like. The plants contain triglycerides with C-18 acetylenic acids, triterpenic sapogenins and silicic acid. It contains 7-8 families.

Family Olacaceae

It occurs widely in the tropics of Africa, Asia and America (from southern USA through Mexico and Central America down to Argentina extending into the Antilles), often in lowland rainforests but is also grown in the deciduous forests and savanna. It contains 28 genera and 200 species (Heywood *et al.*, 2007). The plants are tropical woody autotrophic or hemiparastic. It is the family of olives, jasmines, American hog plum and African walnut.

The plants are usually small to medium evergreen trees, sometimes shrubs (*Ximenia*), rarely lianas. Some genera are known to be root semi-parasites (*Ximenia*). The leaves are simple or compound, opposite or rarely alternate, spiral or distichous, without stipules. The lamina margin is entire or sinuate dentate. The inflorescence is axillary, racemose or paniculate. The flowers are actinomorphic, bisexual or rarely unisexual in *Ximenia* (plants dioecious), usually small and greenish. The calyx is small, copular and 3 to 6-lobed. The petals are 3-6, free or connate below or forming a tube and valvate. The stamens are as many or twice as many (or more) as the petals in a single whorl. The filaments are free or connate. An intra- or extra-staminal nectary disk is sometimes present. The ovary is superior, semi-inferior or inferior, of 2-5 carpels, with 1-5 locules. The placentation is free, central or axile and with one ovule per locule. The fruit is a drupe or nut, often surrounded or enclosed by the fleshy accrescent calyx and usually single-seeded (Heywood *et al.*, 2007).

Genus *Anacolosa* (Blume) Blume

The genus contains 15 to 22 species.

Galo nut
(***Anacolosa frutescens*** Blume syn. ***A. luzoniensis*** Merr.)

It occurs in Myanmar, Andaman and Nicobar Islands, eastern Thailand, peninsular Malaysia, Sumatra, Java, Borneo, northeastern Sulawesi, Moluccas and the Philippines.

It is an erect shrub or tree growing up to 5-25 m tall. The leaves are elliptic or elliptic oblong to lanceolate and 7-15 x 4-6.5 cm in size. Flowers are 5-15 per fascicle borne on short warts or tubercles. Calyx is cup-shaped with 5-7 short lobes and is 3 mm in diameter. Petals are 6, ovate lanceolate, of 2-3 x 1-1.5 mm and greenish white. Stamens are 5-7. Fruit is a drupe, obovoid ellipsoid to oblongoid, 1.8-2 cm long, 1.5-2 cm in diameter, yellow to orange borne on stout, up to 8 mm long pedicel. The pulp is 3.5-5.9 mm thick. Fruit has one seed, which contains copious albumen (Sleumer, 1984; Verheij and Coronel, 1991).

Genus *Coula* Baill.

It has one to three species of trees native to tropical Africa. Genetic evidence suggests that it should be transferred to family Strombosiaceae.

African walnut
(***Coula edulis*** Baill. syn. ***Coula catrea*** Wildem and Th. Dur.; gabon nut; almond wood)

It is a native of western tropical Africa (Martin *et al.,* 1987). Its trees are medium or large, 20-25 m in height having a large and dense crown casting a deep shade. The bark is thick, rather smooth lenticellate, greenish brown, slightly fissured and white and resinous in young trees. Leaf is alternate, simple, exstipulate, oblong or elliptic, coriaceous and 10-30 x 4-9 cm in size. The secondary nerves are depressed on the upper surface and prominent on the lower surface. The young shoots and leaves are covered with rust coloured stellate hair. Flowers are small, tetramerous or pentamerous with rather thick glabrous and greenish yellow petals. The fruit is an ellipsoid drupe, 3-4 cm long having extremely hard kernel. The flesh surrounding the kernel is green or reddish, 5-6 mm thick and smooth. The oily kernel has a taste comparable to that of a chestnut or hazelnut (FAO, 1982; Duke, 1989).

Genus *Ximenia* L.

The generic name was given to honour a Spanish priest, Francisco Ximenez.

Tallow wood plum
(*Ximenia americana* L.)

Tallow wood plum is a bush forming densely branched spinous shrub or small tree growing up to 2-7 m in height with pale gray and smooth bark. Branchlets are smooth, armed with stout, axillary and purple red spines covered with a waxy bloom. The leaves are alternate, coriaceous, glabrous, ovate to elliptic or oblong lanceolate, 3-8 x 1.5-4 cm in size and obtuse or emarginate at the apex. Petioles are short, slender, up to 6 mm long and canaliculate. Flowers are yellowish green or whitish in colour, fragrant, borne on shortly pedunculate axillary racemes or umbels. Pedicels are 3-7 mm long. Fruits are globose to ellipsoidal drupes that are about 3 cm long, 2.5 cm thick, glabrous, greenish when young becoming yellowish when ripe and contain a juicy pulp and one seed. Seed is woody and light yellow, up to 1.5 cm long and 1.2 cm thick having a fatty kernel (FAO, 1983).

Family Santalaceae

The family contains 40 genera and 490 species (Heywood *et al.*, 2007). It is widely distributed in tropical, subtropical and temperate regions concentrated in relatively dry areas.

The plants are herbs, shrubs and trees, mostly root parasites, but a few are epiphytic branch parasites. The leaves are usually alternate or in some genera (*Santalum, Buckleya)* opposite and without stipules. Scale like leaves are found in several genera (*Anthobolus*, *Exocarpos*, *Phacellaria*) and the plants resemble *Cytisus* or *Cupressus* superficially, while the other species have flattened branches (cladodes) that imitate true leaves. The flowers are borne in spikes, racemes or heads and are actinomorphic, bisexual or unisexual (plants monoecious or dioecious), generally small and inconspicuous. The perianth comprises a single united whorl of 3-6 segments (tepals) that may be greenish or coloured. The stamens are as many as and opposite the tepals and adnate to them at the base. The ovary is inferior or semi-inferior, of 2-5 carpels with a single locule containing 1-5 naked ovules suspended from a central placental column. The style is more or less simple. The fruit is a nut or drupe containing a single seed that has no testa but has copious endosperm.

Genus *Santalum* L.

It has approximately 25 species that occur across the Indo-Malaya, Australasia and Oceania ecozones, from India through Malesia to Pacific Islands, as far as Hawaii and the Juan Fernandez Islands off the south coast of South America. The plants are woody trees or shrubs, mostly root parasites. The best known and valuable tree is the Indian sandalwood (*Santalum album*) that produces highly aromatic wood used in scents, perfumes and herbal medicines.

Quandong

(***Santalum acuminatum*** (R.Br.) A.DC.; native peach)

It is a native of Australia where it has disjointed distribution throughout the southwest extending into desert areas of central Australia. A related species *S. lanceolatum* (plum bush) also produces bluish edible fruits. It grows in areas with an annual rainfall of 125-275 mm. The tree is highly drought and salt tolerant (Wickens, 1995).

The shrub or small tree grows up to 10 m in height. It is a semi-root parasite. The leaves are opposite, lanceolate, 5-7.5 cm long and coriaceous with the lateral veins becoming prominent with age. Flowers are numerous borne in terminal pyramidal panicles. The perianth is 5 mm in diameter having somewhat concave lobes even when open. Margins of the disk are prominent and broadly rounded between stamens, which curve over the notches. Anthers are short and the style is exceedingly short and conical or scarce with 2 or 3 deeply lobed stigmas. The fruit is a globular drupe, 2-4 cm in diameter with succulent epicarp and bony pitted endocarp. The kernel accounts for 40% of the fruit (Wickens, 1995). The flesh layer is thin (3-4 mm) and red or yellow. No standard cultivar is known. Some genotypes have been selected which yield up to 20 kg fruit per tree.

⑧ ⑧ ⑧

Chapter 33

CARYOPHYLLALES

The plants are perennial or annual herbs, shrubs or trees, rarely climbing, often more or less succulent. Anomalous secondary thickening of the stems is frequent. Vessels are with simple perforations, very rarely also reticulate. Fibres are with simple or bordered pits. Axial parenchyma is diffused or vasicentric. Sieve element plastids are of P-type with a peripheral ring of protein filaments and with or without a central protein crystal. Nodes are unilacunar or rarely trilacunar and very rarely multilacunar. Leaves are alternate or opposite, seldom verticillate, simple and mostly entire, sometimes reduced to spines, often showing Kranz anatomy, exstipulate or rarely with small stipules. Stomata are of various types. Flowers are borne in various types of inflorescences or are solitary in the axils, bisexual or less often unisexual, usually actinomorphic, mostly 4-5-merous and entomophilous or less often anemophilous. Sepals are (l)2-5(-10), free or more or less connate below and sometimes petaloid. Petals are two to many or lacking and free or more or less connate. Stamens are one to many, often as many or twice as many as the sepals or petals, in multistaminate flowers often developing centrifugally, especially in Aizoaceae and Cactaceae. Anthers are usually opening longitudinally.

Pollen grains are 3-celled or rarely 2-celled, 3-colpate to pantocolpate and pantoporate, tectate columellate with spinulose annular perforate/punctate tectum. Tapetum is secretory. Microsporogenesis is simultaneous. Gynoecium is syncarpous (eusyncarpous, paracarpous, or lysicarpous) or less often more or less apocarpous but with short, basal, distinctly fused carpellary sections, of two to many carpels, with distinct stylodia. Ovary is superior, half inferior or inferior. Nectaries are present or absent. Ovules are hemitropous to campylotropous, rarely nearly anatropous, bitegmic and crassinucellate. Female gametophyte is usually of *Polygonum*-type. Endosperm is nuclear. Fruits are of various types. Seeds are with peripheral and mostly curved or coiled embryo, bordering or encircling the more or less copious perisperm or nearly without perisperm. Perisperm is completely absent in *Monococcus* of Phytolaccaceae. Endosperm is lacking or scanty. Seed coat is exotestal endotegmic or exotestal mostly producing betalains (betacyanins and betaxanthins) but not anthocyanins (only in the Molluginaceae and Caryophyllaceae producing anthocyanins and not betalains). At least some species of *Phytolacca* contain glucosilonates (Daxenbichler *et al.*, 1991). It evolved from betalain containing ancestors and has lost betalains and require anthocyanins, the two changes possibly being biochemically interrelated. It contains 34 families.

Family Polygonaceae

It is a buckwheat, rhubarb and dock family. It contains 43-49 genera and 1100 species (Heywood *et al.,* 2007). The plants are annual or perennial herbs, shrubs, trees, scramblers or climbers often with swollen nodes, usually containing tannins, often with oxalic acid. The leaves are usually basal or alternate, opposite or occasionally whorled (some *Eriogonum*), simple (but highly reduced in *Calligonum*), with or without a petiole and in subfamily Polygonoideae with a characteristic ochrea or membranous sheath of connate stipules around the stem uniting the stipules. The inflorescence is terminal or axillary, simple or a branched thyrse, appearing paniculate, racemose, spike-like or umbel-like. The flowers are usually hermaphrodite, occasionally unisexual (plants generally dioecious, sometimes monoecious), small, white, greenish or pinkish and subtended by a sheathing membranous tube of fused bracteoles. The tepals are 3-6, usually in 2 whorls of 3 or 1 whorl of 5 (by fusion of 2 tepals), often becoming enlarged and membranous in fruit. The stamens are (2-)6-9 with 2-locular anthers opening lengthwise. The ovary is superior, of (2)3(4) fused carpels with a single locule containing a single basal ovule. The styles are 2-4, usually free. The fruit is a triangular achene or nut, sometimes attached to or enclosed in a fleshy expanded perianth (e.g. *Antigonon*, *Coccoloba*).

Genus *Coccoloba* P. Browne

The genus consisting of about 120-150 species is native to tropical and subtropical regions of South America, the Caribbean and Central America, with two species extending into Florida. The plants are shrubs and trees, mostly evergreen. The leaves are alternate, often large (very large in some species), with the leaves on juvenile plants often larger and of different shape than those of mature plants. The flowers are produced in spikes. The fruit is a three-angled achene, surrounded by a brightly coloured fleshy perianth, edible in some species, though often astringent.

Seagrape
(*Coccoloba uvifera* L.)

It is an evergreen shrub or small tree and can attain as much as 12 m height. Bark is smooth and gray and can peel off in small scales. Leaves are alternate, entire and shortly petiolate with a sheathing reddish brown stipular growth at the base of the petiole. It is dioecious. The racemose inflorescence bears numerous small greenish white scented flowers. The male flowers have rudimentary pistil while the female flowers contain non-functional stamens (staminodes). Fruit is 2 cm in diameter, ellipsoid or ovoid and purple and occur in clusters resembling a bunch of grapes and is composed of a 1 cm long achene surrounded by edible acidulous sweet flesh.

Family Cactaceae (Cactus family)

The plants are perennial xerophytic trees, shrubs, climbers and epiphytes, recognized by their more or less succulent parts (except spines). It contains 90-110 genera and 2000 species (Heywood *et al.*, 2007). Four subfamilies have been recognized, Pereskioideae (shrubs, trees or vines, leaves absent eg. *Pereskia*),

Maihuenioideae (leaves present on these cushion forming plants. eg. *Maihuenia*), Cactoideae (leaves absent or minute eg. *Cereus*) and Optuntioideae (leaves and glochids present eg., *Optunia*).

The plants mostly bear spines. The spines, branches and flowers arise from special sunken cushions or areoles, which may be regarded as condensed lateral branches and are either set singly on tubercles or serially along the raised ribs. Tufts of short barbed hairs (glochids) may also be present in the areoles. The young green shoots undertake photosynthesis but with age these become corky and in the arborescent species develop into a hard, woody and unarmed trunk as in the conventional trees. The vascular system forms a hollow, cylindrical and reticulated skeleton and lacks true vessels. The roots are typically superficial and in the larger species are widely spreading and adapted for rapid absorption of water and nutrients near the soil surface. The flowers are solitary and sessile (except *Pereskia*), bisexual (with rare exceptions) and regular to oblique-limbed. Colour ranges from red and purple to various shades. Stamens are numerous and spirally arranged. The ovary is inferior borne on an areole and is commonly covered in the hairs, bristles or spines consisting of 2 to numerous carpels and forming one locule with numerous ovules on parietal placentas. The style is simple in *Opuntia*. The detached fruit (pseudocarp) grows roots and shoots to form a new plant. The fruit is a berry, which is juicy but may be dry and leathery splitting open in various ways to release the seeds. The seeds have straight to curved embryo and little or no endosperm.

The genera that yield edible fruits are *Opuntia, Stenocereus, Cephalocereus, Escontria, Hylocereus, Lemaireocereus, Myrtillocactus, Nopales, Pereskia, Poreskiopsis* and *Selenicereus*.

Genus *Opuntia* Miller

It is a large genus of 200 species occurring throughout America from South Canada to Patagonia. The plants are trees and shrubs, some low and caespitose having usually segmented stems. The stem segments are cylindric, clavate and subglobose or more or less flattened, sometimes tuberculate, very rarely ribbed with acicular and subulate or papery spines, usually one to many and rarely none, sometimes sheathed. Flowers are yellow, pink, red or off-white in colour and are lateral or sub-terminal having rotate or spreading perianth, rarely erect or slightly zygomorphic (subg. *Nopalea*). Stamens are sometimes touch sensitive and the style is often more or less expanded near the base. Fruits are fleshy or dry and umbilicate. Seeds are circular to broadly oval, 3-9 mm with aril surface bony, trichomatous or alveolate, rarely almost fleshy (Kubitzki *et al.*, 1993). Various species are grown for forage or edible fruits. Some of these have become noxious pests in Australia and South Africa.

The species of *Opuntia* that yield edible fruits are *O. ficus-indica* Mill. (*Verde apastillada), O. robusta* Wendl., *O. lindheimeri* Engl., (Camusea), *O. streptacantha* Lema. (Cardona), *O. leucotricha* DC, *O. tuna* Mill., *O. inermis* DC., *O. caribaea* Britt. and Rose, *O. macrocentra* Engelm., *O. clavata* Engelm., *O. phaeacantha* Engelm., *O. vulpina* Web., *O. amyclaea* Tan. (Fafayuca), *O. megacantha* Salm-Dyck. (Amarilla), *Opuntia* spp. (Blanca I) and *Opuntia* spp. (Blanca II) (Mizrahi *et al.*, 1997).

Prickly pear

(***Opuntia ficus-indica*** Mill.; ***O. gymnocarpa*** Web.; Indian fig; tuna)

It is a native of Mexico. Prickly pear is a bushy tree of 5 m height having oblong or elliptic, 30-50 cm long joints and dark green areoles disposed in 8-9 spiral series with or without spines. Flowers are yellow orange, 7-9 cm long. Pericarpel is 2-2.5 times longer than the perianth, tuberculate with approximately 8 spiral series of areoles. The fruit is an oblong or pear shaped, 4-9 cm long, dark purple or red berry, botanically an accessory fruit, formed from an inferior ovary adhering to the receptacle (Fig. 33.1). The edible portion is made up of a number of funicles intermixed with juicy papillary hairs. The funicles are outgrowths from the internal fruit wall having many soft black seeds of 3-4 cm diameter. The sour fruits of some species are called xoconoxtles and the sweet fruits of the other species are called tunas. The fruit is generally pear shaped and has number of small prickles all over. The prickles fall off easily on rubbing. The joints (segments) either lack spines or have very few (Barbera *et al.*, 1995).

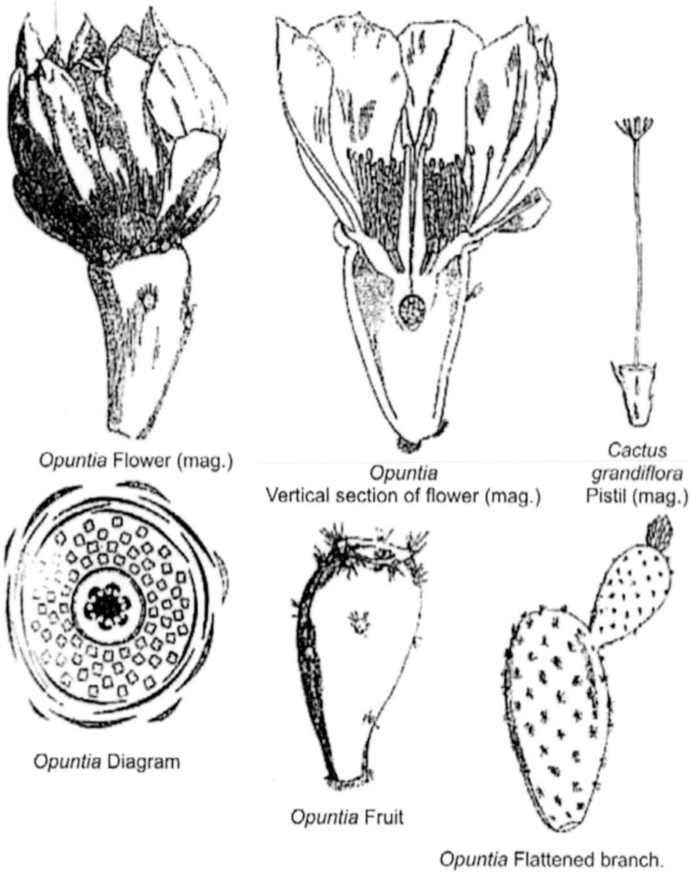

Fig. 33.1. Leaf, flower and fruit of *Opuntia* (Source: Le Maout, 1877).

Cultivars

From the germplasm available at the Centre for Semi-arid Forest Resources, Texas A & M University, Kingsville Campus, Texas, USA (Russel and Felker, 1987) several promising clones have been identified. The important ones are 1270 (high protein forage, fairly sweet pale green fruits), 1271 (fast growing, fairly sweet fruits), 1280 (thornless, pretty good yellow fruits), 1287 (thorny, fruit cultivar) and 1308 (very fast growing vegetable cultivar) (Singh and Felker, 1998). Several cultivars are grown in Mexico, Italy, Israel, Argentina and South Africa (Table 33.1).

Table 33.1. Fruit characteristics of cactus pear cultivars.

Cultivar	Fruit colour	Fruit weight (g)	% flesh	Ripening time	TSS (°B)
		Mexico			
Alfajayucan[a]	Green clear	187	61	July-September	14
Amarilla Huesona[b]	Yellow brown	143	58	August-September	15
Blanca de Castilla[a]	Green clear	148	55	August-September	14
Burrona[a]	Green clear	205	65	September-October	13
Chapeada[a]	Green clear	128	60	July-September	14
Crstalina[a]	Green clear	240	65	August-September	14
Naranjona[a]	Yellow orange	114	55	July-August	15
Papanton[a]	Green clear	130	58	August-September	14
Raja Pelona[b]	Red purple	116	55	July-August	14
		Italy			
Bianca[c]	Pale cream	123	68	August-November	15
Gialla[c]	Yellow orange	125	65	August-November	13
Rossa[c]	Red	127	63	August-November	15
Argentina					
Amarilla[b]	Yellow	160	-	-	15
Naranja[b]	Yellow	130	-	-	12
Rosada[b]	Red	140	-	-	13
Israel					
BS 1	Yellow	103	30	July-August	-
Ofer	Yellow	116	55	July-August	14
		South Africa			
Algeria	Dark pink	110	56	January-February	12
Bakenslip	Light green	84	37	January-February	13
Blue Motto	Pale yellow	140	51	February-March	12
Castillo	Light brown	114	52	February-March	14

Corfu	Pale green	91	35	January-February	14
Direckteur	Light yellow	152	50	January-March	13
Fresno	Yellow white	147	55	January-February	14
Fusicaulus	Pale green	104	45	January-February	14
Fusicaulus O P	Light green	144	52	February-March	12
Guaya Quil	Yellow white	89	41	January-February	14
Gymno Carpo	Yellow	126	52	February-March	12
Hardy Bred	Green white	102	26	January-February	12
Mexican	Light green	139	54	January-February	14
Morado	Pale yellow	114	48	January-February	13
Muscatel	Pale green	161	42	January-February	13
Niagara	Yellow white	147	47	January-February	13
Nudosa	Brownish green	175	47	January-February	11
Oos Kaap	White green	151	49	January-March	14
Santa Rosa	Pink	131	52	January-February	12
Signal	Red	111	34	January-February	14

[a]Hybrids of *O. ficus-indica* with different wild cactus pear forms;
[b]*O. ficus-indica;* [c]Artificially induced late fruits of *O. ficus-indica.*

Genus *Pereskia* Miller

Sixteen species occur in tropical America from South Mexico and the Caribbean region to North Argentina (Kubitzki *et al.*, 1993). The plants are trees, shrubs and woody climbers and some with tuberous roots. Stems are not conspicuously succulent, terete and elongated. Leaves are broad, thin, deciduous or sub-persistent with usually numerous spines. Inflorescence is paniculate or corymbose up to 50-flowered or flowers are clustered or solitary. The flowers are stalked or sessile, pericarpel areoles with wool, often hairs and rarely spines, tube none, perianth rotate, spreading or rarely erect, red, orange, purplish, pink, yellow or white. Fruits are sometimes with persistent scales and sometimes proliferous.

Barbados gooseberry

(***Pereskia aculeata*** Plum. ex Mill. syn. ***P. pereskia*** Karst.; West Indian gooseberry; Spanish gooseberry; leafy cactus; lemon vine; blade apple)

It is native of tropical America. When young the erect woody deciduous shrub grows up to 10 m long into a scrambler or climber having viselike branches. Spines on the trunk are long, slender and in groups. The spines on the branches are short, recurved, usually in pairs and rarely solitary. Leaves are alternate, 3.2-10 cm long, short-petioled, elliptic oblong or ovate and shortly pointed at the apex and sometimes fleshy. Flowers are white, yellowish or pink tinted and 2.5-4.5 cm across. The calyx tube is prickly. Fruit is round, oval or pyriform, lemon or orange yellow or reddish, 1-2 cm wide with thin, smooth and somewhat leathery skin. Seeds are 4 mm long, thin, brown or black and soft (Morton, 1987).

Cultivars

Two cultivars well known in ornamental plants trade are:

Godseffiana: It has broad leaves, basically yellow green variegated with scarlet and coppery on the upper surface.

Rubescens: It has leaves variegated with red (Morton, 1987).

Genus *Hylocereus* (A. Berger) Britton & Rose

Hylocereus species originate principally from Latin America (probably from Mexico and Colombia), and possibly from the West Indies (Britton and Rose, 1963). The genus has 16 species endemic to Latin America and distributed in Central America, the West Indies and North and South America (Kubitzki *et al.*, 1993). *H. undatus* and *H. costaricensis* are the two most widely cultivated species. These are rarely cultivated on a large scale outside Colombia, Costa Rica and Nicaragua. However, *H. undatus,* the most cosmopolitan species, has undergone extensive development in Vietnam. *Hylocereus* spp. are diploid ($2n$ = 22). These are characterized by climbing plants with aerial roots that bear a glabrous berry with large scales. The plants are scandent or epiphytic shrubs. The stems are often 5 m or more long, usually 3-winged or angled, segmented, green or glaucous often having horny margins, aerial roots and areoles with short spines or rarely spineless. Flowers are usually very large, funnel form, nocturnal, white or rarely red having stout pericarpel and tube, usually broad and triangular scales, usually naked or nearly so areoles of pericarpel and tube, continuous series of stamens, thick style and sometimes bifid stigmas. The flowers of *H. undatus* and *H. costaricensis* appear under the areoles. The flowers are 25-30 cm long and 15-17 cm wide. The ovary is located at the base of a long tube carrying the foliaceous scales to the exterior. There are numerous stamens on a slender anther stalk. The tubular style is 20 cm in length and 0.5 cm in diameter, the stigmas have 24 slender lobes, creamy green in colour. Pollination of these two allogamous species is carried out by bats at night. Manual cross pollination is needed to ensure fruit set and development and may be carried out before anthesis. Fruits are large, globose, ovoid or oblong and fleshy with broad scales. Fruit characteristics of some *Hylocereus* species (Le Bellec and Vaillant, 2011) are given in Table 33.2.

Table 33.2. Fruit characteristics of some *Hylocereus* species.

Species	Common name	Weight (g)	Peel colour	Flesh colour
H. costaricensis	Red pitaya	250-600	Red	Red purple
H. megalanthus	Yellow pitaya	120-250	Yellow	White
H. ocamponis	Red pitaya	150-400	Red	Red
H. monocanthus	Red pitaya	200-400	Purple	Red purple
H. undatus	Dragon fruit	300-800	Rosy-red	White
H. undatus subsp. *luteocarpa*	-	100-480	Clear yellow	White

Key to species of the genus *Hylocereus* (Britton and Rose, 1963)

A. Stems bluish or whitish or greyish, areoles spiniferous; margin foliaceous, scales red

Margins of joints horny; spines short, conic; oblong fruit, red flesh .... 1. *H. purpusii*

Margin of joints not horny; spines few, conic

Branches slender, oblong fruit .... 2. *H. polyrhizus*

Branches stout, ovoid fruit, purple red flesh .... 3. *H. costaricensis*

AA. Stems bright green; areoles spiniferous; margin foliaceous, scales green

Margin of joints horny; ribs of stem crenate; oblong fruit, white flesh .... 4. *H.undatus*

Margin of joints not horny; oblong obovoid fruit becoming nearly smooth, white flesh .... 5. *H. trigonus*

Pitahaya

(***Hylocereus*** spp.; pitaya, pitajaya, pitajuia or pitalla)

All pitahaya are grouped into four main genera; *Stenocereus* Britton & Rose, *Cereus* Mill., *Selenicereus* (A. Berger) Riccob and *Hylocereus* Britton & Rose. The generic term 'pitahaya' includes several different species. Yellow pitahaya, *S. megalanthus* (Schum.) Britt & Rose is a fruit with a yellow skin and white pulp and red pitahaya (*Hylocereus* spp. Britt & Rose) has fruits with a red skin and white or red pulp, depending on the species (Le Bellec *et al.*, 2006). The fruits are attractive in shape and colour and have very good properties for the food industry. The red flesh species are also rich in betalains, meeting the increasing trade interest for antioxidant products and natural food colorant. Pitahaya fruits are not very fragile and are easy to keep fresh.

Strawberry pear or Pitaya

(***Hylocereus undatus*** Britt. and Rose syn. ***Cereus undatus*** Haw.) (2n=22)

It is a terrestrial or epiphytic vine cactus. The stem is heavy, 3-sided, green, fleshy, much branched having flat wavy wings with horny margins that may reach 3 m in length. The stem clings to a wall by means of numerous strong aerial roots. There are 2 to 5 short spines on each areole. The flowers are magnificent, night blooming, very fragrant, bell shaped, white, up to 35 cm long, 22.5 cm wide having a thick tube bearing several linear, 4-7.5 cm long green scales above which is a circle of recurved, greenish yellow, linear, 10 cm long and 3.2-4 cm wide segments. Numerous cream coloured stamens form a showy fringe in the centre and at the apex of the thick perianth tube. The non-spiny fruit is oblong oval, 10 x 6.25 cm in size and is coated with bright red fleshy or yellow and ovate bases of scales. White, juicy and sweet pulp contains innumerable tiny black and partly hollow seeds. In Costa Rica and Nicaragua the important cultivars are Lisa, Rosa and Cebra.

Family Simmondsiaceae

The plants are rigid and divaricately branching xerophytic shrubs. Leaves are evergreen, small (about 2 cm long), opposite, flat, leathery, shortly petiolate, simple (but jointed at the base), oblong or ovate, pinnately veined, cross venulate and exstipulate with entire margins. Flowers are small, regular, (4–)5(–6) merous, cyclic, solitary or aggregated in inflorescences in racemes (female) or in heads (male). Free hypanthium and hypogynous disk is absent. Calyx are (4–)5(–6), 1- whorled, polysepalous, regular, persistent, accrescent (foliaceous) and imbricate (fringed). Androecium (8)10(–12), free of one another and exclusively of fertile stamens. Stamens are (8–)10(–12), diplostemonous and filantherous. Anthers are basifixed (or ventrifixed), dehiscing via longitudinal slits and conspicuously extrorse. Microsporogenesis is simultaneous. Tapetum is glandular. Gynoecium is 3 carpelled, superior, syncarpous and stylate. Carpels are reduced in number relative to the perianth. The pistil 3-celled. Ovary is 3-locular. Styles are 3 and free and apical. Stigmas are papillate. Placentation is axile to apical. One ovule per locule is pendulous, non-arillate, anatropous, bitegmic and crassinucellate. Embryo sac development is *Polygonum*-type. Endosperm formation is nuclear. Fruit is non-fleshy dehiscent capsule. It is 1-seeded (two of the locules empty). Seeds are non-endospermic, large, glandular and conspicuously hairy.

Jojoba
(*Simmondsia chinensis* L.)

Jojoba is a shrub that may reach 4-5 m in height. Its flat gray green leathery leaves and deep root system make it well adapted to withstand drought and the desert heat. The shrub has a long life, perhaps up to 200 years. These may either be male (staminate) producing pollen or female having pistillate flowers. The flowers have no smell or petals to attract insects. Pollination usually takes place by wind during late March. The pistillate flowers develop into fruits in August. During the summer heat, the green fruits may dry resulting in shriveling and peeling of outer skin exposing wrinkled, brown, soft skinned and small olive sized nuts. The nuts contain oil that is clear and odourless. Half the weight of the nut is oil. It is felt less oily than the traditional edible oils. Its chemical structure is unique among all known vegetable oils. It is a polyunsaturated liquid wax of a type not easily synthesized commercially. The only other source is the sperm whale, which has now nearly vanished as a result of its killing to meet the oil demand.

⑧ ⑧ ⑧

ASTERIDS

The plants are mostly sympetalous. Ovules are with single integument and thin-walled megasporangium although both are homoplasius. Number of stamens equal the number of petals. The stamens are epipetalous with an obviously sympetalous corolla. Gynoecium is nectary and the endosperm is cellular. Iridoids are widespread.

Chapter 34

CORNALES

The plants are woody. The leaves are alternate or more often opposite, simple undivided. Hydathodes are present in the teeth. Flowers are often 4-merous having persisting calyx much smaller than corolla, epigynous intrastaminal nectary disk, ovules with single integument and often a thin megasporangium and inferior ovary. Fruit is drupaceous. Diverse iridoid compounds are present. It has 6 families and over 650 species.

Family Cornaceae (dogwood family)

It contains about 7 genera and 105 species (Heywood *et al.*, 2007), mostly occurring in northern temperate regions and tropical Asia and less commonly in South America, Africa and Australia.

The plants are mostly small trees, but are sometimes large trees and occasionally small perennial herbs with woody rhizomes. The leaves are alternate or opposite, usually entire but rarely serrate (*Davidia*). The inflorescence is usually a terminal but sometimes axillary thyrse, which may occasionally be capitate and subtended by showy white bracts. The flowers are epigynous and actinomorphic and sometimes unisexual. The calyx has 4-5 or rarely 10 small teeth. The petals are 4-10, usually short and broad but conspicuously long and narrow in *Alangium*. The stamens are usually as many as or twice the number of petals, often in 2 whorls or up to 40 in one whorl as in *Alangium*. The ovary is inferior and usually 1 or 2-locular but up to 6 or l0-locular in *Davidia* with one pendulous ovule in each locule. The fruit is drupaceous with a fleshy exocarp enclosing a stone dehiscing while germination except in *Alangium*.

Genus *Cornus* L. (n = 9-11)

It has about 60 species mainly in temperate regions of the northern hemisphere, 1 or 2 species in South America and 1 in Africa (Kubitzki, 2004). The important species is *C. mas* L. (cornelian cherry or dogwood). The other species are *C. florida* (flowering dogwood), which is a small tree, 7 m in height and produces showy white bracts during May, *C. kousa*, which is a bushy plant bearing 4 creamy white bracts and fleshy strawberry like red fruits, *C. macrophylla*, which is a tree of 10-16 m height bearing yellowish white flowers and globose and blue fruits when ripe and *C. nuttallii* (Pacific dogwood) that has magnificient white showy bracts.

The plants are evergreen or deciduous trees or shrubs or rarely rhizomatous herbs having 2-armed and calcified trichomes. Leaves are usually opposite (alternate in *C. alternifolia* and *C. controversa*). Inflorescences are terminal or axillary and thyrso-paniculate or in heads surrounded by petal-like involucral bracts. The flowers are hermaphrodite, rarely dioecious and tetramerous having distinct or obsolete calyx, valvate petals, alternipetalous stamens, fleshy disk, dorsifixed anthers and style with capitate or truncate stigma and (1)2(3)-locular ovary. Drupe is areolate at the apex having fleshy, blue, white or red pericarp, bony or crustaceous and 2(1)-locular endocarp opening with valves. Seeds are oblong and compressed having elongated embryo and fleshy endosperm.

Cornelian cherry
(***Cornus mas*** L.; dogwood)

It is native to Europe and is a deciduous shrub or small tree attaining a height of 5-8 m. Its brown and gray bark flakes off toward the base of the bole from where numerous shoots emerge. Leaves are opposite, ovate to elliptic having acute to acuminate apex, broad cuneate base, lustrous above showing 3-5 pairs of veins. Flowers are yellow and fragrant and appear early in clusters on the bare branches. Flower has 4 greenish external bracts, smaller in size having 4 smooth petals and 4 stamens. Fruits are 1.2-1.5 cm long, oblong, pendulous, brilliantly red and glossy berries. Each berry contains a hard ellipsoidal nut (Bailey, 1971).

⑧ ⑧ ⑧

Chapter 35

ERICALES

The plants are shrubs or less often small trees, perennial herbs or woody lianas. Raphide cells are absent. Vessels are with scalariform perforations, sometimes with many bars (up to 50 in Clethraceae and 65-70 in Cyrillaceae) but often with both scalariform and simple perforations or only with simple perforations. Lateral pitting is from scalariform to alternate. Fibres have bordered pits. Rays are typically heterogeneous. Axial parenchyma is diffused, often scanty or wanting. Sieve element plastids are of S- or Pcf-type (Cyrillaceae). Nodes are unilacunar, trilacunar (some Ericaceae and some Styphelioideae) or multilacunar (some Styphelioideae). Leaves are alternate, less often opposite or verticillate, simple, entire, exstipulate, sometimes without chlorophyll and reduced to mere scales. Stomata are anomocytic, paracytic, tetracytic or of other types. Flowers are borne in racemose inflorescences or sometimes solitary (axillary or terminal), bisexual or rarely unisexual and usually actinomorphic. Sepals are (3-) 5(-7), free or basally connate, imbricate or valvate. Petals are isomerous with sepals, free or more often connate into sympetalous corolla with usually imbricate or contorted, more rarely valvate lobes and rarely petals are wanting. Stamens are twice as many as the petals or less often the same number and then alternate, rarely only two (*Ceratiola* in Empetroideae) or up to 20. Filaments are free or more or less adnate to the corolla. Anthers are tetrasporangiate or disporangiate, usually inverted, opening longitudinally or more often by apical (morphologically basal) pores or slits, often with two or more slender appendages, which are rarely borne on the upper part of the filament instead of on the anther. Tapetum is secretory. Microsporogenesis is simultaneous. Pollen grains are 2-celled or sometimes 3-celled, very often in tetrads and mostly 3-colporate. Nectary disc is usually present and intrastaminal. Gynoecium is of 2-10 united carpels having simple style. Ovary is superior or inferior, 1-10-locular having numerous or rarely several or one ovule per carpel. Ovules are anatropous or hemitropous, unitegmic, tenuinucellate. Female gametophyte is usually of *Polygonum*-type. Endothelium is present. Endosperm is usually cellular (nuclear in *Rhododendron* and *Vaccinium*) with terminal haustoria that are usually micropylar and chalazal. Fruits are septicidal or loculicidal capsules, berries or drupes. Seeds are small. Nonhydrolyzable tannins, ellagic acid and hydroquinones are present.

Family Lecythidaceae (lecythis family; Brazil nuts and cannonball tree)

It contains 25 genera and 310 species (Heywood *et al.*, 2007). The plants are small to very large trees or occasionally shrubs with spirally arranged leaves in clusters at the tips of twigs or borne in 2 ranks. Leaves are large, simple and usually without gland dots, sometimes toothed and asymmetrical at the base, with or without stipules. The inflorescence is up to 1 m long spike, as in *Courouptia guianensis* or is a panicle borne terminally or racemes in the axils of leaves on side shoots of young stems or on the older parts of the stems. The flowers are actinomorphic or zygomorphic, bisexual, often large and showy in shades of red, pink and yellow or white and have an attractive and fluffy appearance due to numerous stamens (subfam. Lecythidoideae) or with long pedicels (subfam. Scytopetaloideae). They seldom last long in *Barringtonia* and remain open only one night before the stamens and petals fall. The flowers are regular or irregular having (2-) 3-6 sepals fused into a cup-shaped structure and 3-6(-18) petals (absent in *Foetidia* and subfam. Scytopetaloideae), free (subfam. Lecythidoideae) or united at the base or into a ribbed tube (subfam. Napoleonaeoideae). In subfam. Scytopetaloideae, the petals are replaced by staminodes that are sometimes thick and become either reflexed as the anthers mature or do not separate and fall off the flower as a cap. The stamens are joined to each other and to the petals or staminodes at their base are in one or more rings. The ovary of 2-6 fused carpels is inferior, totally fused with the top of the receptacle or partially sunk into it or is superior (subfam. Scytopetaloideae) having 2-6 or more locules, each having 1 to many ovules, a long simple style and lobed or disk-shaped stigma, or 3-8 locules and 2 or more pendulous ovules per locule borne on axile placentas, a simple style and simple stigma (subfam. Scytopetaloideae). Pollination is frequently by bats that are attracted by the sweet smelling flowers. The usually large fruits have indehiscent fleshy outer layers, hard and woody inner layers having a lid through which the seed leaves the fruit or woody capsule, but is drupe-like in a few cases. The seed is large and woody and lacks endosperm except in subfam. Scytopetaloideae where the seed contains a narrow embryo surrounded by copious endosperm which may have a rough or mottled appearance.

The family consists of a group of large evergreen or deciduous trees of which two genera *Lecythis* and *Bertholletia* produce fruits of commercial importance.

Genus *Bertholletia* Humb. & Bonpl. (2n = 34)

It has only one species, *B. excelsa* Humb. & Bonpl. confined to Amazonia (Kubitzki, 2004). The trees are large, up to 55 m tall. Leaves are not clustered at the ends of branches. Inflorescences are terminal and axillary spikes or panicles of spikes. Flowers are zygomorphic having 2-lobed calyx, 6 petals, androecium with strap-like hood arching over the summit of ovary and tightly pressed against staminal ring, all sterile hood appendages and 80-135 stamens. Ovary is (3)4(-6)-Iocular with 16-25 ovules inserted at the base of septum of each loculus. Fruit is a large, globose woody circumscissile capsule with small, internally detaching operculum and is thus secondarily indehiscent. The seeds, triangular in cross section are 10-25 in number having hard and woody testa, not embedded in pulp. The embryo is undifferentiated and cotyledons are represented by two small scales.

Brazil nut

(*Bertholletia excelsa* Humb. et Bonpl.; nigger-toe; castanha-do-brasil; pavory nut; cremenut; castanas; para nut; neuz de para)

The species is native of the forests of southeastern Amazonia in South America. The tree is deciduous, large, 35-50 m in height with cylindrical trunk having grayish bark. Leaves are simple, alternate, oblong or elliptic oblong with wavy margins and 25-35 x 8-12 cm size. The light yellow flowers are borne on spike like racemes. The bisexual flower has two sepals, 6 pale yellow petals, numerous stamens and short style. Brown and woody ripe fruits (pyxidum) are about 10-15 cm in diameter, globose, heavy (3 kg) with 12-25 or more triangular nuts (4-7 cm long) packed inside the thick woody shell and a thin brown testa with an aperture at one end, which is closed by a woody plug and must be broken open to extract the nut. The nut has an oily endosperm (FAO, 1986; Wickens, 1995).

Genus *Lecythis* Loefl (2n = 34)

The genus has 26 species that are neotropical in Central America to São Paulo, Brazil (Kubitzki, 2004). The species such as *L. pisonis* Cambess. syn. *L. urnigera* Mart. ex Berg, *L. minor* Jacquin. syn. *J. elliptica* H.B.K., *L. zabucajo* Aubl., *L. lanceolata* Poir syn. *L. ollaria* Loefling, *L. usitata* Miers., *L. tuyrana* Pitt., *L. grandiflora* Aubl. and *L. ampla* Mier produce edible nuts. These are native of South America, Brazil in particular (Martin *et al*., 1987).

The trees are small to very large. Leaves are not clustered at ends of branches. Inflorescences are terminal or axillary racemes, spikes or panicles, rarely on young stems below the leaves. Flowers are zygomorphic having 6-lobed calyx, the lobes often with mucilage ducts and 6 petals. Androecium has strap-like hood arching over the summit of ovary, the hood appendages are sterile or with anthers and 70-1000 ring stamens. Ovary is (3)4(-6)-locular having ovules inserted on the lower part of septum. Fruits are small to very large, woody circumscissile capsules, sometimes indehiscent and falling to the ground with seeds inside, generally anulate or urceolate, the base usually remaining on the tree after dehiscence. The seeds are often with a basal swollen aril projecting beyond the micropylar end, rarely with lateral aril or are exarillate.

Monkey pot (paradise nut or sapucaia nut)

The *Lecythis* spp. are commonly known as monkey pot or paradise nut (English), olla de mono or sapucaia (Portuguese) and zabucajo (Spanish). The name is said to be derived from baiting an empty fruit with food and fixing it to a low branch for a monkey to easily insert its paw through the opening, but cannot withdraw it once it has grasped the contents. The other local names are tocari, marmite-de-singe, monkey nut and cream nut.

***L. minor* Jacquin. syn. *L. elliptica* H.B.K.:** The tree is found wild in the Orinoco and Amazonas basin and has good potential for commercial planting (Martin *et al.,* 1987). It attains 18-22 m height and has large, coriaceous, broadly oblong oval leaves. Hermaphrodite sessile flowers are borne on paniculate inflorescence. Of the 6 petals, two are longer. Androecium covers the ovary with many sterile appendages. The dark brown fruit is globose, roughly verrucose and 10-15 cm in diameter with convex margin at the orifice enclosing 8 seeds. The seed kernels are of dark brown colour, 3-5 cm in size enclosing the small cotyledons with big hypocotyls containing the reserve substances required for germination (Brücher, 1989).

Lecythis pisonis **Cambess. syn.** ***L. urnigera*** **Mart. ex Berg:** It is distributed throughout the Brazilian Amazonia, Colombia, Orinoco Basin of Venezuela and the Guyanas. It is also cultivated on small scale in Brazil, Guiana, West Indies and Malaysia. This is a deciduous tree growing up to 30-40 m tall with 60-80 cm trunk diameter. Leaves are simple, petiolate and ovate to widely elliptic in shape, 2.5-12 x 1.5-5 cm in size, herbaceous to coriaceous. The bisexual flowers are small, borne in terminal racemes. The flower consists of 6 unequal and ovate sepals, 6 unequal petals and numerous stamens present within a hooded androecium. The androecium hood is flat with proximal appendages bearing anthers and the distal ones being antherless. The pollen turns yellow or white or black after 24 hours. Ovary is inferior and the styles are short with an annular expansion towards the apex. Fruit is a large woody, dehiscent, bell-shaped and pendent capsule up to 25 cm long with 1-2 cm thick woody pericarp borne in groups of 30-40. Large operculum (circular lid) detaches on maturity leaving the capsule dangling by a slender fleshy funicle that decays and allows the seeds to fall. The 30-40 irregularly wrinkled seeds are oblong, 5 cm long with creamy white texture (Prance and Mori, 1979; FAO, 1986; Duke, 1989; Mori and Prance, 1990).

Lecythis zabucajo **Aubl.:** It is distributed throughout South America (Martin *et al.*, 1987) and is common in the rainforests of Venezuela, Colombia, Brazil and Panama. The tree is evergreen and large, growing up to 50 m high with a straight trunk. Leaves are simple, elliptic obovate to oblong lanceolate, 4.3-14.3 cm long and 2.6-6.7 cm wide, acuminate and shed towards the end of the dry season. The white flowers are borne on terminal racemes. The fruits are urn-shaped, 15-20 cm long, 22-26 cm in diameter having a thick woody lid containing nuts of about 5 x 2.5 cm size with less sharply angled, thin and brown shells than those of Brazil nut having longitudinal ridges. The kernel is ivory white, 5 cm long and almost creamy in texture. It is sweeter than Brazil nut. It bears fruits all over the canopy unlike the Brazil nut which bears fruits only on its topmost branches.

Lecythis lanceolata **Poir syn.** ***L. ollaria*** **Loefling:** It is found growing in forests in Brazil and Venezuela. The tree is large having spreading branches. Fruit is a large woody capsule with a lid which bursts open when the seeds are ripe. Monkeys are said to put their hands inside the capsule in order to extract the seeds which have a fleshy interior rich in oil and a woody outer casing. The nuts are edible and have pleasant flavour. The tree is considered to be a selenium accumulator. Part of the element is bound to selenium rich proteins. The tissues of the bark, leaves, capsule and seeds contain selenium. The highest concentration (about 5 g/kg) has been observed in the nuts, half of which is soluble in water.

Family Sapotaceae (chicle, gutta, percha, masticand, sapodilla)

It contains about 35-75 genera and 800 species that occur pantropically mainly in low and lower montane rainforest. The plants are trees reaching 30 m tall and 2 m in girth, shrubs, lianas or geophytes with extensive underground woody stems containing white (sometimes yellow or blue) latex in the twigs and fruit, usually also in the bark. The leaves are simple, entire or rarely with spiny teeth, spirally arranged or distichous, often crowded in false whorls and sometimes with stipules that soon drop. The inflorescence is fasciculate or rarely a solitary flower, axillary, often behind the leaves or on the trunk. The flowers are bisexual or unisexual (plants monoecious or dioecious), regular, scented, often white or cream and often nocturnal and bat pollinated. The sepals are free and overlapping or sometimes not, in 2 whorls of 2-4 or one whorl of (4)5(6) or 6-11 in a spiral. The petals are usually equal in number to the sepals but usually in one whorl and are fused at the base, entire or divided, sometimes into 3 lobes, the lateral ones sometimes laciniate. The stamens are 4-35(-43), epipetalous, rarely free, either equal in number opposite the corolla lobes or numerous, sometimes alternating with staminodes or in clusters. Staminodes (0-8)-12) occur in one whorl and are sometimes petaloid. The ovary is superior, usually of many fused carpels, with 1-15)(-30) locules each containing a single axile or basal ovule, sometimes 2-5 ovules. The style is simple. The fruit is a berry and is not articulated. The seeds are 1 or few with an oily endosperm and bony testa and a large embryo.

Genus *Manilkara* Aubreville

The pantropical genus has 30 species in America, 20 in Africa and Madagascar and 15 in Asia and the Pacific (Kubitzki, 2004). The species that produce edible fruits are *Manilkara hexandra* (Roxb.) Dubard (khirni), *M. kauki* (L.) Dubard (wongi, kauki), *M. huberi* (Ducke) Chev. (macaranduba), *M. rufula* (Miq.) H. J. Lam. (macaranduba) and *M. salzmannii* (A. DC.) H. J. Lam. (macaranduba). These are native of Brazil and are grown in homesteads (Anon., 1962).

The plants are trees, rarely shrubs, nearly always with sympodial branching and rarely dioecious. Venation is nearly always brochidodromous, secondaries often looping to form a submarginal vein (eucamptodromous and then with convergent secondaries), tertiaries parallel to the secondaries often descending from the margin and the higher order venation is usually reticulate or areolate. Stipules are small, caducous or absent. Flowers are solitary or fasciculate. Corolla is nearly always glabrous, occasionally carnose having usually much shorter tube than the lobes, rarely equaling or exceeding them. The lobes are 6(-9), usually spreading and divided to the base into 3 segments with the median segment usually erect, often clawed and clasping the stamen. Two lateral segments are spreading, shorter, equaling or exceeding the median segment, entire or deeply divided or laciniate or less frequently corolla lobe only partially divided or 2-3-lobed at the apex or rarely entire. Stamens are 6(-12), very rarely fixed within the tube with filaments, sometimes fused with the staminodes. Staminodes are (0-)6(-12), bifid, laciniate, truncate or irregularly divided or vestigial, erect or rarely incurved but not forming an envelope round the gynoecium and nearly always glabrous. Disk is small and annular. Ovary is 6-14-locular, hairy or glabrous with axile or basi-ventral placentation and exserted style. Fruit is 1 to several-seeded. Seed is ellipsoid to obovoid, laterally compressed with a hard, shining and woody testa, nearly always narrowly elongate hilum, basi-ventral or less frequently extending along most of the adaxial face and very rarely broad. Embryo has foliaceous cotyledons and exserted radicle and copious endosperm.

Sapodilla

(*Manilkara zapota* (L.) van Royen; sapota; chiku)

The plants are tender, evergreen, subtropical or hardy tropical. It is native of tropical America. The tree has straggly branches, dense dark foliage, medium height and milky latex. The leaves are lanceolate, shiny above and pubescent below. The flowers have a white tubular corolla. The fruit is a berry of medium size having 2-4 cells, russeted, brown and rough skin. The pulp is yellowish brown, astringent before ripening but not as in the persimmon. The seeds are 1-4 or even more.

Cultivars

In India, the cultivars are grouped under the following four types on the basis of tree growth, nature of branching and foliage colour:-

1. Trees with erect habit; branches appear in whorls having deep green, broad and oval foliage. Fruits are large with smooth and yellow skin and butter-like and sweet pulp.
2. Trees with drooping habit; branches appear in whorls having light green, narrow and elliptical foliage. Fruits are small with rough brownish skin and pulp of inferior quality.
3. Trees with spreading habit; branching irregular having deep green, broad and oval foliage. Fruits have smooth and yellow skin with butter- like sweet pulp.
4. Trees with spreading habit and the fruits have rough skin and inferior pulp quality.

Each group may be subdivided into 2 sections having round or oval fruits.

Kali Patti: It is an early and high yielding (160 kg per tree) cultivar of Maharashtra, Karnataka and Gujarat. The tree is medium tall and round topped having spreading branches. The foliage is dark green, broad ovate and thick with entire margin. The fruits are sweet, of medium size and excellent quality and varied shape (oval in monsoon and round in other seasons) and occur singly. The flesh is sweet and mellow. having 1-4 seeds.

Chhatri: It is the second important cultivar of Maharashtra. The trunk remains exposed. The tree form resembles an umbrella with horizontal and drooping branches that appear in whorls at different levels of the trunk. The leaves are light green and ovate lanceolate with entire margin. The fruit is similar to Kali Patti but is less sweet and poorer in quality. It is a fairly high yielding cultivar and produces fruits regularly.

Cricket Ball (Calcutta Large): The cultivar does well in dry weather and at elevations below 300 m. It is grown mostly in West Bengal, Andhra Pradesh, Karnataka, Tamil Nadu and Maharashtra States of India. The tree is round topped having spreading branches. The leaves are light green and shiny. The trees do not bear heavily. The fruits are very large and round having rusty brown surface. Its flesh is pulpy, crisp, granular/gritty and is moderately sweet.

Oval: It is a shy bearing cultivar like Cricket Ball. The fruits are small to medium in size and oval in form having gritty or granular and less sweet flesh.

Thagarampudi: The cultivar is grown mainly in Tamil Nadu. The tree has round top and spreading branches. The leaves are green and ovate to ovate lanceolate. The fruit is round oval and medium sized having thin and rough skin and dull colour, very sweet, melting and juicy flesh having 1-4 seeds.

Dwarapudi: The fruit is smaller, sweet and very popular.

Pala: It is popular in Tamil Nadu and Andhra Pradesh. The fruit is small to medium in size and roundish oval similar to egg in shape with broadly pointed apex and is very sweet.

Baramasi: It is popular in Bengal, Bihar, Uttar Pradesh and Karnataka.

CO 1 (Cricket Ball x Oval): Fruit is long oval, medium sized, bigger than Oval and sweeter than Cricket Ball, weighs about 125 g having reddish brown and granular flesh and 18°B TSS. Fruit yield is 175-200 kg/tree.

CO 2: It is a clonal selection from Baramasi. Fruit is round, medium sized with outer surface cinnamon brown and inner surface yellowish green in colour. The flesh is soft, juicy, slightly gritty and light brown in colour having slight aroma and sweet taste. The seed is medium sized, black, obovate and beaked having distinct suture and slightly adhering to the flesh. Fruit yield is 175 kg/tree or 11.8 tonnes/ha.

CO 3 (Cricket Ball x Vavilavalasa): The cultivar is suitable for growing in plains up to an elevation of 100 m above mean sea level under wide range of soil conditions. The trees bear throughout the year with two peaks during February-June and September-October. It gives 40-50 tonnes fruit yield per hectare under high density planting.

PKM 1: It is a clonal selection from cultivar Guthi. It is suitable for southern and central districts of Tamil Nadu. The trees are dwarf and high yielding. Fruit is usually oval and rarely round, medium sized and weighs 100 g. Fruit yield is 3547 kg per hectare.

PKM 2 (Guthi x Kirthibarthi): The cultivar gives 43% higher fruit yield than PKM 1. The trees start producing from fifth year after planting with mean fruit yield of 80 kg/tree. The fruit weighs 95 g and has 26.30°BTSS.

PKM 3 (Guthi x Cricket Ball): The cultivar is adapted to the tropical plains of Tamil Nadu and is tolerant to leaf spot and leaf webber. The vertical growth habit of tree makes it suited for high density planting. Fruits are large and oval and mature earlier than the other cultivars. The fruits bear in clusters.

PKM (Sa) 4: It is a selection from open pollinated plants of PKM 1. The fruit is spindle-shaped suited for making dry flakes. The pulp is attractive, light pinkish honey brown having 24-25°B TSS. Fruit yield is 20.8 tonnes/ha.

PKM (Sa) 5: It is a selection from open pollinated seedlings in Virudunagar, Tamil Nadu. The fruit is attractive oval having light brown skin and 25.5°B TSS. The flesh is crisp and retains coppery brown colour. It is suitable for making dry flakes, milk shake powder and jam. Fruit yield is 18.7 tonnes/ha.

Dhola Diwan: It has light green leaves and whitish oval fruits which are harvested in summer. Fruits are superior in quality.

Kirthibarthi: The fruits are small to medium in size and oval similar to egg in shape. Fruit skin is rough and medium to thick. Pulp is very sweet and is good for transportation to distant places.

The other cultivars known in India are **Calcutta Special** (large fruit, late), **Pilipatti** (small fruit, mid season to late), **Bhuripatti** (small fruit, mid season), **Jumakhia** (small fruits in clusters, late), **Mohan Gooti** (small fruit, mid season, not very sweet), **Bangalore** (large, ridged), **Vavivalasa** (oval, medium-sweet, bear poorly), **Jonnavalosa-I** (medium size, pale-fleshed, sweet), **Jonnavalosa-II** (medium size, ridged, yellowish pink flesh, sweet, not agreeable flavour), **Jonnavalosa Round** (large, ridged, with cream-coloured flesh, very sweet), **Gauranga** (small, lopsided, ridged, very sweet, bears heavily), **Ayyangar** (large, very thick skin, sweet, rose-scented), **Oaka** (small, rounded to oval, good flavour), **Badam, Bhuri, Calcutta Round, Dhola diwani, Fingar, Gavarayya, Guthi, Kali** and **Vanjet.**

Siamese cultivars

The main cultivars are Makok, Tikal, Alano, Oxkutzcab, Hasya, Morena and Molix).

Makok: The cultivar is native to Thailand. It is high producer and is excellent for home gardens because the tree is a small and compact. The fruit is long, pointed having smooth and brown pulp with good taste and sweet aroma. It ripens from May to November.

Hasya: The cultivar is native of Mexico. Its fruits are football shaped having a reddish hue throughout the pulp and is of excellent eating quality. The tree is a large upright and prolific producer of large fruits (370 g). The fruit ripens from November to June.

Alano: It is native of Thailand. It bears oval and sweet fruits having the texture of an ultrafine pear. The trees are heavy and regular producers of medium sized (257 g) fruits of superior quality. The tree tends to be less brittle and is much smaller than other cultivars. The fruit ripens from November to June.

Tikal: The cultivar originated in Mexico and was selected in USA. The fruits are ovoid, but are fat at one end like a top. The fruit size varies and can be as large as 314 g. The fruit ripens from December to March.

Molix: The cultivar is native of Mexico and bears football-shaped fruits similar to Hasya but the fruits tend to be darker brown outside and less red inside and the tree has curly leaves. The pulp is exceptionally sweet with a fine pear texture and pleasant aroma. The fruits are large (370 g). The fruits ripen from February to May.

American cultivars

Silas Woods: The tree has dwarf growth habit (under 6 m). This cultivar is highly productive. Trees bear fruit year round and the branches often require support as they get very heavily loaded with fruits. The cultivar is suitable for container culture.

Brown Sugar: The fruit is medium small and is 5-6.25 cm long. The tree is tall and bushy. The fruit is granular, sweet, juicy and extremely fragrant.

Prolific: It has round conical fruit with a scurffy brown skin. The flesh has a pinkish tan and is mildly fragrant. It is sweet, juicy and extremely delicious. The tree bears fruits early, consistently and heavily.

Russel: The cultivar bears large, almost round fruit, which is 7.5 to 12.5 cm in size. It is mildly fragrant and has a granular texture and rich and sweet flavour.

Oxkutzcab (gigantic sapodilla): It is originally from Oxkutzcab (Yucatan). Fruits are very large, roundish with speckled, light-coloured skin. Fruit flesh is brown with excellent taste. The tree is heavy bearing and gives high yields of very good and extremely large fruit (800 g) from May to September.

Malaysian clones (Sarip, 1991)

The 7 recommended clones are C4, C54, C55, C60, C61 (Betawi), C62 (Jantung) and C 63 (Subang).

Jantung (C62): The trees establish early and grow into larger size and stem girth compared with those of Betawi and Subang. The canopy is broad and conical similat to Betawi. It is less prolific in flowering compared with Subang and Betawi, but fruit set is high (21.1%), fruit yield is about 33.2 tonnes/hectare/year similar to Subang. The fruit is oval tapered towards the end assuming heart shape with reddish brown skin. Average fruit size is 6.6 x 5.2 cm with weight of 110 g. The pulp is sweet (19.8ºB TSS), medium to high grittiness (60-80%). The fruit contains 1-4 seeds. Like Subang and Betawi, this clone is susceptible to stem gall.

Subang (C63): The tree is less vigorous in early growth resulting in smaller plant size and stem girth similar to Betawi. The tree has a pyramid-shaped canopy. It is most prolific in flowering compared with Betawi and Jantung. The fruit set is low (10-11%) giving fruit yield of about 32.2 tonnes per hectare. The fruit is oval, elongated and tapered towards the end with brown skin and contains 1-3 small seeds. Fruit size is 6.2-8.0 x 3.1-5.2 cm. The pulp is very sweet (21.6ºB TSS), less gritty (20-40%) and has smooth texture. Multiple fruit set in an inflorescence encourages infestation by pests and diseases like fruit borer (*Nephothetrix pirates*) and sooty mold (*Anthaloderma setosum*) and results in uneven fruit growth and ripening.

Mega (CM 19): The tree is vigorous and developing sturdy trunk and fairly small stature suited for high density planting (336 trees/ha). The stems are free from galls unlike in other cultivars. It has low to medium flowering intensity, but bears 1.5 times more flowers than other cultivars. It is self compatible (5-20% fruit set through self pollination), but pollination with Subang pollen gives higher fruit set of 55-80%. The fruit is medium to large, three times the size of Jantung or Subang and weighs about 320-400 g. It is round and symmetrical with flat shoulder and facilitates packing and reduces fruit injury during handling and transportation. The fruit skin is not astringent. The pulp is as good and sweet (20-21ºB TSS) as that of Jantung and Subang with low grittiness. The fruit contains 2-5 seeds. Its fruit matures in 32-35 weeks compared to those of Jantung and Subang requiring 28-35 weeks. The fruits are borne singly. The incidence of fruit borer (*Nephothetrix pirates*) and sooty mold (*Anthaloderma setosum*) is less.

Betawi (C61): It is an Indonesian clone. The tree is small, similar to Subang and of low vigour. The tree has a broad conical canopy. The cultivar has medium flowering intensity and gives low fruit set of about 10.4%. Consequently, it is much lower yielding clone bearing about 21 tonnes fruits per hectare. The fruit is round and symmetrical, weighs about 86.2 g and measures 5.1 x 5.4 cm and is seedy (2-5 seeds/fruit). The flesh is gritty (80%) and sweet (19°B TSS). The cultivar is susceptible to stem gall.

Khirni

(***Manilkara hexandra*** (Roxb.) Dubard syn. ***Mimusops hexandra*** Roxb.; ralu; rayan)

It is believed to be native of India and is found growing in central India and Deccan Peninsula and is widespread in the tropics. It is cultivated throughout India for fruit and is also grown as an ornamental tree.

The tree is evergreen and long lived, of small to medium size, 10-12 m in height with a spreading crown and straight and massive bole. The bark is gray and deeply furrowed. Leaves are elliptic obovate or oblong and coriaceous. White or pale yellow solitary flowers are borne in fascicles in the axils of leaves. The berry is ovoid or ellipsoid, 1.5-2 cm long and reddish yellow. It contains usually one or rarely 2 seeds which are ovoid, 1-1.5 cm long, reddish brown or black and shining. Lot of variability occurs in India with regard to size, shape and quality of the fruit.

Genus *Sideroxylon* L.

Forty nine species occur in the neotropics and about 25 are present elsewhere (6 Africa, 6 Madagascar, 8 Mascarenes Islands, 4 Asia, 1 each in North West Pakistan, Afghanistan, Oman, Somalia, Ethiopia and Djibouti).

The plants are spinous or unarmed trees or shrubs. Leaves are spirally arranged, rarely opposite, often fascicled borne on short lateral shoots. Venation is variable. Stipules are absent. Inflorescence is axillary, sessile or very rarely pedunculate. Flowers are solitary or fasciculate, bisexual or rarely unisexual (dioecious). Sepals are in one whorl of 5(-8), quincuncial and free. Corolla is cyathiform and usually glabrous. Tube is nearly always shorter than the lobes, rarely equaling or exceeding it. The lobes are (4)5(-8), imbricate or quincuncial, spreading, entire or divided into a larger median segment and two smaller lateral segments. Stamens are (4)5(-8), exerted with well developed filaments, extrorse anthers, usually glabrous and stamens sometimes converted into sterile staminodes in male flowers. Staminodes are (4)5(-8), usually well developed and often lanceolate, erose, infolded and incurved against the style, usually glabrous and less frequently hairy. Ovary is (1-)5(-8)-locular, hairy or glabrous with basi-ventral or basal placentation and exerted or included style. Fruit is 1(-2)-seeded, fleshy and usually glabrous. Seed is globose, ovoid, oblong or ellipsoid, not laterally compressed and rarely plano-convex. The fruit has 2 seeds having smooth, shining testa, free from the pericarp, often thick and woody, often sculptured on the adaxial surface with several prominent thickened wrinkles. The hilum is nearly always basal or basi-ventral, small, circular, lanceolate or elliptic, rarely adaxial and broad. Embryo is vertical, oblique or horizontal having thin foliaceous cotyledons and copious endosperm or plano-convex cotyledons and a thin sheath of endosperm or no endosperm with exserted radicle.

Miracle berry
(***Synsepalum dulcificum*** (Schumach. & Thonn.) Daniell.; miraculous berry; sweet berry)

It is native of West Africa. The plant is known for its berry which contains miraculin. As a result after it is eaten it makes the subsequently consumed sour foods to taste sweet. The plant grows 150 to 450 cm high and has dense foliage. The leaves are glabrous below and are 5-10 x 2-3.7 cm in size. The flowers are white and are produced in flushes. The fruit is 2-3 cm long bright red, ellipsoid berry containing a single seed. A form called 'Hirsutus' has hairy leaves and bears oval red fruits that are larger than those of the common smooth leaf-type (CRFG, 1996c).

Genus *Argania* Roemer & Schultes

The plants are spinous shrubs or small trees. Leaves are spirally arranged becoming fascicled in short lateral shoots. Venation is eucamptodromous to brochidodromous with convergent secondaries.

Argan tree
(***Argania spinosa*** (L.) Skeels syn. ***A. sideroxylon*** Roem. and Schultes, ***Sideroxylon spinosum*** L.; arga; argania; Morocco ironwood; arganier)

It is native of southwestern Morocco where it grows wild. The argan has been introduced into other Mediterranean countries, Australia, South Africa, Haiti and USA (Morton and Voss, 1987; Bouachrine, 1994).

It is a thorny multi-stemmed evergreen tree or shrub growing up to 4.5 to 9 m in height. The rough bark is grooved longitudinally and transversely. Although the tree is evergreen but sheds leaves during dry season. Leaves are simple, alternate, clustered, bright green when young and dark green on both surfaces when mature, oblong spathulate to oblanceolate, 2-2.5 x 0.8 cm in size, leathery, hairless and sessile or nearly so and have conspicuous veins. Flowers are bisexual, greenish, 5 mm wide, each subtended by two bracts. The corolla is bell shaped and deeply five lobed. The five hairy sepals are slightly connate at the base. The five slightly protruding stamens alternate with five staminodes, which are sometimes fertile. The superior and hairy ovary is surmounted by a single style as long as or longer than the stamens. The flowers are partially dichogamous and cross pollination improves fruit set (Nerd *et al.,* 1998). The olive sized fruit is round or ovoid to conical and turns bright yellow on ripening. The thick, acrid, bitter, gummy and unpleasant peel (epicarp) contains a layer of pulp (mesocarp) having milky latex. There may be 1-3 seeds, often only one, as the others abort. The seeds are oval brown, smooth and 2 cm long. The testa is composed of thick sclerified cells. The very brittle, oil rich white kernel is surrounded by a very thin, tangled network of loose and woody vascular wrinkles imprinted on its outer surface. The seeds are polyembryonic (Baumer, 1983; Morton and Voss, 1987; Bouachrine, 1994).

Genus *Vitellaria* Gaertner

Possibly two species occur in western tropical Africa and Cameroon (Kubitzki, 2004). The plants are trees or shrubs. Leaves are in a dense terminal cluster having craspedodromous venation with a prominent vein, tertiaries parallel to the secondaries or reticulate and quaternaries finely areolate. Stipules are small and caducous. Inflorescences are densely clustered at the shoot apex. Corolla is glabrous and tube is much shorter than the lobes. The (6-) 8 lobes are entire, contorted in bud and spreading. Stamens are (6-) 8 with glabrous anthers. Filaments are 7-12 mm long. Staminodes are (6-)8, erect or inflexed forming an envelope round the gynoecium. Style is 8-12 mm long. Ovary is (5)6- locular with axile placentation and slightly exerted style. Fruit is 1-2-seeded and fleshy. Seed is globose or broadly ellipsoid, not compressed laterally having a rather thin and shining testa and broad adaxial hilum. Embryo is with thick, fleshy and fused cotyledons and exserted radicle and without endosperm.

Shea butter tree
(*Vitellaria paradoxa* Gaertn. f. syn. ***Butyrospermum paradoxum*** (Gaertn. f.) Hepper.; karate; shea nut; butter seed)

It is a native of tropical West Africa and is widely distributed from West Africa (Liberia, Gold Coast, Nigeria, Tongo, Dahomey, Senegal, and Sierra Leone) to Sudan and Uganda and from southern to eastern Congo (Booth and Wickens, 1988; FAO, 1988). It is found growing in dry to wet subtropics and dry to moist tropical zones. It holds good potential in areas where oil palm does not grow well (Martin *et al.,* 1987; Hall *et al.*, 1996).

It is a stout, much branched, 20 m tall tree with a spreading crown. Its bark is usually gray or blackish, deeply fissured splitting into square or rectangular corky scales. The leaves are oblong to ovate oblong, 10-25 x 4.5-14 cm in size, rounded at apex, acute to broadly cuneate at base with both surfaces either pubescent or glabrescent. Flowers are fragrant and are borne in dense clusters at the tips of branchlets. The outer sepals are lanceolate, 9-14 x 3.5-6 mm in size and pubescent. The inner sepals are slightly smaller. The corolla is creamy white, glabrous externally and 7-11 x 4.5-7 mm in size with broadly ovate lobes. The filaments are 7 to 12 mm long, anthers are 4.5 mm long, staminodes are 8 mm long and the style is 8 to 15 mm long. The ovary is superior having 5-10 locules each containing a single anatropous ovule. The fruit is ellipsoid, greenish up to 6.5 cm long, 4.5 cm diameter and has sweet pulp surrounding the seeds. Seed is 5 cm long and 3.5 cm in diameter and is usually solitary but sometimes up to three per fruit. It is shining and dark brown with a large white scar on one side (Booth and Wickens, 1988; FAO, 1988).

Two subspecies have been recognized. These are *Vitellaria paradoxa* subsp *parkii* (G. Don) Hepper. and *Vitellaria paradoxa* subsp. *niloticum* (Kotschy) Hepper. The former has less dense plants with shorter indumentum and smaller flowers whereas the latter has densely ferrugineous parts with externally pilose corolla tube having 9.5 to 11 m long lobes.

Genus *Chrysophyllum* L.

Forty three species occur in the neotropics, 15 in Africa, 10 in Madagascar and 2-3 in the region from India to Malesia and Australia (Kubitzki, 2004). The plants are trees or shrubs, very rarely lianas. Leaves are distichous or spirally arranged. Venation is brochidodromous or eucamptodromous, tertiary veins often parallel to the secondaries and descending from the margin, or oblique and closely parallel to reticulate. Stipules are absent. Inflorescence is axillary, ramiflorous or cauliflorous. Flowers are unisexual or bisexual, fasciculated or rarely solitary. Sepals are (4)5(6), imbricate or quincuncial, sometimes accrescent in fruit and frequently ciliate. Corolla is globose, campanulate or cylindrical, shorter tube, equaling or exceeding the lobes. The lobes are (4)5(-8), simple. Stamens are (4)5(-8) fixed in the lower or upper part of the corolla tube. Anthers are extrorse in the bud and hairy or glabrous. Staminodes are rarely present. Small lanceolate or subulate structures are present in the corolla lobe sinuses alternating with the stamen. Disk is absent. Ovary is (4)5(-12)-locular with axile placentation. Fruit is 1 to many-seeded. Seed is laterally compressed with a narrow adaxial hilum, sometimes extending around the base of the seed or not laterally compressed and then the hilum is broader, basi-ventral or adaxial. The testa is smooth and shining or when rough it is adherent to the pericarp. Embryo is vertical with thin foliaceous or thick flat cotyledons and exserted radicle and abundant endosperm nearly equaling the thickness of the cotyledons.

Star apple

(***Chrysophyllum cainito*** L.; caimito; pitaya; golden leaf tree; pomme surette)

The star apple or caimito is native to the West Indies and Central America (Martin *et al.*, 1987). The other related edible species is *C. oliviforme* L. (caimitillo or satin leaf) from tropical America. It is cultivated in South America, Mexico, Florida and to a limited extent in Hawaii. It also occurs in Sri Lanka, Singapore, southern India, tropical Africa, Queensland (Australia), Thailand, Vietnam and the Philippines (Morton, 1987; Verheij and Coronel, 1991).

The tree is strikingly handsome, ornamental and evergreen growing up to a height of 30 m. The leaves are alternate, oval, 5-16 x 3-6 cm in size, shiny, dark green above and silky coppery gold beneath. The young shoots are coppery gold in colour and are heavily pubescent. The small yellow purplish white flowers are borne in axillary umbellate clusters scattered along the mature twigs. The flower has 5 circular to ovate sepals, 5 tubular and lobed corolla, 5 stamens and 7-11 lobed stigma. The fruit is smooth skinned, round, oblate or ellipsoid, 5-10 cm in diameter and the core is star shaped. The white pulp is sweet containing 3-10 glossy and flattened dark seeds loosely placed in the cavities. The unripe fruit contains viscous latex of unpleasant taste. The flavour of properly ripened fruit is agreeable (Page, 1984; Morton, 1987; Verheij and Coronel, 1991).

Genus *Pouteria* Aublet

About 200 species occur in the neotropics, 120 in Asia, Malesia, Australia and the Pacific and 5 in Africa (Kubitzki, 2004). The plants are trees or shrubs, rarely geoxylic suffrutices. Leaves are spirally arranged, rarely opposite. Venation is eucamptodromous or brochidodrornous, usually without a submarginal vein, never finely striate (except *P. keyensis*). Stipules are absent (present in *P. congestifolia*). Inflorescence is axillary or ramiflorous, fasciculate with fascicles single or occasionally arranged along the short leafless shoots. Flowers are often unisexual (plant dioecious). Sepals are 4-6, free, imbricate or quincuncial or 6-11 in a closely imbricate spiral. Corolla is cyathiform to tubular, rarely rotate having tube shorter to longer than the lobes. The lobes are 4-6(-9), usually erect, rarely spreading, simple, sometimes fringed ciliate or papillose. Stamens are 4-6(-9), fixed inside the corolla tube or rarely at the base of lobes, rarely free, usually included, less frequently exerted. Filaments are generally short. Anthers are usually extrorse or laterally dehiscent and usually glabrous. Staminodes are usually in same number as corolla lobes, less frequently lacking, inserted in the corolla sinus or inside the tube. Disk is present or absent. Ovary is 1-6(-15)-locular with axile placentation with style included or exerted. Fruit is a 1 to several-seeded berry. Seed is broadly ellipsoid, plano-convex, shaped like the segment of an orange or laterally compressed with smooth, wrinkled or pitted testa, usually adaxial full length with hilum; narrow, broad or sometimes covering almost all the seed surface. Embryo is vertical with plano-convex or thin foliaceous cotyledons with radicle exerted or included and endosperm present or absent.

The genus is widespread throughout the tropical regions of the world and includes the canistel (*P. campechiana*), the mamey sapote (*P. sapota*) and the lucuma (*P. splendens*). The largest fruit of the genus is mamey sapote.

Mamey sapote

(***Pouteria sapota*** (Jacq.) H.E. Moore and Stearn syn. ***P. mammosa*** (L.) Cronquist, ***Calocarpum sapota*** (Jacq) Merr., ***C. mammosum*** (L.) Pierre, ***Lucuma mammosa*** Gaertn.; sapote; mamey Colorado; marmalade plum)

The mamey sapote is native of Central America and Mexico. It is found growing in the tropical forests of Guatemala and is cultivated in Central America, Mexico, Nicaragua and West Indies (Brücher, 1989; Verheij and Coronel, 1991).

It is an attractive tree of medium size growing up to 50 m in height and tends to be uniform and hemispherical. The leaves are entire, obovate or oblanceolate, 20 to 45 cm long with dark green and shiny upper surface, yellowish veins and light green undersurface. These are grouped in whorls of 8 or 10 near the ends of young branches. The branches are covered with a brown pubescence. The small, hermaphrodite and sessile flowers develop in large numbers along the bare branches below the young leaves. The calyx consists of 8-10 imbricate sepals. The corolla has 5, white petals united in a tube enclosing five fertile stamens, five staminodes and a conical and pilose pistil terminating into a simple stigma. The ovary has five carpels with single ovule in each cell. The fruit is ovoid or ellipsoid in shape having persistent calyx at the base and sometimes with the remnants of the pistil at the apex. It is large, (10-25 x 18-12 cm) in size with a thin, brown, lignified and hard epicarp and a rough and rusty brown surface. The pulp is salmon to reddish brown, thick and very sweet and aromatic. The fruit contains one ellipsoid or spindle-shaped seed, which is glossy brown, 5 to 6 cm or more long having oily kernel and a hard, shiny testa and a large cream coloured or opaque hilum. In some cultivars, 2 or 3 seeds per fruit are seen (Martin *et al.,* 1987; Morton, 1987; Brücher, 1989; Verheij and Coronel, 1991). Important cultivars are Magna, Tazumal, Pantin, Pace, Copan, Cuban No.1, Mayapan, Progreso, Valiante and Rivera.

Canistel

(***Pouteria campechiana*** (Kunth.) Baehni syn. ***P. campechiana*** var. ***salicifolia*** Baehni., ***Lucuma campechiana*** Kunth., ***L. salicifolia*** HBK, ***L. nervosa*** A.DC., ***L. sphaerocarpa*** A.DC., ***L. palmeri*** Fern., ***L. laeteviridis*** Pittier, ***Richardella nervosa*** Pierre, ***B. salicifolia*** Pierre; yellow sapote; egg fruit)

The canistel is a native of Central America and West Indies (Martin *et al.*, 1987). It is commonly grown in Central and South America. Canistel is also cultivated in the Philippines. It has been introduced in Florida, Puerto Rico, Hawaii, Bahamas and USA (Brücher, 1989; Verheij and Coronel, 1991).

The canistel is a small, evergreen and open tree growing up to a height of 25 m if not injured by cold weather. The leaves are oblong obovate to oblanceolate, 10-20 cm long, glabrous and bright green in colour. The flowers are small borne in the leaf axils of young branches in clusters of 2-5 having 5-lobed calyx consisting of 4-6 sepals, of which the inner ones are rounded at the apex, whitish corolla with ovate lobes, 5 stamens, 5 staminodes, columnar style, slightly dilated stigma and 5-6 celled ovary. The fruit is orange yellow when ripe, globose or ovoid, 5-10 cm long, frequently pointed at the apex having bright orange and soft flesh which is mealy like the yolk of a hardboiled egg and usually contains 2-3 shiny dark brown seeds (Martin *et al.*, 1987; Morton, 1987; Brücher, 1989; Verheij and Coronel, 1991).

Abiu

(***Pouteria caimito*** (Ruiz & Pav.) Radlk. syn. ***Achras caimito*** Ruiz. and Pav., ***Lucuma caimito*** Roem. & Schultes, ***Labaitia caimito*** Martius, ***Guapeba caimito*** Pierre; caimito)

Abiu is native of Peru particularly the Amazon headwater area (Clement, 1989a). The fruit was well known in the pre-Columbian civilizations of South and Central America. It is well distributed throughout the tropics, though not grown to any large extent commercially (Nakasone and Paull, 1998). Its trees grow wild in the Amazon lowlands from Peru to the State of Para in Brazil and on the lower eastern slopes of the Andes from southwestern Venezuela to Peru. A botanical variety *P. caimito* var. *strigosa* is reported from Venezuela. It is cultivated in Peru, Ecuador and Venezuela (Morton, 1987). It has also been introduced in Australia.

The tree is evergreen and grows 4.5-6 m tall and has pyramidal or round crown. Leaves are dark green, alternate, entire, obovate to lanceolate but variable in shape and 10-20 x 3-6 cm in size. Flowers are borne solitary or in clusters of 2-5 in the leaf axils or above the leaf scars and are practically sessile and hermaphrodite having usually 4, ovate, small sepals, cylindrical, white or greenish white corolla, 4-5 stamens on the throat of the corolla, 4-5 staminodes, pistil little longer than corolla and a small globose, 10-12-celled ovary. Fruit is bright yellow, globose to cylindrical berry, 4-5 cm in diameter, 5-10 cm long and tomentose when young having whitish, translucent and jelly like flesh. The fruit skin is leathery, 3-5 mm thick and produces sticky latex. Seeds are 1-4, cylindrical ovoid and slightly compressed on the sides (Menninger, 1977; FAO, 1986; Morton, 1987)

Genus *Madhuca* Hamilton ex Gmelin

About 100 species occur in the region from India through Malesia and South China to Guinea (Kubitzki, 2004). Leaves are spirally arranged, very rarely distichous borne on horizontal shoots with eucamptodromous or brochidodromous venation, oblique, horizontal or reticulate tertiaries, rarely intersecondaries and tertiaries descending from the margin and parallel to the secondaries. Stipules are present, sometimes large and persistent. Flowers are bisexual in 2 to many-flowered fascicles, usually on young branches in the axils of fallen leaves. Sepals are 2 + 2, free, rarely a mixture of 2+2 and 5 imbricate sepals in a single whorl. Aestivation of the outer pair is valvate, imbricate or open. Corolla lobes are (6-)8-12(-17), imbricate usually spreading or reflexed, nearly always equaling or exceeding the tube. Corolla tube is usually hirsute or barbate at the throat. Stamens are (12-) 14 -36(-43), exerted (uppermost whorls only) in 1-3 alternating whorls inserted in the throat of the corolla tube or occasionally the lowermost inserted near the base of the tube. Filaments are usually free, rarely partially or completely fused in pairs or into a short tube, not geniculate or the anthers are sessile. Anthers are extrorse or laterally dehiscent, usually large and tapering gradually up to the acute hairy or glabrous apex. Nectary is sometimes present, poorly developed and annular. Ovary is usually glabrous, less frequently hairy, (5-)8-9(-15)-locular and exerted style is often long tapering gradually up to the apex. Fruit is 1-seeded. Seed is broadly ellipsoid oblong, plano-convex or laterally compressed, usually the long narrow adaxial hilum is frequently broader and cover the seed surface. Embryo is with plano-convex, thick but flat rarely foliaceous cotyledons with exerted radicle.

Mahua
(***Madhuca longifolia*** Macbr. syn. ***M. indica*** J. F. Gmel., ***M. latifolia*** Macb., ***Bassia latifolia*** Roxb.)

Mahua or illipe tree or Indian butter tree is native of India and Sri Lanka (Martin *et al.,* 1987). The trees are distributed from India to Australia and Polynesia. The related species found in India are *M. malabarica* (Bedd.) Parker (madhuca) and *Diploknema butyracea* (Roxb.) H.J. Lam. syn. *Madhuca butyracea* Macb. (phulwara, bassia fat). In India, it is cultivated in Uttar Pradesh, Bihar, Andhra Pradesh, Karnataka, Bengal, Maharashtra, Madhya Pradesh and Orissa.

It is a medium to large, much branched and deciduous tree up to 18 m in height usually with a short bole 0.8 m in diameter and large and round crown. The bark is dark and cracked. Leaves, clustered near the ends of branches, are elliptic or elliptic oblong, 7.5-23 x 3.8-11.5 cm in size, coriaceous, pubescent when young and almost glabrous when mature. Flowers appear in dense fascicles near the ends of branches. Calyx is coriaceous and corolla is tubular, fleshy, cream coloured, 1.5 cm long and caducous. Fruits are fleshy, ovoid, up to 5 cm long and greenish turning reddish yellow or orange when ripe and contain 1-4 brown, ovoid, shining and 2.5-3.7 cm long seeds (Anon., 1990; Duke, 1989; Hocking, 1993).

Two botanical varieties are recognized, viz. *M. longifolia* (Koeng.) Macbr. var. *longifolia* and *M. longifolia* var. *latifolia* (Roxb.) Chevalier. The former grows in southern India while the latter is common throughout India. Some early, mid season and late maturing selections have been made.

Genus *Diploknema* Pierre

It is found in the tropical and subtropical Asia. Seven species have been recognized to be growing in Nepal, India and South East Asia. These are *Diploknema butyracea* (Roxb.) H. J. Lam., *D. butyraceoides* (M. B. Scott) H. J. Lam., *D. oligomera* H. J. Lam., *D. ramiflora* (Merr.) H. J. Lam., *D. sebifera* Pierre, *D. siamensis* Fletcher and *D. yunnanensis* D. D. Tao, Z. H. Yang & Q. T. Zhang.

Indian butter tree

(*Diploknema butyracea* H. J. Lam. syn. ***Madhuca butyracea*** (Roxb.) Pierre ex Dubard, ***Aesandra butyracea*** (Roxb.) Baehni**, *Bassia butyracea*** H. J. Lam.)

Indian butter tree or phulwara or chiura is native of India. It occurs wild in the Himalayan foothills at higher elevations than that suitable for mahua (between 300 to 1500 m altitudes) with the greatest frequency in Kumaon and Gharwal and in Sikkim and Bhutan. It is distributed throughout tropical Asia especially India. It is also found in the Andamans in evergreen forests, chiefly along the ravine sides (FAO, 1982; Anon, 1990). It is a large deciduous tree having rapid growth increasing in circumference by about 4-5 cm per year. Leaves are petiolate, obovate or obovate oblong, 20-35 x 7.5-15 cm in size, coriaceous, silky or tomentose on the lower surface when young. Flowers are white with sticky fragrance. Flower is composed of 4-5 sepals, campanulate corolla surrounded by petaloid lobes slightly longer than the tube numbering 8-12 and 30-45 stamens with glabrous filaments equaling the lanceolate and acuminate anthers. The ovary is 10-locular surrounded by a linear style. Fruit is a 1-3-seeded berry with a more or less fleshy pericarp, about the size of a plum. Each fruit contains an ellipsoid seed, which occupies about 28% of the volume of fruit. The hilum is long and rather broad. The very fleshy cotyledons fill the entire seed and the endosperm is reduced to a thin easily separable layer (FAO, 1982).

Family Ebenaceae

The plants are shrubs or trees having hard wood and are more important for timber than for wood. The leaves are alternate, entire and exstipulate. The flowers are solitary or cymose in the axil of the leaf, regular, seldom perfect but are usually dioecious. The calyx is 3-7-lobed and persistent. Corolla is fleshy with 3-7 lobes that are twisted or convolute in the bud. Stamens are short, as many as petals or twice their number inserted at the base of petals, the filaments being free or united in pairs. Ovary is superior with 2-16 carpels each having 1-2 ovules. Styles are 2-8, free or united at the base. The fruit is a berry or drupe, usually fleshy but are rarely dehiscent. Embryo is small, straight with large foliaceous cotyledons. Endosperm is copious.

Genus *Diospyros* L. (x = 15; 2n = 60, 90)

About 500-600 species are known; 200-300 in Asia and the Pacific area, 95 in Madagascar, 94 on the African mainland, 100 in the Americas and 15 in Australia (Kubitzki, 2004). A number *of Diospyros* species produce ebony wood and edible fruits of which the main are *D. kaki* L.f. (Japanese or Oriental persimmon), *D. lotus* L. (date plum or Central Asian persimmon), *D. virginiana L.* (American or common persimmon), *D. ebenaster* Retz. (black sapote), *D. discolor* Willd. syn. *D. mabola* Roxb. (mabola persimmon, butter fruit), *D. ebenum* Koenig syn. *D. assimilis* Bedd. (Ceylon ebony persimmon), *Diospyros digyna* Jacq. (black persimmon), *Diospyros mespiliformis* Hochst. ex A. DC (West African ebony) and *Diospyros blancoi* A.DC (velvet apple). *D. ebenester* and *D. virginiana* are neotropical species originated in North and Central America.

The plants are tropical and subtropical, deciduous or evergreen trees or shrubs. Winter buds are ovoid with about 3 outer scales and the terminal bud is lacking. The leaves are usually alternate with entire margins and are exstipulate. Inflorescences are usually cymose or fasciculate if male and uniflorous if female (less frequently few-flowered). The flowers are whitish, dioecious and sometimes monoecious. The pistillate flower is pale yellow to white, usually solitary. Staminate one is cymose and axillary. Male flowers are usually smaller than the female ones and differ in shape. Calyx and corolla are usually 4-lobed, rarely 3-7-lobed. Stamens are 2 to 100, usually 8-16. Ovary is 2-8-carpellate, 4-12-celled with 2-6 styles. Calyx is variable, 3-8-lobed or cup-shaped and entire, persistent and usually accrescent on fruits. The fruit is a large juicy berry, 1 to 10-seeded usually with an enlarged calyx at base having large flattened seeds.

Japanese persimmon

(*Diospyros kaki* L. f. syn. ***D. chinensis*** Blume; Oriental persimmon; sharon fruit; date plum; keg fig; kaki) (2n=90)

It is native of China and Japan and is the national fruit of Japan. It grows in the foothill areas. In India, it is cultivated in Kulu, Jammu and Kashmir and Coonoor in Tamil Nadu. The plant is hardy deciduous and subtropical, moderately resistant to winter cold but not very resistant to summer heat. The tree is up to 12 m high with rounded or pyramidal shape. The bark is fairly rough. The leaves are large, 7-25 x 5-10 cm in size, tough, thick, alternate, entire, short petioled, glossy green and glabrous above and sparsely pubescent on the lower surface and ovate to elliptic varying in form according to cultivar. Male and hermaphrodite flowers occur on the same tree in varying proportions on current season's growth. Male flowers emerge in leaf axils in 3-flowered cymes with four lobed calyx and corolla, 16-24 stamens arranged in 2 rows on the petals. Sometimes, the central flower in the cyme is hermaphrodite. Female flowers are borne singly in the leaf axils with a lobed and leaf-like calyx, light yellow corolla which is tubular at the base and divided in 4 lobes, 8 abortive stamens, globular and more or less flattened 4-cellular ovary each divided into two forming 8 compartments each having a single ovule. The fruit sets parthenocarpically in some cultivars. Pollination influences the fruit characters. The fruit is a large berry of variable shape with a fine membranous skin of variable colour. The flesh is orange yellow, more or less soft and almost fluid when fully ripe, consistency being influenced by pollination. Flattened and dark brown seeds constitute up to 8 per cent of the fruit but many cultivars have only 4 or less number of seeds or may be completely seedless.

Cultivars

As many as 1000 cultivars are known. The cultivars have been divided into non-astringent and astringent groups each divided into 2 subgroups based on the effect of pollination on flesh colour and fruit shape. The first group, called pollination constant, consists of the cultivars showing no change in the colour of flesh under the influence of pollination and bear light coloured and seedless fruits. The second group, called pollination variant, consists of the cultivars in which the flesh colour depends on pollination. These bear low coloured fruits when seedless and dark reddish brown fruits when seeded. Thus the Japanese persimmon cultivars fall into the following 4 groups:

a. Non-astringent and pollination constant (PCNA), e.g. Fuyu, Jiro, Gosho and Suruga.

b. Non-astringent and pollination variant (PVNA), e.g. Zenjimaru, Shogatsu, Mizushina and Amahyakume.

c. Astringent and pollination constant (PCA), e.g. Yokono, Yotsumizo, Shakokushi, Hagakushi, Hachiya and Gionba.

d. Astringent and pollination variant (PVA), e.g., Aizumishirazu, Emon, Roshuhyakume and Hiratanenashi.

The cultivars have also been classified into 5 groups on the basis of flowering pattern:

1. Pistillate– the commonest type.
2. Pistillate but sporadically monoecious.
3. Dioecious.
4. Dioecious but sporadically monoecious.
5. Staminate.

Hachiya: The cultivar is pollination constant and astringent type. The astringency is due to the tannins that turn soluble at fruit maturity. The cultivar is widely cultivated. The fruit is large, oblong and glossy orange red having pointed apex and persistent calyx. The fruit remains astringent till it becomes soft and is usually seedless and of very good quality.

Hykume: The cultivar is pollination variant. Fruit is medium to large in size, angular, four sided, pale orange in colour and is russeted or mottled.

Izu: The cultivar is non-astringent type and is susceptible to fruit fly. The plant has medium upright growth. The fruit matures early and is orange red in colour and flat in shape having attractive appearance.

Jiro: The cultivar is non-astringent type. The tree is upright and vigorous. The fruit ripens in mid season and is large, orange red, oblate with eight conspicuous lines and large calyx. The fruit quality is excellent.

Fuyu: It is a late ripening and non-astringent cultivar. The plant is medium in vigour with spreading growth habit and rounded crown. The fruit is medium sized and oblate like tomato having orange to red skin with four distinct lines and light orange flesh. The quality and storage ability of the fruit are excellent.

Jambu: The fruits are non-astringent.

Tamopan: The fruits remain astringent till fully ripe. These are very large, oblong, orange red and quadrangular but are equatorial in form and have a tough skin. The fruit quality is good.

Tanenashi: It is an astringent type. The fruit is broadly conical, large, light orange and has mealy or pasty consistency. The fruit quality is not very high.

American persimmon
(*Diospyros virginiana* L.)

It is native to the eastern United States from Florida to Connecticut. Its range has spread up to Oregon and southern parts of Canada. The spreading tree is deciduous and grows up to 10-15 m in height with a round top. The bark is rough and shows the characteristic checking. The leaves are less shiny and more pubescent on the lower surface. The plants are dioecious. Flowers are borne on one year old wood near the branch tip. Flowers are greenish whitish, the pistillate flowers are usually solitary where as the staminate flowers are generally cymose in group of three having 4-lobed calyx and corolla, 8-16 stamens and 4-12-celled ovary with 2-6 styles. The fruit is a large and juicy berry having 5 or more flattened seeds and are very astringent till they ripen. Cross pollination is necessary for setting seeds. Some selected cultivars are John Rick, Garretson, Killen, Craags, Beavers, Wabash, Florence, Early Golden, Ruby and Miller.

West African ebony
(*Diospyros mespiliformis* Hochst. ex A. DC; African ebony; swamp ebony)

It is native of tropical Africa. It has widespread distribution in tropical Africa, south of the Sahara from Senegal to Sudan and down to southern Africa. The tree is evergreen or semi evergreen growing up to 20 m in height having grayish black bark. Leaves are simple, alternate and oblong to elliptic in shape, leathery with dull shine and 4-17 x 1.5-5.5 cm in size. The tree is dioecious. The flowers are pentamerous, white, hairy and fragrant. Male flowers are borne in clusters whereas female flowers are solitary, shortly pedicellate and axillary. Fruit is round, 3 cm in diameter, greenish and hairy when young, yellow or purplish and smooth on ripening and contains 4-6 dark brown seeds (FAO, 1983; FAO, 1988).

Velvet apple
(*Diospyros blancoi* A.DC syn. *D. discolor* Willd., *D. philippensis* (Dear.) Guerke; mabolo; mabola; butter fruit; *pommier velours*)

It is native of the Philippines where it is commonly found at low altitudes (Martin *et al.*, 1987; Verheij and Coronel, 1991). It is cultivated in Taiwan and Indonesia. It has been introduced throughout the eastern tropics. In India, it is grown in southern parts of the Peninsula and in Bihar and Assam.

It is an evergreen tree growing up to 7-15 m in height and about 50 cm bole diameter and has a conical crown. The leaves are alternate, oblong, 8-30 x 2.5-12 cm in size, entire, coriaceous, usually with rounded base, pointed apex, silvery hairy upper surface and 1.7 cm long petioles. The tree is dioecious. Male flowers are borne in 3-7 flowered axillary cymes having short pedicels, tubular but deeply 4-lobed, about 1 cm long calyx, slightly larger tubular and 4-lobed corolla, 24-30 creamy white stamens which are united in pairs at the base. The female flowers are solitary, axillary and sub-sessile, slightly larger than male flowers having 4-5 staminodes. The fruit is a globose or depressed berry, 5-12 x 8-10 cm in size, velvety and brown reddish, capped at the base with the persistent stiff calyx having thin skin densely coated with golden brown hairs giving a strong cheese-like odour. The flesh is whitish, firm, rather dry, sweet astringent and aromatic having 4-10 wedge-shaped brown seeds of 4 x 2.5 x 1.5 cm size (Morton, 1987; Verheij and Coronel, 1991).

Cultivars

Seedling trees exhibit a high degree of variability in leaf and fruit characters (shape and size, flesh colour and seedlessness). White-fleshed and sweet cultivars are popular in the Philippines. Superior types such as Manila and Valesca have been selected in the Philippines (Morton, 1987; Verheij and Coronel, 1991).

Black sapote

(***Diospyros digyna*** Jacq. syn. ***D. ebenaster*** Retz.; black persimmon; sapote negro)

It is native to tropical forests of Central America (Guatemala and Mexico). It grows wild in the Antilles and is a popular fruit of Mexico and Guatemala (Morton, 1987; Brücher, 1989).

It is an evergreen tree with a dominant trunk and grows up to 10-12 m in height. Leaves are alternate, elliptic oblong to oblong lanceolate, 10-30 x 3-5 cm in size with a dark green, shining and leathery lamina. Flowers are axillary, unisexual or hermaphrodite borne singly or in groups of 3-7 in leaf axils. Male flowers occur usually in clusters of 3 and female flowers are solitary. Calyx consists of 5 sepals which are green and persistent. Corolla consists of 5, white and tubular petals. The male flowers have 16-60 stamens and female flowers have 8, 10 or 12-celled ovary with one ovule per cell. Fruit is a flattened globose to slightly lobed berry of 5-15 cm diameter, dark olive green, seated on a persistent, 4-6 lobed calyx which reflexes at maturity. Pulp is soft and brown to black. The dark chocolate colour of the pulp is considered unattractive (Martin *et al.*, 1987; Verheij and Coronel, 1991). The seeds are 0-12, flat, 2 cm long, smooth and brown (Morton, 1987; Verheij and Coronel, 1991).

Cultivars

A few promising selections of black sapote have been made. Bernecker and Maher are selections made in Australia (Morton, 1987). More selections, especially those having hermaphrodite flowers can be made from the wide range of genetic diversity. The cultivated types are generally seedless.

Date plum
(***Diospyros lotus*** L.; Central Asian persimmon)

The plants are dioecious. The tree is of medium size. The leaves are large, elliptic, glabrous on the upper surface. The fruits are small, globular, yellowish turning to chocolate in colour. Parthenocarpy is found. Cross pollination also takes place.

Mabola persimmon
(***Diospyros discolor*** Willd. syn. ***D. mabola*** Roxb.; butter fruit)

It is found growing wild in South India, Bihar and Assam. The fruits are pale orange in colour.

***Diospyros tomentosa* Roxb.**

It is found growing wild. It is a hairy species. This species gives the well known ebony wood. The fruits are small.

Family Actinidiaceae (kiwi vine, Chinese gooseberry)

This is a family of trees and shrubs, previously thought to be related to Theaceae. It contains 3 genera and 350 species (Heywood *et al*., 2007) and occurs mainly in tropical and subtropical America and South East Asia with another centre in temperate Asia. *Saurauia* is the largest genus with 250 species and is found in both tropical America and in the region from the Himalayas to East Asia, Fiji and Queensland. *Actinidia* having 62 species occurs from Indonesia to Siberia with the centre of distribution in China.

The plants are trees, shrubs or woody lianas frequently having hairy young growth. The leaves are alternate, simple having entire or serrate margin and are exstipulate. The inflorescences are axillary and cymose having few to numerous flowers (up to 500 in *Saurauia*). These are unisexual (plants monoecious in *Actinidia* and monoecious or functionally dioecious in *Saurauia*) or hermaphrodite, actinomorphic and usually 5-merous. The sepals are 3-4 5(-9), free or adnate to the petals. The petals are usually white, sometimes red or brown yellow and free or connate. The stamens are as few as 10-15 or numerous in fascicles. The ovary is superior with 3 to 5 carpels, usually 5-locuar, sometimes 3 or up to 20 or more. Placentation is axile. The fruit is a berry or a loculicidal capsule in some *Saurauia* species. Seeds are usually numerous.

Genus *Actinidia* Lindl. (n = 29 or multiples)

About 60 species occur in East Asia, mostly in West to East China, north to Sakhalin and Kuril Islands, south to Taiwan, Himalayas, North East India, Indochina and Malaysia (Kubitzki, 2004). All *Actinidia* species are perennial, climbing or scrambling plants, mostly deciduous although a few from warmer areas are evergreen. The plants are woody climbers and may or may not be pubescent. The genus includes shrubs growing to 6 m tall, and vigorous, strong growing vines, growing up to 30 m in tree canopies. They mostly tolerate temperatures down to

−15°C, and some are much hardier. The leaves are alternate and simple, with a dentate margin and a long petiole. These may be dioecious or polygamous. The flowers are solitary or in axillary cymes, usually white, with five small petals. These have sepals that are distinct or somewhat fused at the base, white or yellow to reddish petals, numerous stamens, longitudinally dehiscent anthers, pubescent or glabrous and many-locular ovary having distinct and persistent stylar branches. The flowers on male vines produce viable pollen but lack a properly developed ovary, ovules or styles. The flowers of female vines appear perfect but the pollen they release is shrivelled and non-viable. The fruit can occur singly, in small bunches of three to five fruits, or sometimes in larger bunches or infructescences containing up to 30 or more fruits. The fruit is globose to oblong berry, sometimes pubescent having numerous seeds embedded in the pulp. They vary in size, shape, hairiness and external colour and do not split open at maturity. Some change colour as they ripen. The flesh can also vary in colour, juiciness, texture and composition.

Distribution of sections and genera of *Actinidia*

Section *Leiocarpae* (Dunn) Li: Leaves are glabrous or only slightly pubescent. The fruit is not spotted.

Series *Lamellatae* C. F. Liang: Pith is lamellate.

Actinidia arguta (Sieb. et Zucc.) Planch. ex Miq. var. *arguta*: distributed in Heilongjiang, Jilin, Liaoning, Hebei, Shandong, Shanxi, Shaanxi, Henan, Anhui, Hubei, Zhejiang, Jiangxi, Yunnan, Eastern Siberia, Korea and Japan (4x).

A. arguta (Sieb. et Zucc.) Planch. ex Miq. var. *cordifolia* (Miq.) Bean: distributed in Jilin, Liaoning, Shandong, Shaanxi, Henan, Hubei, Zhejiang, Korea, Japan

A. arguta (Sieb. et Zucc.) Planch. ex Miq. var. *giraldii* (Diels) Voroshilov: distributed in Hebei, Shaanxi, Henan and Hubei.

A. arguta (Sieb. et Zucc.) Planch. ex Miq. var. *nervosa* C.F. Liang: distributed in Henan, Zhejiang, Sichuan and Yunnan.

A. arguta (Sieb. et Zucc.) Planch. ex Miq. var. *purpurea* (Rehd.) C. F. Liang: distributed in Shanxi, Shaanxi, Henan, Hubei, Sichuan, Guizhou, Hunan, Jiangxi, Zheiang, Yunnan and Guangxi.

A. globosa C. F. Liang: distributed in Hunan and Guangxi.

A. henanensis C. F. Liang: distributed in Henan.

A. hypoleuca Nakai: distributed in Japan

A. kolomikta (Maxim. et Rupr.) Maxim.: distributed in Heilongjiang, Jilin, Liaoning, Hebei, Shaanxi and Hubei, Sichuan, Yunnan, Eastern Siberia, Korea and Japan (4x).

A. maloides Li var. *maloides:* distributed in Gansu, Sichuan, Hubei and Yunnan (4x).

A. maloides Li var. *cordata* C. F. Liang: distributed in Sichuan.

A. melanandra Franch. var. *melanandra*: distributed in Gansu, Shaanxi, Henan, Sichuan, Hubei, Anhui, Yunnan, Guizhou; Hunan, Jiangzi, Zhejiang and Fujian (4x).

A. melanandra Franch. var. *cretacea* C. F. Liang: distributed in Hubei (4x).

A. melanandra Franch. var. *glabrescens* C. F. Liang: distributed in Hunan (4x).

A. melanandra Franch. var. *kwangsiensis* (Li) C. F. Liang: distributed in Guangxi (4x).

A. melanandra Franch. var. *subconcolor* C. F.Liang: distributed in Zhejiang (4x).

A. rufa (Sieb. et Zucc.) Planch. ex Miq.: distributed in Korea and Japan.

A. tetramera Maxim. var. *badongensis* C. F. Liang: distributed in Hubei (4x).

A. tetramera Maxim. var. *tetramera:* distributed in Gansu, Shaanxi, Henan, Hubei and Sichuan (4x).

Series *Solidae* C. F. Liang: Pith is not lamellate (2n==4x=116).

A. polygama (Sieb. et Zucc.) Maxim.: distributed in Jilin, Liaoning, Shandong, Gansu, Shaanxi, Henan, Hubei, Zhejiang, Hunan, Sichuan, Yunnan, Guizhou, Siberia, Korea and Japan (4x).

A. macrosperma C. F. Liang var. *macrosperma*: distributed in Jiangus, Anhui, Hubei, Zhejiang, Jiangxi and Guangdong (4x).

A. macrosperma C. F. Liang var. *mumoides* C. F. Liang: distributed in Jiangsu, Anhui, Zhejiang and Jiangxi (4x).

A. valvata Dunn var. *valvata*: distributed in Anhui, Jiangsu, Zhejiang, Hubei, Jiangxi, Hunan and Guangdong (4x).

A. valvata Dunn var. *boehmeriifolia* C. F. Liang (originally published as *boehmeriaefolia):* distributed in Jiangxi and Zhejiang (4x).

A. valvata Dunn var. *longipedicellata* L. L. Yu: distributed in Zhejiang (4x).

Section *Maculatae* Dunn: Leaves are glabrous or only slightly pubescent. The fruit is spotted.

A. callosa Lindl. var. *acuminata* C. F. Liang: distributed in Hunan.

A. callosa Lindl. var. *callosa*: distributed in Yunnan, Hunan, Zhejiang, Taiwan, Burma, Northern India, Indo-China, Malaysia and Indonesia (2x).

A. callosa Lindl var. *discolor* C. F. Liang: distributed in Sichuan, Yunnan, Guizhou, Anhui, Zhejiang, Jiangxi, Hunah, Guangdong, Guangxi and Taiwan (2x).

A. callosa Lindl var. *ephippioidea* C. F. Liang: distributed in Yunnan

A. callosa Lindl. var. *formosana* Fin. et Gagn.: distributed in Taiwan.

A. callosa Lindl var. *henryi* Maxim.: distributed in Gansu, Shaanxi, Anhui, Sichuan, Hubei, Hunan, Jiangxi, Zhejiang, Fujian, Yunnan, Guizhou, Guangdong and Guangxi (2x).

A. callosa Lindl var. *strigillosa* C .F. Liang: distributed in Guizhou, Hunan and Guangxi (4x).

A. chrysantha C. F. Liang: distributed in Hunan, Guangxi and Guangdong (4x).

A. cylindrica C. F. Liang var. *cylindrica*: distributed in Guangxi (2x).

A. cylindrica C. F. Liang var. *obtusifolia* C. F. Liang: distributed in Guangxi (4x).

A. cylindrica C. F. Liang var. *reticulata* C. F. Liang: distributed in Guangxi (4x)

A. fasciculoides C. F. Liang var. *fasciculoides*: distributed in Yunnan.

A. fasciculoides C. F. Liang. var. *cuneata* C. F. Liang : distributed in Guangxi (2x).

A. fasciculoides C. F. Liang var. *orbiculata* C. F. Liang: distributed in Guangxi.

A. glaucocallosa C. Y. Wu (originally published as *A. glauco-callosa):* distributed in Yunnan.

A. glaucophylla F. Chun. var. *glaucophylla:* distributed in Hunan, Guangxi, Guangdong and Guizhou (2x)

A. glaucophylla F. Chun var. *robusta* C.F. Liang: distributed in Guangxi (2x).

A. glaucophylla F. Chun var. *rotunda* C.F. Liang: distributed in Guangxi (2x).

A. glaucophylla F. Chun var. *symmetrica* (F. Chun) C. F. Liang: distributed in Guangxi and Guangdong (2x).

A. gracilis C .F. Liang: distributed in Guangxi.

A. indochinensis Merr.: distributed in Yunnan, Guangxi, Guangdong and Indochina (2x).

A. laevissima C. F. Liang: distributed in Guizhou.

A. leptophylla C. Y. Wu: distributed in Sichuan, Yunnan and Guizhou.

A. rubricaulis Dunn var. *rubricaulis:* distributed in Sichuan, Hubei, Yunnan, Guizhou, Hunan, Guangxi

A. rubricaulis Dunn var. *coriacea* (Fin. et Gagn.) C. F. Liang: distributed in Sichuan, Hubei, Yunnan, Guizhou, Hunan, Jiangxi and Guangxi.

A. sabifolia Dunn (originally published as *A. sabiaefolia)*: distributed in Anhui, Jiangxi, Hunan and Fujian (4x).

A. trichogyna Franch.: distributed in Sichuan, Hubei and Jiangxi.

A. venosa Rehd. var. *venosa*: distributed in Sichuan, Yunnan, Tibet and India.

A. venosa Rehd. var. *pubescens* Li: distributed in Sichuan and Yunnan.

A. ulmifolia C. F. Liang: distributed in Sichuan.

A. umbelloides C. F. Liang var. *umbelloides*: distributed in Yunnan.

A. umbelloides C. F. Liang var. *flabellifolia* C. F. Liang: distributed in Yunnan.

Section Strigosae Li: Leaves and branches are very hairy or woolly with coarse, hard and simple hairs.

A. arisanensis Hayata: distributed in Taiwan

A. benryi Dunn var. *henryi:* distributed in Yunnan.

A. benryi Dunn var. *polyodonta* Hand.-Mazz.: distributed in Yunnan.

A. benryi Dunn var. *glabricaulis* (C. Y. Wu) C. F. Liang: distributed in Yunnan.

A. carnosifolia C. Y. Wu var. *carnosifolia*: distributed in *Yunnan.*

A. carnosifolia C. Y. Wu var. *glaucescens* C. F. Liang: distributed in Hunan, Guangxi, Guangdong, Guizhou and Yunnan

A. *chengkouensis* C. Y. Chang: distributed in Shaanxi, Sichuan and Hubei.

A. fortunatii Fin. et Gagn.: distributed in Guizhou.

A. hemsleyana Dunn *var. hemsleyana*: distributed in Zhejiang, Jiangxi and Fujian (2x).

A. *hemsleyana* Dunn var. *kengiana* (Metc.) C. F. Liang: distributed in Zhejiang and Fujian (2x).

A. holotricha Fin. et Gagn.: distributed in Sichuan and Yunnan.

A. melliana Hand.-Mazz.: distributed in Jiangxi, Hunan, Guangxi, Guangdong and Hainan (2x).

A. petelotii Diels: distributed in Indochina.

Actinidia rudis Dunn: distributed in Yunnan.

rubus Levl.: distributed in Yunnan.

A. strigosa Hook. f. et Thoms.: distributed in Sikkim.

A. *vitifolia* C.Y Wu: distributed in Sichuan and Yunnan.

Actinidia rudis Dunn: distributed in Yunnan.

A. *holotricha* Fin. et Gagn.: distributed in Sichuan and Yunnan.

A. *rubus* Levl.: distributed in Yunnan.

A. *chengkouensis* C. Y. Chang: distributed in Shaanxi, Sichuan and Hubei.

A. *vitifolia* C.Y Wu: distributed in Sichuan and Yunnan.

A. *fortunatii* Fin. et Gagn.: distributed in Guizhou.

A. melliana Hand.-Mazz.: distributed in Jiangxi, Hunan, Guangxi, Guangdong and Hainan.

A. carnosifolia C. Y. Wu var. *carnosifolia*: distributed in *Yunnan.*

A. carnosifolia C. Y. Wu var. *glaucescens* C. F. Liang: distributed in Hunan, Guangxi, Guangdong, Guizhou and Yunnan.

A. henryi Dunn var. *henryi:* distributed in Yunnan.

A. henryi Dunn var. *polyodonta* Hand.-Mazz.: distributed in Yunnan.

A. henryi Dunn var. *glabricaulis* (C. Y. Wu) C. F. Liang: distributed in Yunnan.

A. *petelotii* Diels: distributed in Indochina.

A. *strigosa* Hook. f. et Thoms.: distributed in Sikkim.

A. *hemsleyana* Dunn *var. hemsleyana*: distributed in Zhejiang, Jiangxi and Fujian (2x).

A. *hemsleyana* Dunn var. *kengiana* (Metc.) C. F. Liang: distributed in Zhejiang and Fujian (2x).

A. *arisanensis* Hayata: distributed in Taiwan.

Section *Stellatae* Li: Leaves and branches are very hairy or woolly with soft, fine and stellate hairs on the under surface.

Series *Perfectae* C. F. Liang: Lower surface of leaf is thickly covered with persistent stellate hairs.

A. chinensis Planch. var. *chinensis*: distributed in Shaanxi, Henan, Anhui, Jiangsu, Hubei, Jiangxi, Zhejiang, Hunan, Fujian, Guangdong and Guangxi (4x).

A. chinensis Planch. var. *jinggangshanensis* (C. F. Liang) C. F. Liang et A. R. Ferguson: distributed in Jiangxi (2x).

A. chinensis Planch. var. *rufopulpa* (C. F. Liang et R. H. Huang) C. F. Liang et A. R. Ferguson: distributed in Hubei.

A. cinerascens C. F. Liang var. *cinerascens*: distributed in Guangdong.

A. cinerascens C. F. Liang var. *tenuifolia* C .F. Liang: distributed in Guangdong.

A. cinerascens C. F. Liang var. *longipetiolata* C. F. Liang: distributed in Hunan.

A. deliciosa (A. Chev.) C. F. Liang et A. R. Ferguson var. *deliciosa*: distributed in Gansu, Shaanxi, Henan, Hubei, Hunan, Sichuan, Yunnan, Guizhou and Guangxi (6x).

A. deliciosa (A. Chev.) C. F. Liang et A. R. Ferguson var. *longipila* (C.F. Liang et R.Z. Wang) C.F. Liang et A. R. Ferguson: distributed in Shaanxi and Sichuan.

A. deliciosa (A. Chev.) C.F. Liang et A. R. Ferguson var. *chlorocarpa* (C. F. Liang) C. F. Liang et A. R. Ferguson: distributed in Sichuan, Yunnan and Guangxi (4x).

A. eriantha Benth. f. *eriantha:* distributed in Guizhou, Hunan, Zhejiang, Jiangxi, Fujian, Guangdong and Guangxi (2x).

A. eriantha Benth. f. *alba* C.F. Gan: distributed in Zhejiang.

Acinidia farinosa C. F. Liang: distributed in Guangxi (2x).

A. fulvicoma Hance var. *fulvicoma*: distributed in Hunan, Jiangxi, Fujjan and Guangdong (2x).

A. fulvicoma Hance var. *pachyphylla* (Dunn) Li: distributed in Jiangxi and Guangdong.

A. fulvicoma Hance var. *lanata* (Hemsl.) C. F. Liang f. *lanata*: distributed in Jiangxi, Fujian, Hunan, Guangdong, Guangxi and Guizhou.

A. fulvicoma Hance var. *lanata* (Hemsl.) C. F. Liang f. *hirsuta* (Fin. et Gagn.) C. F. Liang: distributed in Hunan, Guizhou, Yunnan and Guangxi.

A. fulvicoma Hance var. *lanata* (Hemsl.) C. F. Liang f. *arahchnoidea* C. F. Liang: distributed in Guangxi.

A. lanceolata Dunn: distributed in Anhui, Zhejiang, Jiangxi, Fujian, Hunan and Guangdong.

A. latifolia (Gardn. et Champ.) Merr. var. *latifolia:* distributed in Anhui, Zhejiang, Jiangxi, Fujian, Hunan, Sichuan, Yunnan, Guizhou, Guangxl, Guangdong, Hainan, Taiwan, Indochine, Malaysia and Indonesia.

A. latifolia (Gardn. et Champ.) Merr. var. *mollis* (Dunn) Hand.-Mazz.: distributed in Yunnan.

A. liangguangensis C. F. Liang: Hunan, Guangdong, Guangxi

A. rufotricha C. Y. Wu var. *rufotricha:* distributed in Yunnan (2x).

A. rufotricha C. Y. Wu var. *glomerata* C. F. Liang: distributed in Guizhou and Guangxi

A. setosa (Li) C.F. Liang et A. R. Ferguson: distributed in Taiwan.

A. styracifolia C. F. Liang: distributed in Hunan and Fujian (2x).

A. suberifolia C. Y. Wu: distributed in Yunnan.

Series *Imperfectae* C.F. Liang: Lower surface of leaf has imperfectly stellate hairs or scattered deciduous hairs.

A. grandiflora C. F. Liang. (originally published as *A. stellato-pilosa)*: distributed in Sichuan (4x).

A. guilinensis C.F. Liang: distributed in Guangxi (4x).

A. lijiangensis C.F. Liang et Y.X. Lu: distributed in Guangxi (4x).

A. obovata Chun ex C. F. Liang: distributed in Gulzbou.

A. pilosula (Fin. et Gagn.) Stapf ex Hand.-Mazz.: distributed in Yunnan and Burma.

A. stellatopilosa C.Y. Chang: distributed in Sichuan.

A. sorbifolia C. F. Liang: distributed in Guizhou.

A. zhejiangensis C.F. Liang: distributed in Zhejiang and Fujian (2x).

Kiwifruit

x = 29; 2n = 58-116

Kiwifruit is a common name for all commercially grown fruit from the genus *Actinidia*. The most common kiwifruit is the fuzzy kiwifruit, from the species *A. deliciosa*. It is native to northern China. Other species of *Actinidia* are native to India, Japan and southeastern Siberia. Its cultivation spread in the early 20th century in New Zealand when seeds were introduced by Isabel Fraser from China. The fruit was called *yang tao*, Chinese gooseberry and melonette. It was renamed as kiwifruit because of its resemblance to the New Zealand kiwi bird. Other species that are commonly eaten include golden kiwifruit (*A. chinensis*), Chinese egg gooseberry (*A. coriacea*), baby kiwifruit or hardy kiwi (*A. arguta*), Arctic kiwifruit or Arctic beauty (*A. kolomikta*), red kiwifruit (*A. melanandra*), silver vine (*A. polygama*), purple kiwifruit (*A. purpurea*).

In China, kiwifruit is spread over the entire southeast and is grown mainly in the mountainous area upstream of the Yangtze River, as well as Sichuan. It can be grown in temperate climates with adequate summer heat. Where fuzzy kiwi (*A. deliciosa*) is not hardy, other species can be grown as substitutes. Kiwifruit is deciduous, woody and vigorous climber reaching up to 9 m in length. Pith is solid or lamellate. Winter buds are small enclosed in swollen base of the petiole. Leaves are alternate, long-petioled, serrate, rarely entire, exstipulate and oval to nearly circular and 7.5-12.5 cm long. Young leaves and shoots are coated with red hair. Mature leaves are dark green and hairless on the upper side and downy white with prominent light coloured veins beneath. The trees are dioecious. The flowers are borne singly or in 3's in cymes in the leaf axils. These are cup-shaped, 2.5-5 cm broad, 5-6 petalled, white at first and changing to buff yellow having central tufts of many stamens. Pistillate flowers appear to be perfect but their stamens produce non-viable pollen. The staminate flowers have a greatly reduced ovary and poorly developed style. Pistillate flowers are generally larger than the staminate ones and often contain more perianth parts. Both staminate and pistillate flowers have a distinctive fragrance. The sepals are imbricate, 5, rarely less. The petals are 5, rarely 4, thin and convolute. Anthers are versatile. Ovary is superior and many-celled with many radiating styles. The fruit is oval berry, 50-100 g in weight, 6-15 cm long and ovoid or oblong with russet brown skin, which is densely covered with short, stiff and brown hair. The young fruit is capped at the base with a prominent 5 pointed calyx which shrivels and abscises when the fruit matures but 5 small sepals persist. The flesh is firm, glistening, juicy, bright green to yellow except for the white succulent centre from which radiate many fine and pale lines. Minute dark purple or nearly black albuminous seeds having large embryos are scattered between these lines. In some inferior types, the central core is fibrous or even woody (Morton, 1987; Ferguson, 1990a, b).

Fuzzy kiwifruit or mangüeyo
(***Actinidia deliciosa*** (A. Chev.) C. F. Liang & A. R. Ferguson) (2n=2x=58)

It is a vigorous, woody, twining vine or climbing shrub reaching 9 m length. It is native to southern China and has been declared the national fruit of the country. This species grows naturally at altitudes between 600 and 2000 m. The leaves are alternate, long-petioled, deciduous, oval to nearly circular, cordate at the base, and 7.5–12.5 cm long. Young leaves are coated with red hairs. Mature leaves are dark green and hairless on the upper surface and downy white with prominent, light-coloured veins beneath. It is dioecious and is pollinated by bees. Bee hives are normally used in commercial orchards. The flowers are fragrant, unisexual, borne singly or in threes in the leaf axils, are five to six-petalled, white at first, changing to buff yellow, 2.5–5 cm broad, and both sexes have central tufts of many stamens, though those of the female flowers have no viable pollen. The flowers also lack nectar. Kiwifruit is notoriously difficult to pollinate, because the flowers are not very attractive to bees. Sometimes, the more labour intensive hand pollination is employed. Male flowers are gathered and processed to extract their pollen which is then sprayed on the female flowers. The fruits are oblong, up to 6.25 cm long. The russet brown skin of the fruits is densely covered with short, stiff, brown hairs. The flesh is firm until fully ripened and it is glistening, juicy and luscious. The colour of the flesh is bright green or sometimes yellow, brownish or off-white, except for the white, succulent centre from which radiate many fine, pale lines. The flavour is subacid to quite acid similar to that of the gooseberry or strawberry. Firm fruits ripen after a few days to a week when stored at room temperature, but should not be kept in direct sunlight. The two botanical varieties are *Actinidia deliciosa* var. *chlorocarpa* and *Actinidia deliciosa* var. *deliciosa.*

Cultivars

In China, the following four main classes of cultivars are recognized:

1. Zhong Hua (Chinese gooseberry) is most commonly grown and has three subclasses; yellow flesh, green flesh and yellow green and green yellow flesh.
2. Jing Li (northern pear gooseberry).
3. Ruan Zao (soft date gooseberry) is good for jam and is usually grown in hills.
4. Mao Hua class constitutes selections made by growers for fresh fruit market, such as Quing Yuan # 17 rated as superior quality, Quing Yuan # 22 of high quality, Quing Yuan # 28 of medium quality, Quing Yuan # 18 good as fresh fruit and for processing, Quing Yuan # 20 of excellent quality and Long Quan # 3 of above average quality. Hayward, Blake and Saanichton 12 belong to this group. They have a fuzzy, dull brown skin, and bright-green flesh.

Numerous selections have been made in China from the wild growing populations of *A. deliciosa*. The best known of these is Qinmei. Some more selections made by growers for processing purposes are Qing Yuan # 27, Qing Yuan # 29, Quing Yuan # 6 and Huang Yan. The important pistillate cultivars are Abbott (Greens: Rounds), Allison (large fruited), Bruno (Mc Loughlin: Longs; Long fruited; Te puke), Wilkins Super, Constricted, Elmwood, Gracie, Hayward (Giant Hooper's Giant; Mc Whannels; Chico), Monty (Montgomery) and Greensill. Hayward is markedly superior to the other pistillate cultivars.

For areas with warmer winters, some of the older New Zealand selections, e.g., Abbott or Allison, Tewi from the Canary Islands or Vincent from Orange County, California are good, but these generally have fruit of mediocre quality (Meyer 1992; Ferguson, 1999).

Hayward: The cultivar was developed by Hayward Wright in Avondale, New Zealand. It is grown all over the world. It has high chilling requirement. It is a shy bearer and has tendency of alternate bearing and requires pollinizer for good yields. Solitary flowers emerge on the fruiting canes. The fruits are large, broad and flat, each of 80-120 g having very good keeping quality and storage life (4-6 months), superior flavour and high sugar and fairly high ascorbic acid content. Core is larger than in the fruit of any other cultivar.

Allison: The cultivar has performed well in the mid hills of Himachal Pradesh in India. It very closely resembles Abbott except that both apices of the fruit are more flat and it weighs 50-75 g. The corolla overlap and are crimpled over the margins. The cultivar bears very heavily and the fruits ripen early. The fruits are sweet in taste with somewhat low acid and ascorbic acid contents.

Bruno: The cultivar requires less chilling hours and is suitable for low hill areas. In contrast to Allison, its fruits have very high titratable acid and ascorbic acid contents. It is a good cropper and the fruits are smaller than those of Hayward but larger than Abbott and Allison. Fruits are dark brown in colour with very dense, short and bright hairs and taper towards the stem end and are longer.

Monty: It is a late flowering type but fruits ripen in mid season. The fruit apices are flat as in Allison and Abbott and are distinct and broader at the blossom end. The cultivar is very prolific and productive and requires thinning to get good fruit quality. The fruit is from small to medium, highly acidic and has medium sugar and ascorbic acid contents.

Qinmei: The cultivar was selected in China from wild populations from the Qinling Mountains, Shaanxi. It has high adaptability to both low and high temperature conditions. It is regarded to be superior to Hayward in most of the characters. It is very productive and starts bearing early. Fruits are large (110 g) and oblong in shape but have only a mild flavour and a comparatively short shelf life.

Saanichton 12: It is from British Columbia. The fruit is somewhat more rectangular than that of Hayward and comparably sweet, but the inner core of the fruit can be tough.

Blake: The cultivar can self pollinate. The fruit is smaller, oval and has inferior flavour.

Koryoku: It is a selection from open pollinated population of Hayward. It is grown on a limited scale in Japan, mainly because the fruit is sweeter and reaches harvest maturity earlier.

Top Star: It is a bud mutant of Hayward having smooth, essentially hairless fruit. The reduced vigour of the vine allows for easier management and less summer pruning.

Tomua: It means "early" in Maori. This is the first female kiwifruit cultivar from New Zealand developed from a cross between Hayward and males from an accession of *A. deliciosa* collected from Qinling Mountains. This accession was notable for flowering early in the season and the fruit, although small, also mature early. The fruit is large having good, sweet flavour but has limited storage life. Depending on location, the fruit can be harvested about one month earlier than Hayward.

Donné: In some areas with warmer winters, winter chilling requirements are not satisfied. Old cultivars such as Abbott or Allison or selections such as Tewi (from the Canary Islands) or Vincent (from Orange County, California) may be suitable but these generally have fruit of mediocre quality. Therefore, Donné was specifically selected for the warmer climatic conditions of South Africa.

Staminate cultivars

Male cultivars have been selected for weather conditions of different areas so as to coincide their flowering with that of Hayward to ensure effective seed set and full fruit size. A series of males have been selected in New Zealand for Hayward. Autari has been selected in Italy and Californian Male or Chico Male in California. King and Ranger have been selected for Tomua as it flowers several weeks ahead of Hayward. Matua selections from M-series, Chico and Timouri males are good pollinizers (Morton, 1987; Ferguson, 1990a; Pandey and Joshi, 1997).

Tomuri: Tomuri means "late" in Maori language. This cultivar flowers earlier and is considered to be a very effective pollinizer for Hayward, but in New Zealand it flowers late. It is a good pollinizer for early and mid season pistillate cultivars. The flowers appear in clusters of five (1-7) with short hairs on the peduncle.

Matua: The cultivar is a good pollinizer for early and mid season pistillate cultivars as it profusely flowers over a long period. The flowers have pubescence or hairs on the peduncles that are usually short and are borne in clusters of 1-5 (usually 3).

Golden kiwi

(***Actinidia chinensis*** Planch., Chinese gooseberry; Yang tao) (2n=2x=58)

Its origin is supposed to be the northern Yangtse Valley in China. In its native habitat it grows in thickets, thick *Quercus* forests and light secondary forests and bushland. Of all *Actinidia* species, *A. chinensis* is the most similar to *A. deliciosa*, the cultivated kiwifruit. These two species were earlier classified together. But their fruits are distinctly different. In wild plants of *A. chinensis*, the fruits are generally much smaller, more rounded and less cylindrical than those of the cultivated kiwifruit. However, from the considerable variation in fruit size and shape, good selections of *A. chinensis* can approach or even exceed the average size of 'Hayward' fruit. The fruit is almost hairless at maturity, and what hair remains is usually much shorter and finer than kiwifruit hair, more like the fuzz of a peach. Flesh colour can vary greatly from bright green, shading through lime green to a clear, intense yellow. Possibly the most attractive fruits are those in which the inner pericarp flesh is red `iris' in centre surrounded by the outer yellow. The flavour of good selections of *A. chinensis* is much better than that of Hayward. The fruits are sweeter, more aromatic with a flavour reminiscent of some subtropical fruit. The cultivar Zaperi Gold has been developed in *A. chinensis*. The cultivar has sweet fruits with 12°B TSS.

Cultivars

The female selections available include Lushanxiang (syn. 79-2, ACC 226, Elizabeth, First Emperor, K189, Yellow Joy); Jiangxi 79-1 (syn. Koushin, Kosuei 79-1, Lushan 79-1, Red Princess); Kuimi (syn. F.Y. 79-1, ACC 211, Apple Sensation, Kamitsu, Turandot and possibly K191); Jinfeng (syn. FT79-3, Golden Yellow, Kinpo). These selections have been grown in Italy, New Zealand and USA. The only commercial production outside of China appears to be in California, with Lushanxiang, Jiangxi-79-1.

Hort16A (Zespri Gold): It was identified from the seedling progeny of a cross of two accessions of *A. chinensis*. It is diploid unlike the kiwifruit which is hexaploid and flowers almost a month earlier. Therefore, diploid males of *A. chinensis* have to be selected as pollenizers. Two such plants are Meteor and Sparkler. The fruit has a very characteristic pointed shape, quite different in appearance to Hayward fruit. The skin is covered with very soft downy hair, which is easily rubbed off. The flesh is a bright yellow when fruit are harvested at the right maturity and the flavour is much preferred to that of Hayward fruit. The fruits need to be handled differently, especially as the fruit skin is more tender and more easily damaged than that of Hayward. It is susceptible to PSA virus.

China Belle: The cultivar was selected in France. It is a tetraploid. The fruit is large averaging between 80 and 100 g and it has yellow flesh. A pollenizer, PolliChina, has also been selected.

Enza Red: It is a red-ringed cultivar of the Chinese *hong yang*.

Gold 3 (Sun Gold): The cultivar is more disease resistant. It is not quite as sweet as Hort16A and has tanginess and lacks the slightly pointed tip usually seen in Hort16A.

Issai: It is a hybrid of hardy kiwi and silver vine. It can self pollinate. The fruit is relatively large. It is less hardy than most hardy kiwi.

Kiwi berries

These are composed of three species of kiwifruit, hardy kiwi (*A. arguta*), Arctic beauty (*A. kolomikta*), and silver vine (*A. polygama*). These are fast growing, climbing vines. The fruit is edible berry or grape-sized fruit similar to the fuzzy kiwi in taste and appearance with thin and smooth skin. These are referred to as kiwi berry, baby kiwi, dessert kiwi, grape kiwi or cocktail kiwi.

Hardy kiwi
(***Actinidia arguta*** (Siebold and Zucc.) Planch. ex Miq.; tara vine; wild fig)

It is native of northern China, Korea, Siberia and possibly Japan (Reich, 1991). The related edible species is *A. kolomikta* Maxim.

The plants are climbing vines (lianas), sometimes climbing 30 m high on the trees (CRFG, 1996a). In cultivation, the vines must be supported by a trellising system. The plant has a more delicate appearance than the common kiwifruit. Leaves are elongated, generally 1-2 cm long and attached to the stem on red petioles. They are usually serrated and far less leathery and fuzzy than the kiwifruit. The flowers are about 4 cm in diameter, white to cream in colour and somewhat fragrant emerging from the leaf axils singly or in triplets. Flowering period extends over several weeks from early May to June depending on climatic conditions. Plants are dioecious. However, self fruiting females are known to exist.

The fruit is about the size of a European gooseberry or grape, with a skin that is smooth and polished, completely hairless and generally palatable. The fruits are generally green in colour and contain small black seeds of emerald green colour and with typical rayed pattern. The flesh is sweet but the best feature is the unique, sophisticated flavour. It is suitable for areas that are too cold for kiwifruit cultivation.

Cultivars

Selections have been made in Canada, France, Germany, Italy, New Zealand and USA for size, colour (red or green skins), sweetness and flavour. The female cultivars are Ananasnaja Minchurian (Anna), Clara Zetkin, Poozdniani (late), Raniaia, Urezhainaia (late), Cordifolla, Dumbarton Oaks, Geneva, Issai (2 distinct self fruitful cultivars from Japan), Ken's Red, Michigan State, 119-40B (claimed to be self fruitful), Red Princess and seedling selections made by Professor Meader. Various male cultivars are also known. The cultivar Anna (Anansnaya) is possibly a hybrid of *A. arguta* and *A. kolomikta* (Evreinoff, 1949).

Arctic kiwifruit
(***Actinidia kolomikta*** (Maxim. et Rupr.) Maxim.)

It is a particularly cold hardy species with small but sweet fruit, very rich in vitamin C. It is of very limited commercial potential except for very cold areas.

Zhong hua (Chinese gooseberry), *jing li* (northern pear gooseberry), *ruan zao* (soft date gooseberry) and *mao hua* (may be tight or loose haired) are the four main cultivars of this species in China.

Family Ericaceae (bilberries, blueberries. cranberries, heathers, heaths, rhododendrons, wintergreens)

The plants are evergreen or deciduous shrubs, sometimes scramblers or climbers (some *Agapetes*), herbs or trees (*Arbutus*) and epiparasites without chlorophyll (*Monotropa*) having simple leaves. The leaves are alternate and spirally arranged or opposite or whorled, entire or serrate, with or without a petiole, sometimes strongly revolute and forming narrow ericoid needles and without stipules. The inflorescence is terminal or axillary racemes usually with paired bracteoles (prophylls) and rarely single-flowered. The flowers are regular, usually hermaphrodite having (2-)4-5(-9) sepals fused at the base. The petals are (3-)4-5(-9), equal and usually fused except for the pointed tips. The stamens are (2-)5)-5(or 10 (-16), free or rarely adnate to the petals. The ovary is usually superior but is inferior in the Vaccinieae, of (1-)4-5(-14) fused carpels, containing one to many ovules. The style is usually as long as the corolla tube, hollow and is without lobes. The fruit is a capsule, berry or drupe having small seeds.

The family has 8 subfamilies (Enkianthoideae, Monotropoideae, Arbutoideae, Cassiopoideae, Ericoideae, Harrimanelloideae, Styphelioideae and Vaccinioideae) and 20 tribes, 124 genera and 4050 species.

Subfam. Vaccinioideae

The plants are evergreen or deciduous trees, shrubs or climbers having the apical bud of aborting vegetative shoots. It has five tribes:

1. **Oxydendreae:** It contains deciduous trees with terminal inflorescences, anthers without an appendage and splitting for half their length. It has 1 genus, *Oxydendrum* with 1 species.
2. **Lyonieae:** It comprises trees, shrubs, occasionally vines with woody stems, varied flowers and stamens with the bent filaments. It has 4 genera and 48 species of which *Lyonia* has 35 species.
3. **Andromedeae:** It consists of small shrubs with hairs only on the leaf margins, terminal inflorescence and flowers with awned anthers. It has 2 genera and 3 species.
4. **Gaultherieae:** It consists of shrubs with racemose inflorescences, usually urn-shaped, stamens with swollen filaments and anthers with short slits. It has 6 genera and 246 species of which *Gaultheria* has 130 and *Diplycosia* has 100 species.
5. **Vaccinieae:** It comprises evergreen or deciduous shrubs or small trees, sometimes lianas, often epiphytic having leathery leaves, axillary inflorescences and the flowers with a jointed pedicel and fleshy fruits. It has 35 genera and 1275 species including 400 of *Agapetes,* 140 of *Vaccinium,* 130 of *Cavendishia* and 80 of *Dimorphanthera* (Heywood *et al.*, 2007).

The Vaccinieae contains whortle berries, billberries, cranberries and blueberries. The true blueberries belong to the ancient genus *Vaccinium* and subgenus *Cyanococcus*. There are several types of blueberries. The species range in height from 1.5 to 7 m and are classified as highbush blueberries. The cultivated highbush blueberry or swamp berry has been developed primarily from two species, *V. corymbosum* L. (highbush blueberry) and *V. australe* Small, although several other species such as *V. angustifolium* have been used in the breeding programme. The less common species of Vaccinieae are *V. myrtillus* (biliberry) and *V. reticulatum* (Hawaiian ohelo).

Genus *Vaccinium* L. (x=12; 2n=24, 48, 72)

It contains 140 species in arctic and north temperate (few species), Central and South America, South East Pacific, Africa and Madagascar (Kubitzki, 2004). The plants are shrubs, rarely trees, lianas or rhizomatous having leaves rarely thin, deciduous and penninerved, inflorescences are racemes, rarely terminal or rarely fascicles or single axillary flowers. Flowers are (4)5(6)-merous, rarely the pedicel is continuous, calyx limb is erect, rarely spreading, often shorter than the tube and lobes are variable. Corolla is <1.5 cm long, cylindric, urceolate or campanulate having imbricate lobes, rarely valvate and rarely parted to the base, short to long filaments, anthers with or without spurs. Disintegration tissue is rarely present, thecae is smooth or papillate, tubules is long and slender with terminal pores, rarely with slits. Ovaries are 4-5, rarely falsely if 10-locular. Fruit is purple or black, rarely red or white.

Cranberry and blueberry
(***Vaccinium*** spp.) x = 12

The plants are deciduous or evergreen shrubs, rarely trees having small ovoid winter buds with 2 or several outer scales. The leaves are alternate, short petioled and entire or serrate. The flowers are axillary or terminal, solitary or in racemes having 4-5 calyx lobes, rarely obsolete, cylindric, urceolate or campanulate, 4-5 lobed or sometimes parted corolla, 8 or 10 stamens, awned or awnless anthers, prolonged into terminal tubes with an opening at the apex and 4-16 celled inferior ovary. The fruit is a many seeded false or epigynous berry crowned by the persistent calyx lobes. It contains about 130 species in the Northern Hemisphere from Alaska to the tropics.

The genus *Vaccinium* L. has the following species yielding edible fruits (Janick and Paull, 2008):

Section *Cyanococcus* A. Gray: *V. corymbosum* L (highbush blueberry) distributed in Eastern North America (transcontinental with *V. myrtilloides* Michx.). *V. angustifolium* Ait. (lowbush blueberry).

Section *Hemimyrtillus* Sleumer: *V. arctostaphylos* L. (Caucausian bilberry) distributed in Caucacus mountains, Azores, Madeira and eastern Asia.

Section *Myrtillus* Dumort: *V. myrtillus* L. (Billberry) with circumpolar distribution (Central America in Rocky Mountains). *V. membranaceum* Dougl. ex Hook. (thin-leaved huckleberry).

Section *Oxycoccus* (Hill) Koch.: *V. macrocarpon* Ait. (Cranberry) Circumpolar distribution (south to mid Atlantic and south Appalachians). *V. oxycoccus* L. (small cranberry).

Section *Polycodium* (Raf.) Rebder: *V. stamineum* (Deerberry): distributed in Eastern North America (disjunct populations in Mexico).

Section *Pyxothamnus* (Nutt.) Sleumer: *V. floribundum* H.B.K. (Mortina) distributed in Andes, Mexico, Central America, California to British Columbia.

Section *Vaccinium* L.: *V. uliginosum* L. (Bogberry) with circumpolar distribution.

Section *Vitis-idaea* (Moench) Koch.: *V. vitis-idaea* L. (Lingonberry) with circumpolar distribution.

Five major classes of blueberry are grown i.e. lowbush, highbush, half highbush, southern high bush (low chilli) and rabbit eye.

Highbush blueberry
(*Vaccinium corymbosum* L.) 2n = 48, 72

It is a woody perennial plant, very variable in size. The shrubs are normally 1-4 m high having branches spreading up to 4 m. It hybridizes freely with lowbush blueberry. Growth of individual shoots of the blueberry are sympodial and episodic. Leaves are deciduous, ovate to ovate lanceolate, 3.75-7.5 cm long, entire, glabrous or slightly pubescent beneath. Flower bud differentiation occurs in the basipetal fashion, proceeding from the uppermost bud on a shoot. Flowers are borne at the tips of the previous season's growth. These epigynous blueberry flowers are white or pinkish appearing in short dense racemes and are generally urn-shaped and inverted. The five-lobed calyx is glaucous and glabrous and adnate to the ovary. Petals are usually five and are fused into a five lobed corolla, 8 mm in length, which is generally white but may have some pink colouration along the ribs. The 8 or 10 stamens correspond to the four or five corolla. Fruit is blue black and 1 cm across in size (Bailey, 1971; Eck *et al.*, 1990).

Cultivars

Important cultivars of highbush blueberries are Aron, Collins, Earliblue, Concord, Stanley, Berkeley, Bluejay, Bluecrop, Bluecross, Bounty, Blueray, Ivanhoe, Dixi, Jersey, Atlantic, Burlington, Coville, Darrow, Weymouth, Wolcott, Herbert and Rubel. New cultivars developed by USDA breeding programme are Bluetta, Spartan Patriot, Darrow, Duke, Toro, Lateblue and Elliott. Hybrids between highbush cultivars and species native to the southern USA (particularly *V. darrowi* and *V. ashei*) have produced a series of highbush types adapted to the South as they require low chilling. These are cultivars Flordablue, Sharpblue, Avonblue, Georgiagem, Cooper, Gulfcoast, Cape Fear, Blue Ridge and O'Neal (Eck *et al.*, 1990; Thakur and Rathore, 1991). The cultivars listed in order of their ripening time are:-

Earliblue: The bush is vigorous, erect and productive. Fruit cluster is medium loose. Berry is medium large, light blue, firm, resistant to cracking having medium scar, good flavour, sweet and aromatic pulp. The fruit ripens early and does not drop quickly, and ships well. It is suitable for fresh market and U-pick.

Bluetta: The bush is moderately vigorous, compact, somewhat spreading consistent in production and is more frost and winter hardy than Earliblue. Fruit cluster is Loose. Berry is medium, light blue and firm having medium broad scar and fair to good flavour.

Duke: The bush is vigorous, erect, open and consistent in production. Berry is medium, light blue, firm having small scar and fair flavour and becomes more aromatic on storage. The fruit has good shipping qualities.

Spartan: The bush is vigorous, erect, open and productive. Berry is very large light blue, firm having medium scar and excellent flavour. It blooms late and thus avoids frost injury, but ripens early concentrated to two main pickings. It prefers light well drained soil.

Patriot: The bush is vigorous, moderately erect, open and very productive. Fruit cluster is tight. Berry is very large, slightly flat, medium blue, firm having small scar and excellent flavour. Ripening is concentrated to two main pickings. The berries are red black when immature and can be soft in very hot weather. Plants are tolerant to heavy and wet soils, cold weather and frost. It has some resistance to *Phytophthora* root rot.

Meader: The bush is vigorous, erect, very hardy and very productive. Fruit cluster is medium loose. Berry is large, firm having small scar and good flavour. The plant tends to overbear requiring heavy pruning.

Collins: The bush is medium in vigour, slightly spreading and productive. Berry is large, light blue, firm having small scar and excellent flavour and resists cracking. The fruit does not drop quickly once ripe and is suitable for fresh market. The plants tolerate spring frosts.

Bluejay: The bush is vigorous, fast growing, erect, open and moderately productive. Fruit cluster is loose. Berry is medium large, light blue, very firm having small scar and pleasant, mild and slightly tart flavour and resists cracking. Fruit does not drop quickly once ripe and retains quality. Clusters are borne at the periphery of the plant on long stems. It has exceptional fruit retention and suits machine harvesting. Keeping quality of the berry is good and it ships well.

Blueray: The bush is vigorous, erect, open and productive. Fruit cluster is small and tight. Berry is large to very large, light blue, firm, resistant to cracking having small to medium scar and excellent, aromatic flavour. It has good shipping quality. The plants perform well in hot climates. Its tight fruit clusters discourages machine harvesting.

Toro: The bush is vigorous, erect and productive. Berry is medium large, firm and light blue having small scar and good flavour. Ripe fruits are harvested in two pickings and are suitable for processing and fresh market. The plants have good cold hardiness.

1613-A: The bush is moderately vigorous, erect, open and very productive. Fruit cluster is medium and loose. Berry is medium, light blue and very sweet and does not drop quickly once ripe, but tends to shrivel when fully ripe. The fruits can be harvested in one or two pickings and can be machine harvested. It is not suitable for long distance shipping and is excellent for processing.

Ivanhoe: The bush is very vigorous, erect and is not consistent in production. Fruit cluster is medium and loose. Berry is large, light blue, firm having medium scar and good, tart and aromatic flavour and is resistant to cracking.

Olympia: The bush is vigorous, spreading and productive. Berry is medium large, medium blue having thin skin, small scar and very good flavour and aroma and is resistant to cracking. Fruit does not shatter and has uniform size and high sugar content. It is a popular fresh market fruit well suited for processing in frozen packs. It can be shipped fresh on first and second picking if the weather is not too hot. It can be machine harvested if the bush is properly pruned.

Bluecrop: The bush is vigorous, erect, open and very productive. Fruit cluster is loose. Berry is very large, light blue, firm, having small scar and very good, slightly aromatic flavour and is resistant to cracking. Fruit can be tart or becomes red black (only half fruit blue) if picked too early. Plant tends to overproduce if not pruned correctly.

Berkeley: The bush is vigorous, open, spreading and is productive to very productive. Fruit cluster is large and loose. Berry is large to very large, light blue, firm, having small to medium scar and fair and slightly aromatic and mild flavour, less acidic than most cultivars and is resistant to cracking. The fruit stores well and is popular for U-pick and for fresh markets and processing. The plant does the best in light well drained soil.

Rancocas: The bush is vigorous and consistent in production. Berry is medium- sized having poor quality, large picking scar and tendency for ripe fruit to crack during heavy rains. Berries ripen about 3 days after Bluecrop. It is suitable only for processing. The plants have good cold hardiness.

Pemberton: The bush is vigorous, erect and very productive. Fruit cluster is loose. Berry is medium large, dark blue, firm having large scar and fair and slightly aromatic flavour. The plants are easy to grow and prune, but picking is more difficult than in most other cultivars and is thus not suited for machine harvest.

Herbert: The bush is vigorous, open, spreading and consistent in production. Fruit cluster is medium and loose. Berry is large to very large, medium blue having medium firmness (fruit can be tender), medium scar and very good, aromatic flavour and is resistant to cracking. The fruit does not shatter from the bush.

Rubel: The bush is moderately vigorous and moderately productive. Berry is small, firm having small picking scar and tends to retain the stem if drought stressed or not picked on time. The fruit ripens about 2 weeks after Bluecrop and is suitable for processing. The plants have good cold hardiness.

Coville: The bush is vigorous, open, spreading and productive. Fruit cluster is loose. Berry is very large, medium blue, firm, having medium scar and good, tart and highly aromatic flavour and is resistant to cracking. Fruit does not drop quickly once ripe. The plants are moderately resistant to *Godronia* canker.

Jersey: The bush is very vigorous, erect and productive. Fruit cluster is loose. Berry is medium, medium blue, firm, having medium scar, fair flavour and no aroma and is resistant to cracking. The fruits are well suited for processing.

Dixi: The bush is vigorous, open, spreading and productive. Fruit cluster is medium and tight. Berry is large to very large, medium blue, firm, having large scar and good to excellent and aromatic flavour but is subject to cracking.

Darrow: The bush is vigorous, erect and consistent in production. Fruit cluster is medium and loose. Berry is large to very large, light blue firm having small medium scar and excellent, tart flavour and is resistant to cracking. The fruit is best suited for fresh market or U-pick.

Lateblue: The bush is vigorous, erect and consistent in production. Fruit cluster is moderately loose. Berry is medium large, light medium blue and firm having small to medium scar and good to very good flavour. Fruit ripening is simultaneous. It is suitable for machine harvest. However, high temperatures at harvest time may cause stemminess.

Elliott: The bush is vigorous, erect and consistent in production. Fruit cluster is loose. Berry is medium, light blue, firm having small scar and mild to good, slightly tart flavour. Fruit ripening is simultaneous.

Sunrise: The bush is moderately vigorous, upright and moderately productive and early. Berry is medium, firm and medium blue having small scar and good flavour.

Nelson: It is a mid to late season cultivar. The bush is vigorous, upright, productive and cold hardy. Berry is very large, light blue and firm having medium scar and good flavour.

Sierra: It is an early to mid season cultivar. The bush is upright, vigorous and productive. Berry is medium, light blue and firm having small scar and good flavour. It is an interspecific hybrid of 4 species (*V. darrowi, V. corymbosum, V. ashei* and *V. constablaei*) and requires high chilling like the northern highbush cultivars grown in the Pacific Northwest.

Bluegold: It is a late season cultivar ripening with Jersey. The bush is vigorous, but low growing and is extremely productive. Berry is medium, light blue and firm having small scar and good flavour. It is closely related to Northland but has a more upright growth habit and the fruit has a longer shelf life. It is likely to be well suited for fresh market.

Half high blueberries are derivatives of low bush - high bush hybrids usually *V. angustifolium* x *V. corymbosum.*

Half high bush cultivars

These have been developed by crossing highbush blueberry and lowbush blueberry. The plants are short statured, winter hardy and have fruit characters similar to those of highbush cultivars. Some of the cultivars may be suitable in very cold winter temperature areas. These have survived low temperatures between -37 and -40°C. Survival is best until the snow cover is less than 30 cm thick.

Northsky: It is a mid to late season cultivar. The bush is attractive, compact, short statured (25-45 cm), dense and highly branched with dark green leaves in summer and bright red in the fall. It withstands temperatures between 1.7 and 4.4°C and is moderately productive (0.5-1.0 kg per plant). Berry is medium, light blue having more of a good "wild" blueberry flavour. The fruit stores well.

Northblue: It is a mid to late season cultivar. The bush is vigorous, low statured (60-90 cm), withstands temperatures between 1.7 and 4.4°C, consistently productive (1.4-3.2 kg per plant). Berry is large, dark blue, firm having small scar and good flavour. Fruits store well and have more of a "wild" blueberry flavour and are good for processing, U-pick or local sale. The bush has large, glossy and dark green leaves and is attractive and ornamental in home gardens.

Northcountry: It is an early to mid season cultivar ripening about 10 days before Bluecrop. The bush is moderately vigorous, less than 90 cm high and productive. Berry is medium sized and very light blue having sweet and mild "wild" flavour. Its attractive bushy foliage and bloom makes it an ornamental plant.

St. Cloud: It is an early cultivar ripening with Bluetta. The bush is hardy and productive (1.4-3.2 kg per plant). The bush attains 90-120 cm height at maturity. Berry is large, medium blue, firm having small scar and good flavour. Ripening is simultaneous.

Northland: It is an early cultivar ripening before Bluecrop. The bush is hardy, vigorous, moderately spreading and attains a height of about 120 cm at maturity and is very productive. Berry is medium, medium blue, moderately firm having small to medium scar and good and sweet flavour. Fruit ripening is concentrated. The fruit is good for processing or for local sale. The plants must be pruned to upright stature for machine pick.

Southern highbush blueberries are similar to highbush types but have a low chilling requirement as a result of crosses with *V. darrowi* and other species.

Rabbit eye Blueberry
(*Vaccinium ashedi* Reade (2n=72)

Rabbiteye blueberries are considered suitable for planting in drought susceptible areas. These are essentially self sterile and therefore, compatible cultivars must be interplanted for effective pollination. Plants reach height of 1.8-4.5 m depending on cultivar. Most of these are not as winter hardy as the highbush cultivars and often have shiny blue black fruit due to the absence of the waxy bloom found on highbush types. Fruit ripening is promoted by frequent harvests. The important cultivars are Tifblue, Baldwin, Bluebelle, Climax, Alliceblue, Brightwell and Choice (Thakur and Rathore, 1991).

Climax: It is an early cultivar. The bush is moderately vigorous, upright, spreading. Berry is medium and dark blue having good flavour. Ripening is simultaneous.

Premier: It is an early to early mid season cultivar. The bush is vigorous and productive. Berry is large and light blue having excellent flavour.

Brightwell: The bush is vigorous and upright. Berry is large, firm and light blue having small scar and good flavour.

Bluebelle: The bush is moderately vigorous, upright and productive. Berry is large and light blue having excellent flavour.

Tifblue: The bush is vigorous and erect. Berry is medium large, very light blue and firm having small scar and high dessert quality. The harvest season is long.

Centurion: The bush is vigorous, productive and is adapted to heavy soils. Berry is medium blue and firm having excellent dessert quality. It is, however, late blooming and late ripening.

Lowbush Blueberry
(*Vaccinium angustifolium* Ait.) (2n = 2x = 48)

The term lowbush is usually applied to those species whose mature stems are generally less than one metre in height. Maine (USA) is a principal producer of lowbush berries from species *V. myrtilloides* Michx., *V. angustfolium* Ait., *V.*

brittonii Porter ex Bricknell (Westwood, 1993). Lowbush Blueberry is native to eastern and central Canada and northeastern USA, growing as far south as West Virginia and west to the Great Lakes region, Minnesota and Manitoba. It is a low spreading deciduous shrub growing 60 cm tall, but is usually only 35 cm or less. The leaves are glossy blue green in summer turning purple in the fall. The leaf is broad to elliptical. Buds borne in stem axils are brownish red. The flowers are white, bell shaped and 5 mm long. The berry is small, sweet and dark blue to black. The plant grows the best in wooded or open areas in well drained acidic soils. Important cultivars are Auguste Brunswick and Cumberland.

Cranberry

(***Vaccinium macrocarpon*** Ait.) (x=12; 2n=24)

Cranberry is evergreen creeping shrub with slender stems of 1 m having upright ends. The leaves are alternate, elliptic oblong, 0.8-1.3 cm long, obtuse, flat or revolute and lighter beneath having entire margins. The flowering shoots bear 1-7 solitary flowers at successive axils of the leaves. The flower has 4 sepals, 4 reflexed petals and 8 stamens with anthers opening through a terminal pore, a single inferior ovary with 4 cells and few to many ovules. Fruit is oblong, globose or nearly pyriform berry, 0.8-1.9 cm in diameter and red at maturity (Dana, 1990). Principal producing States in the USA are Massachusetts, New Jersey, Wisconsin, Washington and Oregon.

Cultivars

The cultivars grown commercially are the selections from the native populations. These are Early Black (dark red fruit), Mc Farlin (large berry), Searles (bright red berry) and the late maturing Howes (Dana, 1990), Jersey and Stevens. The other important cultivars are Beever, Beckwith, Bergmen, Franklin, McFarlin, Pilgrim, Stevens and Wilcose (Rathore and Thakur, 1991).

Lingon berry

(***Vaccinium vitis-idaea*** L.; cowberry; mountain cranberry; foeberries; moss cranberry; mountain cranberry; partridge berry; red whortle berry; alpine cranberry)

It is native to Scandinavia and is grown in Sweden, Norway, Denmark, Poland, Germany, Finland, eastern Canada and the former Soviet Union (Hjalmarsson and Ortiz, 2001).

It is an evergreen dwarf and prostrate shrub with semi-woody stems bearing numerous shoots of 1-2 mm diameter. The leathery leaves are simple, petiolate and obovate and are borne in an alternate spiral. The upper surface of leaf is dark green and the lower surface is pale green and waxy with black glandular dots. White flowers are produced singly or in clusters in terminal racemes and have 4 sepals, a bell-shaped corolla, 4 -5 petals, 8 stamens and inferior ovary having four locules. The fruit is a globose berry of carmine colour and up to 1 cm in diameter (Stang *et al.*, 1990; Stang, 1994).

Cultivars

Two subspecies of lingon berry are subsp. *minus* and subsp. *vitis-idaea*. The subsp. *minus* is shorter and has smaller leaves and berries than subsp. *vitis–idaea.* Unlike in subsp. *minus,* the venation is conspicuous in subsp. *vitis–idaea*. The important cultivars are Ida, Linnea, Splendor, Regal, Red Pearl, Koralle, Sussi, Sanna, Scarlet, Erntekrone, Erntesegen, Erntedank Masovia and Koralle. Two races have been reported. The smaller North American form is *V. vitis-idaea* var. *vitis-idaea*. Lot of variability has been observed in open pollinated seedlings (Stang, *et al*., 1990; Stang, 1994).

⑧ ⑧ ⑧

Bael (Dr Sanjay Singh, CIAH, Godhra)

NB 5

NB 9

Goma Yashi

Pant Urvashi

Pant Sujata

Durian (TFNet, 2008)

D 99 (Kob)

D 123 (Chanee or Gibbon)

Cacao (CPCRI, Kasargod)

G VI 56

M 16.9

Papaya (Scheldeman, 2002; TNAU, Coimbatore; TFNet, 2008; IIHR, Bengaluru)

CO 1

CO 3

CO 6

CO 8

Sun-up GMO-papaya

Sunrise

Eksotika II 6

Babaco
(*Vasconcellea x heilbornii*)

Coorg Honey Dew

Arka Surya

Sapodilla (TFNet, 2008; TNAU, Coimbatore)

Kali Patti

Makok

Molix

Sapodilla (TFNet, 2008; TNAU, Coimbatore)

PKM 1

PKM 4

PKM 5

Karonda (CHES, Chethali)

CHESK-II-7

CHESK-V-6

Kiwifruit (Ferguson, 1999)

Actinidia chinensis 'Hort 16A'

A. deliciosa 'Hayward'

A. rufa

Blueberry (Strik *et al.*, 2014)

Draper

Duke (Northern highbush)

Darrow

Elliot

[LAMIIDS: EUASTERID I]

Frequently against constant and ganzrandige (primel) sheets, early sympetalie, frequently upper-constant ovary and cap fruits, ellagic acid is missing.

Chapter 36

RUBIALES (GENTIANALES)

The plants are trees, shrubs, subshrubs and perennial or annual herbs. Intraxyllary (internal) phloem is present. Interxyllary (included) phloem of mate type is present in Antoniaceae and some species of *Strychnos* and *Logania*. Vessels are with simple perforations or rarely scalariform perforations and mostly lateral pitting is alternate. Fibres are with bordered or more often simple pits, sometimes septate. Vestured pits are present or rarely absent (*Gelsemium* and *Logania*). Rays are heterogeneous to homogeneous or wanting. Axial parenchyma is of diverse types or absent. Sieve element plastids are of S-type or Ss-type. Nodes are trilacunar, unilacunar or multilacunar. Leaves are opposite or sometimes verticillate, rarely alternate, simple and usually entire, exstipulate or less often with interpetiolar stipules. Leaf mesophyll cells sometimes contain oil bodies. Colleters are often present. Stomata are mostly paracytic or anomocytic.

Flowers are borne in cymose or less often racemose inflorescences, sometimes solitary, bisexual or rarely unisexual, mostly actinomorphic or nearly actinomorphic with double perianth. Corolla is sympetalous having usually contorted lobes, less often imbricate or valvate. Stamens are usually as many as the corolla lobes in a single whorl and alternate with them or very rarely stamens are 16-26, in two whorls usually attached to the corolla tube (*Dialypetalanthus*). Anthers are tetraspora-ngiate opening longitudinally or very rarely by apical pores (as in *Exacum*). Tapetum is secretory. Microsporogenesis is simultaneous or successive (some Apocynaceae). Pollen grains are 2-celled or 3-celled, mostly 3-colporate. Intrastaminal nectary disc is present or absent. Gynoecium is of two or less often 3-5 (up to eight) united carpels. Stylodia are more or less united into a terminal style. Ovary is superior to inferior, mostly 1-locular with one to many ovules in each locule or on each placenta. Ovules are anatropous to hemitropous, unitegmic, tenuinucellate, sometimes with endothelium. Female gametophyte is usually of *Polygonum*-type. Endosperm is nuclear or rarely cellular (*Mitrasacme, Voyriella*), very rarely (in some Apocynaceae) with chalazal haustorium. Fruits are capsules or of other types. Seeds are exotestal with more or less straight embryo and copious endosperm and rarely the endosperm is nearly wanting. It usually produces iridoid compounds and often also alkaloids, but without cardiotonic glycosides. It has 5 families.

Family Rubiaceae (coffee or madder family)

The family comprises 615 genera and 13150 species (Heywood *et al.*, 2007). The plants are mostly small trees or shrubs, sometimes armed but nearly all life forms are found including large trees, annual and perennial herbaceous plants, woody monocaul dwarfs, lianas, epiphytes, geofrutices (herbaceous stems from a woody rootstock) and rarely succulent or aquatic occasionally associated with hollow stems or special chambered tubers. The leaves are opposite or sometimes whorled, simple and nearly always entire. A few genera have bacterial nodules and often domatia (pits or hairy tufts in the axils). Colleters, which are modified hairs secreting exudates, are found on the inside surface of the stipule or otherwise associated with the stipule and sometimes present on inflorescence and floral parts. Stipules are always present, most commonly interpetiolar and often are fused above the axils, sometimes forming a sheath or less often are entirely free or exactly leaf-like (*Galium*), rarely intrapetiolar.

The inflorescence is highly variable. The flowers are usually bisexual but are also unisexual (plants usually dioecious), often heterostylous or with pollen presentation and commonly 4 or 5-merous. The calyx is adnate to the ovary, extending to a free limb tube and/or lobes, occasionally reduced to a rim, sometimes with calycophylls (or semaphylls: large, leaf-like, white or coloured calyx lobes). The corolla is always tubular (gamopetalous), although in a few taxa the tube is almost obsolete or rarely absent with radial symmetry (actinomorphic), although some species are secondarily zygomorphic. The stamens are usually as many as the corolla lobes and alternate with them and are fixed to the internal surface of the corolla tube (epipetalous). Occasionally the filaments are free to the base, sometimes fused in a ring (monadelphous as in Chiococceae). The anthers are usually dorsifixed but are occasionally basifixed having commonly introrse dehiscence or infrequently porate. A disk usually functions as a nectary positioned above the ovary inside the calyx limb, most commonly cushion-like or annular. The ovary is inferior of usually 2 and sometimes of 5 or more carpels, with axile or sometimes parietal placentation or a mixture of axile and parietal. In some members of the tribe Gardenieae the ovules are anatropous and one to numerous per locule. The style is simple and the stigma is lobed or capitate. Morphologically elaborate stigmas are present in the groups having distinct secondary pollen presentation mechanism, e.g. in tribe Vanguerieae. The fruit is small or quite large, indehiscent (berries or drupes with usually 2 or more single-seeded pyrenes, less frequently a multiseeded stone) or dehiscent (capsules, mericarps) sometimes united into syncarps. The seeds may be small or large and sometimes winged. Endosperm is present and usually conspicuous, sometimes ruminate and absent in tribe Guettardeae.

Genus *Vangueria* Juss.

The genus is named for Voa vanguer as *V. madagascariensis* is known in Malagasy.

Spanish tamarind

(***Vangueria madagascariensis*** J. F. Gmel. syn. ***V. acutilobe*** Robyns; voo vanga)

It is native of Africa and Madagascar and is widely distributed in Tanzania. It is a profusely branched shrub or small tree growing up to 5 m in height with smooth gray bark. The leaves are of 7-20 x 2-11 cm size, opposite, elliptic ovate, dark green above and paler beneath, glabrous or rarely slightly pubescent with acuminate or rarely obtuse or acute apex. Flowers are greenish, fulvous pubescent and borne in dense axillary cymes. The fruit is globose, 3-4.5 x 2.5-4.2 cm in size having a hard seed coat.

Borojo

(***Borojoa patinoi*** Cuatr. syn. ***Alibertia patinoi***; borojó)

Borojo is native of Colombia and is distributed in tropical humid rainforests in western Colombia along the Pacific Coast (geographic zone of Choco) and in the Esmeraldas Province of Ecuador. Borojó is an Embera word meaning head shaped fruit or round and globulous fruit (*boro* = head, *ne-jo* = fruit).

The small tree grows up to 3-5 m in height. Fruit is globose in shape, 7-12 cm in diameter, 750 g to 1 kg in weight having green to brown pulp with 90-640 seeds.

Genus *Genipa* L.

The genus is closely allied to *Gardenia*. Several *Gardenia* species were originally placed in *Genipa*. It is not to be confused with Genip (*Melicoccus bijugatus*), a completely unrelated eudicot. The name is derived from *genipapo*, the Guianan name for *G. americana* (commonly known as Huito). The genus is also commonly called *jagua*, *xagua* or *xaqua*.

Genipap

(***Genipa americana*** L. syn. ***G. americana*** L. var. ***caruto*** (Kunth.) Schumann; genipapo; genipa; marmalade box)

Genipap is a native of West Indies and South America (Martin *et al*., 1987). It is also distributed in Central America, western Latin America and the Caribbean region. It has been introduced in South East Asia (Morton, 1987).

Genipap is an erect tree growing up to 20-30 m in height. Leaves are shiny, dark, opposite, green and elliptic obovate. Inflorescence is corymbose with 3-12 flowers. The flowers are bisexual, actinomorphic having 5-lobed calyx, 14-18 mm long, yellow, gamopetalous and 5 lobed corolla, 5 epipetalous stamens, 1 pistil and inferior ovary. Fruit is an ellipsoid to ovoid and yellowish brown berry, 9-15 x 7-9 cm in size and 5-8 cm in diameter with sweet aromatic flavour and weighs 200-400 g. The flesh is whitish filled with numerous flat and yellow brown seeds of 0.8 x 0.7 cm size (Martin *et al*., 1987; Brücher, 1989).

Family Loganiaceae (Strychnaceae)

It contains 15 genera and 420 species (Heywood *et al*., 2007). The plants are annual to perennial herbs, shrubs, small to large trees and high climbing lianas. The leaves are opposite with the petiole bases usually joined by a raised line and the lamina is simple with an entire margin. The inflorescence is usually cymose and terminal or axillary, often many-flowered and sometimes conspicuously cincinnate. The flowers are actinomorphic, bisexual, sometimes unisexual and 4 to 5-merous. The sepals are fused at least at the base and sometimes very unequal. In *Phyllangium*, the calyx is absent and replaced by a 2-lobed foliaceous involucre. The corolla usually forms a narrow tube with short lobes and the stamens are inserted on the tube. The ovary is superior and if semi-inferior, it is of 2 fused carpels having (1)2(3) locules. Placentation is axile often with a peltate placenta usually bearing many ovules. The fruit is a capsule dehiscing either apically or basally, a drupe, or berry like or fleshy and coriaceous and sometimes globose (*Strychnos*) and up to 15 cm in diameter or pear-shaped and up to 20 cm long. The seeds are 2 to numerous and sometimes winged.

Genus *Strychnos* L.

It includes about 190 species of trees and lianas distributed around the tropics. The strychnine tree, *S. nux-vomica*, native to tropical Asia, is the source of strychnine. *S. ignatia* (St. Ignatius bean) is a closely related Asian shrub/tree. *S. toxifera* is one of the two plant sources of the poison curare. The drought tolerant corky barked monkey orange or suurklapper (Arikaans), *S. cocculoides* Bak, the Natal orange or green or spiny monkey orange, *S. spinosa* Lam. (kaffir orange) and the black or spiny leaved monkey orange *S. pungens* produce edible fruits. Another related species is *S. innocua* Del. The ripe seeds of *S. potatorum,* known as Therran or Nirmal is ground and used as a coagulant to purify water and is rubbed against the inside walls of the earthen water containers.

Kaffir orange

(***Strychnos spinosa*** Lam.; monkey orange; elephant orange; monkey ball)

It is native of Africa and is distributed throughout the drier parts of tropical Africa from Senegal and Sierra Leone, eastward throughout the southern Sahel to southern Ethiopia and from East to South Africa (Martin *et al*., 1987; FAO, 1988).

Kaffir orange is a spiny semi-evergreen shrub with gray brown scaly bark growing up to 9 m height. Branches are flexible to fairly stiff, smooth or hairy armed with about 7 mm long paired spines and slightly curved backwards having a black tip. Leaves are opposite, broadly ovate to round, up to 10 cm long, hairy and 3-5 nerved. Flowers are small and whitish green borne in terminal cymes. Fruit is spherical, 5-12 cm in diameter, green, light brown at maturity, sometimes prickly and has hard shell containing many seeds. The seeds are compressed, up to 2 cm in diameter and embedded in the fleshy pulp (FAO, 1988).

Two forms are distinguished. The sweet form bears fruits clustered on short stalks whereas the bitter form has long fruit stalks and narrow leaves (FAO, 1988).

Family Apocynaceae

The family comprises about 180 genera and 1500 species. It contains herbs, shrubs, trees or climbers. The plants have milky sap containing poisonous drugs. The leaves are opposite, verticillate or alternate, simple and entire and exstipulate. The flowers are regular, bisexual, hypogynous or sometimes epigynous. The calyx is deeply 5 parted, free almost to the base, often glandular on the inside. The corolla is sympetalous and mostly tubular. Stamens are as many as the corolla lobes and attached at different levels of corolla tube. Filaments are usually free. Anthers are usually sagittate and acute. Disc is usually present and is entire or lobed. Ovaries are 2, superior or partly inferior, 1 or 2-celled with few to many ovules. Style is one, simple or divided. The fruit is entire or indehiscent or of two carpels, berry like or follicular. Seeds are often with tufts of hair.

Genus *Carissa* L.

It contains about 8 species of shrubs or small trees native to the tropical and subtropical regions of Africa, Australia and Asia. Some of the species are *C. bispinosa* (L.) Desf. Ex Brenan, *C. boiviniana* (Baill.) Leeuwenb, *C. carandas* L., *C. macrocarpa* (Eckl.) A. DC., *C. pichoniana* Leeuwen, *C. spinarum* L. and *C. tetramera* Stapfl.

Karonda

(***Carissa carandas*** L. syn. ***C. congesta*** Wight.; Christ's thorn)

It is a native of India. Spines are usually simple. The leaves are lighter coloured, not so thick and elliptic. The flowers are small, not as fragrant as those of Natal plum and the base of the corolla is purplish borne in clusters of 2-3. In bud stage, the corolla is twisted to the left. The fruit is small and yellowish with a red blush.

Standard cultivars are not known but three distinct types having green, pink and white fruits attaining red blush on ripening are known which have variability in shape and size from which superior types could be selected. Selections 3, 9, 12 and 13 identified at Mahatma Phule Krishi Vidyalaya, Rahuri are considered to be good in quality.

Pant Manohar: It is a seedling selection made in 1998 at Pantnagar (Uttar Pradesh), India. The bushes are dense and medium sized. Fruit is dark brown, 2.13 x 1.69 cm in size having 88.27% flesh, 3.92°B TSS and 1.82% acidity.

Pant Sudarshan: It is a selection made at Pantnagar (Uttar Pradesh), India. It is a bush of medium size. The fruit is light pink, of 2.16 x 1.69 cm size and 3.46 g weight having 4.68 seeds, 88.47% pulp, 11.8% dry matter, 3.45°B TSS and 1.89% acidity.

Pant Suvarna: It is a selection made at Pantnagar (Uttar Pradesh), India. The bushy plant has upright growth. The fruit is green with light maroon blush and attain deep maroon colour on maturity. The fruit is of 2.55x.1.66 cm size and 3.62 g weight having 88.27% pulp, 12.39% dry matter, 3.83°B TSS and 2.30% acidity. The cultivar gives on an average 25 kg fruit yield per bush.

Maroon Coloured: It is a selection made at Faizabad (Uttar Pradesh), India. The fruit has attractive maroon colour and 3.2 g weight. The pulp has 7.1°B TSS and 4.1% acidity and contains 17.9 mg vitamin C, 0.7 mg anthocyanin and 3.8 mg iron per 100 g pulp. The cultivar gives on an average 3.7 kg fruit yield per bush.

White-Pink blush: It is a selection made at Faizabad (Uttar Pradesh), India. The fruit is 2.9 g in weight and has lighter maroon colour than that of cultivar Maroon Coloured. The pulp has 6.6°B TSS and 4.7% acidity and contains 15.4 mg vitamin C, 0.2 mg anthocyanins and 4.15 mg iron per100 g pulp. The cultivar gives 2.8 kg fruit yield per bush.

CHESK- II-7: This is a seedling selection from Konkan region in India. The plants are medium bushy, 2-3 m tall, spreading with 2-3 stems. The leaf is 6-7 cm long and 4-5 cm wide and green. Young leaves are red. Each node harbours two thorns. Flowering takes place in February–March. The flowers are white. The fruits mature during May-July. The fruit is oblong, dark blackish violet, of 12-14 g weight and almost seedless (0.3 seeds/fruit) having sweet and red pulp with 11.5°Brix TSS and 0.76% acidity. A 4 year old plant gives yield of 3000-3100 fruits.

CHESK- V-6: This is a seedling selection from Konkan region in India. It has characters similar to CHESK-II-7. The fruit is 13-15 g in weight, dark red and contains no or few seeds. The flesh is dark red having 14°Brix TSS, 0.44% acidity and 7.30 mg/100 g vitamin C. A 4 year old plant gives yield of 1200-1500 fruits.

Natal plum
(*Carissa macrocarpa* (Eckl.) A.DC syn. ***C. grandiflora*** (E. Mey) A.DC.)

It is a native of India. It is an evergreen, straggly and woody climbing shrub or a small tree attaining a height of 3-7 m if allowed to grow. The leaves are dark green, entire, opposite, ovate or elliptic, 2.5 - 7.5 cm long and tapering towards the base. It possesses strong axillary spines which are often forked. The flowers are white and fragrant and are borne in clusters of two or three. The corolla tube is cylindrical and naked, anthers are free from stigma and the ovary is 2-celled. Fruit is an ellipsoid or globose berry having granular pulp, nearly 2.5 cm in diameter and has 2-8 small, flat and brown seeds. On ripening, the fruit turns green to white or pink and later black (FAO, 1982; Morton, 1987; Verheij and Coronel, 1991).

Carissa bispinosa (L.) Desf.

The spines are 5 cm long and are very strong. It is a very hardy species. The flowers are smaller than those of *C. grandiflora*. The fruit resembles that of Natal plum but is of cherry size and is borne in clusters.

Carissa spinarum L.

It is the hardiest among the *Carissa* species. It is native of India. It can grow on rocky as well as alkaline soils.

Genus *Couma* Aublet

The plants grow in northeastern America and are a source of couma rubber, chewing gum base and edible fruit.

Cow tree
(*Couma utilis* (Mart) Muell. Arg. syn. ***C. guianensis*** Aubl., ***C. rigila*** Muell. Arg.; couma; sorva; sorva pequena**)**

Cow tree is native of South America. The related edible species is *C. macrocarpa* Barb. It is found growing throughout the Amazon most frequently in non-flooded areas of Upper Amazon and Rio Negro at lower than 500 m elevation. Usually its fruits are collected from the wild. It is also cultivated.

The fast growing tree is 2-12 m in height. Stem is thick with smooth bark. The leaves are long, elliptical, 8-20 cm in size, glossy and dark green. The flowers are white, pentamerous with 1 cm long tube. The fruits are borne in clusters and fruits are round berries, 5-8 cm in diameter and yellow when ripe. The exocarp is thin and the pulp is white, sweet and aromatic having numerous seeds (2-42).

⑧ ⑧ ⑧

Chapter 37

BORAGINALES

The plants are herbs, shrubs or less often trees, sometimes lianas, rarely herbaceous parasites. In the herbaceous genera, less frequently in the woody forms, the stems and leaves have characteristic harsh unicellular hairs usually having a basal cystolith and often calcified or silicified walls, less often also have gland tipped hairs (Lennoaceae) or only stalked and glandular hairs. Vessels are with simple or very rarely reticulate perforations and alternate lateral pitting. Fibres are with bordered or more often simple pits and are often septate. Rays are heterogeneous to homogeneous. Axial parenchyma is of various types, but mostly apotracheal. Vestured pits are present or absent. Sieve element plastids are of S-type. Nodes are unilacunar, very rarely (*Cordia*) trilacunar with three traces. Leaves are alternate or sometimes partly or wholly opposite, simple, usually entire, sometimes lobed, pinnatisect or pinnate, rarely palmate and exstipulate. Stomata are usually anomocytic.

Flowers are borne in scorpioid cymose inflorescences, rarely solitary, mostly bisexual, actinomorphic or slightly zygomorphic having usually 5-merous perianth and androecium. Sepals are more or less connate, imbricate or rarely valvate. Corolla is sympetalous, imbricate or contorted, rarely valvate. Stamens are as many as and alternate with the corolla lobes, attached to the tube. Anthers are tetrasporangiate, introrse or rarely (*Hoplestigma*) latrorse opening longitudinally. Pollen grains are (2)3-celled, 3-colpate, 3-colporate, 6-colpate, pericolpate, pericolporate or 3-porate. Nectary disc is present around the base of the ovary or is wanting. Gynoecium is of two median carpels, rarely of 4-5 (in *Zoelleria*, Boraginaceae), 6-14 (Lennoaceae) carpels. Stylodia is usually more or less united into terminal or gynobasic style. Ovary is superior or rarely semi-inferior (as in Hydrophyllaceae) having two to many ovules on each placenta or in each locule. Ovules are anatropous, rarely hemitropous (some Boraginaceae) or campylotropous (Lennoaceae), unitegmic, tenuinucellate or rarely (Ehretioideae) pseudocrassinuc-ellate (crassinucellate), sometimes with endothelium (Hydrophyllaceae, *Heliotropium* in Boraginaceae). Female gametophyte is usually of *Polygonum*-type, rarely (some Boraginaceae) of *Allium*-type. Endosperm is cellular or less often (species of *Phacelia* in Hydrophyllaceae and some Boraginoideae) nuclear and rarely (species of *Echium* and *Pulmonaria*) helobial. Often micropylar and/or chalazal haustoria are present. Fruits are of various types. Seeds have straight or nearly straight embryos and copious (many Hydrophyllaceae) or scanty endosperm or often (most of Boraginaceae) without endosperm. Funicle and placenta are present in seed coat. Higher inulins and pyrrolizidin alkaloids are present. It contains only one family Boraginaceae.

Family Boraginaceae (borage and forget-me-not family)

The family contains 150 genera and 2700 species (Heywood *et al.*, 2007). The tree genera are pantropical and subtropical while the herbaceous genera are distributed worldwide although are particularly abundant in the Mediterranean region and warm temperate Asia. They occur in a wide range of habitats from sea level to more than 4000 m altitude.

The plants are annual or perennial herbs to shrubs, occasionally woody climbers or large forest trees. Vegetative parts characteristically have conspicuous stiff hairs that are usually unicellular and mounted on a swollen multicellular base embedded in the epidermis. Glabrous plants are rare (*Cerinthe*, some woody genera). The leaves are alternate or sometimes the lower ones are opposite or sometimes (some *Tournefortia*) simple with entire or sub-entire margins, rarely serrate (some woody species), shallowly lobed (*Coldenia*) or serrate spinulose (*Halgania*) rarely succulent (*Heliotropium curassavicum*).

The inflorescence is characteristically a lax to condensed, 1-sided, terminal or axillary and cincinnate cyme or occasionally the flowers are borne solitary in leaf axils. The flowers are hypogynous, gamopetalous and actinomorphic or sometimes slightly zygomorphic (*Echium*), usually 5-merous. Sometimes unisexual plants are dioecious (*Cordia*), frequently heterostylous. The calyx lobes are free or fused and sometimes conspicuously accrescent as in *Sacellsum, Auxemma, Cordia, Trichodesma*. The corolla is usually tubular trumpet-shaped, rarely ampanulate, sometimes bearing 5 appendages in the throat. The stamens are attached to the corolla tube and the anthers are rarely (*Halgania*) connivent around the style. The ovary is bicarpellate, entire to deeply 4-lobed having 1, 2, or 4 locules and a terminal or gynobasic (subfam. Boraginoideae) style with 1, 2 or 4 stigmas. The ovules are usually 1 per locule and are sub-basal. The fruit may be a drupe with 1-4 pyrenes but is more often a schizocarp splitting into four 1-seeded nutlets or is rarely a dehiscent capsule. Two species from New Guinea, formerly referred to the genus *Zoelleria*, have 8-10 ovules and nutlets but are otherwise typical of *Trigonotis* and are included in that genus.

Genus *Cordia* L.

It contains about 300 species of shrubs and trees, which are found worldwide mostly in warmer regions. Many species are commonly called 'manjack', while several central American species are called 'bocote' in Spanish. The generic name has been given to honour the German botanist and pharmacist Valerius Cordus (1515-1544). A majority of plants have trichomes (hairs) on the leaves.

Gonda

(***Cordia myxa*** L. syn. ***C. dichotoma*** Forsk. f.; lasora; lehsua)

It is found growing in arid and semi-arid regions of North India and seems to be indigenous to the region. A related species found growing is *C. gharaf* Forsk. commonly known as *gondi* having smaller tree with smaller and narrower leaves.

It is evergreen or partially deciduous tree of moderate size attaining a height of 12.2 to 15.2 m and has a crooked trunk. Bark is gray or brown, rough with shallow longitudinal furrows. Young shoots are silvery gray in colour. Leaves are alternate, thinly coriaceous, 3 to 5-nerved, 7-15 cm long, broadly ovate or sub-orbicular rounded cordate. Flowers are small, white, usually pentamerous and borne on large lax terminal and axillary cymes. Fruit is drupe, 1-3 cm long, yellowish brown, pink or black and has persistent calyx when ripe having a viscid, sweet, transparent pulp surrounding the stone. A large fruited type Paras is known in Gujarat. An early bearing and high yielding cultivar, Thar Bold, has been identified at CIAH, Bikaner.

⑧ ⑧ ⑧

Chapter 38

SOLANALES

The plants are herbs (erect or climbing), shrubs or trees, sometimes parasitic. Indumentum of diverse types of trichomes is often stellate or otherwise branched. Intraxylar phloem is usually present in the stem and petioles. Vessels are with simple perforations. Fibres are mostly with simple pores. Rays are heterogeneous to homogeneous. Axial parenchyma is apotracheal or less often scanty paratracheal. Sieve element plastids are of S-type. Nodes are unilacunar. Leaves are alternate, or rarely opposite, simple or sometimes compound, sometimes succulent, rarely scale like and exstipulate. Stomata are of diverse types.

Flowers are borne in various types of inflorescences (cymose or derived from the cymose type) or are solitary and axillary, usually bisexual, commonly 5-merous, actinomorphic to more or less zygomorphic. Calyx is synsepalous usually 5-lobed and persistent. Corolla is sympetalous, usually 5-lobed, lobes mostly plicate in bud, contorted, imbricate or valvate. Stamens are usually 5, attached to the corolla tube and alternate with its lobes, but in Solanaceae-Salpiglossideae one or even 3 of them are transformed into staminodia. Anthers are usually introrse, tetrasporangiate opening longitudinally or by apical pores. Tapetum is secretory. Microsporogenesis is simultaneous. Pollen grains are 2-celled or sometimes 3-celled, 3-colporate, 6-polycolpate or pantoporate. Well developed intrastaminal nectary disc is usually present. Gynoecium is usually of 2, sometimes 5, rarely 3 united carpels. Style is usually terminal or sometimes gynobasic having usually 2-lobed or sometimes peltate stigma. Ovary is superior, (1)2(3-5)-locular with one or more often several or numerous ovules in each locule or on each placenta. Ovules are anatropous, hemitropous, campylotropous or amphitropous, unitegmic, commonly tenuinucellate, usually with endothelium. Female gametophyte is of *Polygonum* or *Allium*-types. Endosperm is cellular, sometimes nuclear and helobial. Fruits are of diverse types, mostly baccate or septicidal capsules. Seeds are exotestal having straight or curved or sometimes spiral embryo and usually copious endosperm. Various kinds of alkaloids, flavonol and flavone glycosides, coumarins, acylated anthocyanins and caffeic acid are commonly produced. Iridoid compounds are absent. It contains 7 families.

Family Solanaceae (nightshade family, potato family)

It contains 90 genera and 1000-4000 species (Heywood *et al.*, 2007). Many species contain alkaloids and are poisonous.

The plants are trees (some *Cyphomandra*, *Solanum*) or shrubs (*Cestrum*), vines, lianas, epiphytes and perennial or annual herbs. The leaves are alternate, usually simple, entire or dissected and exstipulate. The flowers are bisexual or rarely unisexual (plants andromonoecious), usually actinomorphic, sometimes slightly or markedly zygomorphic borne in terminal or axillary, sometimes complex cymose inflorescences or reduced to a single flower (*Solanum*). The sepals are (4(5(-7), in one whorl, partly fused, usually regular and persistent and sometimes enlarged around the fruit (*Physalis, Nicandra*). The petals are (4)5(-10), usually contorted and plicate, imbricate or valvate, variously fused making the corolla round and flat (*Solanum, Lycopersicon*), campanulate (*Nicandra, Withania*) or tubular (*Cestrum, Nicotiana*), rarely 2-lipped as in *Schizanthus* where the abaxial petals form a keel. The stamens are usually 4 or 5, 2+2 or up to 8 or 10. The ovary is superior, of 2 fused carpels (3-5 in *Nicandra* or more in some *Nolana*) with a single style and usually with 2 locules, sometimes 3-5 (*Nicandra, Datura*), generally with numerous axile ovules. The fruits show considerable diversity but are usually bicarpellate and a septicidal capsule or a berry, sometimes a drupe or pyrene (*Goetzea* group, *Capsicum, Lycium*) or a schizocarp with up to 15(-30) single seeded mericarps (*Nolana*).

Genus *Physalis* L.

It is a native of warm temperate and subtropical regions. It is characterized by the small orange fruits similar in size, shape and structure to a small tomato but is partly or fully enclosed in a large papery husk derived from the calyx. Many *Physalis* species are called ground cherries.

Cape gooseberry

(***Physalis peruviana*** L.; golden berry; husk cherry; Peruvian ground cherry; poha; poha cherry)

The plant is annual or herbaceous and perennial belonging to the temperate region. It is 0.5-2 m tall having purplish, ribbed, densely pubescent and spreading branches. The leaves are simple, alternate, sub-opposite, ovate, 5-15 x 4-10 cm in size, cordate at base, acuminate at apex and entire or shallowly toothed with petioles as long as the blades. Flowers are solitary, pendent and axillary with 1 cm long pedicels having 5-toothed, 2 cm wide, thin, papery and slightly 5-lobed calyx, short or campanulate yellow corolla with 5 large dark brown purple spots, 5 stamens, purple anthers and sub-capitate stigma. Fruit is a globose berry, 1-2 cm in diameter, orange yellow and glabrous enclosed in the densely pubescent, ovoid, inflated bladdery calyx (husk) having juicy pulp. Seeds are numerous, yellowish and very small (FAO, 1982; Morton, 1987; Brücher, 1989; Verheij and Coronel, 1991).

Cultivars

The important cultivars Giallo Grosso, Giant Poha Berry, Golden Berry, Long Aston, Golden Nugget and New Sugar Giant are often large fruited but have rather bland taste. The small fruited types have better flavour and are preferred for the preparation of jam and preserve (Verheij and Coronel, 1991; NRC, 1989b).

Genus *Cyphomandra* L.

It may be considered as a clade within the genus *Solanum* rather than as a separate genus uniting the members of the old genus with some other *Solanum*. Probably it is best to consider retaining the taxon, *Cyphomandra* as a section in *Solanum* subgenus *Leptostemonum* (Dunal) Bitter (Bohs, 1995).

Tree tomato

(***Solanum betaceum*** Cav. syn. ***Cyphomandra betacea*** Sendt.; tomarillo)

It is a native of Peru. It is a brittle and semi-woody, semi-perennial shrub, 1.8-4 m in height having up to 6 years life span. The leaves are simple, entire, cordate, ovate, large, 20-30 cm long, short-pointed and prominently pinnately veined and soft pubescent which give unpleasant smell when bruised. The leaves look like those of brinjal. Flowers are small borne in loose axillary clusters near the branch tips and are fragrant, pink to light blue in colour, about 1 cm in diameter and pentamerous. The corolla is bell-shaped and 5-lobed. Stamens are 5 borne on the throat of corolla and the ovary is 2-celled having many ovules with small stigma. The fruit is an egg-shaped, obovoid or ovoid, pendent and long stalked 2-loculed berry, 3-10 x 3-5 cm in size, pointed at both the ends having persistent calyx, thin, smooth, purple reddish and orange red to yellowish and faintly striped skin. It looks like brinjal but has a musky and acidic taste like that of tomato. The flesh is juicy, subacid to sweet and blackish to yellowish with many circular, flat, thin and hard seeds (FAO, 1982; Morton, 1987; Brücher, 1989; Verheij and Coronel, 1991).

Cultivars

Lot of variability in fruit characters has been observed in seedling plants. The skin colour may be purple, red, orange or yellow while the pulp may be red to yellow. Important cultivars are Ecuadorian Orange, Goldmine, Inca Gold, Oratia Red, Rothamer, Ruby Red and Soild Gold (CRFG, 1996d). The cultivar Kaitaia Yellow has been released in New Zealand for canning and a dark red strain has been selected for the fresh fruit market (Morton, 1987; Verheij and Coronel, 1991). The yellow fruit cultivar with its milder but superior flavour is preferred for processing (Martin *et al*., 1987; Verheij and Coronel, 1991).

Genus *Solanum* L.

Besides the potato and tomato, it contains the nightshades and horse nettles as well as numerous plants cultivated for their ornamental flowers and fruits.

Orinoco apple

(***Solanum topiro*** Humb. et Bonpl.; orinoko apfel; cubiu; cocona; topiro; topiru)

The orinoco apple is a robust shrub having many branched bushy growth of 1-2 m height. The whole plant is covered with hairy indument. Leaves are large (50-60 cm), simple and cordiform with strong nervature and slightly lobed margins and are borne on short petioles. Inflorescence has 5-8 spiral shaped flowers with 0.2 to 0.5 cm long floral peduncles. One to three fruits arise from the branches at each group of three leaves. Fruit is ovoid, 7-10 cm in diameter enclosed in a tough, yellow or orange red and slightly pubescent skin. The pulp is creamy yellow and has a pleasant sweet acid taste very similar to that of peaches. The fruit contains 500-2000 oval and flattened seeds.

Naranjilla

(***Solanum quitoense*** Lam.; lulo; the golden fruit of the Andes)

It is native of the Andes in South America. The plant is a large and robust shrub of 1-2 m height having hairy leaves and spines. The leaves are alternate, large, 40 cm in diameter, elliptic to ovate with very prominent veins. The veins are purple in young leaves. The flower buds are densely covered with short and violet hair. The inflorescence is a short axillary cyme of 4-5 flowers. The flowers are hermaphrodite, white in colour, borne on 1-1.5 cm long pedicels. Calyx is campanulate, 1.4-1.6 cm broad with ovate lanceolate and acute lobes. Corolla is 2 cm across, oblong or linear and 6 mm broad with 5 large yellow stamens. It is an excellent dessert fruit and is also used to flavour confections and to prepare drinks, jelly, jam, pies and preserve. Freshly squeezed naranjilla juice is used in Ecuador and Colombia to make "sorbete", a green foamy drink with an appealing sweet sour flavour of pineapple and strawberry.

Some spiny and spineless cultivars are available. The spiny cultivar grown in Colombia bears sour, bright yellow and somewhat flattened fruits. In Ecuador, its two groups, Naranjilla Agaria (acid naranjilla) and Naranjilla Dulce (sweet naranjilla), have been distinguished. The widely grown spineless cultivars produce sweet, reddish yellow and spherical fruits. A spiny form, "berenjena de olor", is common in Costa Rica (Morton, 1987). The related species *S. hirtum* can be used in the breeding programme to develop root knot nematode resistant types and *S. topiro* Humb. and Bonpl. to evolve cultivars suited to lower elevations.

⑧ ⑧ ⑧

Chapter 39

LAMIALES

The plants are herbs or shrubs. Leaves are opposite or sometimes verticillate and nodes are unilacunar or rarely multilacunar. Flowers are mostly monosymmetric, bisexual or rarely unisexual. Androecium is often 2(+2) and the gynoecium is mostly of two united median carpels, sometimes pseudomonomerous, rarely of three or four carpels. Style is terminal or gynobasic with 2-lobed stigma (style sometimes very short and the stigma is essentially sessile). Ovary is superior or less often semi-inferior or inferior with (1)2 or many ovules per locule. Ovules are anatropous, hemitropous, amphitropous or less often orthotropous, unitegmic, tenuinucellate, usually with endothelium. Female gametophyte is usually of *Polygonum*-type. Endosperm is cellular. Endosperm haustoria are developed. Fruits are capsules or capsule-like or sometimes indehiscent and then drupaceous or baccate, sometimes separating into drupe like mericarps. Seeds are exotestal with copious endosperm or endosperm is scanty or wanting. It usually produces various iridoid compounds (ethereal oils in gland-headed hairs; 6 oxygenated flavones, rosmarinic acid) and the phenolic glycoside orobanchin, flavones, but only rarely alkaloids and not secoiridoids. It contains approximately 23 families.

Family Oleaceae (olive family)

It contains about 25 genera and 570 species (Heywood *et al.*, 2007). It is widespread in the temperate and tropical regions. Several genera show an unusual disjunction between southern Europe and the Far East. Most tropical genera have relatively narrow distribution either in the New World or Old World, but *Schrebera* and *Chionanthsus* span both the hemispheres. *Menodora* is in the New World, apart from 3 species in South Africa (one of which has its closest relatives in the USA and Mexico), and *Nestegis* is confined to New Zealand and Norfolk Island, except for one species in Hawaii. *Hesperelaea* was confined to Guadalupe Island off the Pacific coast of Mexico but is now extinct. *Fraxinus* is an important woodland genus in the temperate northern hemisphere but also extends to the tropics in Costa Rica and mountains of Malaysia.

The family comprises mostly trees and shrubs, but *Jasminum* and *Myxopyrum* include woody climbers growing 12 m or more long, and *Menodora* and *Dimetra* may be herbs. The leaves are opposite except in *Jasminum*. The inflorescence is a compound cyme and is usually terminal. The flowers are actinomorphic, occasionally unisexual or the plants are dioecious and usually sweetly scented. The calyx is 4-

lobed, or in *Jasmineae* 5 to 15-lobed or sometimes absent. The corolla is usually gamopetalous and actinomorphic but is absent in *Forestiera* and most *Fraxinus* and usually 4-lobed but 5 to 12-lobed in *Jasmineae*. The stamens are characteristically 2 but rarely 4 in *Fraxinus*. The ovary is 2-carpellate and 2-locular, each locule bearing 1, 4 or more ovules (*Forsythia*). The fruit is a capsule, samara, berry or drupe

Genus *Olea* L. x=23; 2n=46

The plants are shrubs or trees. Leaves are evergreen, opposite, entire or toothed. Inflorescence is terminal or axillary, cymose-paniculate. Flowers are hermaphrodite or unisexual having 4-lobed or 4-toothed calyx, short corolla tube with 4 lobes, valvate and cucullate, 2 stamens and bilocular ovary with 2 pendulous ovules per locule. Fruit is a drupe having hard endocarp. It contains about 32 species occurring in Old world tropics and warm temperate regions (Kadereit, 2004). The only edible species is *Olea europaea* L.

The genus has six subspecies, *O. europaea* subsp. *cuspidata, O. europaea* subsp. *laperrinei, O. europaea* subsp. *maroccana, O. europaea* subsp. *cerasiformis, O. europaea* subsp. *Guanchiea* and *O. europaea* subsp. *europaea.*

Olive

(***Olea europaea*** L.) (x=23; 2n=46)

Olive is native to mediterranean region. Olive has been described in old Egyptian literature. The olive includes (i) *O. europaea* (Miller) Lehr., the widespread Mediterranean olive; (ii) *O. europaea* subsp. *maroccana,* endemic to Morocco; (iii) *O. europaea* subsp. *laperrinei* from Saharan mountains; (iv) *O. europaea* subsp. *cerasiformis* from Madeira; (v) *O. europaea* subsp. *guanchica* from Canary Islands; (vi) *O. europaea* subsp. *cuspidata* distributed from South Africa to China. There are two forms of Mediterranean olive, wild olive or oleaster (*O. europaea* subsp. *europaea* var. *sylvestris*) and the cultivated olive (*O. europaea* subsp. *europaea* var. *europaea.* Only the cultivated sub-species produces edible fruits.

The trees are long lived and evergreen from few to 20 m in height. The root system is generally shallow, spreading to 0.9-1.2 m even in deep soils. The leaves are thick, leathery and broad, opposite, lanceolate, pale green and persisting for 2 or 3 years. Leaves have stomata on their abaxial surfaces only, nestled in peltate trichomes making it relatively resistant to drought. It is polygamo-monoecious. The inflorescences are panicles arising from the axils of leaves on the previous year's growth. Each inflorescence contains 15-30 flowers, depending on the cultivar. The flowers are perfect or staminate depending on cultivar and season. These are small, yellowish white and inconspicuous. Perfect flowers normally consist of a short toothed calyx and short tubular corolla with 4 lobes, 2 stamens with a filament supporting a large pollen bearing anther, and a plum green pistil with a short thick style and a large capitate stigma. Imperfect flowers are staminate with the pistil either lacking or rudimentary. The fruit is a 1-seeded drupe, 1.5-2 cm in diameter. It consists of a carpel and the wall of the ovary having both fleshy and dry portions. It is yellow white (straw) with anthocyanin at the distal or basal end. Fruit shape and size and pit size and surface morphology vary greatly among cultivars. The mature seed is covered with a thin coat that covers the starch-filled endosperm.

Cultivars

Olive cultivars are divided into two interbreeding groups, the oil types having up to 28% oil and table olives that are pickled (Zohary, 1995). Many cultivars are duel purpose. Some of the important cultivars are:

Frontoio: The trees are medium tall and moderately vigorous having rounded crown and drooping branches. The leaves are medium sized, lanceolate and bright green above and silvery below with entire margin. The cultivar is regular and heavy bearer. Inflorescence is pale yellow, quite large and sparse. The flowers are paniculate. Fruits are purple black, oval elongated and of medium size and ripen late. Oil content is 23-25%. Stone is large and oval elongated with furrowed surface. It is susceptible to cold, *Cycloconium* knot and olive fly.

Ascolano: The cultivar originated at Merche, Italy. It is resistant to cold and olive fly. Plants are fast growing, upright in stature with dense and compact crown. The leaves are elliptic, of medium size, bright green above and silvery below with entire margin. Inflorescences are few, small and light yellow and the flowers are borne in paniculate thyrse. Trees are easy to harvest. Fruit matures early and thus frost injury is not a problem, but the tender fruits get bruised easily during harvest and processing. The fruits are large, nearly as large as those of Sevillano, spheroid, somewhat asymmetric and almost black when ripe and have a high flesh:pit ratio. Stone is medium, elliptic elongated and asymmetric having rough surface, relatively high pulp stone ratio and 22-24 oil percentage. The pickling quality of fruits is high, but these are not suitable for Spanish-green pickling because of 'salt shrivel' during fermentation.

Leccino: The cultivar originated at Toscana, Italy. It is resistant to low temperature but is susceptible to sooty mould and *Cycloconium* knot. The tree is vigorous having spreading branches and open crown. The leaves are elliptic lanceolate, light green and silvery below. Inflorescences are few and small and the flowers occur in raceme. The fruits are purple black at picking, spheroid and medium sized with round apex. Stonc shape is similar to that of fruit and asymmetric with rough surface. The cultivar bears regularly giving fair fruit yield having 25-30% oil content.

Coratina: It is an Italian cultivar. The tree is semi-vigorous with open crown and somewhat drooping branches. The leaves are elliptic, dark olive green and silvery white beneath. Inflorescence is short and compact. The flowers are in paniculate thyrse. The fruits are black, big in size, oval and somewhat asymmetric with no nipple and many small lenticels and ripen late. The stone shape is like that of the fruit with furrowed surface. It is heavy yielder with oil content of 23-25%. The cultivar is resistant to drought and susceptible to sooty mould, peacock spot/eye olive fly, knot, etc. It is suitable for pickling.

Picholine: It is a French cultivar. The tree is upright and moderate in growth with open crown. The leaves are pale green, elliptic lanceolate and tomentose below with entire margin. Inflorescence is compact and large with racemose flowers. Fruits are medium in size, green and spheroid elongated with no nipple having pointed apex. The stone is small and oval with smooth surface. It is a late ripening cultivar giving medium fruit yield having 16-18% oil content. Trees are tolerant to cold and resistant to peacock spot.

Pendolino: The tree is moderate in growth having open crown and drooping branches. The leaves are bright green, lanceolate, entire and tomentose below. Inflorescence is like that of Picholine. The fruits are black when ripe, medium sized, asymmetric oval with rounded apex. Stone is spheroidal and medium in size having smooth surface. Oil content is 23-25%. The cultivar is heavy cropper but is susceptible to cold and *Cycloconium*.

Manzanillo: The trees are low, spreading and easy to harvest. Fruits are larger than those of Mission and also have higher flesh:pit ratio. Oil content is high in small, frozen or cull fruits. The fruits mature early and thus are rarely injured by frost. However, the trees of this cultivar are less winter hardy than the other major cultivars.

Mission: The trees grow tall and are difficult to harvest unless they are kept low by pruning. The fruits are easy to process. The fruits have high oil content but are small in size and have the lowest flesh:pit ratio among the main cultivars. Since the fruits mature late in the fall, the danger of frost injury is greater. The trees of this cultivar are hardier than the other cultivars. It has survived under temperatures as low as -13°C.

Sevillano: The trees do not grow tall if properly trained and thus can be easily harvested. The fruits are large and fetch premium price but flesh:pit ratio is only average. Since oil content is too low oil extraction is not feasible. Fruits mature early enough and thus escape frost damage.

Picnal: Trees are moderately vigorous and spreading with dense canopy. Tees are very hardy and highly productive. The fruit shape is oval and has a nipple on the end. Oil content and fruit yield are high. Fruit ripens in mid to late season. It is resistant to drought and is sensitive to wilt and peacock spot/eye.

Arbequina: Trees are hardy, low in vigour, spreading with medium to dense canopy. It is early and consistent in production. The fruit ripens unevenly from early to mid season. The fruits are small in size. The trees are resistant to cold and tolerant to salinity. Oil content is very high.

Barnea: Trees are vigorous with an erect growth habit and sparse canopy. Productivity is moderate to high. The fruits are medium in size, elongated having a nipple and many large lenticels. Fruits ripen early and are suitable for pickling. It is susceptible to olive fly.

Koronoiki: It is the main cultivar for oil. Oil content is 20-21%. Trees are upright and moderately vigorous. Inflorescence is racemose but is short and compact. It comes into production early and its productivity is high and consistent. Fruits are oval and small and bright black in colour. It is resistant to cold, drought and frost but is susceptible to *Cycloconium* knot and olive fly.

The early ripening cultivars from western Australia (Despeissis, 1921) are Verdale (roundish oval fruit and oval kernel), Picholine (oblong oval fruit and spindle shaped kernel), Sevillano (oval fruit and kernel), Gordal (oval oblique fruit and kernel), Pendoulier (oblong conical fruit and kernel) and Manzanillo (roundish fruit and oval kernel).

Family Lamiaceae syn. Labiatae (mint family)

The family contains 233 genera and 6870 species and includes genera such as *Tectona* (teak), *Gmelina, Vitex, Clerodendrum* and *Caryopteris* (Heywood *et al.*, 2007). The family is cosmopolitan, ranging from tropical forests to arctic tundra and from sea level to high altitudes. The genera such as *Phlomis, Salvia, Sideritis, Micromeria* and *Thymus* are aromatic and characteristic of the Mediterranean region.

The plants are annual or perennial herbs and shrubs to small or large trees like *Tectona grandis* growing 30 m high or more or rarely woody climbers. The stems are characteristically square in section in the more herbaceous genera. The leaves are opposite or very rarely alternate (some *Aeollanthus*) or whorled, usually simple and often variously toothed on the margins, but sometimes deeply divided (some *Lavandula*) or 3-foliate (*Cedronella*) or digitately (*Vitex*), pinnately (*Petraeovitex)* or even bipinnately (*Perrierastrum*, *Plectranthus*) compound, sometimes simple and very large (35 x 25 cm in *Tectona grandis*). The inflorescence is cymose, usually dichasial but sometimes monochasial, lax or often contracted into axillary verticillasters, usually with many bracts and bracteoles. The flowers are gamopetalous and strongly zygomorphic or occasionally secondarily actinomorphic. The calyx has a short tube, often with 10-15 conspicuous nerves and 4 lobes or teeth, or 2-lipped (*Scutellaria*), rarely fleshy in fruit (*Hoslundia*). The corolla is usually strongly 2-lipped but sometimes sub-actinomorphic. Stamens are usually 4 (2 long and 2 short) and characteristically exerted from the corolla tube. There is often a nectariferous disk below the ovary. The ovary is superior, of 2 carpels divided into 2 or 4 locules by intrusion of the carpel wall, often deeply 4-lobed and with a gynobasic style having 4 basal or sub-basal ovules. When the ovary is entire, the fruit may be a drupe having 4 pyrenes, but when the ovary is 4-lobed the fruit breaks up into 4 dry nutlets.

Genus *Vitex* L. (2n=12, 16)

The plants are trees or shrubs. Leaves are usually digitately compound with 3-8 leaflets but unifoliate, sometimes in whorls. Inflorescence is compound into a thyrse or panicle. Flower is blue, yellow, white to purple, actinomorphic to weakly 2-lipped having campanulate or tubular, 3-5 lobed or truncate calyx, infundibular, zygomorphic, 5-lobed, 2-lipped corolla, straight or curved tube, stamens 4, didinamous, attached near base of tube, usually exserted, thecae dorsifixed, divergent, 3-coplate pollen, bicarpellate ovary, 2-locular and 2-seeded carpels, terminal style, stigma with equal short lobes. Fruit is drupaceous with hard 4-seeded pyrene. It has approximately 250 species in Old and New World tropics and few in temperate regions (Kadereit, 2004). The edible fruit bearing species are *V. triflora* Vahl. (taruma), found growing in subtropical part of Brazil and Paraguay, *V. flexuosa* Thunb., distributed in Japan, China, Indonesia and Philippines and *V. pentagona* Diels. and Gilg., found growing in China and Indochina.

Black plum

(*Vitex doniana* Sweet syn. *V. cuneata* K. Schum. et Thonn., *V. cienkowskii* Kotschy et Perr.; African oak; prune noire; koro)

It is native of Africa and is distributed in the Sahelian eastern and tropical Africa (Martin *et al.*, 1987). It is a small or medium, semi-deciduous tree of 10-12 m height with stout, obscurely pubescent branches. Leaves are opposite, 4-foliate, sub-coriaceous and glabrous on both sides. Leaflets are obovate cuneate, entire, obtuse and shortly stalked. The white to yellow flowers are borne axillary on compound cymes. Calyx is campanulate, densely villose with minute teeth. Corolla is 2-3 times the length of calyx and very hairy outside. Fruit is a smooth, globose, 2 cm long and black drupe with hard and 4-celled endocarp. The flavour is sweet and resembles that of prune. Seeds are obovate and oblong (FAO, 1988).

⑧ ⑧ ⑧

[CAMPANULIDS: EUASTERID II]

Chapter 40

DIPSACALES

The plants are small trees, shrubs, more or less woody lianas or herbs. Vessels are with simple or less often scalariform perforations having commonly alternate lateral pitting. Fibres are with bordered or simple pits. Sieve element plastids are of S-type. Nodes are trilacunar or sometimes unilacunar, pentalacunar or multilacunar. Leaves are opposite or sometimes verticillate, simple or less often compound, exstipulate or stipulate, but sometimes with stipule like appendages. Leaf mesophyll cells contain oil bodies. Stomata are anomocytic, rarely paracytic or anisocytic.

Flowers are borne in various kinds of cymose or mixed inflorescences, usually bracteolate, mostly bisexual, actinomorphic or zygomorphic and 4-5-merous. Calyx is small, often variously reduced or modified. Corolla is sympetalous, usually 5-lobed or 4-lobed by fusion of two. The lobes are imbricate or sometimes valvate. Corolla tube is usually with a nectary consisting of one-celled epidermal hairs and an underlying nectary tissue. Stamens are as many as or fewer than the corolla lobes and are attached to the corolla tube at the throat alternating with the corolla lobes. Anthers are mostly tetrasporangiate, opening longitudinally. Tapetum is amoeboid or secretory. Microsporogenesis is simultaneous. Pollen grains are 3-celled and mostly 3-colporate. Gynoecium is of 2-5 united carpels with more or less connate stylodia. Stigma is capitate or lobed. Ovary is inferior or semi-inferior, sometimes more or less pseudomonomerous with one to many ovules per locule. Ovules are anatropous, unitegmic, mostly tenuinucellate with well developed endothelium. Endosperm is cellular, usually without haustoria and rarely (*Centranthus*) with chalazal haustorium. Female gametophyte is of *Polygonum*-type, rarely *Allium*- or *Adoxa*-type. Fruits are of various types. Calyx is persistent in the fruit. Seeds have small to large and straight, embryo with copious endosperm or rarely the endosperm is reduced to a single layer of cells (most Valerianaceae). The plants produce iridoid compounds and sometimes various kinds of alkaloids and are somewhat tanniniferous (Takhtajan, 2009).

Family Adoxaceae syn. Sambucaceae (Elder family)

It occurs in all the continents, widespread in the northern temperate region extending south in the Old World to the Canaries, North Africa, the Middle East and China and in the New World through Central America to the Andes.

The plants are robust, somewhat shrubby rhizomatous herbs to many-stemmed shrubs or small deciduous to semi-evergreen up to 10 m tall trees. The stems of herbaceous species are grooved, those of woody species are conspicuously lenticellate with well-developed white or brown pith. Stipule like outgrowths that occur in the whorls of 3-(4)-6 at the nodes may be peg shaped or stipitate glandular or foliaceous and up to 3 cm, found irregularly in all the species, but are sometimes absent, their development being dependent on growth conditions. The leaves are opposite, pinnate up to 7 pairs of leaflets or occasionally bipinnate, sometimes with stipels on the midrib. The leaflets are up to 15 cm long, ovate to lanceolate or oblanceolate having serrate margins and usually an acute apex. The inflorescence is a broad flat or rounded terminal panicle or corymb up to 30 cm across with the primary rays umbellate or alternate. Bracts and bracteoles are usually small. The flowers are sessilc or pedicellate, actinomorphic, unisexual in *S. australis,* usually 5-merous but 3 or 4-merous respectively in the Australian species *S. australasica* and *S. gaudichaudiana* and are often fragrant. The calyx is small and inconspicuous. The corolla has a short tube and broad rounded lobes and is white or cream to greenish yellow or pinkish. The stamens are as many as corolla lobes inserted near the base of the tube, sometimes with purple anthers. The ovary is three-quarters inferior having 3-5 locules each with a single pendulous ovule. The stigma is sessile and 3 to 5-lobed. The fruit is a spherical to ovoid berry, black or blue to purple or red including 3-5 hard pyrenes. The family contains 3 genera and 8-10 species.

Genus *Sambucus* L.

It has about 20 species native of North America including *Sambucus canadensis* L. (American elderberry). The other species are *S. nigra* L. (European elderberry or black elder), found growing wild throughout Europe and *S. racemosa* L. (grape elder) growing in Eurasia (Ritter and Mckee, 1964). *Sambucus adnata* and *S. wightiana* occur in the Himalayan region and *S. lavanica* extends from the eastern Himalayas to Japan and New Guinea. *S. australis* is native in northeastern Argentina, Uruguay, and southern Brazil. In tropical Africa, the European *S. ebulus* has an isolated disjunct subsp. on the high mountains of East Africa. The other 2 spp. are endemic from the Cape York Peninsula to Tasmania in Australia.

Elderberry

(***Sambucus canadensis*** L.; American elderberry; common elder)

($x = 9$; $2n = 36$)

It grows wild in the northeast and midwest of North America. It is stoloniferous and deciduous shrub growing up to 4 m in height with white pith and pale yellowish gray and slightly lenticellate branches. Leaves are bright green, odd pinnate usually with 7 elliptic to lanceolate, 15 cm long, acuminate and sharply toothed leaflets borne opposite on short petioles. Their lower surface is smooth or slightly downy. Flowers are small, bisexual, white to cream-coloured borne in five-rayed cymes (convex umbels) that are 25 cm across and are produced during June-July. There are usually 5 minute sepals, rarely 3 or 4 petals, 5 stamens and 5-chambered ovary. Fruit matures in September and is a purplish black berry, 0.6 cm in diameter and has 3-5 seeds (Bailey, 1949; Stang, 1990). European elderberry (*S. nigra* L.) tree is larger than that of American elderberry and grows up to 8 m in height.

Cultivars

About 12 cultivars of American elder are known of which the important are Adams No. 1 and 2 (very productive), Johns (late ripening), Scotia (productive and having high TSS), New York 21 (very large berries), Nova (early ripening and productive), Kent (large fruit size and early maturing), Victoria (early ripening) and Ezyoff (large and bold berries) (Hill, 1983; Way, 1981). The other important cultivars are Andrews (late), Hahs (mid season) and Wentworth (early ripening). Manitou Pembina (Manitou) is a Canadian selection noted for large fruit size. Fruits of Phillips are free of the musky flavour and odour common amongst the cooking fruit species. Thus, the jelly of its fruits develops equally good flavour and colour as that of red currants (Stang, 1990).

The following cultivars of ornamental elderberry were developed at East Malling Research Station, East Malling, England:-

Gerda (Black Beauty): (Fastigiata x Guincho Purple) x (Fastigiata x Guincho Purple 528-180).

Eva (Black Lace): [(Laciniata x Guincho Purple) x (Laciniata x Aurea)]; Selection 595-35.

Genus *Viburnum* L.

The important species are *Viburnum trilobum* Marsh. (Highbush cranberry), *V. opulus* (European cranberry), *V. sargentii* Koehne (Sargent viburnum) and *V. trilobum* Marsh. (American cranberry), (Wood, 1976). Most *Viburnum* species are grown as ornamentals owing to their attractive foliage and colourful fruit.

Highbush cranberry
(***Viburnum trilobum*** Marsh.; American cranberry bush; pembina)

It is native of North America and is found growing in Canada from New Brunswick to British Columbia in Alaska and in the northern USA from New York to Michigan, South Dakota and Oregon (Stang, 1990).

It is a deciduous shrub of about 3.6 m height with gray, glabrous branches and scaly buds. The leaves are broad ovate, entire, 5-13 cm long, 3-5 lobed with three to five major veins branching palmately from the base, pubescent or nearly glabrous beneath having petioles with shallow groove and small, usually stalked glands. The flowers are small, showy and white borne in large cymes. The outer flowers are enlarged and sterile providing an attractive unusual display at full bloom. The stamens are exerted. The fruits are 0.8 cm long and are similar to those of the true cranberry (*V. macrocarpon*) in size and colour. The bright scarlet colour begins by the end of July and that retains until the following spring (Stang, 1990).

⑧ ⑧ ⑧

REFERENCES

Acedo, A. L. (1992). *Jackfruit biology, production, use and Philippine Research Monograph Number 1.* Multipurpose Tree Species Network Research Series. Forestry/Fuelwood Research and Development (F/FRED) Project, Winrock International Institute for Agricultural Development, Thailand. pp12.

Adanson, M. (1763). *Families des plantes*, Paris, 2 Vols.

Aguiar, J. P. L., Marinho, H. A., Rebelo, Y. S. and Shrimpton, R. (1980). Aspectos nutritivus de alguns frutos da Amazonia., 10: 755-758.

Akpinar, A. E., Kocal, H., Ergul, A., Kazan, K., Selli, M. E., Bakir, M., Aslantas, S., Kaymak, S. and Saribas R. (2010). SSR-based molecular analysis of economically important Turkish apricot cultivars. *Genetics and Molecular Research*, 9, 324-332.

Alexander, D. Mc., E. (1983). *Some Citrus species and varieties in Australia*, CSIRO, Australia.

Almeyda, N., Malo, S. E. and Martin F. W. (1979). *Cultivation of neglected tropical fruits with promise, Part 6: the rambutan.* Agricultural Research and Education Administration Division, US Department of Agriculture, New Orleans, USA.

AL-Saghir, Mohamad G. and Porter, Duncan M. (2012). Taxonomic revision of the genus *Pistacia* L. Anacardiaceae). *American Journal of Plant Sciences,* 3: 12-32.

Anderson, A. B. (1988). Use and management of native forest dominated by acai palm (*Euterpe oleracea* Mart.) in the Amazon Estuary. In: *The palm-tree of life* (ed.) M. J. Balick. *Advances in Economic Botany*, 6: 144-154.

Angiosperm Phylogeny Group I. (1998). An ordinal classification for the families of flowering plants. *Ann. Missouri Bot. Gard.,* 85(4): 531-553.

Angiosperm Phylogeny Group II. (2003). An update of the Angiosperm Phylogeny Group classification for the orders and families of flowering plants: APG II. *Bot. J. Linn. Soc.,* 141: 399-436.

Angiosperm Phylogeny Group III. (2009). An update of the Angiosperm Phylogeny Group classification for the orders and families of flowering plants: APG III. *Bot. J. Linn. Soc.,* 161: 105-121.

Anon. (1952). *A Dictionary of Indian Raw Materials* Vol. III: D-E. CSIR, New Delhi. pp.64-65.

Anon. (1956). *A Dictionary of Indian Raw Materials* Vol. IV: F-G. CSIR, New Delhi pp.101-103.

Anon. (1959). *A Dictionary of Indian Raw materials* Vol. V: H-K. CSIR, New Delhi, pp. 133.

Anon. (1962). *A Dictionary of Indian Raw Materials* Vol. VI: *L-M.* CSIR, New Delhi, pp 429-473, 298-300.

Anon. (1969). *The wealth of India- raw materials*. Publication and Information Directorate, CSIR, New Delhi, India, 8: 250–282.

Anon. (1976). *A Dictionary of Indian Raw Materials Vol. X: SP-W.* CSIR, New Delhi, pp. 20-21, 25-28, 122—123. 164-167, 171-177, 194-195, 274-277.

Anon. (1984). *Cash crop science sub committee report*, MAU, Parbhani (Maharashtra), India.

Anon. (1985). *A Dictionary of Indian Raw Materials Vol. 1: A* (Revised). Council of Scientific and Industrial Research, New Delhi.

Anon. (1987). *IC and RF Society Newsletter.* Fall, 1987, Issue 23, pp. 15.

Anon. (1990). *Plants for reclamation of wastelands*. Publication and Information Directorate. CSIR, New Delhi, pp. 18-21, 530.

Anon. (1998). *Project Report 1996 and 1997*. Ninth Group Workers Meeting AICRP on Arid Zone Fruits, CCS HAU, Hisar, 16-18 October, 1998. NRC for Arid Horticulture, Bikaner, pp. III.

Argent, G. C. G. (1976). The wild bananas of Papua New Guinea. *Notes on Royal Botanic Garden (Edinburgh)*, 35: 77-114.

Armstrong, W. P. (2006). *Sex determination & life cycle of Ficus carica.* <http://Waynesworld.Palomar.edn/p/Jun99b. htm>.

Arnold, T. H. Wells, M. J. and Wehmeyer, A. S. (1985). Khoisan food plants: taxa with potential for future economic exploitation. In: *Plants for Arid Lands*. (eds.) Wickens, G. E. Goodin, J. R. and Field, D. V. George Allen and Unwin, London, UK, pp. 69-86.

Arora, R. K. (1985). *Genetic Resources of less known cultivated Food Plants. NBPGR Sci. Mongr. No. 9*. National Bureau of Plant Genetic Resources, New Delhi, India, 126 p.

Arora, R. K. (1998). Genetic Resources of Native Tropical Fruits in Asia: Diversity, Distribution and IPGRI emphasis on their conservation and use. In: *Tropical Fruits in Asia: Diversity, Maintenance, Conservation and Use*. Proceedings of the IPGRI-ICAR-UTFANET Regional training course on the conservation and use of germplasm of tropical fruits in Asia held at IIHR. 18-31 May, 1997 Bengaluru, India. (eds.) R. K. Arora and V. Ramanatha Rao, pp. 42-53.

Arora, R. K. and Pandey, A. (1996). *Wild Edible Plants of India: Diversity, conservation and use*. National Bureau of Plant Genetic Resources, New Delhi, India, 294 p.

Arora, R. K. and Ramanatha Rao, V. (eds.) (1995). *Proceedings of Expert Consultation on Tropical Fruit Species of Asia.* The Malaysian Agricultural Research and Development Institute, Serdang, Kuala Lumpur, Malaysia, 17-19 May, 1994,114 pp.

Arya, S., Kumar, N., Tokay, O. P. (1995) *Khejri* (*Prosopis cineraria* (L.) Druce), *its value, research and extension*. HRD-ODA Project, India, 131 pp.

Asma, B. M. (2011). *Her yönüyle kayisi.* Uyum Ajans, Ankara– Türkiye, p. 367. (in Turkish)

Azhakiamanavalan, R. S. and Vadivel, E. (1997). Dryland Tamarind PKM-1: Pre- and Post-release status. In: *Proceedings of National Symposium on Tamarindus indica* L.). (ed.) P. S. Rao, 27-28 June, 1997, Tirupati, pp. 7-9.

Badanes, M. L., Asins, M. J., Carbonell, E. A. and Llacer, G. (1996). Genetic diversity in apricot, *Prunus armeniaca*, aimed at improving resistance to plum pox virus. *Plant Breed.*, 115, 133-139.

Badillo, V. M. (2001). Nota correctiva *Vasconcellea* St. Hill. Y no *Vasconcella* (Caricaceae). *Ernstia*, 11(1): 75-76.

Badillo, V. M. (1971). *Monografía de la Familia Caricaceae.* Universidad Central de Venezuela, Maracay, Venezuala, 221 p.

Badillo, V. M. (1993). *Caricaceae. Segundo esquema.* Revista de la Facultad de Agronomía de la Universidad Central de Venezuela, Alcance 43, Maracay, Venezuela, 111 p.

Badillo, V. M. (1999). New records of species of the *Carica* genus for Ecuador. *Ernstia*, 9(3-4): 185-186.

Bailey L. H. (1949). *Manual of Cultivated Plants.* McMillan, New York.

Bailey L. H. (reprinted 1947). Standard Cyclopedia of Horticulture. Vol. 3. The Macmillan Co., New York, 1930.

Bailey, L. H. (1916). Prunus. In: *The standard cyclopedia of horticulture*. V. P. R. Mount Pleasant Press, J. Horace McFarland Co., Harrisburg, PA, pp. 2822- 2845.

Bailey, L. H. (1923). Various cultigens and transfers in nomenclature. *Gentes Herbarum,* 1: 113–136.

Bailey, L. H. (1934). The species of grapes peculiar in North America. *Genetes Herbarium,* 5: 151-244.

Bailey, L. H. and Hough, L. F. (1975). Apricots. In: Janick J, Moore J. N. (eds.) *Advances in fruit breeding*. Purdue University Press, West Lafayette, Indiana, pp. 367- 383.

Baird, J. R. and Thieret, J. W. (1989). The medlar (*Mespilus germanica*), Rosaceae from Antiquity to Obscurity. *Economic Botany*, 43 (3): 328-372.

Baker, J. G. (1893). A synopsis of the genera and species of Musaceae. *Annals of Botany*., 78: 189-229.

Balick, M. J. (1988). *Jessenia* and *Oenocarpus*: neotropical oil palms worthy of domestication. FAO Plant Production and Protection paper 88, FAO, Rome, 191 pp.

Barbera, G., Inglese, P., Pimienta-Barrios, E. (1995). *Agro-ecology, cultivation and uses of cactus pear*. FAO Plant Production and protection paper 132, FAO, Rome.

Barkley, N. A., Roose, M. L., Krueger, R. R. and Federici, C. T. (2006). Assessing genetic diversity and population structure in a citrus germplasm collection utilizing simple sequence repeat markers (SSRs). *Theoretical and Applied Genetics*, 112: 1519-1531.

Barooah, S. and Meitei, W. I. (1977). Tree bean, an economic crop of Manipur. Assam Agrian. *The Journal of the College of Agriculture, A. A. U., Jorhat*, Vol. VII & VIII, December, 1977, p. 51.

Barret, H. C. and Rhodes, A. M. (1976). A numerical taxonomic study of affinity relationships in cultivated *Citrus* and its close relatives. *Systematic Botany*, 1: 105-136.

Barros, L. M. (1988). Melhoramento. In: *a cultura do Cajucivo no Nordeste do Brasil* (lima VPMS Ed.), Banco do Norderte do Brasil/ETENE. Fortaleza, pp. 321-356 (BNB/ETENE). Estudos Economicos e Sociais,

Batlle, I. and Tous, J. (1997). *Carob tree, Ceratonia siliqua* L. In: *Promoting the conservation and use of underutilized and neglected crops. 17*. Institute of Plant Genetics and Crop Plant Research Gatersleben/International Plant Genetic Resource Institute, Rome, Italy, 92 p.

Baumer, M. (1983). Notes on trees and shrubs in arid and semi-arid regions. EMASAR Phase II. FAO, Rome.

Bayer, R. J., Mabberley, D. J., Morton, C. M., Cathy, H., Sharma, I. K., Pfeil, B. E., Rich, S., Hitchcock, R. and Sykes, S. (2009). A molecular phylogeny of the orange subfamily (Rutaceae: Aurantioideae) using nine cp DNA sequences. *Amer. J. Bot.*, 96 (3): 668-685.

Bell, R. L., Quamme, H. A., Layne, R. E. C. and Skirvin, R. (1996). Pears. In: J. Janick and J. N. Moore (eds.), *Fruit breeding. Vol. 1. Tree and Tropical Fruits*. Wiley John Sons Inc., New York. pp. 441-514.

Bellini, E. Watkins, R. and Pomarci, E. (1984). *Descriptor list for Peach* (*Prunus persica*). IBPGR Secretariat and CEC Secretariat, Brussels.

Benson, L. (1979). *Plant classification, 2nd ed.* D. C. Heath, Lexington, MA.

Bentham, G. and Hooker, W. J. (1862-1883). *Genera Plantarum*. London, L. Reeve & Co.

Bergh, B. and Ellstrand, N. (1986). Taxonomy of the avocado. *California Avocado Society Yearbook*, 70: 135-145.

Bergh, B. O. (1975). Avocados. In: *Advances in Fruit Breeding.* (eds.) J. Janick and J. N. Moore. Purdue University Press, West Lafayette, Indiana, USA, pp. 541-567.

Bergh, B. O. (1976). Avocado breeding and selection. In: *The Avocado. Proceeding of the 1st International Tropical Fruit short course*. (eds.) J. W. Sauls, R. L. Phillips and L. K. Jackson. Fruit Crops Department, University of Florida, Gainesville, Florida, USA, pp. 24-33.

Bermejo, Hernandez, J. E. and Leon, J. (1994). *Neglected crops 1492 from a different perspective*. FAO Plant Production and Protection Series No. 26. FAO, Rome.

Bertoni, M. S. (1919). Contributions a l'etude botanique des plantes cultivees. I. Essai d'une monographie du enre. *Ananas*. *Anales Cientificos Paraguayos (Ser.II)*: 4, 250-322.

Bessey, C. E. (1915). Phylogenetic taxonomy of flowering plants. *Ann. Mo. Bot. Gard.*, 2: 109-164.

Bettencourt, E. and Konopka, J. (1989). 6.11. Directory of Germplasm collection. Temperate Fruits and Tree nuts: Actinidia, Amelanchier, Carya, Castanea, Corylus, Cydonia, Diospyros, Fragaria, Juglans, Malus, Mespilus, Morus, Olea, Pistacia, Prunus, Punica, Pyrus, Ribes, Rosa, Rubus, Sambucus, Sorbus, Vaccinium and others. IBPGR, Rome, 296 pp.

Bettencourt, E., Hazekamp, Th. and Perry, M. C. (1992). *Directory of Germplasm collection 6.1. Tropical and subtropical fruits and tree nuts*: Annona, Avocado, Banana and Plantain, Breadfruits, Cashew, Citrus, Date, Fig, Guava, Mango, Passionfruit, Papaya, Pineapple and others. IBPGR, Rome, 334 pp.

Bhansali, A. K. (1975). *Monographic study of the family Rhamanaceae of India.* Ph.D. Thesis, Univ. of Jodhpur.

Bhattacharya, S. C. and Dutta, S. (1956). *Classification of Citrus fruits of Assam.* Scientific Monograph No. 20. ICAR, New Delhi, 110 pp.

Bohs, Lynn. (1995). Transfer of *Cyphomandra* (Solanaceae) and its species to *Solanum.* Taxon, 44(4): 583-87.

Bompard, J. M. and Schnell, R. J. (1997). Taxonomy and systematics. In: Litz, R. E. (ed.) *The Mango: Botany, Production and Uses.* CAB International, Wallingford, Oxon, U.K. pp. 21-47.

Bonavia, E. (1888-90). *The cultivated oranges and lemons, etc. of India and Ceylon.* W. H. Allen and Co. London, 384, pp.

Booth, F. E. M. and Wickens, G. E. (1988). *Non-timber uses of selected arid zone trees and shrubs in Africa.* FAO Conservation Guide 19. FAO, Rome, 176 p.

Bortiri, E., Oh, S. H.; Gao, F. Y. and Potter, D. (2002). The phylogenetic utility of nucleotide sequence analysis of sorbitol 6-phosphate dehydrogenase in *Prunus* (Rosaceae). *Am. J. Bot.,* 89: 1697-1708.

Bouachrine, J. E. (1994). Multipurpose trees and shrubs– focus on the argan trees. *Agroforestry Today*, 6 (1) : 11.

Bremer, K. and Wanntorp, H. E. (1981). The cladistic approach to plant classification. In: *Advances in Cladistics. 1.* V. A. Funk and D. R. Brroki (eds.) 87-94, New York Botanical Garden, New York.

Brenan, J. P. M. (1963). *Flora of Tropical East Africa*: *Trapaceae*. Crown Agents for the Colonies, London.

Brickell, C. D. (1995). *International Code of Nomenclature for Cultivated Plants—1995.* Int. Comm. for the Nomenclature of Cultivated Plants, Wimborne, UK.

Brickell, C. D., Alexander, C., David, J. C., Hatter Seheid, W. L. A., Leslie, A. C., Malecot, V., Xiaoba, J. and Cubeg, J. J. (2009). *International Code of Nomenclature for plants (ICNAP or Plants Code),* Eighth Edition Scripta Horticulture, ISHS Secretariat, Belgium. 10 :1-168.

Britton, N. L. and Rose, J. N. (1963). *Descriptions and illustrations of plants of the cactus family*, Vol. I and II. Dover Publ., Inc., New York, USA, pp. 183–195.

Broadley, R., Rigden, P., Chay-Prove, P., Daniells, J. (2004). *Subtropical Banana Grower's Handbook.* Queensland Department of Primary Industries, pp 1-206.

Brücher, H. (1989). *Useful Plants of Neotropical origin and their wild relatives.* Springer-Verlag, Berlin, Heidelberg, New York, pp. 128-142, 215-289.

Burgos L., Pérez-Tornero, O., Ballester, J., Olmos, E. (1998). Detection and inheritance of stylar ribonucleases associated with incompatibility alleles in apricot. *Sex Plant Reprod,* 11: 153–158.

Burns, W. and S. H. Prayag (1921). *The book of the mango*. Department of Agriculture, Bombay, Bulletin No. 103.

Callaway, M. B. (1992). Current Research for the commercial development of Pawpaw (*Asimina triloba* (L.) Dunal). *HortScience*, 27 (2): 90, 191.

Calzada Benza, J. (1993). *143 frutales nativos*. Ediciones Universidad Nacional Agraria La Molina, Lima, Peru, 366 p.

Campbell, C. W. (1963). Promising new guava cultivars. *Proceedings of the Florida State Horticultural Society,* 76: 363-365.

Carbonneau, A. (1983). Sterilites male et femelle dars le genre. Vitis I. Modelsation de leur heredite. *Agronomis*, 5: 635-44.

Carreel, F. (1994). *Etude de la diversite des bananiars (genera Musa) al aide des marquers RFLP*. Thesis. Institut National Agronomiques. Paris-Grignon. 90 pp.

Carreel, F. S., Faure, D., Gonzalez, de Leon, P. G. L, Lagoda, X., Perrier, F., Bakry, et al. (1994). Evaluation de la diversite genetique chez les bananiers diploides a BIRFA-CIRAD. Fruit (special issue): 25-40.

Chadha, T. R. (2001). *Textbook of temperate fruits*. DIPA, ICAR, New Delhi, pp. 26-28.

Chase, M. W. and Reveal, J. L. (2009). A phylogeneric classification of the land plants to accompany APG-III. *Botanical Journal of Linnean Society*, 161: 122-127.

Cheesman E. E. (1947). Classification of the Bananas: The Genus *Musa* L. *Kew Bulletin*, 2 (2):106-117.

Cheesman, E. E. (1947-1950). *The classification of the bananas. Kew Bulletin*. 1. 2:97-106. 1947; 11.2:106-117. 1947; III a. 3:11-17. 1948; III b. 3:17-28. 1948; III c. 3:145-153. 1948; III d. 3:154-157. 1948; III e. 3:323-325. 1948; III. 3:325-328. 1948; III. 4:23-24. 1949; III h. 4:24-28. 1949; III i. 2:133-135. 1949; III J. 2:135-137. 1949; III k. 3:265-267. 1949; III l. 3:267-272.1949; IIIm. 4:445-449. 1949; III n. 4:450-452. 1949; IIIo. 5:27-28. 1950; III p. 5:28-29. 1950; III. q. 5:29-31. 1950; III r. 2:151-152. 1950; III S. 2152-155. 1950.

Cheesman, E. E. (1948). Classification of the bananas. III. Critical notes on species a. *Musa balbisiana* Colla. *Kew Bull.*, 3: 11-17; b. *Musa acuminata* Colla. *Kew Bull.*, 3: 17-28; and c. *Musa paradasiaca* and *Musa sapientum.* Kew Bull., 3: 145-153.

Chhabra, A. K. (2006). *Practical manual on floral biology of fruit plants*. Deptt. of Plant Breeding, CCS HAU, Hisar.

Clamot, G., Linden, N. and Borg van der (1986). *Strawberry* (*Fragaria* L.) *Descriptors*. ECSC, EEC, EAEC, Brussels and Luxembourg; and International Board of Plant Genetic Resources, Rome.

Clay, J. W. and Clement, C. R. (1993). Selected species and strategies to enhance income generation from Amazonian Forests. FO: Misc/93/6 Working paper. FAO, Rome.

Clement, C. R. (1988). Domestication of the pejibaye palm (*Bactris gasipaes*): past and present. *Advances in Economic Botany*, 6: 155-174.

Clement, C. R. (1989a). A center of crop genetic diversity in Western Amazonia, a new hypothesis of indigenous fruit crop distribution. *BioScience*, 39 (9): 624-631.

Clement, C. R. (1989b). The potential use of the pejibaye palm in agroforestry systems. *Agroforestry Systems*, 7 (3): 201-212.

Colla, L. A. (1820). Memoria sul genera *Musa*. *Mem. Reale Accad. Sci. Torino.,* 25: 384-394.

Collins, G. N. (1903). *The mangoes in Puerto Rico*. United States Department of Agriculture Bulletin No. 28.

Condit, I. J. (1947). *The Fig*. Walthams, Mass., Pub. USA.

Constantine, D. and Rossel, G. (2001). *The Musaceae: An annotated list of the species of Ensete, Musa* and *Musella*. http://www.users.globalnet.co.uk/~drc/index.htm.

Coppens d' Eeckenburugge, Geo and Leal, Freddy (2003). Morphology, Anatomy and Taxonomy. In: *The Pineapple: Botany, production and uses* (eds.) D.P. Bartholomew, R. E. Paull & K. G. Rohrabach, CAB International, UK.

Cowan, A. M. and Cowan, J. M. (1929). The trees of Northern Bengal, including shrubs, woody climbers, bamboos, palms and tree ferns. Bengal Secretariat Book Depot, Calcutta, 178 p.

CRFG (1996a). *Hardy Kiwi Fruit Facts.* California Rare fruit Growers, Inc. California, USA.

CRFG (1996b). *Jaboticaba* (*Myrciaria* spp.) California Rare fruit Unit Inc, California, USA, pp. 1-4.

CRFG (1996c). *Miracle Fruit Facts.* California Rare fruit Growers, Inc. California, USA, pp. 1-2.

CRFG (1996d). *Tomarillo Fruit Facts*. California Rare fruit Growers, Inc., California, USA.

CRFG (1999). Pawpaw Fruit Facts. California fruit Growers, Inc. California, USA.

Cronquist, A. (1988). *The evolution and classification of flowering plants*, *2nd ed.* New York Botanical Garden, Bronx, NY.

Crouch, J. H., Crouch, H. K., Constandt, H., Van Gysel, A., Breyne, P., Van Montaga, M., *et al.* (1999). Comparison of PCR- based molecular marker analyses of *Musa* breeding poputation. *Mol. Breed.*, 5: 233-244.

Dalziel, J. M. (1937). *The useful plants of west Tropical Africa*. Whitefriars Press, London, p.164.

Dana, M. N. (1990). Cranberry management. In: *Small Fruit Crop Management.* (eds.) Gene Galletta and D. G. Himelrick. Prentice-Hall, Englewood Cliffs, New Jersey. USA, pp. 334-362.

Daniells, J., Jenny, C., Karamura, D. and Tomekpe, K. (2001). *Musalogue: A catalogue of Musa germplasm: Diversity in the genus Musa*. p. 213. INBAP. Montpellier, France.

Darrow, G. M. (1975). Minor temperate fruits. In: *Advances in fruit breeding*. (eds.) Jules Janick and J. N. Moore. Purdue University Press, West Lafayette, Indiana, pp. 271-284.

Darwin, Charles (1868). Peach and almond stones of natural size viewed edgeways in the variation of Animals and Plants under Domestication, Vol. I & II. Albemarle street, London

Das, B., Ahmed, N. and Singh, P. (2011). *Prunus* diversity- early and present development: A review. *International Journal of Biodiversity and Conservation,* 3(14): 721-734.

Dawes, S. N. and Pringle, G. J. (1983). Subtropical fruits from South and Central America. In: *Plant breeding in New Zealand.* (eds.) G. S. Smith and H. C. Wratt. Butterworths, Wellington, New Zealand.

Daxenbichler, M. E., Spencer, F. G., Carlson, D. G., Rose, G. B., Brinker, A. M., Powell, R. G. 1991). Glucosinolate composition of seeds from 297 species of wild plants. *Phytochemistry,* 30: 2623-2638.

De Candolle, A. (1806-1893). *Origin of cultivated plants*. Kagan Paul, London.

De Langhe, E. A. L. (1996). Banana and plantain: The earliest fruit crop? Focus Paper No. 1. In: *INIBAP annual report 1995*. Montpelier, France: INIBAP.

De Langhe, E. and De Maret, P. (1999). Tracking the banana: Its significance in early agriculture. In: *Prehistory of food: Appetite for change*. Gosden, C. and Hather, J., eds., 377-396. New York: Routledge.

De Moraes, A. P., Fiho, W. and Guerra, M. (2007). Karyotpe diversity and the origin of grapefruit. *Chromosome Research,* 15: 115-121.

De Nettancourt, D. (2001). *Incompatibility and incongruity in wild and cultivated plants,* 2nd ed. Springer-Verlag, New York.

De Zerpa, D. M. (1980). Comportamiento meiotíco de la descendencia híbrida producida al transferir el carácter bisexual de *Carica pubescens* a *Carica stipulata. Revista de la Facultad de Agronomía* (Maracay), XI (1-4): 5-47.

Desai, A. R. (2009). *Glimpses of Cashew Research in Goa*. Technical Bulletin No. 17, ICAR Research Complex for Goa, Indian Council of Agricultural Research, Ela, Old Goa-403402, Goa, India.

Desai, A. R. and Singh, N. P. (2011). *Cashew Germpalsm of Goa: Characterization and Cataloguing*, Technical Bulletin No. 23. ICAR Research Complex for Goa, Indian Council of Agricultural Research, Old Goa-403402, Goa, India.

Descourtilz, M. E. (1829). *Juglans fraxinifolia. Fl Pitt Med Antill*, 7: 5–8.

Despeissis, 1921. The Handbook of Horticulture and Viticulture, Australia.

Dessauw, D. (1988). Etude des facteurs de la stérilité du bananier (*Musa* spp.) et des relations cytotaxinomiques entre *M. acuminata* Colla et *M. balbisiana* Colla. *Fruits,* 27:539–558.

Dickinson, T. A., Evans, R. C. and Campbell, C.S. (2002). *Rosaceae classification and phylogeny introduction and overview*. http/www.botany2002.org/symposl3/abstracts/5.shtml.

Ding Hou (1978). Anacardiaceae. In: *Flora Malesiana, Series* 1. (ed.) *J. Van Steenis*, *Vol. 8 (C. G. G.):* 437-438.

Dodds, K. S. and Simmonds, N. W. (1948). Genetical and Cytological studies of *Musa.* IX. The origin of edible diploid and the significance of interspecific hybridization in bananas complex. *J. Genet.,* 48: 285-296.

Dode, L. A. (1909a). Contribution to the study of the genus *Juglans* (English translation by R. E. Cuendett). *Bulletin of the Society of Dendrology, France,* 11: 22-90.

Dode, L. A. (1909b). Contribution to the study of the genus *Juglans* (English translation by R. E. Cuendett). *Bulletin of the Society of Dendrology, France,* 12: 165-215.

Drew, R. A., O'Brien, C. M., Magdalita, P. M. (1998). Development of interspecific *Carica* hybrids. *Proceedings of the International Symposium on Biotechnology of Tropical and Subtropical Species*, part II, Brisbane, Queensland, Australia, 29 September to 3 October 1997, 285-291.

Duke, J. A. (1989). *CRC Hand book of Nuts*. CRC Press, Boca Raton, Florida, USA, 347 pp.

Dumbre, R. B. (1997). Konkan Amruta– The first variety of Kokum released in India. *ICAR News*, Oct.-Dec., 1999, pp. 14-15.

Dunemann, F., Kahnau, R. and Schmidt, H. (1994). Genetic relationships in *Malus* evaluated by RAPD fingerprinting of cultivars and species. *Plant Breeding*, 113: 150-159.

Eck, P., Gough, R. E., Hall, I. V. and Spiers, J. M. (1990). Blueberry management. In: *Small Fruit Crop Management*. (eds.) Gene J. Galletta and David, G. Himelrick. Prentice Hall, Englewood Cliffs, New Jersey, USA, pp. 273-333.

Egea, J. and Burgos, L. (1996). Detecting cross-incompatibility of three North American apricot cultivars and establishing the first incompatibility group in apricot. *J. Amer. Soc. Hort. Sci.,* 121: 1002-1005.

Eichler, A. W. (1875). *Bluĭhendiagramme*. Leipzig.

Elzebroek, A. T. G. and Wind, K. (2008). *Guide to cultivated plants*. Wellingford, CAB International. P.27.

Engler, A. (1896). Rutaceae, von A. Engler. In: *Dienaturlichen Pfanzenfamilien, begr.* von A. Engler und K. Prantl. Englemann, Leipzig. Teil 3, Abt. 4, pp. 95-201.

Engler, A. (1931). Rutaceae. In: Engler, A. and Prantl, K. *Die Naturlichen Pflanzenfamilien*. Second edition. Engelmann, Leipzig. 19a: 187-359.

Engler, Adolf. (1892) Syllabus der Pflanzenfamilien, Berlin.

Entani, T., Iwano, M., Shiba, H., Che, S. F., Isogai, A. and Takayama, S. (2003). Comparative analysis of the self-incompatibility (S-) locus region of *Prunus mume*: Identification of a pollen-expressed F-box gene with allelic-diversity. *Genes Cells*, 8: 203-213.

Espino, R. R. C., Jamaluddin, S. H., Silayoi, B., Nasution, R. E. (1992). *Musa* L. (edible cultivars). In: Verheij, E. W. M., Coronel, R. E., eds. *Plant Resources of South-East Asia No. 2 Edible Fruits and Nuts*. PROSEA, Bogor, Indonesia.

Evreinoff, V. A. (1949). Notes sur les varieties d'*Actinidia*. *Revue Hort*., 121: 155-158.

Evreinoff, V.A. (1964). Notes sur le jujubier (*Ziziphus jujuba* Gaertner). *Journal d'agriculture et de botanique appliquee,*11:177-187.

FAO (1982). *Fruit bearing forest trees*. FAO Forestry paper No. 34. FAO, Rome, 177 pp.

FAO (1983). *Food and fruit bearing forest species*. 1: Examples from Eastern Africa. FAO Forest Paper No. 44/1. FAO, Rome, 165 pp.

FAO (1984). Food and fruit bearing forest species. 2: Examples from Southeastern Asia. FAO Forestry Paper No. 44/2. FAO, Rome.

FAO (1986). Food and fruit bearing forest species. 3. Examples from Latin America. FAO Forestry paper No. 44/3. FAO, Rome.

FAO (1988). Traditional Food plants- A resource book for promoting the exploitation and consumption of food plants in arid, semi-arid and sub-humid lands of East Africa. FAO Food and Nutrition Paper 42. FAO, Rome, 593 pp.

FAO. (1996a). Report on the state of the world's plant genetic resources for food and agriculture. FAO, Rome, Italy, 511 p.

FAO. (1996b). Global plan of action for the conservation and sustainable utilisation of plant genetic resources for food and agriculture and the Leipzig declaration. FAO, Rome, Italy, 63 p.

Faust, M., Suranyi, D. and Nyujto, F. (1998). Origin and dissemination of apricot. *Horticultural Reviews*, 22: 225-265.

Ferguson, A. R. (1990a). Kiwifruit (*Actinidia*). In: Moore, J.N. and Ballington, J.R. Jr. (eds.) *Genetic Resources of Temperate Fruit and Nut Crops*. International Society for Horticultural Science Wageningen, The Netherlands, pp. 603-653.

Ferguson, A. R. (1990b). Kiwifruit management. In: *Small Fruit Crop Management* (eds.) Gene, J. Galletta and David, G. Himelrick. Prentice Hall, Englewood Cliffs, New Jersey, USA, pp. 472-503.

Ferguson, A. R. (1997). Kiwifruit. In: *the Brooks and Olmo register of fruit and nut varieties* (pp. 319-323) (3rd ed.). Alexandria: ASHS Press.

Ferguson, A. R. (1999). New temperate fruits: *Actinidia chinensis* and *Actinidia deliciosa*. In: Janick, J. (ed.). *Perspectives on new crops and new uses*. ASHS Press, Alexandria, USA.

Galang, F. G. (1955). *Fruit and nut growing in the Philippines*. AIA Printing Press, Malabon, Rizal, Philippines.

Galet, P. (1979). *A Practical Ampelography: Grapevine identification.* Comstock Publishing Associates, Cornell University Press, Ithaca, New York.

Galet, P. (1988). *Cepages et Vigndoles de France*. Tome, 1 Les Vignes americaines. Imp. Dehan, Montpellier, France.

Galet, Pierre (1998). *Grapes varieties and rootstocks varieties*. Oenoplurimedia sarl Chateau de Chaintre, Chaintre, France. 315 p.

Gangolly, S. R., Singh, Ranjit, Katyal, S. L. and Singh, Daljit (1957). *The mango*. Indian Council of Agriculture Research, New Delhi, India 320 p.

Gardiazabal, F. and Rosenberg, G. (1993). *El cultivo del chirimoyo.* Universidad Católica de Valparaíso, Facultad de Agronomía, Valparaíso, Chile, 145 p.

Ge, X. J., Liu, M. H., Wang, W. K., Schaal, B. A. and Chiag, T. Y. (2005). Population structure of wild bananas, *Musa balbisiana*, in China determined by SSR fingerprinting and cpDNA PCR-RFLP. *Molecular Ecology*, 14:933-944.

Geuna, F., Toschi, M. and Bassi, D. (2003). The use of AFLP markers for cultivar identification in apricot. *Plant Breed.*, 122, 526-531.

Ghora, C. and Panigrahi, G. (1984). *Fascicles of flora of India. Rosaceae*: genus *Prunus*. Botanical Survey of India, Calcutta, India. 18.

Ghora, C. and Panigrahi, G. (1995). *The family Rosaceae in India, vol. 2*. Bishan Singh, Mahender Pal Singh, Dehra Dun, India.

Goren, M., Tomer, E. and Gazit, S. (2001). Litchi cultivation in Israel. *Proc. 1st Intl. Symp. Litchi and Longan*, Guangzhou, China. p. 129-133.

Griesbach, Jurgen (2005a). *Avocado growing in Kenya*. World Agroforestry Centre, Nairobi, Kenya.

Griesbach, Jurgen (2005b). *Mango growing in Kenya*. World Agroforestry Centre, Nairobi, Kenya.

GRIN. (2007). *USDA, ARS, National Genetic Resources Program. Germplasm Resources Information Network–* (GRIN) [Online Database]. National Germplasm Resources Laboratory, Beltsville, Maryland.

Groff, G. W. (1921). *The Lychee and Longan*. Orange Judd Co. New York.

Guerriero, R. and Watkins, R. (1984). *Revised descriptor list for apricot* (*Prunus armeniaca*). AGPG: IBPGR/84/91. IBPGR Secretariat, Rome, CEC Secretariat, Brüssels.

Gülcan, R. (1985). *Descriptor List for almond* (*Prunus amygdalus*) (Revised). AGPG, International Board of Plant Genetic Resources Secretariat, Rome.

Gwankar, M. S., Shingrae, D. V., Patil, B. P., Sagvekar, V. V. and Joshi, G. D. (2004). Evaluation of seedling progenies of kokam (*Garcinia indica* Choicy). *Indian J. Arecanut, Spices and Medicinal Plants*., 6(3): 83-86.

Häkkinen, M. and De Langhe, E. (2001). *Musa acuminata in Northern Bornea*, Montpellier, France: INIBAP.

Häkkinen, M. and Väre, H. (2008). Typification and check list of *Musa* L. names (Musaceae) with nomenclatural notes. *Adansonia*, 30:63–112.

Häkkinen, M. and Wallace, R. (2011). Genetic Resources for banana improvement. In: *Banana Breeding Progress and Challenges* (eds.) M. Pillay and A. Tenkouano CRC Press, Bosc Raton, London and New York, pp. 41-51.

Häkkinen, M. and Wang, H. (2007). New species and variety of *Musa* (Musaceae) from Yunnan, China. *Novon*, 17(4): 440-446.

Häkkinen, M. and Wang, H. (2008). *Musa yunnanensis* (Musaceae) and its intraspecific taxa in China. *Novon J. Bot*., 26: 317-324.

Häkkinen, M. and Wang, H. (2008). *Musa zaifui* sp. nov. (Musaceae) from Yunnan, China. *Nordic Journal of Botany,* 26: 42–46.

Häkkinen, M., Wang, H. and Ge, X. J. (2008). *Musa itinerans* (Musaceae) and its intraspecific taxa in China. *Novon,* 18(l):50—60.

Häkkinen, Markku (2013). Reappraisal of sectional taxonomy in *Musa* (Musaceae). *TAXON*, 62 (4): 809–813.

Halasz, J. (2007). *Molecular background of the S-locus controlled self-incompatibility in apricot*. Ph. D. Diss., Corvinus University of Budapest, Budapest, Hungary.

Halasz, J., Hegedus, A., Herman, R., Stefanovits-Banyai, E. and Pedryc, A. (2005). New self-incompatibility alleles in apricot (*Prunus armeniaca* L.) revealed by stylar ribonuclease assay and S-PCR analysis. *Euphytica*, 145: 57-66.

Halasz, J., Pedryc, A. and Hegedus, A. (2007). Origin and dissemination of the pollen-part mutated Sc-haplotype that confers self-compatibility in apricot (*Prunus armeniaca*). *New Phytol*., 176, 793-803.

Halasz, J., Pedryc, A., Ercisli, S., Yilmaz, K. U. and Hegedus, A. (2010). S-genotyping supports the genetic relationships between Turkish and Hungarian apricot germplasm. *J. Amer. Soc. Hort. Sci.,* 135(5): 410-417.

Hall, J. B. and Walker, D. H. (1991). *Balanites aegyptiaca– A monograph*. School of Agricultural and Forest Sciences, University of Wales, Bangor, UK, 65 pp.

Hall, J. B., Aebischer, D. P., Tomlinson, H. F., Osei-Amaning, E. and Hindle, J. R. (1996). *Vitellaria paradoxa– A monograph*. School of Agricultural and Forest Sciences, University of Wales, Bangor, UK, 105 pp.

Hall, John, B., O'Brien, E. M. and Sinair, Fergus L. (2002). *Sclerocarya birrea– A monograph.* School of Agricultural and Forest Sciences Publication Number 19, University of Wales, Bangor, UK, 157 pp.

Hall, John, B., Tomlinson, H. F., Oni, P. I., Buchy, M. and Aebischer, P. (1997). *Parkia biglobosa– A monograph*. School of Agricultural and Forest Sciences Publication Number 9. University of Wales, Bangor, UK, 107 pp.

Harms, H. (1930). *Pflanzenfamilien. Bromeliaceae*. Engler Prantl. 65-159.

Hartless, A. C. (1913). The flowering of the mango. *Agriculture Journal of India,* 8: Encyclopedia 90.

Hartmann, H. T. and Kester, D. E. (1975). *Plant Propagation*, 3rd ed. Prentice Hall, Englewood, Cliffs, Jersey.

Hassler, E. (1919). *Bromeliaceraum paraguariensium conspectus*. Annuaire du Conservatoire et du Jardin Botaniquie de Geneve, 268-341.

Haston, E., Richardson, J. E., Stevens, P. F., Chase, M. W., Harris, D. J. (2009). The Linear Angiosperm Phylogeny Group (LAPG) III: a linear sequence of the families in APG III. *Botanical Journal of the Linnean Society,* 161: 128–131.

Hayashi, K., Shimazu, K., Yaegaki, H., Yamaguchi, M., Iketani, H. and Yamamoto, T. (2008). Genetic diversity in fruiting and flower- ornamental Japanese apricot (*Prunus mume*) germplasms assessed by SSR markers. *Breeding Science*, 58, 401-410.

Hedrick, U. P. (1911). *The plums of New York*. New York (Geneva) Agric. Exp. Sta. Rep.

Hedrick, U. P. (1925). *Systematic Pomology*. Macmillan, New York.

Henderson, A., Galeano, G. and Bernal, R. (1995). *Field guide to the palms of the Americas*. Princeton, NJ: Princeton University Press, New Jersey, USA.

Hendrickson, Robert (1981). *The Berry Book*, Doubledy, 1981.

Herzog, F., Gutier-Beguin, D. and Miller, K. (1996). Uncutlivated plants for human nutrition in Cote d' ivoire. In Domstication and commercialization of non-timber forest products in Agroforestry systems *Proceedings of an International Conference held in Nairobi*-Kenya from 19-23, Feb. 1996 (eds.) R. R. B. Leakey, A. B., Melnyk, M. & Vantomme, P. (Editors). FAO, Rome, Italy. pp. 40–50.

Heslop-Harrison, J. S. and Schwarzacher, T. (2007). Domestication, genomics and the future for banana. *Ann. Bot.* (Lond), 100: 1073–1084.

Heywood, V. H. (1985). *Flowering plants of the world*. Croom Helm, London, UK and Sydney, Australia, 336 p.

Heywood, V. H., Bramrritt, R. K., Kulham, A. and Seberg, O. (2007). *Flowering plant families of the world*. Royal Botanic Garden Kew Publications.

Higgins, J. E. (1906). *The mango in Hawaii*. Hawaii Agricultural Experiment Station, Bulletin No. 12.

Hill, Lewis (1983). Vintage elderberries. *Horticulture,* LXI (6): 44-47.

Hilton, R. J. (1982). Registration of *Amelanchier* cultivar names. *Fruit Var. J.,* 36 (4): 108-110.

Hjalmarsson, Inger and Ortiz, Rodomiro (2001). Lingonberry: Botany and Horticulture. *Horticultural Reviews*, 27: 79-123.

Hocking, Drake (1993). *Trees for Drylands*. Oxford and IBH Publishing Co. Pvt. Ltd., New Delhi, Bombay and Calcutta, 370 pp.

Hodgson, R. W. (1961). Taxonomy and nomenclature in citrus. *International Org. Citrus Virol. Proc.* , 2: 1-7.

Hodgson, R. W. (1965). Taxonomy and nomenclature in citrus fruits. In: Advances in Agricultural Science and their applications. The Madras Agr. J., Coimbatore: 317-331.

Hooker, J. D. (1875). *The Flora of British India (Rutaceae)*. L. Reece and Co., Henrietta Street, Convent Garden, London. pp. 632-641.

Horiuchi, S., Yoshida, M., Kariya, H., Nakamura, T., Hasebe, H., Suzaki, T. and Sakitani, T. (1996). Nihonnoume Sekainoume. Yokendo.

Horovitz, S. and Jiménez, H. (1967). Cruzamientos interespecíficos e intergenéricos en Caricaceas y sus implicaciones fitotécnicas. *Agronomía Tropical (Maracay)*, 17(3): 323-343.

Horry, J. P. and Jay, M. (1990). A evolutionary background of bananas as deduced from flavanoid diversification. In: R. L. Jarret (ed.) *Identification of genetic diversity in the genus Musa*. Proceedings of an International Workshop, Las Banos (1989), the Philippines. INIBAP, Montpellier, France, pp. 41-55.

Horry, J. P. and Jay, M. (1988). Distribution of anthocyanins in wild and cultivated banana varieties. *Phytochemistry*, 277: 2667-2672.

Horry, J. P., Oritz, R., Arnaud, E., Crouch, J. H. Ferris, R. S. B., Jones, D. R., Mateo, N. Picq, C. and Vuylsteke, D. (1997). Banana and Plantain. In: *Biodiversity in Trust. Conservation and use of plant genetic resources in CGIAR centres* (Fuccillo, D., Sears L. and Stapleton, P. eds.). Cambridge University Press. pp. 67-81.

Howell, E. C., Newburry, H. J., Swennen, R., Withers, L. A. and Ford Lloyd, B. V. (1994). The use of RAPD for identifying and classifying *Musa* germplasm. *Genome*, 37:328-332.

Huckins, C. A. (1972). *A revision of the sections of the genus Malus* Miller. Ph. D. dissertation, University of Ithaca, New York.

Hume, Harold H. (1957). *Citrus Fruits*. The Macmillan Co., New York. 444 p.

Hummer, K. E. and Janick, J. (2009). 1. Rosaceae: Taxonomy, Economic Importance, Genomics, In: K. M. Folta, S. E. Gardiner (eds.), *Genetics and Genomics of Rosaceae*, *Plant Genetics and Genomics: Crops and Models*. Springer's and Verlag, Germany.

Hunger, F. W. T. (1920). *Cocos nucifera*, Handbook voor de kennis van den Cocos palm in Nederlandesh—Indie, Zinjne Geschiedenis, Jeschrijving, cultuur en production. Groesbeck & Nijhoff, Amsterdam.

Hutchinson, J. (1948). *British Flowering Plants*. London.

Hyam, R., Pankhurst, R. (1995). *Plants and Their Names: A Concise Dictionary*. Oxford University Press, Oxford.

IBPGR (1986a). *Cashew Descriptor*. International Board of Plant Genetic Resources, Rome, Italy. 33p.

IBPGR (1986b). *Genetic Resources of tropical and subtropical fruits and nuts (excluding Musa)*. International Board for Plant Genetic Resources, Rome, Italy, 162 pp.

IBPGR (1988). *Descriptors for Papaya*. International Board of Plant Genetic Resources, Rome, Italy. 34 p.

Ibrahim, Omar (2000). Business proposal for commercial cultivation of Pummelo (Citrus maxima). Horticulture Research Centre, MARDI.

Iinuma, F. Y. (1874). *Musa basjoo*. Sintei Somoku Dzusetsu (Illustration and description of plants), 2nd ed. 3 : pl. 1.

Ingram, C. (1948). *Ornamental cherries*. Charles Scribner's Sons, New York, 259, pp.

INIBAP (2000). *Bananas. International Network for the Improvement of Banana and Plantain*, International Plant Genetic Resources Institute, Rome, Italy.

IPGRI (1997). *Descriptors for pistachio* (*Pistacia vera* L.). International Plant Genetic Resources Institute, Rome, Italy.

IPGRI (2003). *Descriptors for mangosteen* (*Garcinia mangostana*). International Plant Genetic Resources Institute, Rome, Italy.

IPGRI (2006). *Descriptors for mango* (*Mangifera indica* L.). International Plant Genetic Resources Institute, Rome, Italy.

IPGRI (2006). Food security- Plant Genetic Resources for Food Security– factsheets. IPGRI, Rome, p. 1.

IPGRI and CIHEAM (2003). *Descriptors for fig*. International Plant Genetic Resources Institute, Rome, Italy, and International Centre for Advanced Mediterrannean Agronomic Studies, Paris, France.

IPGRI, INIBAP, CIRAD (1996). *Descriptors for banana* (*Musa* spp.). International Plant Genetic Resources Institute, Rome, Italy; INIBAP, Montpellier, France; Centre de Cooperation Internationale en recherché agronomique pour le development, Montpellier, France. 55 pp.

IPGRI, UPOV, OIV (1997). *Descriptors for grapevine* (*Vitis* spp). International Plant Genetic Resources Institute, Rome, Italy, 62 p. International Union for the Protection of New Varieties of Plants, Geneva, Switzerland. Office International de la Vigne et du Vin, Paris, France.

IPGRI. (2000). *Annual report 1999*. International Plant Resources Institute, Rome, Italy, 36 p.

Isawumi, M. A. (1993). The common edible fruits of Nigeria- Part II. *The Nigerian Field*, 58 (1, 2): 157-171.

IUCN (2003). *The IUCN Red List of Threatened Species*. The World Conservation Union. IUCN Species Survival Commission. IUCN, Gland, Switzerland and Cambridge, UK.

Jackson, David (1986). Hazelnut. In: *Temperate and Subtropical Fruit Production*. (eds.) David Jackson. Butterworths Horticultural Books, New Zealand, pp. 274-275.

Jackson, R. S. (2014). *Wine Science- Principles and Applications*. Elsevier Press, Inc., Amsterdam. p 25.

Jacob H. E. (1950). *Grape growing in California,* Circ. 166, Calif. Agric. Ext. Serv. Coll. Agric. Univ. California, Berkeley, California

Jacob, K. C. (1952). *Madras Bananas, a monograph*. Government Press, Madras, pp. 228.

Janick, J. and Paul, R. E. (eds.) (2008). *The Encyclopedia of Fruits and Nuts*. CAB International, UK.

Janick, J., Cummins, J. N., Brown, S. K. and Hemmat, M. (1996). Apples. In: *Fruit Breeding, Vol. 1. Tree and Tropical Fruits*, ed. J. Janick and J. N. Moore, pp. 1-77. John Wiley and Sons, New York.

Jarret, R. L. (1990). Molecular methods for detecting genetic diversity in *Musa*. In: Jarret, R. L. (ed.) *Identification of Genetic Diversity in the Genus Musa*, pp. 56-66. INIBAP, Montpellier.

Jarret, R. L. and Gawel, N. (1995). Molecular markers, genetic diversity and systematics in *Musa*. In: Gowen, S. R. (ed.) *Bananas and Plantains*, pp. 66-83. Chapman and Hall, London.

Jarret, R. L. and Litz, R. E. (1986a). Isozymes as genetic markers in banana and plantain. *Euphytica,* 35:539-549.

Jarret, R. L. and Litz, R. E. (1986b). Enzyme polymorphism in *Musa acuminata* Colla. *J. Heredity,* 77:183-188.

Jarret, R. L., Gawel, N., Whittemore, A. P. and Sharrock, S. (1992). RFLP based phylogeny of *Musa* species in Papua New Guinea. *Theor. Appl. Genet,* 84: 579-584.

Jarrett, F. M. (1959). Studies in *Artocarpus* and allied genera. III A Revision of *Artocarpus* subgenus *Artocarpus*. *Journal of Arnold Arboretum*, 40: 329-334.

Jiménez, Y., Romero, J. and Scheldeman, X. (1998). Colección, caracterización y descripción de *Carica × heilbornii* nm. *pentagona* B.; *Carica pubescens* (A.DC.) Solms-Laub. y *Carica stipulata B.*, en la provincia de Loja. *Revista de Difusión Técnica y Científica de la Facultad de Ciencias Agrícolas* (Universidad Nacional de Loja), 29(1-2): 43-54.

Jintanawong, S. H., Dunagpikul Hiranpradit, P. and Polprasid, P. (1992). Group characterization of Thai mango, *Mangifera indica* Linn. *Acta Horticulturae*, 321: 245-261.

Johnston, M. C. (1972). Rhamnaceae. In: *Flora of Tropical East Africa,* eds. Milne-Redhead, E. and Polhill, R.M. Crown Agents for Overseas Governments and Administration, London, pp. 23-31.

Jordán, M. and Botti, C. (1992). Tropical and subtropical small fruits. In: Hammerschlag, F. and Litz, R. (eds.) *Biotechnology of perennial fruit crops*. Biotechnology of Agriculture Series No. 8, CAB International, Wallingford, U.K., pp. 513-531.

Judd, W. S., Campbell, C. S. Kellogg, E. A., and Stevens, P. F. (1999). *Plant Systematics: A phylogenetic approach*. Sinauer Associates, Inc., Sunderland, MA. 290–306.

Jussieu, Antoine Laurent de. (1774). Exposition d'un nouvel ordre de plantes adopté dans les démonstrations du Jardin royal In: *Histoire de l'Académie Royale des Sciences.* Année 1774. Paris 1777, p. 175–197

Kadereit, J. W. ed. (2004). The Families and Genera of Vascular Plants VII. Flowering Plants- Dicotyledons- Lamiales (except Acanthaceae) including Avicenniaceae). Springer Verlag, Berlin, Heidelberg, Germany.

Karale, A. R. (1998). Status report on Improvement of Tamarind by selection. *Research Workers Group Meeting of AICRP on Arid Zone Fruits*, CCS HAU, Hisar.

Kay-Ming, Chau (1990). Rambutan, In: *Fruits: Tropical and Subtropical.* (eds.) T. K. Bose and S. K. Mitra. Naya Prokash, 206 Bidhan Sarani, Calcutta, pp. 732-739.

Keller, M. (2010). *The Science of grapevines: Anatomy and physiology*. Elsevier and Academic Press, U.K. pp. 4-19.

Kessler, P. J. A. (1993). Annonaceae. In: Kubitzki, K., Rohwer, J. G. and Brittrich, V. (eds.). *The families and genera of vascular plants*. Vol. II. Flowering Dicotyledons. Springer-Verlag, Berlin, Germany, pp. 93-104.

Knight, R. J. Jr. (1980). Origin and World Importance of tropical and subtropical fruit crops. In: *Tropical and Subtropical Fruits: Composition, Properties and Uses.* (eds.) S. Nagy and P. W. Shaw. AVI Publishing, Westport, Connecticut, USA, pp. 1-120.

Knight, R. L. (1969). *Abstract Bibliography of fruit breeding and genetics to 1965*: *Prunus*. Commonwealth Agricultural Bureau, Buckinghamshire, UK. Tech. Communication 31, 649 pp.

Kokwaro, J. O. (1986). *Flora of Tropical West Africa Ed. 2, vols. 1, 2*. Crown Agents for Overseas Governments and Administrations, London.

Kostermans, A. J. G. H. and Bompard, J. M. (1993). *The Mangoes- Their Botany, Nomenclature, Horticulture and Utilization.* Academic Press, New York, pp. 167-171.

Kostina, K. F. (1969). The use of varietal resources of apricots for breeding. *Trud. Nikit. Bot. Sad.* 40: 45-63.

Kramer, K. U. and Green, P. S., eds. (1990). *The Families and Genera of Vascular Plants* I. *Pteridophytes and Gymnosperms*. Narosa Publishing. House and Springer-Verlag, Berlin, Germany.

Kubitzki, H., Bayer, C. and Stevens, P. F., eds. (2007). *The Families and Genera of Vascular Plants IX. Flowering plants- Eudicots, Berberidales, Buxales Crossosomatales, Fabales, Geraniales, Gunnerales, Myrtales, Proteales, Saxifragales, Vitales, Zygophyllales, Clusiaceae Alliance, Passifloraceae Alliance, Dilleniaceae, Huaceae, Picramniaceae, Sabiaceae.* Springer Verlag, Berlin, Heidelberg, Germany.

Kubitzki, K. and Bayer, C., eds. (2003). *The Families and Genera of Vascular plants V. Flowering Plants- Dicotyledons- Malvales, Capparales and non-betalain Caryophyllales.* Springer Verlag, Berlin, Heidelberg, Germany.

Kubitzki, K., ed. (2004). *The Families and Genera of Vascular plants VI. Flowering Plants- Dicotyledons Celastrales, Oxalidales, Rosales, Cornales, Ericales.* Springer Verlag, Berlin, Heidelberg, Germany.

Kubitzki, K., ed. (2011). *The Families and Genera of Vascular plants X. Flowering Plants. Eudicots- Sapindales, Cucurbitales, Myrtaceae.* Springer Verlag, Berlin, Heidelberg, Germany.

Kubitzki, K., Huber, H., Rudall, P. J., Stevens, P. S. and Stutzel, T., eds. (1998). *The Families and Genera of Vascular plants III. Flowering Plants- Monocotyledons Lilianae (except Orchidaceae).* Springer Verlag, Berlin, Heidelberg, Germany.

Kubitzki, K., Huber, H., Rudall, P. J., Stevens, P. S. and Stutzel, T., eds. (1998). *The Families and Genera of Vascular Plants IV. Flowering plants- Monocotyledons Alismatanae and Commelinanae (except Gramineae).* Springer Verlag, Berlin, Heidelberg, Germany.

Kubitzki, K., Rohwer, J. G. and Bittrich, V., eds. (1993). *The Families and Genera of Vascular plants II. Flowering Plants- Dicotyledons- Magnolid, Hamamelid and Caryophyllid Families.* Springer-Verlag, Berlin, Germany.

Kurz, S. (1877). The banana: Pomological contribution. *J. Agric. Soc.*, Ind., N.S.V: 112—168.

Kyndt, T., Romeijn-peeters, E., Van Droogenbroeck, B., Romero-Motochi, J. S., Gheysen, G. and Goetghebeur, P. (2005). Species relationships in the genus *Vasconcellea* (Caricaceae) based on molecular and morphological evidence. *American Journal of Botany*, 92(6): 1033-1044.

Ladipo, D. O., Fondoun, J. M. and Ganga, N. (1998). Domestication of the bush mango (*Irvingia* sp.): Some exploitable intraspecific variations in West and Central Africa. In: *Domestication and Commercialization of non-timber forest products in agroforestry systems. Proc. International Conference held in Kenya*, 19-23 February, 1996. (eds.) R. R. B. Leaky, A. B. Temu, M. Melnyk and P. Vantomme. FAO, Rome, Italy, pp. 193-205.

Lamarck, J. B. P. (1778). *Flore Francaise Imprimerie Ruyale*, Paris, 3 Vols.

Landrum, L. R. and Donoso, Z. Claudio (1990). *Ugni molinae* (Myrtaceae), a potential fruit crop for regions of mediterranean, maritime and subtropical climates. *Economic Botany*, 44: 536-539.

Lane, I. E. (1955). Genera and generic relationships in Musaceae. *Mitt. Bot. Staatssamm. Munchen*, 13: 114-31.

Langdon, R. (1993). The bananas as a key to early American and Polynesian history. *J. Pacific Hist.,* 28: 15-35.

Layne, R. E. C., Bailey, C. H. and Hough L. F. (1996). Apricots. In: Janick J. and Moore J. N. (eds.), *Fruit breeding*, Vol 1. p 79-111. John Wiley & Sons, Inc., New York Wiley, New York

Le Bellec, F. and Vaillant, F. (2011). Pitaya, Pitahaya (Hylocereus spp.). In: *Postharvest Biology and Technology of Tropical and Subtropical Fruits,* vol. IV. Woodhead Publishing Limited, Cambridge, UK, 247-271.

Le Bellec, F., Vaillant, F. and Imbert, E. (2006). Pitahaya (*Hylocereus* spp.): a new fruit crop, a market with a future. *Fruits,* 61: 237–250.

Le Maout, E. (1877). *A general system of Botany, Descriptive and Analytical. Part-II. Descriptions and Illustrations of the Orders.* Longmans and Green, London.

Leaky, R. R. B. and Newton, A. C., eds. (1994). *Tropical trees: The potential for domestication and the rebuilding of forest resources.* Proceedings of a conference organized by the Edinburgh Centre for Tropical Forests, Heriot-Watt University, Edinburgh, 23-28 August, 1992.

Leal, F. J. and Soule, J. (1977). Maipure, a new spineless group of pineapple cvs. *HortScience*, 12: 301-305.

Leal, F., Coppens, d' Eeckenbrugge, G. and Holst, B.K. (1998). Taxonomy of the genera *Ananas* and *Pseudananas,* a historical Review. *Selbyana.,* 19: 227-235.

Ledbetter, C. A. (2008). Apricots. In: *Temperate Fruit Crop Breeding*: *Germplasm to genomics*, Hancock, J. F. (ed.), Chapter 2, pp. 39-82. Springer Science-Business Media B.V.

Ledin, R. B. (1957). A note on the fruiting of the Mauritius variety of Lychee. *Yearbook and Proceedings of the Florida Lychee Growers Association*, 4: 57.

Lee, S. and Wen, J. (2001). A phylogenetic analysis of *Prunus* and the Amygdaloideae (Rosaceae) using ITS sequences of nuclear ribosomal DNA. *Amer. J. Bot.*, 88: 150–160.

Leenhouts, P. W. (1978). Systematic notes on the Sapindaceae- Nephelieae. *Blumea.*, 24: 395-403.

León, J. (1987). *Botánica de los Cultivos Tropicales* [Spanish]. Instituto Interamericano de Cooperacion para la Agricultura, San José, Costa Rica. 445 pp.

Levadoux, L. (1956). *Les populations Sauvages et cultivees de Vitis vinifera* L. Annales de I. Institut National de la Recherche Agonomique. Serie, B., Annales de 1, Amelioration de Plantes: *Genetique, Selection, Ecologie, Technique Cultiurales,* 6: 59-118.

Li, X., Xie, R., Lu, Z. and Zhou, Z. (2010). The origin of cultivated citrus as inferred from internal transcribed spaces and chloroplast DNA sequence and amplified fragment length polymorphism fingerprints. *J. Am. Soc. Hort. Sci.*, 135: 340-350.

Lingdi, L. and Bartholomew, B. (2003). Armeniaca. In: Wu, C. Y., Raven, P. H. (eds.) *Flora of China*, vol 9 (*Pittosporaceae* through *Connaraceae*). Science Press, Beijing, and Missouri Botanical Garden Press, St. Louis. pp 396-401.

Linnaeus, C. (1751). *Philosophia Botanica*. Stockholm.

Linnaeus, C. (1753). *Species Plantarum*, Stockholm, Sweden, 2 vol., pp. 1043.

Linnaeus, C. (1754). Genera plantarum eorumque characteres naturalessecundum numerum, figuram, situm, et proportionem omnium fructificationis partium (editio quinta ab auctore reformata et aucta). Stockholm: Salvius.

Linnaeus, C. (1754). *Genera Plantarum*, 5th edition, pp. 466.

Liu, M. J. and Cheng, C. Y. (1995). A taxonomic study of the genus *Ziziphus. Acta Horticulturae,* 390:161-165.

Loomis, H. F. (1955). Bengal, a promising large clustered Indian Lychee. *Proceedings of the Florida Lychee Growers Association,* 2: 9-12.

Loudon, J. C. (1838). The genus *Armeniaca, Arboretum Et Fruticetum Britannicum*. Vol. II. London: Longman, Orme, Brown, Green and Longmans. pp. 681–684.

Lushington, A. W. (1910). The genus *Citrus*. *Indian Forester*, 36: 323-353.

Luther, H. E. and Sieff, E. (1998). *An Alphabetical list of Bromeliad.* Bionomials, 6th edn. The Bromelial Society, Newberg, Oregoon, 138 pp.

Mabberley, D. J. (2008). *Mabberley's plant-book: A portable dictionary of plants*, 3rd ed. Cambridge University Press, Avon, UK.

Macia, M. J. and Barfod, A. S. (2000). Economic Botany of *Spondias purpurea* (Anacardiaceae) in Ecuador. *Economic Botany*, 54 (4): 449-458.

Maghuly, F., Borroto Fernandez, E., Ruthner, S., Pedryc, A. and Laimer, M. (2005). Microsatellite variability in apricot (*Prunus armeniaca*) reflects their geographical origin and breeding history. *Tree Genetics & Genomes,* 1, 151-165.

Manning, W. E. (1960). The genus *Juglans* in South America and West Indies. *Brittonia,* 12: 1-26.

Manning, W. E. (1978). The classification within the Juglandaceae. *Annals of the Missouri Botanical Garden*, 65: 1058-1087.

Manshardt, R. M. and Moore, P. H. (2003). Natural history of papaya and the Caricaceae, *Abstract # 792 in "Plant Biology 2003". Proceedings of the American Society of Plant Biologists* held 25-30 July 2003 at Honolulu, USA.

Manshardt, R. M. and Wenslaff, T. F. (1989a). Zygotic polyembryony in interspecific hybrids of *Carica papaya* and *C. cauliflora. Journal of the American Society for Horticultural Science*, 114(4): 684- 689.

Manshardt, R. M. and Wenslaff, T. F. (1989b). Interspecific hybridization of papaya with other *Carica* species. *Journal of the American Society for Horticultural Science*, 114(4): 689-694.

Maries, C. (1902). Indian mangoes. *Journal Royal Horticulture Society*, 26: 755-770.

Martin, F. W., Campbell, C. W. and Ruberte, R. M. (1987). *Perennial Edible Fruits of the Tropics– An Inventory*. United States Department of Agriculture. Agri. Research Services, Agriculture Handbook No. 642. Washington, DC., USA, 247 pp.

Maydell, H. J. von. (1986). *Trees and shrubs of the Sahel- their characteristics and uses.* Deütsche Gesellschaft fürtechnische Zusammenirbeit GTZ No. 196, Eschborn, Germany, 525 pp.

Mbida, C. A., Van Neer, W. Doutrelepont, T. H. and Vrydaghs, L. (2000). Evidence of banana cultivation and animal husbandry during the first millennium BC in the forest of South Cameroon. *J. Arch. Sci.*, 27:151-162.

McClure, B. A., Haring, V., Ebert, P. R., Anderson, M. A., Simpson, R. J., Sakiyama, F. and Clarke, A. E. (1989). Style self-incompatibility gene products of *Nicotiana alata* are ribonucleases. *Nature*, 342, 955-957.

McGranahan G, Leslie, C. (1991). Walnuts (*Juglans*). In: Moore J. N., Ballington J. R. Jr. (eds.) *Genetic resources of temperate fruit and nut crops*. International Society for Horticultural Science, Wageningen, pp. 907–951.

McGuire, C. M. (1999). *Passiflora incarnata* (Passifloraceae): A new fruit crop. *Economic Botany*, 53 (2): 161-176.

McNeil, J. (2004). Nomenclature of cultivated plants: a Historical Botanical Stand point. *Proc. XXVI IHC-IV. International Symposium Taxonomy of Cultivated Plants.* (ed.) C. G. Davidson and P. Trehane, *Acta Hort.* 634: ISHS.

McNeill, J., Barie, F. R., Burdet, H. M., Demoulin, V., Hawkswork, D. L., Marhold, K., Nicolson, D. H., Prado, J., Silva, P. C., Skog, J. E., Wiersema, J. H., and Turland, N. J., eds. (2006). *International code of Botanical Nomenclature (Vienna Code): Adopted by the Seventeenth International Botanical Congress* Vienna, Austria, July 2005. Regnum Vegetabile 146. Ruggell. Liechtenstein: Gantner:(i)-xviii, 1-568.

Mega, K., Tomita, E., Kitamura, S., Saito, S. and Mizukami, S. (1988). Ume. In: Aoba *et al.* (eds.) *The Grand Dictionary of Horticulture*, Shogakukan, Tokyo, pp. 289-300.

Mehlenbacher, S. A. (1995). Progress in breeding New Hazelnut cultivars in Oregon. *FAO-IHEAM Nucis Newsletter*, Number 3, January, 1995, pp. 8-9.

Mehlenbacher, S. A. Cociu, V. and Hough, L. F. (1991). Apricots (*Prunus*). In: Moore J. N., Ballington J. R. (eds.) *Genetic Resources of Temperate Fruit and Nut Crops. Acta Hort.*, 290, 63-109.

Mehlenbacher, S. A., Cociu, V. and Hough, L. F. (1990). Apricot. In: Moore, J. N. and Ballington, J. R. (eds) *Genetic Resources of Temperate Fruit and Nut Crops. Acta Horticulturae,* 290: 65-107. International Society for Horticultural Sciences, Wagenningen. The Netherlands,

Mendiola, N. B. (1941). Introduction and trial culture of Java Nephelium in the U. P. College of Agriculture. *Philippines Agriculture*, 30: 168-196.

Menninger, E. A. (1977). *Edible nuts of the world.* Horticultural Books Inc., Stuart, Florida.

Menzel, C. M. (1991). Litchi. In: *Plant Resources of Southeast Asia*. 2. *Edible Fruits and Nuts* (Eds. E. W. M. Verheji and R. E. Coronel). Pudoc. Wageningen, The Netherlands, 191-195 pp.

Menzel, C. M., Watson, B. J. and Simpson, D. R. (1990). Longan. In: *Fruits: Tropical and Subtropical*. (eds.) T. K. Bose and S. K. Mitra. Naya Prokash, 206, Bidhan, Sarani, Calcutta, pp. 522-546.

Mez, C. (1892). Bromeliaceae; *Ananas. Martius, Flora Brasiliensis*: vol. 3(3). Reprinted 1965 Verlag von J. Cramer, Weinheim, Codicote (Hertfordshire), Wheldon and Wesley, New York, pp 288-294.

MEZ-ZIEGENSPECK. (1926). Der Königsberger serodiagiiostisehe Stammbaum. *Bot. Arch.,* 13.

Miller, P. (1768). *Garden Dictionary*. 8th edn. Henrey, Staflen and Cowan, London.

Milne, D. L. (1999). Lychee production and research in South Africa. In: Menzel, C. M., Waite, G. K. and Bronson, R. (eds.). *Proceedings of the Fifth National Lychee Confe-rence. Australian Lychee Growers Association*, Twin Waters, Queensland. pp. 25-31.

Mitra, S. K. and Maity, C. S. (2002). A summary of the genetic resources in Jackfruit (*Artocarpus heterphyllus* Lam.) in West Bengal, India. *Acta Horticulturae*, 575p.

Mizrahi, Y., Nerd, A. and Nobel, P. S. (1997). Cacti as Crops. *Horticultural Reviews*, 18: 291-321.

Mohd. Shamsudin, O., Ahmad Redzuan, P. A., Zainal Abidin, M. and Zaharah, T. (2000). Use of wild durian, *Durio lowianus* as rootstock. *Prosiding Seminar Durian*, (Zainal Abidin, M., Mohd. Shamsudin, O., Ahmad Tarmizi, S., Zabedah, M. and Salma, I., eds.), 1-3 August, Ipoh, p. 56-64. Serdang: MARDI.

Moore, H. E. Jr. (1973). The major groups of palms and their distribution. *Gentes Herbarium*, 11: 27-141.

Moore, J. N. and Ballington, J. R. Jr. eds. (1990). *Genetic Resources of Temperate Fruit and Nut Crops. Acta Horticulturae*, 290: 235-287. International Society for Horticultural Sciences (ISHS), Leuven, Belgium.

Mora-Urpi, J. (1984). El pejibaye (*Bactris gasipaes* H. B. K.): Origen, biologia floral y manejo agronomico. In: *Palmeras Poco Utilizadas de America Tropical.* Food and Agriculture Organization/Centro Agronomica Tropical de investigacion y Ensenaza. CATIE, Turrialba, Costa Rica, pp. 118-160.

Mora-Urpi, J. (1996). *Peach Palm (Bactris gasipaes* H. B. K.). School of Biology, University of Costa Rica.

Mora-Urpi, Jorge, Weber, J. C. and Clement, C. R. (1997). *Peach palm. Bactris gasipaes* Kunth. *Promoting the conservation and use of underutilized and neglected crops* 20. Institute of Plant Genetics and Crop Plant Research, Gatersleben/International Plant Genetic Resources Institute, Rome, Italy, 83 pp.

Mori, S. A. and Prance, G. T. (1990). *Lecythidaceae Part 2, Flora neotropica. Monograph 21*, 2. New York Botanical Garden, New York.

Morley-*Bunker,* Michael (1986a). Medlar. In: *Temperate and Subtropical Fruit Production*. eds. David Jackson. Butterworths Horticultural Books, New Zealand, pp. 265.

Morley-*Bunker,* Michael (1986b). Quince. In: *Temperate and Subtropical Fruit Production*. eds. David Jackson. Butterworths Horticultural Books, New Zealand, pp. 267.

Morton, J. F. (1987). *Fruits of Warm Climates*. Creative Resource Systems, Inc. Winterville, N. C., USA, 503 p.

Morton, J. F. and Voss, G. L. (1987). The argan tree (*Argania sideroxylon*, Sapotaceae), A desert source of edible oil. *Economic Botany*, 41 (2): 221-233.

Mueller, F. J. H. von. (1864). *Musa banksii*. Fragmenla Phytographiae Australiae, 4:132.

Mukherjee, S. K. (1948). The varieties of mango (*M. indica* L.) and their classification. *Bulletin of the Botanical Society of Bengal*, 2: 101-133.

Mukherjee, S. K. (1949). A monograph on the genus *Mangifera*. *Lloydia,* 12: 73.

Naik, K. C. and Gangolly, S. R. (1950). *A monograph on classification and nomenculture of South Indian mangoes*. Superintendent, Govt. Press, Madras, 311p.

Nakasone, H. Y. and Paull, Robert, E. (1998). *Tropical Fruits.* CAB International, Wallingford, Oxford, UK, 445 pp.

Nakasone, H. Y., Paull, R. A. and Ito, P. J. (1967). Evaluation of Introduced cultivars of guava. University of Hawaii. *Hawaii Farm Science,* 16: 4-6.

Nakasone, H. Y., Yamane, G. M. and Miyashita, R. K. (1968). *Selection, evaluation and naming of acerola* (*Malphigia glabra* L.) *cultivars*. Circular 65, Hawaii Agricultural Experiment station, University of Hawaii, Honolulu.

Nambiar, M. C., Bhaskara Rao, E. V. V. and Thankamma Pillai, P. K. (1990). Cashew. In: *Fruits : Tropical and Subtropical.* (eds.) Bose, T. K. and Mitra, S. K. Naya Prokash, 206 Bidhan Sarani, Calcutta, pp. 386-419.

Nasution, R. E. (1991). A taxonomic study of the species *Musa acuminata* Colla with its intraspecific taxa in Indonesia. *Memoirs of Tokyo University of Agriculture, Vol. 23*: 1-122.

Negrul, A. M. (1938). Evolution of cultivated forms of grapes. Comptes Rendus (Doklady) de l Academie des Sciences de 1 URSS, 18: 585-588.

Negrul, A. M. (1946). Origin and evolution of the cultivated grape, In: Baranov, A. K. Y., Lazarevski, M. A., Palibin, T. V. and Prosmoserdov, N. N., eds. *The Ampelography of the USSR*, Moscow: Pischepromizdat, 1: 159-216.

Nerd, A., Irijimovich, V. and Mizrahi, Y. (1998). Phenology, breeding system and fruit development of Argan (*Argania spinosa*, Sapotaceae) cultivated in Israel. *Economic Botany*, 52(2): 161-167.

Ngatchou, J. N. and Kengue, J. (1989). Review of the African plum tree (*Dacroydes edulis*). In: *New crops for Food and Industry.* (eds.) G. E. Wickens, N. Haq and P. Day. Chapman and Hall, London and New York, pp. 265-271.

Nicolosi, *E.* and Deng, Z. N., Gentile, A., La Mafia, S., Continella, G. and Tribulato, E. (2000). *Citrus* phylogeny and genetic origin of important species as investigated by molecular markers. *Theo. Appl. Genet.*, 100: 1156-1166.

Noltie, H. J. (1994). *Musa griersonii. Edinburg J. Bot.*, 5:171.

NRC (1989). Lost crops of the Incas: little-known plants of the Andes with promise for worldwide cultivation. National Academy Press, Washington DC, USA, 415 p.

Nwakanma, D. C., Pillay, M., Okoli, B. E., Tenkouana, A. (2003). Sectional relationships in the genus *Musa* L. inferred from the PCR-RFLP of organelle DNA sequences. *Theoretical & Applied Genetics,* 107: 850-856.

Oke, O. L. (1995). *Leaf protein research in Nigeria Ibadan*, University of Idadan Press, Nigeria.

Okolia, B. E. and Mgbeogu, C. M. (1983). Fluted pumpkin, *Telfairia occidentalis,* West African Vegetable Crop. *Economic Botany,* 37: 145-149.

Orban, I. and Bouharmont, J. (1998). Megagametophyte development of *Nymphaea nouchali* Burm. f. (Nymphaeaceae). *Bot. J. Linn. Soc.,* 126: 339-348.

Ortiz, R. (2008). *Musaceae*, *Musa* sp. (banana and plantain). In: Janick, J. and Paull, R. E. (eds.) *The Encylopedia of Fruits and Nuts*, CAB International, Wallingford , U.K, pp.

Ottman, Yvonne (1987). Rediscovering the Realm of Fruiting Mulberry varieties. *Fruit Varieties Journal*, 41 (1) : 4-7.

Page, P. E., Compiler. (1984). *Tropical tree fruits for Australia.* Queensland Department of Primary Industries Information series Q 183018, Brisbane, Queensland, Australia, pp. 20-217.

Pandey, G. and Joshi, B. D. (1997). *Kiwi– A future fruit for the Indian Hills*. National Bureau of Plant Genetic Resources, Regional Station, Phagli, Shimla, 119 pp.

Pandey, R. M. and Misra, K. K. (1992). *Phal Utpadan.* Publication Directorate, G. B. Pant Krishi and Technology University, Pantnagar, pp. 428-33.

Pandey, S. N. and Misra, S. P. (2008). *Taxonomy of Angiosperms*. Ane Books India, New Delhi.

Pang, X. M., Hu, C. G. and Deng, X. X. (2007). Phylogenetic relationships within *Citrus* and its related genera as inferred from AFLP markers. *Genetic Resources and Crop Evolution,* 54: 429–436.

Pareek, O. P. (2001). *Ber*. International Centre for Underutilized crops. Univ. of Southampton, Southampton, UK, 290 pp.

Pareek, O. P. and Vishal Nath (1996). *Co-ordinated fruit research in Indian arid Zone– A two decades profile (1976-1995)* of research conducted under All India Coordinated Research Project on Arid Zone Fruits of ICAR. National Research Centre for Arid Horticulture, Bikaner, India, 104 pp.

Parmar, C. and Kaushal, M. K. (1982). *Wild fruits of the sub-Himalayan region*. Kalyani Publishers, New Delhi, pp. 49-53.

Pascual, L., Perfectti, F., Gutierrez, M. and Vargas, A. M. (1993). Characterizing isozymes of Spanish cherimoya cultivars. *HortScience*, 28(8): 845-847.

Pathak, R. K. (2003). *Status Report on Genetic Resources of Indian gooseberry-Aonla (Emblica officinalis* Gaertn.) *in South and Southeast Asia*. Bhag Mal, Ramanatha Rao, V. and Arora, R. K. (eds.), IPGRI South Asia Office, New Delhi, India, pp. 99.

Paull, R. E. and Duarte, O. (2011). *Tropical Fruits*, Vol. I. CABI, Wallingford, UK.

Paull, R. E. and Duarte, O. (2012). *Tropical Fruits*, Vol. II. 384 pp. CABI, Wallingford, UK.

Paulos, D. (1975). Tamarind. *Agril. Agro. Ind. J.*, 8: 35-37.

Payne, J. A. and Krewer, G. W. (1990). Mayhaw– A new fruit crop for the South. In: *Advances in New Crops. Proceedings of the first National Symposium in New Crops: Research, Development, Economics*, October 23-28, 1988. (eds.) Jules Janick and James E. Simon. Timber Press, Portland, Oregon, USA, pp. 317-321.

Pedryc, A., Ruthner, S., Herman, R., Krska, B., Hegedus, A. and Halasz, J. (2009). Genetic diversity of apricot revealed by a set of SSR a markers from linkage group G1. *Sci. Hort.*, 121: 19-26.

Pereira, F. M. (1984). Rice e Paluma: novas cultivation de goiabeira. In: *Congresso Brasileiro de Fruiticultura*, 7, Florianopolis, Anais Florianopolis 2, 524-528.

Pereira-Lorenjo and Fernandez-Lopez, J. (1995). Chestnut in Spain: An old culture with future. *FAO-IHEAM Nucis Newsletter*, Number 4, December 1995, pp. 12-15.

Perez Vela, J. M. (1985). *El pijuayo en el Peru: manejo Y usos*. Yurimaguas, Peru.

Peters, C. R. (1987). *Ricinodendron rautanenii* (Euphorbiaceae): Zambezian Wild Food Plant for All Seasons. *Economic Botany*, 41:494-502.

Pickersgill, B. (1994). From descriptors to DNA. New tools and new tasks in the evaluation of genetic resources. In: *Evaluation and exploitation of genetic resources pre-breeding*. Balfourier, F. and Perretant, M. R. (eds.) Clermont-Ferrand, France.

Pillay, M., Tenkouano, A., Hartman, J. (2002). Bananas and Plantains: Future Challenges in *Musa* Breeding. Chapter 8. In: *Crop Improvement, Challenges in the Twenty-First Century*. Food Products Press, New York. pp 223-252.

Pillay, M., Tenkouano, A., Ude, G. and Ortiz, R. (2004). Molecular characterization of genomes in *Musa* and its applications. Chapter 23, In: Jain, S. Mohan and Swennen, R. (eds.) *Banana Improvement*: *Cellular Molecular Biology and induced Mutations*, Science Publishers, Inc., Enfield NH, USA, pp. 271-286.

Pinto, A. C. Q., Cordeiro, M. C. R., de Andrade, S. R. M., Ferreira, F. R., Filgueiras, H. A. C., Alves, R. E., *et al.* (2005). *Annona species*. International Centre for Underutilized Crops, University of Southampton, Southampton. pp. 263.

Planchon J. E. (1887). Monographie des Ampélidées vrais. In: A. F. P. P. De Candolle, and C. De Candolle [eds.] *Monographiae Phanaerogamarum,* 5: 305-654 G Masson Paris, France.

Ploetz, R. C., Kepler, A. K., Daniells, J., Nelson, S. C. (2007). Banana and plantain- an overview with emphasis on Pacific island cultivars. In: C. R. Elevitch, ed. *Species Profiles for Pacific Island Agroforestry*. Permanent Agricultural Resources, Holualoa, Hawaii.

Popenoe, W. (1920). *Manual of Tropical and Subtropical fruits*. Macmillan Co., New York, pp. 100-126.

Popenoe, W. (1970). The cherimoya. *California Avocado Society Yearbook*, 54: 133-139.

Popenoe, W. (1974). *Manual of tropical and subtropical fruits, excluding the banana, coconut, pinapple, citrus fruits, olive and fig*. A Facsimile of the *1920 Edition*. Haffner Press, New York, USA. Collier-McMillan Publishers, London, UK, 474 p.

Potter, D., Eriksson, T., Evans, R. C., Oh, S., Smedmark, J. E. E., Morgan, D. R., Kerr, M., Robertson, K. R., Arsenault, M., Dickinson, T. A. and Campbell, C. S. (2007). Phylogeny and classification of Rosaceae. *Pl. Syst. Evol.*, 266: 5-43.

Prain, D. (1904). *Musa nagensium*. *Journal of the Asiatic Society of Bengal*, 73(2):22.

Prance, G. T. and Mori, S. A. (1979). *Lecythidaceae Part-I, Flora Neotropica Monograph 21, 1*. New York Botanical Garden, New York, USA.

Py, L., Lacoeuihe, J. J. and Teisson, C. (1987). *The pineapple, cultivation and uses*. G. P. Maisonluve et Larsose, Paris.

Rajan, S. and Lal, B. (1993). Fruit colour and quality of Dashehari mango in relation to canopy light penetration. *Golden Jubilee Symposium on Horticultural Research-Changing Scenario*, Bengaluru, India (Abs. 8.5).

Ramachandran, C., Peter, K. V. and Gopalakrishnan, P. K. (1980). Drumstick (*Moringa oleifera*): A multipurpose Indian vegetable. *Economic Botany,* 34 (3): 276-283.

Rao, A. N. and Ramanatha Rao, V. (1998). Genetic Resources and Breeding patterns of Five Humid Tropical Fruit Tree species. In: *Tropical Fruit in Asia: Diversity, Maintenance, conservation and use*. Proceedings of the IPGRI-ICAR-UTFANET Regional Training Course on the conservation and use of germplasm of Tropical Fruits in Asia, (eds.) R. K. Arora and V. Ramanatha Rao, pp. 54-69.

Rao, G. G. (1996). *Salvadora persica* – A promising species for saline wastelands, *Salinity Newsletter,* 7 (1-4) : 6.

Rao, R., Bencivenni, M., Basile, B. and Forlani, M. (2008). Molecular characterization of the apricot cultivars of protected geographical indication (PGI) "Albicocca Vesuviana". XIV. *International Symposium on Apricot Breeding and Culture*, 16-20 June 2008, Matera (Italy), abstract books.

Rathore, D. S. (1991). Pecan. In: *Temperate fruits.* (eds.) S. K. Mitra, T. K. Bose and D. S. Rathore. Horticulture and Allied Publisher, Calcutta, pp. 415-450.

Rathore, D. S. and Thakur, K. K. (1991). Cranberry. In: *Temperate Fruits*. (eds.) S. K. Mitra, T. K. Bose and D. S. Rathore. Horticulture and Allied Publisher, Calcutta, pp. 702-720.

Ray, P. K. (2007). *Litchi Production technology and post harvest handling for export.* Rajendra Agricultural University, Pusa, Bihar (India).

Reckin, J. (1983). The Orinoco nut– a promising tree crop for the tropics. *International Tree Crops Journal*, 2: 105-119.

Rehder, A. (1940). *Manual of cultivated trees and shrubs hardy in North America, exclusive of the subtropical and warmer temperate regions*, 2nd revised and enlarged edition. XXX. 996 pp. Macmillan Co., New York.

Reich, Lee (1991). *Uncommon fruits worthy of Attention.* Reading, Mass., Addison-Wesley. pp. 15-33.

Reid, William (1990). Eastern Black Walnut: potential for commercial nut producing cvs. In: *Advances in New Crops. Proceeding of the first National Symposium New Crops: Research, Development, Economics.* (eds.) J. Janick and J. E. Simons, Timber Press, Portland, Oregon, pp. 327-331.

Rendle, A. B. (1925). *The Classification of Flowering Plants. 2 vols.* Cambridge University Press, Cambridge

Rendle, A. B. (1959). *The Classification of Flowering Plants.* Vol. II. Cambridge University Press, Cambridge, UK.

Reveal, J. L. and Chase, M. W. (2011). APG III: Bibliographical introduction and Synonymy of magnoliidae. *Phytotaxa*, 19:71-134.

Richardson, D. L., Hamilton, K. S. and Hutchinson, D. J. (1965). Notes on bananas. 1-Natural edible tetraploids. *Tropical Agriculture*, *Trinidad*, 42:125-137.

Ridley, H. N. (1922). *The Flora of the Malay Peninsula. Vol. I. Polypetalae.* L. Reeve & Co., Ltd., 6 Henrietta Street, Convent Garden, London.

Ritter, C. M. and Mckee, G. W. (1964). The elderberry: History Classification and Culture. *Pa. Agric. Exp. Stn. Bull.*, 709.

Robertson, Smith, P. G. K. R. and Rohrer, J. R. (1990). A checklist of the family Maloideae (Rosaceae). *Can. J. Bot.*, 68: 2209-2269.

Robinson, J. C. (1996). Bananas and Plantaine. CAB International, Wallingford.

Robinson, J. P., Harris, S. A. and Juniper, B. E. (2001). Taxonomy of genus *Malus* Mill. (Rosaceae) with emphasis on the cultivated apple, *Malus domestica* Borkh. *Plant Syst. Evol.*, 226: 35-58.

Rodrigo, P. A. (1976). *Rambutan and its relatives*. Agrix How-to Series, 41: 11-16.

Rolfs, P. H. (1915). *Mangoes in Florida*. University of Florida Agriculture Experimental Station Bulletin, 127, 105.

Romero, C., Pedryc, A., Munoz, V., Llacer, G. and Badenes, M. L. (2003). Genetic diversity of different apricot geographical groups determined by SSR markers. *Genome,* 46, 244- 252.

Romero, C., Vilanova, S., Burgos, L., Martinez-Calvo, J., Vicente, M., Llacer, G. and Badanes, M. L. (2004). Analysis of the S-locus structure in *Prunus armeniaca* L.: Identification of S-haplotype specific S-RNase and F-box genes. *Plant Mol. Biol.*, 56, 145-157.

Rongsen, Lu. (1992). Seabuckthorn- a multi-purpose plant species for fragile mountains. *ICIMOD Occasional Paper No. 20*, Kathmandu, 62 pp.

Ronning, C. M., Schnell, R. J. and Gazit, S. (1995). Using randomly amplified polymorphic DNA (RAPD) markers to identify *Annona* cultivars. *Journal of the American Society for Horticultural Science*, 120(5): 726-729.

Rosengarten, F. (1984). *The Book of Edible Nuts*. Walker and Company, New York.

Roy, S. K. (1990). Bael. In: *Fruits: Tropical and Subtropical*. (eds.) T. K. Bose and S. K. Mitra. Naya Prokash 206 Bodhan Sarani, Calcutta, pp. 740-745.

Russell, C. E. and Felker, P. (1987). The prickly pears (*Opuntia* spp., Cactaceae): a source of human and animal food in semi-arid regions. *Economic Botany*, 41: 433-445.

Ryugo, Kuy (2008). Pomegranate. In: Janick, J. and Paull, R. E. (eds.) *Encyclopedia of Fruits and Nuts*. CABI, Wallingford, UK. p. 611.

Salem, M. A. (2007). *Cashew Business Line*, November 19, 2007.

Salma, I. (2000). *Kepelbagaian durian di Malaysia (Diversity of durian in Malaysia*). Prosiding Seminar Durian, (Zainal Abidin, M., Mohd. Shamsudin, O., Ahmad Tarmizi, S, Zabedah, M. and Salma, I., eds.) 1-3 Aug. 2000, Ipoh, p. 20-25. Serdang: MARDI.

Sanewski, G. (1991). *Custard apples, cultivation and crop protection*. Queensland Department of Primary Industries, Brisbane, Australia, 103 pp.

Sant Ram and Rajan, S. (2003). *Status report on genetic resources of mango in Asia-Pacific Region*. IPGRI Office for South Asia, New Delhi. India. 196 p.

Sarip, J. (1991). The powder chiku clones (*Manikara achras*) (Mill) Fosberg in Malaysia. *Proc. of Third National Fruit Symposium*, p. 263-268, Serdang: MARDI.

Saunt, J. (1990). *Citrus Varieties of the World*. Sinclair International, UK.

Savolainen, V., Corbaz, R., Moncousin, C., Spichiger, R. and Manen, J. F. (1995). Chloroplast DNA variation and parentage analysis in 55 apples. *Theor. Appl. Genet.*, 40: 1138-1141.

Sawumi, M. A. (1993). The common edible fruits of Nigeria- Part-II. *The Nigerian Field*, 58 (1, 2): 157-171.

Scheldeman X. (2002). *Distribution and potential of cherimoya* (*Annona cherimola* Mill.) *and highland papayas* (*Vasconcellea* spp.) *in Ecuador*. Ph.D. dissertation, Faculty of Agricultural and Applied Biological Sciences, Ghent University (UGent), Ghent, Belgium.

Schmidt, H., Vittrup-Christensen, Watkins, R. and Smith, R. A., eds. (1985). *Cherry Descriptors*, International Board of Plant Genetic Resources, Rome and CEC, Brussels.

Scora, R. W. (1975). *On the history and origin of Citrus*. Bulletin of the Torrey Botanical Club, 102: 369-375.

Scora, R. W. (1988). Biochemistry, taxonomy and evolution of modern cultivated Citrus. *Proceedings of the International Society of Citriculture*, 1, 277-289.

Sehgal, R. N. and Chauhan, Vaidya (1991). Chilgoza tree in imminent danger of extinction in Western Himalaya. *Wasteland News,* 6 (2): 32.

Sharma, S. D. (1991). Filberts. In: *Temperate fruits*. (eds.) S. K. Mitra, T. K. Bose and D. S. Rathore. Horticulture and Allied Publishers, Calcutta, pp. 473-497.

Shepherd, K. (1964). A new species of banana. Kew Bulletin. 17: 461-463.

Shepherd, K. (1999). *Cytogenetics of the genus Musa*. 160 , International Network for the Improvement of Banana and Plantain, Montpellier, France.

Shepherd, K. and Ferreira, F. R. (1982). The PNG Biological Foundation's Banana Collection at Laloki, Port Moresby, Papua New Guinea. *IBPGR/SEAN Newsletter*, 8(4): 28–34.

Shivanandan, V. N. (1980). *Studies on systematic selection of productive types of tamarind*. M.Sc. (Agri.) Thesis Univ. of Agri. Sciences, Bangalore.

Sidibe, M. and Williams, J. T. (2002). *Baobab, Adansonia digitata* L. International Centre for Underutilised Crops, Southampton, UK, 96 p.

Silayoi, B. and Chomchalow, N. (1987). Cytotaxonomic and morphological studies of Thai banana cultivars. In: Persley, G.J. and De Langhe, E. A. (eds.), *Banana and Plantain Breeding Strategies: proceedings of an International of an International workshop*, Cairns, Australia, 1986. ACIAR Proceedings no. 21: 157-160.

Simmonds, N. W. (1954a). Varietal identification in the Cavendish group of bananas. *Journal of Horticultural Science,* 29: 81-87.

Simmonds, N. W. (1954b). Anthocyanins in bananas. *Annals of Botany*, 72: 471-482.

Simmonds, N. W. (1954c). Isolation in *Musa*, sections Eumusa and Rhodochlamys. *Evolution*, 8: 65-74.

Simmonds, N. W. (1959). *Bananas. 1st edition*. Longmans, Green and Co. Ltd., London.

Simmonds, N. W. (1962). *The Evolution of the Bananas*. Longmans, London. 170 pp.

Simmonds, N. W. (1966) *Bananas*. 2nd edition. Longmans, London.

Simmonds, N. W. (1995). Banana: *Musa* (Musaceae). p. 370-375. In: J. Smart and Simmonds, N. W. (eds.), *Evolution of crop plants. 2nd ed.* pp. 370-375. Longman Scientific & Technical. Essex, England..

Simmonds, N. W. and Shepherd, K. (1955). The taxonomy and origin of the cultivated bananas. *J. Linn. Soc. Lond. Bot.*, 55: 302-12.

Simmonds, N. W. and Weatherup, S. T. C. (1990a). Numerical taxonomy of the wild bananas (*Musa*). *New Phytol.*, 115: 567-571.

Simmonds, N. W. and Weatherup, S. T. C. (1990b). Numerical taxonomy of the cultivated bananas. *Tropical Agriculture (Trinidad),* 67: 90-102.

Simmonds, N. W. (1952a). Experiments on the pollination of seeded diploid banana. *J. Gent.,* 51: 32-40.

Simmonds, N. W. (1952b). The germination of banana seeds. *Trop. Africa, Trinidad*, 29: 35-49.

Singh, G. and Felker, P. (1998). Cactus: A new World Food. *Indian Horticulture,* April-June, pp. 27-31.

Singh, H. P. and Uma, S. (1996). Genetic diversity of banana in India. In: *Proceedings of the Conference on "Challenges for banana Production and Utilization in 21st century,* Trichy, Sept. 24-25.

Singh, H. P., Balasubramanian, P. P. and Hubballi, V. N. (Eds.) (2003). Indian Cashew-issues and strategies. *Proceedings of 5th National Seminar on Indian Cashew in the New Millennium*, Visakhapatnam, 12-13 April, 2002. Directorate of Cashew and Arecanut Development, Kochi/Ministry of Agriculture, New Delhi, pp. 761-770.

Singh, L. B. (1960). *The Mango: Botany, Cultivation and Utilization.* Leonard Hill, London.

Singh, L. B. and Singh, R. N. (1956). *A monograph on the mangoes of Uttar Pradesh. Vols I and II*, Superintendent, Printing and Stationery Lucknow, UP. Vol. I, 150 p.

Singh, L. B. and Singh, U. P. (1954). *The litchi* (*Litchi chinensis* Sonn.). Superintendent of Printing and Stationary, Lucknow, UP, India.

Singh, P., Bakhshi, J. C. and Singh, R. (1971). Identification of ber (*Zizyphus mauritiana* Lam.) cultivars through vegetative characters. *Punjab Hort. J.*, 11 (3-4): 176-87.

Singh, P., Bakhshi, J. C. and Singh, R. (1972). Identification of ber (*Z. mauritiana* Lam.). *Purjab Hort. J.,* 12(2-3): 120-33.

Singh, R. and Nath, N. (1969). Practical approach to the classification of *Citrus*. *Proc. First International Citrus Symposium.* p. 435-440.

Singh, Sanjay and Hoda, M. N. (1996). Makhana (Gorgan nut). In: *A text book Pomology (Tropical fruits) Vol. II.* (ed.) T. K. Chattopadhyay. Kalyani Publishers, Ludhiana, pp. 321-328.

Singh, Sham., Krishnamurthi, S. and Katyal, S. L. (1963). *Fruit culture in India*. Indian Council of Agricultural Research, New Delhi, pp. 307-310.

Singh, V. (1998). Seabuckthorn: A Wonder Plant of dry temperate Himalayas. *Indian Horticulture*, July-September, 1998, pp. 6-9.

Sleumer, H. (1984). Olacaceae. In: *Flora Malesiana, Ser. 1, 10, 1: 1-29*. Kluwer Academic Publisher, Dordrecht.

Smith, L. B. and Downs, R. J. (1979). *Bromeliodea (Bromeliaceae). Monograph 14, pt. 3, Flora Neotropica.* The New York Botanical Gardens, New York, 2142pp.

Smith, N. J. H., Williams, J. T. Plucknett, D. L. and Talbot, J. P. (1992). *Tropical forests and their crops*. Cornell University Press, New York, 568 pp.

Sneath, P. H. A. and Sokal, R. R. (1973). *Numerical Taxonomy.* W. H. Freeman, San Francisco.

SOFRI (Southern Fruit Research Institute) (2002). *Fruits in Vietnam*. Agriculture Publishing House, Hochi Minah Citry, Vietnam.

Spiegel-Roy, P. and Goldschmidt, E. E. (1996). *Biology of Citrus*. Cambridge University Press.

Sporne, K. R. (1976). Character correlation among angiosperms and the importance of fossil evidence in assessing their significance. In: *Origin and Early Evolution of Angiosperms,* C. B. Beck (ed.). Columbia Univ. Press, New York.

Srivastava, A. K., Singh, H. K. and Singh, I. S. (1998). Genetic diversity– Bael (*Aegle marmelos* Correa). *Indian Horticulture,* January – March, 1998, pp. 1.

St. Pierre, R. G. (1992). *Growing Saskatoons– A manual for orchardists.* Deptt. of Horticulture Science, University of Saskatchewan, Saskatoon, Saskatchewan, 40 pp.

Stang, E. J. (1990). Elderberry, Highbush Cranberry and Juneberry management. In: *Small Fruit Crop Management* (eds.) Gene J. Galletta and David G. Himelrick. Prentice Hall, Englewood Cliffs, New Jersey, USA, pp. 363-382.

Stang, E. J. (1994). Lingonberry cultivars building blocks for an industry. *Fruit Var. J.*, 48: 3-6.

Stang, E. J., Weis, Gavin, G. and Klueh, J. (1990). Lingonberry: Potential New Fruit for the Northern United States. In: *Advances in New Crops– Proceedings of the First National Symposium on New Crops Research, Development, Economics*, Indianapolis, Indiana, October 23-28, 1998. (eds.) Jules Jackson and J. E. Simons. Timber Press, Portland, Indiana , USA, pp. 321-322.

Stebbins, G. L. (1974). *Flowering Plants: Evolution above the species level.* The Belknap Press, Harvard Univ. Press, Cambridge.

Stover, R. H. and Simmonds, N. W. (1987). *Bananas, 3rd edition.* Longman, Sci. and Tech. London, UK, 468 pp.

Strand, L. L. (1994). *Integrated Pest Management for Strawberries.* University of California State wide Integrated Pest Management Project Publication 3351. University of California, Berkeley, California, 142 pp.

Strik, B. C., Finn, C. E. and Moore, P. P. (2014). *Blueberry cultivars for the Pacific Northwest.* Pacific Northwest Extension Publication, Oregon State University, University of Idaho & Washington State University, USA.

Sturrock, T. T. (1951). A study of the identification of mango varieties in Florida. *Florida Mango Forum, Mango Studies*, 71.

Sturrock, T. T. and Wolfe, H. S. (1944). A key to Florida mango varieties. *Proc. Fl. Hort. Soc.,* 175-180.

Stushnoff, Cecil. (1991). *Amelanchier* species. In: *Genetic Resources of temperate Fruit and Nut Crops. Vol. II.* (eds.) J. N. Moore and J. R. Ballington. *Acta Horticulturae*, 290: 569-600.

Suranyi, D. (1999). Apricot culture in Hungary: Past and present. *Acta Hort.*, 488: 205-210.

Sussenguth, K. (1953). Rhamnaceae. In: *Die Naturlichen Pflanzenfamilien*, Engler, A. and Prantle, K. Duncker & Humblot, Berlin. Band 20 d. pp. 121-132.

Swennen, R. and Ortiz, R. (1997). *Morphology and Growth of Plantain and Banana.* International Institute of Tropical Agriculture (IITA) Research Guide 66. IITA Ibadan, Nigeria, 32 pp.

Swennen, R. and Vuylsteke, D. (1987). Morphological taxonomy of plantains (*Musa* cultivars AAB) in West Africa. In: Persley, G. and De Langhe, E. (eds.) *Banana and Plantain Breeding Strategies*, pp. 165-171, ACIAR Proceedings 21. ACIAR, Canberra.

Swingle, W. T. (1909). The limitation of the Satsuma orange to trifoliate orange stock. US deptt. Agri. *Bur. Plant Indus. Circ.,* 46: 1.

Swingle, W. T. (1915). A new genus, *Fortunella*, comprising four species of kumquat oranges. *Jour. Wash. Acad. Sci.*, 5: 165-76.

Swingle, W. T. (1948). Botany of *Citrus* and its wild relatives of the orange subfamily In: Webber, H. J. and Batchelor, L. D. (eds.). *The Citrus Industry*. University of California Press, California, pp. 129-174.

Swingle, W. T. and Reece, P. C. (1967). The botany of *Citrus* and its wild relatives. In: Reuther, W., Batchelor, L. D. and Webber, H. J. (eds.). *The Citrus Industry*. University of California Press, California, pp. 190-340.

Szabo, Z. and Nyeki, J. (1991). Blossoming, fructification and combination of apricot varieties. *Acta Hort.*, 293: 295-302.

Tabora, P. C., Balick, M. J., Bovi, M. L. A. and Guerra, M. P. (1993). Hearts of palm (*Bactris, Euterpe* and others). In: *Underutilized crops Pulses and Vegetables.* (Ed.) J. T. Williams. Chapman and Hall, London, pp. 193-218.

Takhtajan, A. (1997). *Diversity and classification of flowering plants.* Columbia University Press, New York, New York, USA.

Takhtajan, A. (2009). *Flowering plants*. Second edition. Springer, Germany.

Tanaka, T. (1936). The taxonomy and nomenclature of Rutaceae- Aurantioideae. *Blumea,* 2: 101–110.

Tanaka, T. (1954). Species problem in *Citrus* (Rension aurantiacearum IX). *Jap. Soc. Prom. Sci., Ueno, Tokyo*, 152 pp.

Tanaka, T. (1961). Semi-centennial commemoration papers on citrus studies. *Citrologia supporting Foundation*, University of Osaka, Japan. *Citrologia*, 114 pp.

Tézenas du Montcel, De Langhe, H. and Swennen, E. R. (1983). Essai de classification des bananiers plantains (AAB). *Fruits*, 38: 461-474.

TFNet (2008). *Tropical Fruit Gallery.* TFNet Secretariat, Selangor, Malaysia.

Thakur, K. S. and Rathore, D. S. (1991). Blueberries. In: *Temperate Fruits*. (eds.) S. K. Mitra, T. K. Bose and D. S. Rathore, Horticulture and Allied Publisher, Calcutta, pp. 672-701.

Thorne, R. F. (2000). The classification and geography of the flowering plants: Dicotyledons of the class Angiospermae (subclasses Magnoliidae, Ranunculidae, Caryophyllidae, Dilleniidae, Rosidae, Asteridae, and Lamiidae). *Botanical Review,* 66: 441–647.

Thorne, R. F. (2003). An updated classification of the class Angiospermae. www.rsabg.org/publication/angiosp.htm.

Tindall, H. D. (1994). Sapindaceous Fruits: Botany and Horticulture. *Horticultural Reviews,* 16: 143-196.

Tippo, O. (1942). A Modern Classification of the Plant Kingdom. *Chronica Botanica*, 7: 203–206.

Tomlinson, P. B. (1959). An anatomical approach to the classification of the Musaceae *J. Linn. Soc. (Bot.),* 55: 779-809.

Tomlinson, P. B. (1962). Phylogeny of the Scitamineae- morphological and anatomical considerations. *Evolution*, 16: 192-213.

Tous, J., Romero, A., Hermoso, J. F., Ninot, A., Plana, J. and Batlle, I. (2009). Agronomic and commercial performance of four Spanish carob cultivars *HortTechnology,* 19: 465-470.

Trehane, P., Brickell, C. D., Baum, B. R., Hetterscheid, W. L. A., Leslie, A. C., McNeill J., Spongberg, S. A. and Vrugtman, F. (1995). *International Code of Nomenclature for Cultivated Plants* 1995. Regnum Vegetabile Vol. 133. Quarterjack, Wimborne, UK.

Ude, G., Pillay, M., Nwakanma, D., Tenkouana, A. (2002). Genetic diversity in *Musa acuminata* Colla and *Musa balbisiana* Colla and some of their natural hybrids using AFLP markers. *Theoretical and Applied Genetics,* 104: 1246-1252.

Uhl, N. W. and Dransefield, J. (1988). Genera Palmarum, a new classification of Palms and its implication. *Advances Economic Botany*, 6: 1-19.

Uma, S. and Sathiamoorthy, S. (2002). *Names and synonyms of bananas and plantains of India.* National Research Centre for Banana (ICAR), Tiruchirapalli, India.

Uzun, A., Gulsen, O., Seday, U., Bircan, M., Yilmaz, K. U. (2010). SRAP based genetic analysis of some apricot cvs. *Rom. Biotech. Lett.* 15: 5396-5404.

Vakili, N. G. (1967). The experimental formation of polyploidy and its effect in the genus *Musa*. *American Journal of Botany,* 54: 24-36.

Valmayor, R. V., Espino, R. R. C and Pascua, O. C. (2002). *The wild and cultivated bananas of the Philippines*. Los Banos, Laguna: PARRFI and BAR.

Valmayor, R. V., Jamaluddin, S. H., Silayoi, B., Kusumo, S., Danh, L. D., Pascua, O. C., Espino, R. R. C. (2000). *Banana Cultivar Names and Synonyms in South East Asia.* International Network for the Improvement of Banana and Plantain- Asia and the Pacific Office, Laguna, Philippines.

Valmayor, R. V., Rivera, F. N. and Lomulji, F. N. (1981). *Philippine Banana Cultivar Names and Synonyms*. IBP. Bull. No. 3, University of the Philippines at Los Banos.

Valmayor, R. V., Silayoi, B., Jamaluddin, S. H., Kusumo, S., Espino, R. R. C. and Pascua, O. C. (1991). *Banana classification and commercial cultivars in South East Asia*. PCARRD Information Bulletin no. 24, Los Banos. The Phillippines, 20 pp.

Van Damme, P. (1992). *Toegepaste plantensystematiek. Course Text.* Faculty of Agricultural and Applied Biological Science- Ghent University. 482 p.

Van Welzen, P. C. and Verheij, E. W. M. (1991). *Nephelium lappaceum* L. In: *Edible fruits and Nuts, Plants Resources of South East Asia*. Prosea Foundation Bogor, Indonesia.

Van Wyk, Ben-Erick. (2005). *Food Plants of The World.* Briza Publications, South Africa, p 72.

Vashishtha, B. B. and Pareek, O. P. (1989). Identification key for the cultivars of Indian jujube (*Zizithus mauritiana* Lamk.). *Indian J. Hort.,* 46: 183-88.

Vavilov, N. I. (1930). Wild progenitors of the fruit trees of Turkestan and Caucasus and the problems of the origin of fruit trees. *Proc. 9th International Hort. Congress*, London, pp. 271-286.

Vavilov, N. I. (1951). The origin, immunity and breeding of cultivated plants. Translated from Cyrillic by K. S. Chester. *Chronic Botanic*, 13: 1-336.

Verheij, E. W. M. and Coronel, R. E. (eds.) (1991). *Plant Resources of South East Asia,* No. 2. *Edible Fruits and Nuts*. Pudoc, Wageningen, The Netherlands, 447 pp.

Viala, P. and Vermorel, V. (1901-1909). *Ampelographie*, Masson, Paris, Tomes 1-VII.

Volkova, N. G. (1978). Etricheskie Protessy Gruzinskoi SSR. Etnicheshjeikult-urno-bytorye, Protessyna Kavkaz'e. Moskva. 3-61.

Von Wettstein, R. R. (1907). Handbuck der Systematischen Botanik. *Bot. J. Linnean Soc.*, 38: 29–80 4. (Von Wettstein, R. R. 1901. Handbuch der systematischen Botanik (2nd ed.) Franz Deuticke, Leipzig).

Walter, Annie and Sam, Chanel (2002). *Fruits of Oceania*. ACIAR Monograph No. 85, Canberra, 329 pp.

Warburg, O. (1900). *Musa celebica*. In: H. G. A. Engler. Ed. *Pflanzens*., 4:45-22.

Watkins, R. (1981). Apples (genus Malus), pears (genus Pyrus) and plums, apricots, almonds, peaches, cherries (genus Prunus) in The Oxford Encyclopedia of Trees of the World, ed. F. B. Hora, Oxford University Press, Oxford and New York.

Watkins, R. (1995a). Apples and pears. In: Smartt, J. and Simmonds, N. W. (eds.). *Evolution of Crop Plants*. Longman, London, pp. 418-422.

Watkins, R. (1995b). Cherry, plum, peach, apricot and almond: *Prunus* spp. (Rosaceae). p. 423-429. In: Smartt J. and N. W. Simmonds (eds.), *Evolution of crop plants.* 2 ed. Longman. Scientific & Technical, Essex, England.

Watson, B. J. (1983). *Lychee cultivar names*. Queensland Dept. Primary Ind. Farm note F 36.

Watson, B. J., George, A. P., Nissen, R. J. and Brown, B. I. (1988). Carambola: a star on the horizon. *Queensland Agric. J.,* 114:45 –51.

Watt, G. (1889). *A Dictionary of Economic Products of India*. Vol. 6. Suptd. Govt. Printing, Calcutta.

Waugh, F. A. (1912). *Fruit culture*. Orange Judd Company, New York.

Way, R. D. (1981). *Elderberry culture in New York State*. N. Y. State, Agric. Exp. Stn. Geneva, N. Y. Food Life Sci. Bull. 91.

Weaver, R. J. (1976). *Grape growing*. John Wiley and Sons, New York, USA.

Webber, H. J. and Swingle, W. T. (1905). *New citrus creations of the Department of Agriculture*. U. S. Dept. Agr. Yearbook, 1904: 221-240.

Wei, J. and Gao, H. S. (2002). The production of Asian pears in China. *Acta Horticulturae*, 587: 71-80.

Wertheim, S. J. (1997). The hazelnut in Northwestern Europe: past, present and prospects. *FAO-IHEAM Nucis Newsletter*, Number 8, pp. 17-19.

Wester, P. J. (1920). A descriptive list of mango varieties in India. *Phillppines Agriculture Review*, 13, 265.

Westwood, M. N. (1978). *Temperate zone pomology*. Timber Press. Portland, Oregon., p. 428.

Westwood, M. N. (1982). Pear germplasm of the New National clonal repository: its evaluation and use. *Acta Hort.*, 124: 57-65.

Westwood, M. N. (1993). *Temperate Zone Pomology– Physiology and Culture. 3rd Ed.*, Timber Press, Portland, Oregon.

Wickens, G. E. (1995). Edible nuts. *Non-wood forest products* 5, FAO, Rome, pp. 198.

Wickens, G. E. (1998). *Ecophysiology of economic plants in arid and semi-arid lands.* Springer Verlag, Berlin, Heidelberg.

Wickens, G. E. and Storey, I. N. J. (1984). *Cordeauxia edulis* Hemsley. *Survey of the economic plants of the arid and semi-arid tropics* (SEPASAT). Dossier No. 5 (revised). Royal Botanic Gardens, Kew (mimeo).

Wiersema, J. H. (1985). *Nomenclature of Malus*. Unpublished mimeograph, 22pp. Plant Exploration and Taxonomy Lab, US Deptt. of Agri. Belstltille, M.D.

Willis, J. C. (1985). *A dictionary of the flowering plants and ferns*, 8th ed. Cambridge University Press, Cambridge, UK.

Wong, C., Kiew, R., Argent, G., Set, O., Lee, S. K. and Gan, Y. Y. (2002). Assessment of the validity of the sections in *Musa* (Musaceae) using AFLP. *Annals of Botany,* 90: 231-238.

Wood, K. W. (1976). *Viburnum Collection. Section Opulus: Cranberry bush Viburnums*. University of Wisconsin– Madison Arboretum Field Book. Madison, Wisc.

Woodhouse, E. (1909). Mangoes in Bhagalpur- A preliminary note. *Quart. Journal Department of Agriculture, Bengal*, 2: 168-187.

Wu, C. Y., Tang, Y. C., Chen, Z. D. and Li, D. Z. (2002). Synopsis of new polyphyletic-polychromic polytopic system of the angiosperms. *Acta Phytotax Sinica*, 40: 289-322.

Wu, J., Gu, C., Zhang, S. L., Zhang, S. J., Wu, H. Q., and Heng, W. (2009). Identification of *S*-haplotype-specific *S-RNase* and *SFB* alleles in native Chinese apricot (*Prunus armeniaca* L.). *J. Hort. Sci. Biotechnol.*, 160: 241–248.

Yaacob, O. and Tindall, H. D. (1995). *Mangosteen cultivation*. FAO Plant Production and Protection Paper No. 129, 100 pp.

Yen, C. R. (1984). Seeded and seedless fruit growth of 'Sah Keng" litchi. *Journal of Agricultural of China.,* 33: 257-264.

Yi, T., Wen,. J., Golan-Goldhirsh, A. and Parfitt, D. E. (2008). Phylogenetics and Reticulate Evolution in *Pistacia* (Anacardiaceae). *American Journal of Botany*, 95(2): 241-251. doi:10.3732/ajb.95.2.241

Yilmaz, K. U. (2008). *Phenological, morphological and pomological characteristics of some local apricot genotypes and determination of their genetic relationships and self incompatibility by molecular markers.* Department of Horticulture, Institute of Basic and Applied Sciences, University of Cukurova, Adana, Turkey, Ph. D. Thesis, (in Turkish with an English abstract).

Yilmaz, Kadir Ugurtan and Gurcan, Kaharaman (2012). Genetic Diversity in Apricot. In: *Genetic Diversity in Plants* (2012) edited by Prof. Mahmut Caliskan. In Tech Europe, Slavka Krautzeka, Croatia and InTech China, Shanghai, pp. 249-270.

Yu, T. T., Lu, I. T., T. C., Ku, C. I. Li and Chen, S. X. (1986). *Amydaloideae, Flora Reipublicae Popularis Snicae.* Vol. 38 Science Press, Beijing, China.

Zee, F. T. (1993). Rambutan and pili nuts: potential crops for Hawaii, p. 46-465, In: Janick, J. and Simon, J. E. (eds.). *New Crops*. Wiley, New York.

Zerega, N.J., Ragone, D. and Motley, T.J. (2005). Systematics and species limits of Breqdfruit (*Arvocarpus,* Moraceae). Systematic Botany, 30(3): 610.

Zeven, A. C. and Zhukovsky, P. M. (1975). *Dictionary of cultivated plants and their centres of diversity*. PUDOC, Wageningen, The Netherlands, 219 p.

Zielinski, Q. B. (1955). *Modern Systematic Pomology.* W.M.C. Brown Company, Dubuque, Iowa, USA

Zohary, D. (1995). Olive, *Olea europaea* (Oleaceae) In: Smartt, J. and Simmonds, N. W. (eds.) *Evolution of crop plants*. Longman Scientific and technical, Harlow, U.K., pp. 271-274.

Zohary, D. and Hopf, M. (1994). *Domestication of Plants in Old World.* Clarendon Press, Oxford.

⑧ ⑧ ⑧

GLOSSARY

Abaxial Surface most distant or away from the axis, the lower or outer surface of organ. Syn: dorsal (position)

ABC model A model of floral development in which gene products of the so-called A, B, and C classes combine to produce the four major floral organs: sepals, petals, stamens, and carpels.

Acaulescent A plant stemless or without visible stem (other than the inflorescence axis).

Accession number A number assigned to each new specimen placed into a permanent herbarium collection.

Accessory bud Bud(s) lateral to or above axillary buds.

Accessory fruit A. portion of the mature fruit that is not directly derived from the ovary or ovaries, may include bract(s), stem axes, receptacle, hypanthium, or perianth.

Accrescent Parts persistent and continuing to grow beyond what is normal or typical, e.g., calyx of *Physalis,* Solanaceae.

Acetolysis A standard acid treatment used to dissolve all but the exine of pollen grains in order to better observe pollen wall structure with the light microscope.

Achene The fruit develops from monocarpellary superior ovary. It is unilocular and single seeded. The pericarp does not fuse with seed coat, e.g., unit fruits of sunflowers and other Asteraceae.

Achenecetum An aggregate fruit of achenes as strawberry. The achenes are on the surface of accessory tissue, an enlarged fleshy receptacle.

Achlamydeous Lacking a perianth.

Achlorophyllous Lacking chlorophyll/chloroplasts.

Acicular Needlelike, often round in cross-section, with margins straight and parallel (as in Pinus), length: width ratio > 12:1.

Acrocaulis Positioned at the apex of the stem.

Acrocidal capsule A capsule dehiscing by means of apical slits.

Acrodromous With two or more primary veins or strongly developed secondary veins running in convergent arches toward the apex but not recurved at the base.

Actinodromous With three or more primary veins diverging from one point, inclusive of ternate or palmate venation.

Actinomorphic Syn: *polysynimetric, radial.* Radially symmetrical, with 3 or more planes of symmetry (parts with similar size and shape)

Aculeate Syn: *prickly* with prickles, sharp nonspine, nonthorn appendages (epidermal excrescellce)

Acuminate Apex drawn out into long tapering tail (abruptly incurved) as in *Ficus religiosa*, apical intersection angle <45'.

Acute Pointed and narrow as in mango. Apicai margins approximately straight, the intersection angle 45'-90'.

Adaptation A structure or feature that performs a particular function and which results in increased survival or reproduction.

Adaxial Syn: *ventral.* Surface toward or nearest the axis, the upper or inner surface of organ.

Adherent With unlike parts joined, but only superficially and easily separable.

Adhesive Climbers. Climbing by the formation of adhesive discs, either by the tendrils or by the climbing roots.

Adnate With unlike parts integrally fused, not easily separable, e.g. adhering to the petiole upto a certain height and making it somewhat winged and filament running the whole length of the anther from the base to the apex.

Adventitious root. A root developing from any other plant part, except the radicle. It may arise from the underground or aerial stem or from leaf margins.

Adventive embryony Development of an embryo from a cell of the surrounding tissue, such as megasporangial or integument tissue.

Aerial roots Adventitious roots that absorb moisture and minerals from the air or runoff with the help of spongy velamen tissues common in some epiphytic plants, e.g., of Araceae and Orchidaceae.

Aestivation Referring to position, arrangement, and overlapping of floral parts (perianth, sepals, and petals).

Affinis Abbr: *aff.* Implying, within a taxon name, some type of close relationship, presumably an evolutionary relationship, but also that the specimen differs from the cited taxon in some way, e.g., beyond the described range of variation for one or more characters.

Agamospermy The production of seeds without fertilization.

Ageotropic Growing upward; used for a plant part that normally grows downward; e.g., roots of some cycads, *Avicennia.*

Aggregate fruits (Etaerio). Fruits developed from multicarpellary, apocarpous ovary.

Albuminous Endospemious.

Albuminous cell A parenchyma cell associated with a sieve cell. Derived from a different parent cell than is the sieve cell.

Alete Spores lacking any evidence of a laesura.

Aleurone grain Syn: proteinoplast. Granular protein deposits in plant cells, functioning as storage compounds; a type of ergastic substance.

Alignment The process of matching homologous nucleotide positions of two or more sequences of DNA in order to code the data for phylogenetic or other types of analysis.

Allautogamy Having both outcrossing and selfing flowers. Adj: atlatitogamous.

Allogamy *Outbreeding.* Adj: allogamous.

Allopatric Occurring in different geographic ranges.

Allopolyploidy Polyploidy occurring between different species.

Allozyme One of two or more different molecular forms of an enzyme, corresponding to different alleles of a common gene.

Alternate One leaf or other structure per node.

Alternation of generations Haplod plontic life cycle.

Alternipetalous Stamens with point of attachment between the petals or corolla lobes.

Alternisepalous Stamens, with point of attachment between sepals or calyx lobes.

Alveolar An exine wall structure having numerous, spherical air pockets within the exine.

Amb The outline shape of a pollen grain, its polar view.

Ameat Catkin.

Amoeboid Syn: plasmodial; periplasmodial. An anther tapetum type in which the tapetal cell walls break down, with release of the cytoplasm of the tapetal cells into the locule.

Amphiphloic siphonostele Syn. Solenostele. A siphonoptele in which a ring of xylem is surrounded by an outer and inner layer of phloem.

Amphisarca. The fruit develops from multicarpellary syncarpous, mu superior ovary. The fruit is many seeded. Epicarp is hard. Mesocarp and endocarp are fleshy on which are found scattered seeds, as in wood apple (Aegle marmelos)

Amphistomal Referring to the micropyle of a bitegmic ovule formed or delimited by both integuments.

Amphitropous A general ovule type in which curvature of the ovule during development displaces the micropyle adjacent to the funiculus base, with the nucellus bent along both upper and lower sides and becomes more than that of campylotropous and the embryo sac assumes a horse-shoe shape. It is commonly found in Alismaceae.

Amplexicaul Sessile and clasping most of stem circumference.

Amplification The replication of numerous copies of DNA.

Amplification fragment length polymorphism Abbr: *AFLP.* The use of amplified fragments that exhibit length polymorphism, enabling the recognition of numerous genetic markers.

Amyloplast Starch grain.

Ana-amphitropous An ovule type in which a vascular strand curves, traversing from the base of funiculus to the chalazal region of the nucellus, with the nucellus bent sharply in the middle along both the lower and upper sides, often with differentiated cells (basal body) at the angle of the bend.

Ana-campyotropous An ovule type in which a vascular strand curves, traversing from the base of funiculus to the chalazal region of the nucellus, with the nucellus bent only along the lower side, with no basal body.

Analysis of variables Abbr: ANOVA. A statistical test to evaluate differences among two or more groups in a single test of significance.

Anastomosing Netted, reticulate.

Anatropous (inverted). A type of ovule in which curvature during development results in displacement of micropyle to a position adjacent to the funiculus base: most common and apomorphic for the angiosperms. The ovule bends at 180° and gets inverted, so that the micropyle comes to lie close to the funicle.

Ancestral Syn: Plesiomorphic; primitive. Referring to a preexisting condition or character state.

Ancestral state reconstruction A method of assessing character evolution by calculating the maximum probability of a state at each node, using a selected model of evolution.

Androdioecious/androdioecy Having male flowers on some individuals and perfect flowers on other individuals.

Androecium The male organ(s) of a flower; collectively all stamens of a flower.

Androgynophore A stalk like structure that bears the gynoecium and androecium; e.g., Passiflora.

Andromonoecious Having both staminate and perfect flowers on the same individual.

Androphore A stalk like structure that bears stamens.

Anemochory Dispersal of propagules by wind. Adj: anemochorous.

Anemophily Pollination by wind. Adj: anemophilous.

Anisophylly. In some opposite phyllotaxy, leaves are of unequal size. Spiral in Opposite decussate Opposite superimposed Verticillate

Anisomerous Having a different number of members in different whorls.

Annotated Phylogenetic *classification. A* classification in which monophyletic groups are ordered by a sequential listing of derivative lineages from the base to the apex of the cladogram.

Annotation label A label affixed to an herbarium specimen that verifies or changes the identity of a specimen or that documents the removal of plant material from the specimen.

Annual A plant completing its life cycle in a single season, e.g. Anagalis.

Annual ring The accumulation of secondary xylem (or phloem) over a single growing season, being evident because of the structural difference between the last cells of the summer wood and ' the first cells of the subsequent spring wood.

Annulated roots—roots appear as if a number of discs are placed one above the other, as in ipecac (*Cephalis ipecacuanha*)

Annulus A single row of specialized cells, having differentially thickened cell walls, on the outer rim of a *leptosporangium,* functioning in its dehiscence.

Antepelatous Antipetalous.

Anterior Referring to the lower lobe or part, especially in a horizontally, oriented structure.

Antesepalous Antisepalous.

Anther The pollen bearing part of a filamentous stamen.

Anther sac Microsporangium.

Antheridial wall Syn: *jacket layer; sterile jacket layer.* The outer, sterile layer of cells of the antheridium.

Antheridiophore A specialized, stalked, generally peltate structure that grows from the gametophyte of some liverworts and bears antheridia.

Antheridium The male gametangium of the gametophyte of land plants, producing sperm cells and surrounded by an outer layer of sterile cells, the antheridial wall or jacket layer.

Antherode The sterile anther of some staminodes.

Anthesis Time of flowering; the opening of flower with parts available for pollination.

Anthocarp A fruit in which one or more flower parts functions as accessory tissues. e.g., *Pontederia* (Pontederiaceae), in which an accrescent perianth surrounds and fuses to the achene.

Antipetalous Syn: *antepetalous, obhaplodstemonous.* Stamens with point of attachment in line with (opposite) the petals or corolla lobes, e.g., Primulaceae, Rhamnaceae.

Antipodal cells In a typical angiosperm female gametophyte, the three haploid cells that are positioned opposite the micropyle. i.e., in the chaiazal region.

Antisepalous Stamens with point of attachment in line with (opposite) the sepals or calyx lobes. Syn: *antesepalous, haplostemonous.*

Antrorse Bent or directed upward, usually referring to small appendages.

Aperture A specially delimited region of the pollen grain wall.

Apetalous Having no petals or corolla.

Apex The anterior tip of the lamina is called apex.

Apical At or near the top, tip or end of a structure, e. g. style arising at the apex of the ovary, placenta at the top of the ovary (*pendulous),* apical bud *(terminal bud),* apical meristem (region of actively dividing cells in the land plants, located at the apex of the thallus, shoot, or root), apical-axile (two or more placentae at the top of a septate ovary).

Apiculate With a flexible, apical process, usually slightly curled, length: width ratio >3: 1.

Apocarpous With carpels distinct, the pistil or ovary simple as in Nymphaea and Ranunculus.

Apolar Pollen polarity in which polar and equatorial regions cannot be distinguished after pollen grain separation from the microspore tetrad.

Apomorphy Adj: apomorphic. A derived condition or character state, representing an evolutionary novelty, e.g. apomorphy-based clade (all members of the group share a given, unique evolutionary event).

Apopetalous Syn: choripetalous, polypetalous. With distinct petals.

Aposepalous Syn: chorisepalous, polysepalous. With distinct sepals.

Apostemorious With distinct stamens.

Apotepalous Syn: choritepalous, polytepalous. With distinct tepals.

Appressed Pressed closely to an axis oriented upward, with a divergence angle of 0-15 from upper axis.

Aquatic Growing in water.

Arachnoid/arachnose With trichomes forming a cobwebby mass.

Arborescent Treelike in appearance and size.

Arboreous A tree with well-developed trunk.

Archegonial chamber A cavity between the megasporangium and female gametophyte of gymnosperm seeds, into which sperm cells are released by the male gametophyte.

Archegonial wall Syn: jacket layer; sterile jacket layer. The outer, sterile layer of cells of the archegonium.

Archegoniophore A specialized, stalked, generally peltate structure that grows from the gametophyte of some liverworts and bears archegonia.

Archegonium The female gametangium of the gametophyte, containing a basal egg cell and surrounded by an outer layer of sterile cells, the archegonial wall, which differentiates into a basal ventral and proximal neck.

Archesporial cell A single, large cell of an immature ovule that either directly becomes the megasporocyte or that divides once to form a parietal cell and a megasporocyte.

Areole A modified, reduced, nonelongating shoot apical meristem bearing spines, e.g., Cactaceae.

Aril Adj: arillate. A fleshy outgrowth of the funiculus, raphe, or integuments (but separate from the integuments), generally functioning in animal seed dispersal, e.g., Sapindaceae. It may be a tuft of hairs. A fleshy white aril is distinctive and is the edible part of litchi.

Aristate With a stiff apical process, usually prolonged and straight, length: width ratio >3:1.

Arrangement Placement with respect to similar parts.

Arylphenalenones A class of chemical compounds common in the Haemodoraceae and also found in related families.

Ascending Directed upward, with a divergence angle of 15°-45° from upper axis.

Ascidiate Referring to a carpel that is not leaf like, but develops from a ring of tissue that grows upward, sometimes assuming a somewhat peltate form.

Asepalous Having no sepals or calyx.

Asexual reproduction The transfer of an exact copy of DNA (barring a somatic mutation) from a parent to offspring.

Asexual species Species reproducing without sex, yet are recognizable morphological or genetic units.

Assimilatory roots Hanging green aerial roots which have the power of assimilation (photosynthesis).

Asterad type A type of embryo development in which the terminal cell divides longitudinally, with both basal and terminal cell derivatives contributing to the mature embryo.

Asymmetric Lacking a plane of symmetry and which cannot be divided equally from any plane.

Atactostele A stele consisting of numerous, collateral vascular bundles positioned throughout the stem tissue.

Atectate A pollen grain that ancestrally lacks a tectum.

Ategmic Lacking an integument.

Ategmic ovule. The ovule without any integument as found in *Loranthus* and *San talum*.

Attenuate Basal margins abruptly incurved (concave), the basal intersection angle <45°.

Attractant An aspect of floral morphology that functions to entice an animal pollinator to the flower, either by vision or by odour.

Auctorum non Abbr: *auct.* non. The misapplication of a name, such that the type specimen of the name does not fall within the circumscription of the taken being referred to by that name.

Auriculate With two rounded, basal lobes, margins above lobes concave.

Autapomorphy An apomorphy that occurs for a single lineage or *taxon.*

Author(s) The name of the person(s) who validly published a scientific name.

Autochory Self-dispersal. Adi: autochorous.

Autogamy Syn: *infrafloral selfing.* The transfer of pollen grains from the anther to the stigma of the same flower (Inbreeding) Adj: *endogamous.*

Autonym An automatically created name for infrafamilial, infrageneric and infraspecific taxa.

Autopolyploidy Polyploidy occurring within a species.

Autummal Appearing in fall.

Awn (a) bristle like, apical appendage, e.g., on the glumes or lemmas of grass spikelets. (b) A unit of a pappus type in the Asteraceae that is narrow, elongate, straight, and stiff.

Axial parenchyma Longitudinally oriented parenchyma cells that occur in some secondary xylem (wood) tissues.

Axil The region at the upper (adaxial) junction of leaf and stem.

Axile Ovules present at or near the centre of ovary on the axis formed by the fusion of septa and usually in vertical rows (ovules are attached on the axis of a compound ovary)

axillary Syn: *lateral* (a) On the side of a structure or at the nodes of an axis. (b) With the inflorescence or bud positioned in the axil of the nearest vegetative leaf or leaf scar.

Baccate Syn: *succulent, carnose.* Fleshy or juicy, often with reference to a fruit.

Baculate A pollen sculpturing with rod-shaped elements, each element termed a baculum.

Baculum Pl: *baculi.* Adj: *baculate.* A rod-shaped element, as in the wall sculpturing of some pollen grains.

Balausta A many seeded fruit developed from multicarpellary syncarpous, multilocular superior ovary having persistent calyx. The epicarp is rough arid leathery, mesocarp is papery and thin and endocarp is hard and it forms chambers to enclose seeds irregularly. Testa is red and fleshy whereas tegmen becomes hard as in pomegranate.

Barbed/barbellate With minute, lateral, sharp appendages (barbs) arising along the surface or margin of a bristle, the barbs typically antrorse or retrorse in orientation.

Bark Tissues external to the vascular cambium in stem (and roots) of woody plants, consisting of secondary phloem (inner bark) and derivatives of the cork cambium (outer bark or periderm).

Basal Syn: *radical* (a) At or near the bottom or base of a structure. (b) With the placenta at the base of the ovary, e.g., Asteraceae, Poaceae. (c) With three or more primary veins diverging from one point at the base of the blade, a subcategory of actinodromous and palinactinodromous.

Basic A type of anther wall development in which both secondary parietal cell layers divide to yield two middle layers.

Basicidal capsule A capsule dehiscing by means of basal slits, as in *Aristolochia* spp.

Basifixed. Filament attached with the base of anther lobe

Basionym The original, but now not accepted, name, part of which has been used in a new combination.

Bayesian analysis/inference A method of phylogenetic inference based upon the posterior probability of a phylogenetic tree.

Beak An extended, usually accrescent, basal stylar region, typically functioning in fruit dispersal, e.g., *Taraxacum,* dandelion.

Beard A tuft, line, or zone of trichomes on a perianth or perianth part. Adj: *bearded.*

Berry The fruit develops from mono- or multicarpellary, syncarpous, superior or inferior ovary with axile or parietal placentation. Epicarp forms the rind of the fruit, mesocarp becomes fleshy and endocarp remains thin or membranous, as in grape, tomato, brinjal, guava, date, papaya, chiku, areca nut, etc.

Bicarpellary Gynoecium formed by fusion of two carpels as in Cruciferae.

Bicellateral bundle A vascular bundle with phloem to both the inside and outside of the xylem.

Bicolour unit An inflorescence unit, possibly an apomorphy of the Malvaceae s.l., consisting of a modified, three-bracted cyme, and the bracts modified into an epicalyx in members of the group.

Bicrenate Margin with tooth and teeth again crenate.

Bidentate Teeth again dentate.

Biennial A plant which completes its life cycle in two seasons, typically forming a basal rosette of leaves during the first year and flowering with an elongate inflorescence stalk in the second year, e.g. radish.

Bifacial Secondary growth in which layers of cells are produced both to the inside and outside of a continuously generated cambium.

Bifid Two lobed to two divided, especially at the apex.

Bifoliate Only two leaflets articulated to the rachis as in *Hardwickia binata* and *Bignonia grandillora.*

Bigeminate A compound leaf with two rachillae, each bearing two leaflets.

Bilabiate Two lipped, with two, generally upper and lower segments, e.g., many Lamiaceae.

Bilabiate ringent Corolla irregular with petals used to form a bilipped structure but the slit in between two remains open or broader.

Bilabiate personate. Corolla irregular with petals fused to form a bilipped structure having compactly closed slit.

Bilabiate. Sepals fuse and form a two lipped structure

Bilateral Zygomorphic, irregular monosymmetric.

Bilocular Ovary with two chambers, as in Cruciferae and Solanaceae.

Bimerous Referring to a whorl with two members.

Binary character A character with only two character states.

Binary combination Binomial. Format for the scientific name of species, composed of two names, the genus name and the specific epithet, italicized or underlined.

Binucleate Having two nuclei, referring to some angiosperm pollen grains at the time of release.

Biodiversity The totality of life within a given region.

Biological species Syn: *isolation species.* Groups of populations that interbreed (actually or potentially) and that are reproductively isolated from other such groups in nature.

Biparous Two branches at each node.

Bipinnate bipinnately compound A compound leaf with two orders of axes, the second axes (rachillae) bearing leaflets.

Bipinnate A compound leaf with main segments again pinnately divided into pinnules, i.e., leaflets (pinnules) borne on the rachis but on its branches.

Bipinnatifid Bipinnately lobed.

Biradial Having two planes of symmetry. Adj. disymmetric.

Biseriate With two whorls of parts (perianth parts, stamens) Syn: dicyclic. Rays in wood that are made up of two vertical rows of cells.

Biserrate Teeth again serrated as in elm (Maple).

Bisexual Flowers having both carpel(s) and stamen(s). Syn: *perfect.*

Bisporangiate Anther with two microsporangia and typically one theca. Cf: *monothecal.*

Bisporic Megasporogenesis in which cytokinesis occurs after the first meiotic division, but not the second, resulting in two cells, each of which contain two haploid nuclei, with one of the binucleate cells contributing to the female gametophyte.

Bitegmic With two integuments, apomorphic for the angiosperms, as in ovule

Bitenate, biternately compound A compound leaf with three axes, each of which is ternately compound.

Bivariate plot A graph of a two variables, for several individuals, samples, or taxa.

Blade The flat, expanded portion of leaf. Syn: *lamina.*

Bladder Leaf segments modified into bladder to catch the insects as in aquatic insectivorous plant, e.g. *Lltricularia*

Bootstrap/bootstrapping A method of evaluating cladogram robustness that reanalyzes the data of the original character of taxon matrix by selecting (resampling) characters at random, such that a given character can be selected more than once.

Bostryx *Helicoid* cyme.

Botany The traditional study of photosynthetic organisms (including the green plants, red algae, brown plants, dinoflagellates and euglenoids, but excepting the photosynthetic bacteria), the true fungi, and groups that used to be treated as fungi, such as the slime molds; inclusive of the plant sciences.

Bract A modified, generally reduced leaf, generally found associated with reproductive organs, e.g., subtending the ovuliferous scale of conifers or the flowers or inflorescence axes of flowering plants. Adj: *bracteate.*

Bracteate When bract is present in a flower.

Bractlet/bracteole Syn: prvphyllum. A smaller or secondary bract, often borne on the side of a pedicel.

Branched stamens Filaments of the stamens found branched as in castor

Bremer support *Decay index.*

Bristle An external hair like plant structure, but stouter than a trichome.

Brochidodromous Pinnate venation in which secondary veins do not terminate at the margin, forming prominent upward loops near the margin, joining other, more distal, secondary veins.

Bud An immature shoot system, often surrounded by protective scale leaves, developing into a lateral branch, a flower, or an inflorescence; may be gametophytic or sporophytic.

Bud primordium An immature bud of the shoot, typically located in the leaf axil.

Bulb A rounded, modified, underground stem (bud) bearing highly reduced (disc-like) stem and fleshy scale leaves which make up the bulk of the organ, e.g. onion.

Bulbel A proliferative bulb arising from existing bulbs at or below ground level.

Bulbil A proliferative bulb arising from shoots above ground, typically within an inflorescence.

Bullate *Rugose.*

Bundle cap An outer patch of sclerenchyma fibres associated with a vascular bundle.

Burr A multiple fruit of achenes or grains surrounded by a prickly involucre, e.g., *Cenchrus*, Xanthium (Asteraceae).

Burl A swollen, woody underground stem from which arises persistent, woody, aerial branches, e.g., fire regenerative stems in some *Arctostaphylos* spp., Manzanita. Syn: *lignotuber.*

Buttress roots Enlarged, horizontally spreading and often vertically thickened roots at the base of trees that aid in mechanical support.

C4 photosynthesis An alternate photosynthetic pathway of some land plants in which carbon dioxide is initially fixed in the mesophyll cells by the enzyme PEP carboxylase, producing a four-carbon molecule, which is transported to bundle sheath cells, where the carbon dioxide is released and fixed by ribulosebisphosphate carboxylase in the typical dark reactions.

Caducous Syn- fugacious. Dropping off very early compared with what is typical, usually applied to floral parts as in *Brassica* and laves as in *Opuntia.*

Callose A polyssaccharide, composed of beta- 1,3-glucose units, which lines the pores of sieve areas and sieve plates of sieve clemeats and is commonly deposited within pollen tubes.

Calyptra An apical region of archegonial tissue that is from and lifted up by the elongating sporpphyte during the latter's development and that may function to protect the young sporophyte apex.

Calyptrate Having calyx and corolla fused into a cap that falls off as a unit, e.g. Eucalyptus.

Calyx The outermost series or whorl of modified leaves in the perianth, the units of which are sepals. Sometimes a whorl of epicalyx may be present outer to it as in china rose.

Calyx lobes The segments of a calyx that is synsepalous (with connate sepals).

Campanulate Bell shaped, i.e., with a basally rounded, flaring tube about as broad as long and flaring lobes, e.g., *Campanula;* may also be used for bell-shaped apopetalous corolla or apotepalous perianth or fused sepals.

Camptodromous Pinnate venation in which secondary veins do not terminate at the margin.

Campylodromous With several primary veins running in prominent, recurved arches at the base, curving upward to converge at the leaf apex.

Campylotropous A general ovule type in which curvature of the ovule during development displaces the micropyle adjacent to the funiculus base, with the nucellus bent only along the lower side, and the body of the ovule lies at right angle to the funiculus as in Capparaceae and Leguminosae.

Canaliculate Longitudinally grooved, usually in relation to petioles or midribs.

Canescent Covered with dense, fine grayish-white trichomes; whitish-pubescent.

Capillary bristle A unit of a pappus type in the Asteraceae that is generally straight, very thin, and threadlike, often barbellate.

Capitate Head shaped; spherical with a short basal stalk.

Capitulum or Head It has a flattened receptacle on which several sessile small florets are arranged in a centripetal manner, i.e., youngest in the centre and oldest towards the periphery. Individual florets are bracteate. The whorl of bracts present is called involucre.

Capsule The spore producing component of the sporophytes of liverworts, hornworts, and mosses or A dry, dehiscent fruit derived from a multicarpellary, syncarpous, superior ovary with several locules and axile placentation.

Carcerulus The fruit develops from bi- or polycarpellary, syncarpous. superior ovary and is mutilocular. Number of locules may increase due to false septation. On maturity, single seeded mericarp splits away, as in Althaea, Ocimum and Malva.

Carinate Keeled, having a sharp, median fold projected on the abaxial side; sharply conduplicate.

Carnose *Baccate* Succulent.

Carpel The unit of the gynoecium of angiosperms consisting of ovary, style and stigma; ancestrally, a modified, conduplicate female (mega-) sporophyll of a flower, at maturity enclosing one or more ovules.

Carpophore The stalk like structure that is attached to the mericarps in a schizocarp of mericarps, e.g., Apiaceae.

Cartilaginous With the texture of cartilage; hard and tough but flexible, usually whitish.

Caruncle A fleshy outgrowth at the base of the seed near the micropyle and hilum, generally functioning in animal seed dispersal, e.g., Viola, castor.

Caryophyllaceous. Petals with comparatively long claws and with limbs placed at right angles to the claws.

Caryophyllid type A type of embryo development in which the terminal cell divides transversely, with only terminal cell derivatives contributing to the mature embryo.

Caryopsis *Grain.* The fruit develops from monocarpellary, superior ovary and it remains unilocular and single seeded. The pericarp remains fused with testa. Such fruits are characteristic to family Gramineae, e.g. wheat, rice, maize etc.

Casparian strip A band or ring of mostly suberin that infiltrates the cell wall of endodermal cells, functioning to force water and mineral solutes to pass through the plasma membrane of these cells.

Cataphyll A scale like, often non-Greek, meristem protecting leaf (e.g., in cycads or palms); a rudimentary scale leaf-dry, papery or fleshy scale leaves or bud scales found in usually hypogeous (cryptocotylar) seedlings.

Categorical characters Qualitative characters.

Catkin *Syn: ament.* A unisexual, typically male spike or elotigate delicate and pendulous floral axis that falls as a unit after flowering or fruiting, e.g., *Quercus,* mulberry. All flowers are sessile and densely set.

Caudate Abruptly acuminate into a long, narrowly triangular (tail-like) apical region.

Caudex A short, thick, vertical or branched perennial stem, underground or at/near ground level (stem shoot type). Main trunk straight, unbranched and columnar, bearing crown of leaves (trunk type).

Caudiciform stem A low (at or above ground level), swollen, perennial storage stem from which arise annual or otherwise non persistent photosynthetic shoots, e.g., some *Dioscorea* spp., *Calibanus.*

Caulescent Having an above ground, vegetative stem.

Cauliflorus Inflorescence growing directly from a woody trunk.

Cauline Positioned along the length of the stem. Leaves borne directly on the main trunk

Cell The structural unit of all life.

Cell differentiation The series of changes that a cell undergoes from the point of inception to maturity, involving the transformation of a meristernatic cell into one that assumes a particular structure and function.

Cell expansion Growth of a cell in size, often involving axial elongation.

Cell theory The postulate that all life is composed of one or more cells, that cells arise only from preexisting cells, and that cells are the units of metabolic processes.

Cell wall A layer of the plant cell that is secreted outside the plasma membrane.

Cellular endosperm An endosperm in which the endosperm cell divides mitotically, regularly followed by cytokinesis, each endosperm nucleus contained within a cell wall.

Cellulose A polymer of glucose sugar units (= polysaccharide) in which the glucose molecules are chemically bonded in the beta 1,4 position (= β1,4-glucopyranoside); a major component of the cell wall of green plants.

Central At or near the middle or middle plane of a structure.

Central cell In a typical angiosperm female gametophyte, the single, relatively large cell in the central region that contains the two polar nuclei.

Centric leaf A leaf that is cylindrical in shape, e.g., *Fenestraria* of the Aizoaceae.

Centrifugal Developing from the center region toward the outside or periphery; can be applied to parts of perianth, calyx, corolla, androecium, or gynoecium.

Centripetal Developing from the outside or periphery toward the center region; can be applied to parts of perianth, calyx, corolla, androecium, or gynoecium.

Cernuous With tip drooping downward, abaxially.

Cespitosekaespitose Referring to a generally short, bunched, much-branched plant forming a cushion.

Chaff Syn: palea. One of the bracts subtending flowers in some Asteraceae, e.g., tribe Heliantheae.

Chalazal Referring to the proximal region of the ovule, opposite the micropyle.

Character A feature or attribute of a taxon.

Character correlation The condition in which one character is interdependent upon and influenced by another character.

Character evolution The sequence of evolutionary changes occurring for a given character.

Character optimization The representation of characters in a cladogram in a most parsimonious way, such that the minimal number of character state changes occur.

Character state One of two or more forms of a character.

Character step matrix A numerical tabulation of the number of changes occurring between all pair wise combinations of character states for a given character.

Character X taxon matrix A numerical tabulation of the characters and corresponding character states for each taxon in a phylogenetic analysis.

Chartaccous Opaque and of the texture of writing paper.

Chasmogamy Adj: chasmogamous. Referring to typical flowers in which the perianth opens and exposes the sexual organs, with subsequent crosspollination common.

Cheiropterophily Adj: cheiropterophilous. Pollination by bats.

Chenopodiad type A type of embryo development in which the terminal cell divides transversely, with both basal and terminal cell derivatives contributing to the mature embryo.

Chlorobionta A monophyletic group of eukaryotes, consisting of the "green algae" and the land plants, united by having chloroplasts with chlorophyll a and b, starch, and thylakoids stacked as grana (green plants).

Chlorophyll a The primary pigment in the light reactions of photosynthesis, found in the chloroplasts of all photosynthetic eukaryotes and some bacteria.

Chlorophyll b An accessory pigment in the light reactions of photosynthesis, an apomorphy of the green plant chloroplast.

Chlorophyllous Having chloroplasts at maturity; green.

Chlorophyllous cell 1 Any general, chloroplast-containing cell. 2. One of the chloroplast-containing cells in the specialized leaves of Sphagnum moss that border and surround a hyaline cell.

Chloroplast A double membrane bound organelle with internal thylakoid membranes (lamellae and grana in the green plants), functioning in the reactions of photosynthesis.

Choripetalous Apopetalous, polypetalous.

Chorisepalous Aposepalous, polysepalous.

Choritepalous Apotepalous, polytepalous.

Chromoplast Carotenoid-containing bodies that function to provide yellow, orange, or red pigmentation for a plant organ, as in petals or fruits; a type of ergastic substance.

Ciliate With conspicuous marginal trichomes.

Ciliolate With minute trichomes protruding from margins; minutely ciliate.

Cincinus (or Cincinnus). Scorpioid cyme. A monochasial cymose inflorescence with the lateral axes arising on alternate sides of the relatively main axis.

Circinate Coiled from apex to base, as with the leaf blade and rachis/rachillae, if present, in young fern and cycads or the inflorescence unit of most Boraginaceae.

Circinotropous type It is a special type of ovule formed due to unilateral growth. Initially, the nucellar mass is in line with the funicle and appears orthotropous, then unilateral growth of ovule takes place and it gets inverted, thus becoming anatropous. The curvature continues till the ovule takes a complete 1800 turn and the micropyle again faces upwards as in Opuntia.

Circumferential At or near the circumference; surrounding a rounded structure.

Circumscissile capsule Syn: pyxis/pyxide. A capsule having a transverse line of dehiscence, e.g. Plantago.

Circumscription The boundaries of a taxon, what is included within it and how it is delimited from other taxa.

Cirrhose Mucronate apex with a fine thread-like structure as in banana. With a flexible, greatly curled apical process

Clade Cf. lineage. A monophyletic group, consisting of a common ancestor and all lineages arising from that common ancestor.

Cladistics *Phylogenctic systematics.* A methodology for inferring the pattern of evolutionary history of a group of organisms, utilizing grouping of taxa by apomorphies

Cladode/cladodophyll Adj: *phylloclade.* A flattened photosynthetic stem consisting of one or two internodes only, functioning as and resembling a leaf. In Asparagus, the ultimate branchlets are of one internode only and modified into cylindrical leaf-like structures, whereas leaves are reduced to scales or spines.

Cladodromass Pinnate venation in which secondary veins do not terminate at !he margin and branch near the margin.

Cladogram Syn: *phylogenetic tree.* A branching diagram that conceptually represents the best estimate of phylogeny.

Cladogram robustness A measure of the confidence for which a cladogram actually denotes phylogenetic relationships, e.g., by bootstrapping.

Cladoptosic Referring to dead foliage falling with an accompanying shoot, rather than as individual leaves, e.g., Cupressaceae such as *Taxodium.*

Clambering Syn: *scandent.* Sprawling across objects, without specialized climbing structures.

Classification The arrangement of taxa (or other entities) into some type of order or grouping.

Clathrate Cf: nonclathrate. Fern scales in which the cell walls of adjacent cells (anticlinal walls) are thick.

Clava PI: *clavae.* A club shaped element, as in the sculpturing of some pollen grains.

Clavate Club shaped; terete with a gradually tapering thickened and rounded end.

Claw An attenuate base of a sepal or petal.

Cleft Sinuses extending (pinnately or palmately) one quarter to one half the distance to the midrib, midvein, or vein junction.

Cleistogamy Adj: *cleistogamous.* Referring to flowers in which the perianth remains closed, such that pollen produced from within the flower pollinates only the stigma(s) of that flower. Flowers do not open at the time of the fertilisation and do not expose their sex organs, as in *Commelina benghalensis*.

Climber A plant with weak stem which climbs by taking support of neighbouring objects.

Climbing Growing upward by means of tendrils, petioles, or adventitious roots.

Climbing roots adventitious roots arising from the node of the stem that help in climbing or twining of the plant.

Clinging roots Roots to cling the plant on the stem of other plant as found in epithytic plants such as orchids.

Crown clade A clade in which both or all branches from the common ancestor contain extant members.

Coherent With like parts joined, but only superficially and easily separable.

Cohesion species The largest or most inclusive group of individuals that maintains genetic and phenotypic cohesion via gene flow, genetic drift, and natural selection, maintaining the fundamental niche through genetically determined environmental tolerances.

Coleoptile A protective sheath surrounding the epicotyl, e.g., in some members of the Poaceae.

Coleorhiza A protective sheath surrounding the radicle, e.g., in some members of the Poaceae.

Collateral A vascular bundle with an internal strand of xylem and an external strand of phloem.

Collaleral bud Bud(s) lateral to the axillary bud.

Collateral bundle A vascular bundle with xylem to the inside and phloem to the outside.

Collenclivina A cell type that is live at maturity, has unevenly thickened, pectic-rich, primary cell walls, and functions in structural support, often found at the periphery of stems or leaves.

Colleter A structure on the inner surface of connate stipules that secretes mucilage, aiding in protection of young, developing shoots, e.g., Rubaceae.

Colour pattern The distribution of colours on an object.

Colporate A pollen grain aperture type that is shaped like a colpus but has a circular region in the centre.

Colpus Pl: colpi. Adj: *colpate.* A pollen grain aperture that is elongate with a length: width ratio of greater that, 2:1.

Columella Pl: *columellae.* (a) A central column of sterile (non-spore-producing) tissue within the sporophyte capsule of homworts (b) one of the middle, radially elongate elements of a tectate-columellate pollen exine wall.

Column (a) The central axis to which septae and/or placentae are attached in axile or free-central placentation (b) gynandrium, gynostegium, *gynostemium*

Combinatio nova Abbr: *comb. nov.* Indication that a taxon has recently been transferred to a new position or rank.

Commemorative A name that is after a person or place.

Common name A vernacular name, used by people within a limited geographic region, not formally published and governed by rules.

Comose With an apical tuft of trichomes.

Companion cell A parenchyma cell associated with a sieve tube member, derived from the same parent cell as is the sieve tube member and functioning to load and unload sugars into the cavity of the sieve tube member.

Complete A flower when it contains all the four whorls (calyx, corolla, androecium and gynoeciurn).

Complex tissue A tissue that contains more than one cell type.

Composite fruits (Multiple) Fruits developed from complete inflorescence.

Compound cone The cone of a conifer, consisting of an axis bearing bracts, each of which subtends a modified branch system, the ovuliferous scale.

Compound corymb A branched corymb, consisting of two or more orders of inflorescence axes bearing flat-topped or convex, pedicellate flowers.

Compound cyme A branched determinate inflorescence, similar to a compound dichasium, but lacking a consistent dichasial branching pattern, often by reduction of internodal axes.

Compound dichasium A many flowered, determinate inflorescence of repeatedly branching simple dichasia units.

Compound leaf The leaf has incised margin, and the incision goes so deep, touching the midrib, so that lamina breaks up into two or more lobes or segments called leaflets.

Compound ovary/pistil An ovary/pistil composed of two or more carpels, the gynoecium, syncarpous.

Compound perforation plate A perforation plate composed of several openings.

Compound raceme or panicle As in *Delonix regia* and Yucca.

Compound receptacle Syn: torus. A mass of tissue at the apex of a peduncle that bears more than one flower.

Compound sieve plate A sieve plate that is made up of two or more aggregations of pores.

Compound spike Spike of spikelets, e.g. wheat.

Compound umbel An umbel with the peduncle bearing rays attached at one point and unit simple umbels attached at the tip of the rays, e.g. Apiaceae.

Conduplicate Longitudinally folded at the central axis, with adjacent adaxial sides facing one another.

Cone Syn: *strobilus.* A modified, determinate, reproductive shoot system of many nonflowering vascular plants, consisting of a stem axis bearing either sporophylls, (in simple cones) or ovuliferous scales subtended by bract (in compound Leaves of conifers).

Confer Abbr: cf. Indication that the identity of a specimen is questionable or uncertain and should be compared with specimens of the taxon indicated for more detailed study.

Conical roots Roots broad at the upper end and gradually tapering towards the lower end.

Configuration Referring to gross surface patterns other than venation or epidermal excrescences.

Connate 'With like parts integrally fused, not easily separable.

Connate-perfoliate Two opposite leaves fused basally, such that the blade base of each leaf completely surrounds the stem.

Connective The tissue or filament extension between the thecae of an anther.

Connivent Convergent apically without fusion.

Consensus tree A cladogram derived by combining features common between two or more cladograms.

Conservation biology Branch of biology dealing with the preservation of biodiversity.

Conservation of names A principle of the International Code of Botanical Nomenclature stating that scientific names that are well known and frequently used may be retained over other, earlier, but less well-known, names.

Consistency index (CI) A measure of the relative amount of homoplasy in a cladogram, equal to the ratio of the minimum possible number of character state changes to the actual number of changes that occur.

Contiguous With parts touching but not fused.

Continuous quantitative characters Cf: *discrete quantitative characters.* Characters in which individual measurements are not necessarily integers and may potentially form a continuum.

Contorted *Convolute* Perianth aestivation.

Contractile roots Special type of roots which can contract or pull the stem deep into the soil.

Convergence Syn: *parallelism.* Homoplasy occurring by the independent evolution of a similar feature in two or more lineages.

Convergent Veins converging towards the tip as in jujube and *Cinnamomum tamala.*

Convolute Syn: *contorted.* Perianth parts of a single whorl overlapping at one margin, being overlapped at the other as in Malvaceae.

Cordate Heart-shaped. With two rounded, basal lobes intersecting at a sharp angle, the margins above lobes smoothly rounded.

Cordate/cordiform Shaped like an upside-down Valentine heart; approximately ovate with a cordate base.

Coriaceous Thick and leathery, but somewhat flexible.

Cork The outermost layer of the periderm, generated by the cork cambium.

cork cambium A sheath or hollow cylinder of cells that develops near the periphery of the stem or root, undergoing tangential divisions to form phelloderm to the inside and cork to the outside.

Corm An enlarged, solid, highly condensed underground storage stem or stem base with a large apical bud and with outer, protective scale leaves on its body and adventitious roots growing either from the base or all over its body.

Cormel A proliferative, corm arising from existing corms.

Corolia The innermost series or whorl of modified leaves in the perianth, the units of which are petals.

Corolla lobe A segment of a sympetalous corolla (with connate petals).

Corona A crown like outgrowth between stamens and corolla; may originate from petals or stamens.

Coronate With a tubular or flaring perianth or staminal outgrowth as in Narscissus, Asclepias.

Correct name A validly published, legitimate name that is accepted by a particular author or authors.

Cortex The outer, mostly parenchymatous tissue inner to the epidermis and external to the vasculture.

Corymb An indeterminate inflorescence consisting of a single short main axis and the lower flowers having much longer pedicels than the upper ones so that all flowers are brought more or less to the same level, e.g. candytuft (*Iberis*) and cherry (lateral axes and/or pedicels bearing flat-topped or convex flowers).

Cosmopoliton Having a worldwide distribution.

Costa (a) Midrib. (b) The nonvascularized conductive tissue found in the gametophytic leaves of some mosses.

Costapalmate Essentially palmately lobed compound but with an elongate, rachis like extension of the petiole as in some palms.

Cotyledon A first (seed) leaf of the embryo often functioning in storage of food reserves.

Cotyledonary leaf Cotyledon green and appearing like foliage leaf as in Tilia and Geranium.

Cotylespermous With the food reserve in the cotyledon, part of the embryo.

Couplet The pair of contrasting leads in a dihotomous key.

Craspedodromous Palmate venation in which secondary veins terminate at the leaf margin.

Crassinucellate An ovule in which the nucellus develops two or more layers of cells, the inner ones from divisions of a parietal cell.

Crassulacean acid metabolism Abbr: CAM. An alternate photosynthetic pathway in some xeric, generally succulent plants and functioning to conserve water, in which initial fixation of carbon dioxide occurs at night (when stomata are open) by the enzyme PEP carboxylase to form malic acid, which is stored within vacuoles of the mesophyll cells; during the day the stomata close and CO2 is released from the vacuoles into the cytoplasm, where it is fixed in the chloroplasts.

Creeper Plants with stem growing along the surface of the ground and rooting at its node.

Cremocarp Fruit developed from bicarpellary, syncarpous, inferior ovary. On maturation, it divides along with carpophore into two mericarps, each single seeded.

Crenate With rounded to obtuse, shallowly ascending teeth, cut from 1/16 to 1/8 the distance to the midrib, midvein or junction of primary veins.

Crenulate Diminutive of crenate, teeth cut $\leq 1/16$ the distance to the midrib, midvein, or junction of primary veins.

Cross pollination (Allogamy) The transfer of pollen grains from the anther of a flower to the stigma of another flower (i.e., pollination between the flowers of two different plants).

Crown clade A clade in which both or all branches from the common ancestor contain extant members.

Crownshaft The collection of overlapping, sheathing leaf bases at the apex of a palm trunk.

Crozier A leaf that is coiled during its development, characteristic of the leptosporangiate ferns (Polypodiales) and Marattiales.

Cruciate With four distinct petals in a cross form as in Brassicaceae.

Crucifer type Syn: *onagrad type.* A type of embryo development in which the terminal cell divides longitudinally, with only terminal cell derivatives contributing to the mature embryo.

Cruciform Petals arranged in a cross-like fashion as in Brassica

Crumpled Having a wrinkled or crinkled appearance, particularly in bud.

Cryptantherous *Inserted.*

Cryptic species Species that are not morphologically distinguishable from another species.

Cryptocoylar *Hypogeous.*

Crystal A deposit of silica or calcium oxalate in plant cells that may function as waste products, as calcium ion sinks, or as an irritant to deter herbivory; a type of ergastic substance.

Cucullate Hooded; with an abaxially concave posterior lip.

Culm The flowering and fruiting stem(s) with distinct nodes and internodes as in grasses.

Cultivar The word is a contraction of 'cultivated variety' and describes a group of cultivated plants within a species that are significant in agriculture, forestry or horticulture and have clearly distinguished heritable characteristics. 'Cultivar' is synonymous with the term 'variety'. However, it is not analogous with the category 'botanical variety' that is used to refer to naturally occurring variants within a species (Hartmann and Kester, 1975).

Cuneate Wedge-shaped. With basal margins approximately straight, intersection angle 45°-90°.

Cup-shaped Concave-convex along entire surface; may be abaxially or adaxially concave.

Cupulate Sepals fused to form a cup-like structure.

Cupule A structure that encloses a cluster of unitegmic: ovules/ seeds, with a small opening near the proximal end, through which pollen grains enter. Hard and woody bracts and bracteoles found at the base of flower, somewhat like an involucre that fuse together to form a cup-like structure (cupule) as the fruit matures.

Curator The person in charge of the day-to-day running of an herbarium.

Cuspidate Abruptly acuminate into a triangular, stiff or sharp apex, spiny as in pineapple, date palm, screw-pine, etc.

Cuticle A protective layer containing cutin that is secreted to the outside of epidermal cells functioning to inhibit water loss; found in all land plants.

Cutin A polymer of fatty acids, functioning as a sealant in the cuticle layer of land plants, inhibiting water loss.

Cyathium An inflorescence bearing small unisexual flowers and subtended by an involucre (frequently with petaloid glands), the entire inflorescence resembling a single flower, e.g., *Euphorbia.* The bracts or involucre becomes fused to form a cup-shaped structure. Few nectaries are found in the involucral cup. This cup completely encloses a centrally placed female flower surrounded by a large number of male flowers. Each male flower is bracteate and achiamydeous and is represented by a single stalked stamen. The female flower is bracteate and achiamydeous and is represented by stalked (gynophore), tricarpellary and syncarpous gynoecium. The inflorescence is a special cymose type in which male flowers are arranged basipetally.

Cyclic phyllotaxy The leaves at each node form a whorl with the leaves placed on a circle in which the angles between adjacent leaves are same.

Cycly Number of cycles or whorls of parts.

Cylindrical or centric leaves With no distinct two surfaces, e.g. *Alium cepa.*

Cymbiform Boat shaped, e.g., glumes of grasses.

Cyme General term for a determinate inflorescence.

Cymose The terminal (apical) bud loses its activity and axillary buds grow vigorously to form a large spreading tree as in mango, neem (*Azadirachta indica*), banyan, etc.

Cymose inflorescence Inflorescence having determinate (definite) growth of the main axis, as the apical bud forms the first flower. Other flowers are produced on lateral branches which also bear terminal flower. Thus, every time new lateral branch is formed to produce new flower. Flowers are arranged centrifugally.

Cymule A diminutive of cyme inflorescence (simple dichasium).

Cypsela *Syn: achene* Fruit (achene) developed from bicarpellary syncarpous, inferior ovary with basal placentation. Fruits are unilocular and single seeded. Persistent hairy calyx (pappus) are found at the apex of fruit.

Cystolith A mass of calcium carbonate attached to a stalk from the cell wall, occurring within specialized cells termed lithocysts.

Cytoplasm Everything inside the plasma membrane but not including the nucleus.

Dark reactions Biochemical reactions of photosynthesis in plants, occurring in the stroma of the chloroplast, where atmospheric carbon dioxide reacts to produce a molecule of glucose, requiring the input of the high-energy compounds ATP and $NADPH_2$.

Data information system/database system. The (computerized) organization for inputting and accessing of information.

Decay index Syn: *Bremer support.* A method of evaluating cladogram robustness by calculating how many extra steps are needed (beyond the number in the most parsimonious cladograms) before the original clade is no longer supported; the greater this value, the greater the confidence in a given clade.

Deciduous Parts persistent for one growing season and then falling off, e.g., leaves of deciduous plants, petals fall off after drying.

Decompound A leaf having leaflets in two or more orders: bi-, tri-, and so on, pinnately, palmately, or ternately, compound; also used for a highly divided leaf.

Decumbent Basally prostrate but ascending apically; trailing on the ground to some distance and then straightens upright.

Decurrent Appearing to extend down the stem from the point of attachment, as if fused to the stem. A winged leaf base which extends down to the stem.

Decussate Opposite leaves or other structures at right angles to the preceding pair.

Decussate tetrad A tetrad in which the four grains are in two pairs arranged at right angles to one another.

Deflexed Bent abruptly downward.

Dehiscent or Capsular fruits Pericarp is ruptured after ripening and seeds are dispersed.

Deliquescent Spreading canopy due to suppressed apical bud and dominant lateral buds. Culms. Grass-like strong stem with distinct nodes (solid) and internodes (hollow), may look-like tree.

Deltate Three- sided, length: width ratio 1: 1.

Dendritic Trichomes treelike, with multiple lateral branches.

Dendrochronology The scientific study of wood anatomy to infer details about past events.

Dentate With sharp, coarse teeth that point outward at right angles to margin outline as in *Nymphaea*. Cut from 1/16 to 1/8 distance to midrib, midvein or junction of primary veins.

Denticulate Diminutive of dentate, teeth cut to 1/16 or less the distance to the midrib, midvein, or junction of primary veins.

Depressed Pressed closely to axis downward with divergence angle of 0°-15° from lower axis.

Derived Syn: *apomorphic, advanced.* Referring to a new condition or character state.

Dermal tissue The outer region of plant organs composed of the epidermis and periderm.

Descending Directed downward with divergence angle of 15'-45' from lower axis.

Descent The sequence of ancestral-descendant populations through time.

Description The assignment of features or attributes to a taxon or other entity.

Determinate (a) A shoot that terminates growth after a certain period, the apical meristem aborting or converting into a flower, inflorescence, or other specialized structure. (b) An inflorescence in which the terminal flower matures first, maturating from apex to base.

Dextrorse Twining helically like a typical, right-handed screw as in Convolvulaceae.

Diachronic species A species corresponding to a group of organisms that span a lineage of ancestral-descendent populations, extending through a period of time.

Diadelphous With two groups of stamens, each connate by filaments as in Fabaceae.

Diandrous Two stamens in each flower.

Dichasial or Biparous cyme The peduncle bearing a terminal flower and stops growing. From the main peduncle arise two lateral branches, each of which also terminates in a flower. From the pedicel of each lateral flower further branching may arise repeatedly as in *Dianthus.*

Dichasium A determinate inflorescence that develops along two axes, forming one or more pairs of opposite. Lateral axes.

Dichlamydeous Perianth composed of a distinct outer calyx and inner corolla, regardless of total number of whorls.

Dichogamy Adj: *dichogamous.* A type of outcrossing mechanism that is the result of differences in timing of maturation of male and female floral parts.

Dichopodial Roots in which the apical meristem. Branches into two roots.

Dichotomous With veins successively branching distally into two veins of equal size and orientation, e.g. *Ginkgo biloba.*

Dichotomous key A key utilizing a series of two contrasting statements, each statement a lead, the pair of leads a couplet.

Dicotyledonous A type of anther wall development in which only the outer secondary parietal cell layer divides to yield. The endothecium and a single middle layer.

Didynamous Four stamens, of which two have long filaments and two short filaments as in Labiatae and Scrophulariaceae.

Discriminate function analysis A multivariate statistical method that transforms numerous data variables into other variables in order to determine the variables that discriminate among taxa.

Dietyostele A dissected amphiphloic siphonostele.

Dicyclic *Biseriate.* Perianth cycly and stamen cycly.

Didymous With stamens in two equal pairs.

Didynamous With stamens in two unequal pairs as in Bgnoniaccae, Lamiaceae.

Diffuse Trailers growing in all directions.

Diffuse-porous Wood in which vessels develop more or less uniformly throughout the growth season.

Dimorphic Having two discrete forms, as leaves of some species.

Dioecious/dioecy Having unisexual flowers, staminate and pistillate on separate individual plants.

Diphthong A two vowel combination in Latin that is treated as the equivalent of a single vowel.

Dipiohapiontic life cycle Haplodiplontic life cycle.

Diplostemonous Stamens arranged in two whorls on the thalamus and stamens of outer whorl alternate with petals and of inner whorl opposite the petals, as in Rhamnaceae and Portulacaceae.

Disc florets These are grouped in the centre and are actinomorphic, tubular and bisexual.

Discoid Disk shaped stigma(s).

Discrete quantitative characters *Syn: meristic characters.* Characters in which measurements are always of integers.

Disjunct A population that occurs well outside the geographic range of other members of that taxon.

Disk A discoid or doughnut-shaped structure arising from the receptacle at the outside and surrounding the stamens *(extrastaminal disk),* at the base of the stamens *(staminal disk),* or at the inside of the stamens and/or base of the ovary *(intrastaminal disk);* may be nectar-bearing *(nectariferous disk).*

Disk flower Having an actinomorphic, tubular corolla with flaring lobes as in Asteraceae.

Dispersal The movement of an organism or propagule from one region to another, such as the transport of a seed or fruit (by wind, water, or bird) from a continent to an island.

Disposition Relative placement of objects or parts of objects, inclusive of position, arrangement, and orientation.

Dissected Divided into very fine, often indistinct segments.

Distal Away from the point of origin or attachment.

Distal pole The intersection of the pollen grain polar axis with the grain surface that is away from the center of the microspore tetrad.

Distichous ½ phyllotaxy. Alternate, with points of attachment in two vertical rows/ranks as in grasses. The leaf of second node is just opposite the leaf of the first node.

Distinct With like parts unfused and separate.

Distyly Adj: distylous. Hercogamy in which two floral morphologies occur: pin flowers, with a long style and short stamens, and thrum flowers, with a short style and long stamens.

Disulculate A pollen grain with two elongate apertures on opposite sides of the grain, parallel to the equatorial plane.

Disymmetric *Biradial.*

Dithecal Cf: *tetrasporangiate.* Anther with two thecae and typically four microsporangia.

Diurnal During the day, typically with respect to when flowers open.

Divaricate Divergent; horizontal; patent.

Divergence Syn: *evolutionary divergence.* Formation of two (or more), separate lineages from one common ancestor.

Divergent Veins spread out from one point to different directions.

Divided Sinuses extending (pinnately or palmately) 3/4 to almost to midrib, midvein, or vein junction.

Dorsal Abaxial.

Dorsal vein Syn: *median vein.* The central vein of a carpel, corresponding to the midvein of a leaf.

Dorsifixed Anther attached dorsally and medially to the apex of the filament. Filament attached to the dorsal surface of anther.

Dorsiventral Having a flattened shape, with an upper (adaxial) and lower (abaxial) surface.

Daublyserrate With large, serrate teeth that have along the margin smaller, serrate teeth.

Double samara Fruit developed from bicarpellary syncarpous, superior ovary. Pericarp develops into two wings. On maturation it is divided into two one seeded parts, as in Acer (maple).

Downward Anther dehiscing toward the ground in a horizontally oriented flower.

Drepanium A monochasiurn in which the axes develop on only one side of each sequential axis, typically appearing coiled at least early in development; sometimes equated with helicoid cymes.

Drupe The fruit develops from mono- or polycarpellary, syncarpous, superior ovary. Fruits are usually single seeded, rarely many seeded. Pericarp comprises three layers.

The epicarp forms the skin of the fruit. Mesocarp is fleshy or fibrous and endocarp is hard and stony, as in mango, coconut, peach, walnul, cherry, almond, etc.

Drupeceturn An aggregate fruit of drupes, e.g., *Rubus:* raspberry, blackberry.

Druse A spherical crystal with protruding spikes, composed of calcium oxalate; a type of ergastic substance.

Dry fruits Pericarp remains dry and undifferentiated into three layers.

Duration The length of life of a plant or plant part.

Dyad A fusion product of two pollen grains.

Ebracteate Lacking bracts.

Echina PI: *echinae. Adj: echinate.* A spike like sculpturing element >1 μm long, as in some pollen grain walls.

Eciliate Without trichomes on the margins, regardless of presence or absence of teeth.

Ectophloic siphonostele A siphonostele in which a ring of xylem is surrounded by an outer layer of phloem only.

Ectozoochory Adj: *ectozoochorous.* Dispersal by animals in which propagules are carried on the outside of an animal.

Egg A no motile, evolutionarily enlarged gamete, the end product of oogamy.

Ektexine An outer layer of the pollen grain exine wall.

Elater (a) One of the hygroscopic appendages arising from the spores of *Equisetum,* functioning in spore dispersal. (b) A nonsporogenous, elongate, hygroscopic cell with spiral wall thickenings that develops within the sporangia of some liverworts and that functions in spore dispersal.

Eligulate Cf. *ligulate.* Lacking a ligule; usually used for taxa whose close relatives are ligulate.

Elliptic Margins curved, widest at the midpoint, length: width ratio 2:1 to 3:2.

Emarginate Deeply notched obtuse apex as in Bauhinia. Having an apical incision, cut from 1/16 to 1/8 the distance to midrib, midvein, or junction of primary veins.

Embryo An immature, diploid sporophyte developing from the zygote of land plants.

Embryo proper That portion of proembryo that will eventually grow into the new sporophyte.

Embryo sac Female gametophyte of angiosperms.

Embryogeny Development of embryo within the seed.

Embryophyta/embryophytes A monophyletic group of eukaryotes united by an outer cuticle, specialized gametangia-antheridia and archegonia and an intercalated diploid phase in the life cycle, including the embryo [land plants].

Emendatio Abbr: emend. A correction or amendment of a name.

Emergent With roots or stems anchored to substrate under water, the aerial shoots above water.

Emersed Occurring under water.

Enantiostyly Adj: *enantiosylous.* A type of hercogamy in which the style of different flowers curves either the left or the right.

Enation A small appendage arising from the stem, resembling rudimentary leaf but lacking vascular tissue.

Endarch An orientation of xylem maturation in which the protoxylem is oriented toward the centre of the stele relative to the metaxylem, as occurs in eusteles and atactosteles.

Endexine An inner layer of the pollen grain exine wall.

Endocarp The innermost wall layer of the pericarp, if the latter is divided into layers.

Endoderwis Adj. *endodermal.* A hollow cylinder of cells in roots and some stems that surround the vasculature and functions to selectively control passage of solutes from the outside, via Casparian strips.

Endogenous Arising from the internal tissues, as in the growth of secondary roots from within a primary root.

Endoplasmic reticulurn A cellular organelle consisting of interconnected phospholipid membranes that may function in material transport and as the site of protein synthesis.

Endosperm The triploid tissue that develops from mitotic divisions of the endosperm cell (the product of double fertilization), ultimately enveloping or abutting the embryo and functioning as the nutritive tissue of angiosperm seeds.

Endospermous Syn: *albuminous.* With endosperm as the food reserve in mature seeds.

Endospory Adj: *endosporic.* The development of a gametophyte within the original spore wall.

Endostomal Referring to the micropyle of a bitegmic ovule delimited by only the inner integument, the outer one being foreshortened.

Endosymbiosis The intracellular cohabitation of one cell within another cell; the process that gave rise to mitochondria and chloroplasts by engulfment of a prokaryote by a eukaryotic cell.

Endothecium The outermost cell layer of an anther, typically of enlarged cells with secondary wall thickenings functioning in anther dehiscence.

Endozoochory Adj: *endozoochorous.* Dispersal by animals in which propagules are eaten but are passed through the gut of the animal unharmed.

Ensiform Sword-shaped, with length: width ratio greater than 12: 1, e.g., leaves of *Iris* spp.

Entire Without teeth on margins; locally smooth. (The surface may be divided.)

Entomophily Adj: *entomophilous.* Pollination by insects.

Epicalyx A series of bracts immediately subtending the calyx as in *Hibiscus,* china rose.

Epicotyl The first shoot of a vascular plant that develops from the embryo of the seed.

Epidermal excrescence Surface patterns from structural outgrowths or secretions of the epidermis.

Epidermis The outermost cell layer of all land plant organs, functioning to provide mechanical protection of inner tissue and to inhibit water loss.

Epigeous Syn: *phanerocotylar.* With cotyledon(s) elevated above the ground during germination.

Epigynous With sepals, petals, and stamens attached at the apex of the inferior (completely sunken) ovary.

Epihypogynous With sepals, petals and stamens attached at the middle of the ovary, the ovary haff-inferior.

Epihypoperigynous Hypanthium present. Attached at middle of ovary, with sepals, petals, and stamens attached to hypanthium, the ovary half-inferior.

Epiperigynous Hypanthium present, attached at the apex of the ovary with sepals, petals and stamens attached to hypanthium rim, the ovary inferior.

Epipetalous Syn: *petalostemorous.* With stamens adnate to (inserted upon) petals or corolla.

Epiphyllous Stamens are attached to tepals, as in Asparagus (may be called epitepalous).

Epiphyte Adj: *epiphytic.* A plant that grows upon another plant.

Episepalous With stamens adnate to (inserted upon) sepals or calyx.

Epitropous An ovule position in which the micropyle points distally.

Epitropous-dorsal An epitropous ovule position in which the raphe is dorsal (abaxial), pointing away from the central floral or ovary axis.

Epitropous-ventral An epitropous ovule position in which the raphe is ventral (adaxial), pointing toward the central floral or ovary axis.

Epitype A specimen (or illustration) that is selected to serve as an "interpretive" type if the holotype, lectotype or neotype is ambiguous with respect to the identification.

Epiphyte A plant growing on another plant to get only mechanical support, but not deriving any food from it as in most orchids.

Ephemeral A plant which completes its life cycle in a very short time, say within some weeks as in radish.

Equator The intersection with the pollen surface of a plane at a right angle to the polar axis and passing through the centre of the grain.

Equatorial view Observing a pollen grain from the equatorial region.

Equitant Leaves with overlapping bases, usually sharply folded along midrib.

Ergastic substance A cellular substance that does not function in metabolism, generally functioning in storage, waste secretion, and protection.

et Latin for "and," used in scientific name combinations.

Etaerio of achenes It is an aggregate of achene fruitlets developed from a single flower as in rose, strawberry, Clematis.

Etaerio of berries It is an aggregate fruitlets of berries developed from a single flower as in custard apple.

Etaerio of drupes It is an aggregate of fruitlets of drupes developed from apocarpous ovaries of a single flower as in raspberry.

Etaerio of follicles Each free (separate) carpel develops into a fruitlet known as follicle. Many follicles of a flower make it etaerio of follicles as in *Calotropis* and *Catharanthus*.

Eucamptodromous Pinnate venation in which secondary veins do not terminate at the margin, curving upward near the margin but not directly joining adjacent secondaries.

Euphyll Adj: *eiphrllous.* Syn: *megaphyll.* The sporophytic leaf of the euphyll group, growing by means of either marginal or apical meristems having multiple branched veins with an associated leaf gap.

Eusporangium/eusporangiate sporangium A relatively large sporanigum that is derived from several epidermal cells and having a sporangial wall composed of more than one cell layer.

Eustele A primary stem vasculature that consists of a single ring of discrete collateral or bicollateral vascular bundles.

Even-pinnate Paripinnate/paripinnately compound.

Evergreen Persistent two or more growing seasons, e.g., leaves of most conifers.

Evolution Descent with modification; the transfer of genetic material from parent(s) to offspring over time, with a corresponding change in the genetic material.

Evolutionary species A single, continuous lineage of ancestral descendent populations retaining its identity from other such lineages and having its own tendencies and fate.

Evolutionary divergence *Divergence.*

ex Latin for "from"; in nomenclature meaning validly published by.

Exalbuminous Syn: *nonendospermous.* Lacking endosperm as the food reserve in mature seeds.

Exarch An orientation of xylem maturation in which the protoxylem is oriented toward the organ periphery relative to metaxylem as occurs in some protosteles.

Exarillate Cf: *aril.* Lacking an aril on the seed.

Exerted Length of stamens longer than petals so these come out of corolla tube.

Exfoliating Bark cracking and splitting off in large sheets.

Exindussiate Referring to a sorus that lacks an indusium.

Exide The hard, outermost, desiccation-resistant layer of a pollen grain wall providing structural support and inhibiting desiccation.

Excurrent Main trunk keeps on growing but bears lateral branches in acropetal order, giving a conical shape to the tree.

Exocarp The outermost wall layer of the pericarp, if the latter is divided into layers.

Exospory Adj: exosporic. The formation of a gametophyte external to the original spore wall.

Exostomal Referring to the micropyle of a bitegmic ovule delimited by only the outer integument, the inner one being foreshortened.

Explosively dehiscent Referring to a dry dehiscent fruit that opens with force, in the process ejecting the seeds some distance away.

Exserted Syn: *phanerantherous.* With stamens protruding beyond the perianth.

Exstipellate Without stipels.

Exstipulate Without stipules.

Extrastaminal disk A discoid or doughnut-shaped structure arising from the receptacle at the outside and surrounding the stamens; may be nectar-bearing (nectariferous disk).

Extrorse Dehiscing outward, away from the flower centre towards the periphery of the flower.

Facultative hemiparasite A chlorophyllous, normally parasitic plant not requiring a host to survive and reproduce.

Falcate/faiciform Lanceolate to linear and curved to one side; scimitar-shaped.

False indusium An extension of the blade margin that overlaps the sorus of ·a leptosporangiate fern.

False septum The membranous, intervening tissue of a cross-wall that persists after fruit dehiscence as in Brassicaceae.

Farinaceous Adj: granular scurfy. Finely mealy, covered with small granules.

fascicle (a) A shoot with very short internodes on which flowers or leaves are borne. Syn: *short shoot, spur, spur shoot.* (b) A raceme like or panicle like inflorescence with pedicellate flowers in which internodes between flowers are very short, with pedicel bases appearing congested. Adj: *fasciculate.*

Fat A type of triglyceride compound that may function as high energy storage compounds or secretion products in plants; a type of ergastic substance.

Female Individual with female reproductive organs only. *Pistillate.*

Female gametophyte Syn: *megagametophyte, embryo sac.* A gametophyte that bears only archegonia, housing the egg cell.

Female sporophyll Syn: *megasporophyll.* A sporophyll that bears one or more megasporangia or seeds.

Fenestrate Having window like holes in the surface as in *Monstera deliciosa.*

Fertile segment The sporangia containing component of the leaf of an ophioglossoid fern.

Fibre A sclerenchyma cell that is long and very narrow with sharply tapering end walls, functioning in mechanical support and often occurring in bundles.

Fibrous roots Roots that are adventitious and typically fine and numerous. A root system that developing from the base of the plumule (hypocotyl) in which all roots are equally prominent.

Filement A stamen stalk, generally terete in shape.

Filamentous (a) With a more or less terete stalk as opposed to a laminar body. (b) Filament present and anther sessile.

Filiferous Bearing coarse, fibre like structures.

Filifant Long, thin, and typically flexuous. threadlike, filamentous.

Fissured Bark split or cracked into vertical or horizontal, usually coarse grooves. Fistulose.

Fistular Cylindrical and hollow within.

Flabellate With three or more primary veins diverging from one point and several, equal, fine veins branching toward the leaf apex, a subcategory of actinodromous.

Fleshy leaves In some xerophytes and halophytes leaves become fleshy or succulent to store water due to presence of special storage tissue. Such plants are commonly called succulents, e.g. *Portulaca oleracea*, *Basella rubra*, *Swaeda rnaritima* and species of Aloe, Agave, Bryophyllurn, Sedum, etc.

Flexuous Central axis and tip alternately curved up and down.

Floating Occurring at the water surface.

Floating roots Special adventitious spongy and whitish roots filled with air to facilitate aquatic floating plants.

plant floating on the water. These also help in respiration.

Floccose With dense trichomes in several patches or tufts.

Floral Syn: *manual.* A listing of the plant taxa of a given region, usually accompanied by keys and descriptions.

Floral diagram A diagammatic, cross-sectional view of a flower bud, showing the relative relationship of perianth, androecial, and gynoecial components and illustrating things such as stamen position, placentation, perianth, calyx or corolla aestivation.

Floral formula A symbolic representation of floral morphology, including cycly (number or whorls or series), merosity (number of parts per whorl), fusion of parts, and ovary position.

Floral leaves or sporophylls stamens and carpels are called sporophylls in common with lower plants.

Floral receptacle *Receptacle.*

Floral tube *Hypanthium.*

Floret A unit of a grass (Poaceae) spikelet, consisting of a short, lateral axis bearing two bracts (lemma and palea) that subtend a terminal, reduced flower.

Floristics The documentation of all plant species in a given geographic region.

Flower The reproductive organ of flowering plants; a modified determinate shoot bearing sporophylls (stamens and/or carpels) type) with or without outer modified leaves, the perianth.

Flower bract A modified, generally reduced leaf subtending a flower.

Flower bud A bud that develops into a flower.

Foliaceous stipules as in *Pisum.*

Foliaceous or Leafy bract Bract appears like ordinary green leaf, as in china rose.

Foliage leaves Common green leaves.

Follicetum An aggregate fruit of follicles, e.g. *Magnolia.*

Follicle A dry, dehiscent fruit derived from one carpel that splits along one suture, e.g. *Asclepias,* milkweed, larkspur, Calotropis, Michelia, Vinca etc. Such fruits develop from superior and unilocular ovary and are usually found in pairs or groups.

Food storing roots These accumulate reserve food material to carry out physiological functions.

Foot-layer The inner layer of a tectate-columellate pollen exine wall.

Form genus A genus that corresponds to a particular organ of a fossil plant.

Fossulate A pollen sculpturing with longitudinal grooves.

Foveolate A pollen sculpturing with a pitted surface caused by pores in the surface.

Fragmentation Whereby an originally broadly distributed species with more or less continuous gene exchange and clonal intergradations becomes split into two or more populations.

Free With unlike parts unfused, separate.

Free central. The ovules are borne on a central column with rio septa present, the ovary remains unilocular as in Caryophyllaceae (Superficial). The placenta develop all round the inner surface of a multilocular and multicarpellary gynoecium, as in

Nymphaea (Basal). A single ovule borne at the base of a unilocular ovary as in Compositae.

Free lateral Free on the two sides of the leaf base as in china rose.

Free venation Open venation.

***Fritillaria* type** A type of tetrasporic female gametophyte in which three of the four megaspores fuse to form a triploid nucleus, followed by two sequential mitotic divisions of the haploid and triploid nuclei, resulting in an 8-nucleate female gametophyte in which the three antipodals and one of the polar nuclei are triploid, the other polar nucleus and the cells of the egg apparatus remaining haploid.

Frond Specialized term for a fern leaf.

Fruit The mature ovary of flowering plants, consisting of the pericarp (mature ovary wall), seeds, and (if present) accessory parts.

Frutescent Having the habit of a shrub with numerous, woody, aerial trunks.

Fugacious Caducous.

Fumatory A substance that is smoked by humans, usually for its pleasing or euphoric effects, e.g. tobacco.

Funiculus A stalk that attaches the ovule to the placenta.

Fusiform Spindle-shaped; narrowly ellipsoid with two attenuate ends.

Fusiform roots Swollen in the middle and tapering at the two ends.

Galeate Hooded; with an abaxially concave posterior lip.

Gamete A specialized, haploid cell that fuses with another gamete (in sexual reproduction) to form a diploid zygote.

Gametophyte The haploid phase in the life cycle of all land plants.

Gamopetalous *Sympetalous.* Petals are fused as in Solanaceae.

Gamosepalous *Synsepalous.* Sepals are fused with each other.

Gamotepalous *Syntepalous.*

Gap-coding A method of dividing more or less continuous variation into discrete states.

Geitonogamy Adj: *geitonogamous.* Inbreeding occurring between different flowers derived from one individual. Transfer of pollen grains from the anther to the stigma of another flower of the same plant.

Geminate A compound leaf with two leaflets arising from a petiole and no rachillae.

Geminate-pinnate A compound leaf with two rachillae, each bearing a pinnate arrangement of leaflets.

Gemma (a) An asexual propagule produced within the gemmae cups of some thalloid liverworts. (b) One of the globose or ellipsoid elements of a geminate pollen grain. PI: *gemmae.*

Gemma cup A cup shaped organ on the upper surface of the gametophytes of some thalloid liverworts, containing gemmae propagules.

Gemmate A pollen sculpturing with globose or ellipsoid elements, each element termed a gemma.

Gender The designation of masculine, feminine, or neuter in Latin names.

Gene flow The transfer of the genetic material from parent to offspring.

Gene genealogy/gene lineage Cf: *tokogenetic.* The tokogenetic history of an allele or gene copy, from a particular gene.

Genealogical species Species in which all individuals of the group are more closely related to one another than to any organisms outside the group.

Generative cell One of the two initial, haploid cells in the male gametophyte of angiosperms that mitotically divides to form two sperm cells.

Genetic drift Random genetic modification of a population or species, not the result of natural selection.

Genet Cf: ramet. A genetically different individual of a population.

Geniculate Having a zig-zag posture, e.g., the inflorescence rachis of some grasses..

Genus name The first component of a binomial, always capitalized.

Genus novum Abbr: *gen. n*ov. Meaning that a taxon name, at the rank of genus, is new to science.

Geophyte A perennial herb, typically with a bulb, corm, rhizome or tuber underground stem.

Girdling A type of anther endothecium in which the secondary wall thickenings form rings with cross bridges between them.

Glabrate Nearly glabrous or becoming glabrous with age.

Glabrous Smooth and without any hair or trichomes.

Gland dotted Glands found on the leaf surface as in lemon

Glandular (tapetum type) (a) Covered with minute, blackish to translucent glands (epidermal excrescence). (b) Trichomes secretory or excretory, usually having an apical glandular cell.

Glaucous Syn: Glaucus. Covered with a smooth, usually whitish, waxy coating, which can be rubbed off as in Calotropis and lotus.

Globose Spherical in shape.

Glochidiate With apical, clustered barb like structures.

Glochidium *PI: glochidia; glochids.* A very small leaf spine with numerous, retrorse barbs along its length, produced in the areoles of opuntioid cacti.

Glomerule An inflorescence of sessile or sub sessile flowers in which internodes between flowers are very short, with flowers appearing congested. It is distinguished from a head inflorescerice by blooming of the central flower first.

Glucosinolate A secondary chemical compound found in many Brassicales that functions to deter herbivory and parasitism and also serves as a flavouring agent.

Glume One of usually two scaly bracts occurring at the base of a grass spikelet.

Glutinous Viscid.

Golgi body A cellular organelle comprised of parallel stacks of flattened membranes, functioning in transport and modification of compounds.

Gradate Syn: *sequential.* Development in which the sporangia of a fern sorus mature in succession from the base (periphery) toward the apex (acropetalous) or from the apex toward the base (basipetalous).

Grain Syn: caryopsis. A one seeded, dry, indehiscent fruit with the seed coat adnate to pericarp wall.

Granular Farinaceous.

Granum PI: grana. A pancake like aggregation of thylakoid membranes within the chloroplast of green plants.

Grass spikelet The inflorescence unit of the grass family, consisting of an axis (rachilla) bearing distichous parts: two basal bracts (glumes, sometimes modified or absent) and one or more florets, each floret consisting of a minute lateral axis with two additional bracts (lemma and palea) plus the flower.

Green plants A monophyletic group of eukaryotes, consisting of the "green algae" and the land plants united having chloroplasts with chlorophyll a and b, starch and thylakoids stacked as grana [Chlorobiontal].

Ground meristem The nonvascular, usually parenchymatous tissue between and among the vascular bundles of an atactostele.

Grouad tissue Tissue that is inside the epidermis and not part of the vascular tissue, composed of parenchyma, sclerenchyma and collenchyma cells.

Guard cell One of the two cells that together make up a stomate.

Gynandrium Syn: *column, gynovtegium, gynostemium.* A fusion product of androecium and gynoecium, e.g., Aristolochiaceae, Orchidaceae.

Gynandrous Stamens are fused with the gynoeciurn as in Calotropis.

Gynobasic With style arising at the base and centre of a lobed ovary as in Boraginaceae, Lamiaceae.

Gynodioecious/gynodisecy Having female flowers on some individuals and perfect flowers on other individuals.

gynoecium The female organ(s) of a flower, collectively all carpels of a flower.

gynomonocious/gynomonoecy Having both pistillate and perfect flowers in the same individual.

Gynophore Syn: *stipe.* A stalk of the pistil, usually absent.

gynostegium *Column, gynandrium, gynostemium.* Anthers crowning the column are adnate by the connective to the pentagonal stigma and form gynostegium as in Calotropis.

Habit The general external appearance of a plant (e.g. herbs, shrubs or trees).

Habitat The immediate environment occupied by plant (e.g. aquatic or terrestrial).

Half-inferior With sepals, petals, stamens and/or hypanthium attached at the middle of the ovary.

Halophyte A salt adapted plant.

Hapaxanthi A determinate shoot that completely transforms into a flower or inflorescence.

Haplodiplontic life cycle Syn: *alternation of generations; diplobiontic life cycle.* A life cycle having both haploid and diploid phases, occurring in all land plants.

Haplomorphic Appearing radially symmetric but not having strict mirror image halves because the parts are numerous and/or spirally inserted.

Haplontic life cycle Syn: *haplobiontic life cycle.* A type of sexual life cycle in which the mature, adult phase is haploid, which produces gametes (egg and sperm) that fuse to form a diploid zygote, the latter undergoing meiosis to produce haploid spores developing into new haploid adults.

Haplostemonous Syn: *antesepalous, antisepalous.* Stamens uniseriate, equal in number to the petals, and opposite the sepals.

Haplotype A unique allele of a chromosome or organelle (mitochondrial or chloroplast DNA).

Hardwood Wood derived from a non-monocotyledonous angiosperm, generally (but not always) harder than a softwood because of a greater concentration of fibre cells.

Harmomegathy Volume changes of the pollen grain with changes in water content, e.g., humidity, functioning to inhibit desiccation.

Hastate With two basal lobes, more or less pointed and oriented outwardly approximately 90° relative to central axis.

Hastula An appendage or projection at the junction of petiole and blade, as in some palms.

Haustoria Parasitic roots that penetrate the tissues of a host plant.

Head Syn: *capitulum.* A determinate or indeterminate, crowded group of sessile or sub sessile flowers on a compound receptacle, often subtended by an involucre, e.g., Asteraceae.

Helicoid cyme Syn: *bostryx.* A monochasium in which the branches develop on only one side of each sequential axis, appearing coiled at least early in development; may intergrade with scorpioid cyme.

Helobial An endosperm in which the first mitotic division is followed by cytokinesis, delimiting two cells, with the nucleus of one of the cells dividing without cytokinesis and that of the other cell dividing with cytokinesis.

Hemiparasite A chlorophyllous, parasitic plant.

Hemispheric Half-sphere-shaped.

Hemianatropous type (also called hemitropous). An ovule somewhat intermediate in curvature between anatropous and orhotropous types. It is a modified type of anatropous ovule in which the body of ovule lies at right angle to the funicle and micropyle and chalaza lie in a straight line. The micropyle is not found near the funicle as in Ranunculus and members of Primulaceae.

Herb A plant with annual above-ground shoots, including a flower or inflorescence, the plant itself being annual, biennial, or perennial. It has no persistent parts above the ground. e.g *Chenopodium, Ranunculus, Fumaria* etc.

Herbaceous Having a soft or slightly succulent texture.

Herbarium specimen A pressed and dried plant sample that is permanently glued and/or strapped to a sheet of paper along with a documentation label.

Hercogamy/herkogamy Adj: *hercogamousl herkogamous.* The spatial separation of anthers and stigmas, generally enhancing out breeding. Presence of some morphological barriers which make self pollination impossible, as in Calotropis.

Hermaphroditic A plant with bisexual flowers.

Hesperidium A septate berry with a thick-skinned, leathery pericarp wall and fleshy modified trichomes (juice sacs) arising from the inner walls, e.g. orange, lemon, grapefruit, etc. The fruit develops from multicarpellary, syncarpous, superior ovary with axile placentation. These are multilocular. Epicarp is firm and leathery and has several oil glands. Mesocarp is in the form of white, fibrous part fused with epicarp.

Membranous endocarp projects inwards forming distinct chambers. Many juicy unicellular hairs are found on the inner side of endocarp.

Heteroblasty Adj: *heteroblastic.* The condition in which the juvenile leaves are distinctly different in size or shape from the adult leaves as in Araceae.

Heterochrony An evolutionary change in the rate or timing of development.

Heteromorphy The phenomenon of bearing two types of morphologically dissimilar flowers.

Heterophylly Presence of different types of foliage leaves as in *Ranunculus aquatilis* and *Limnophila heterophylla* where leaves submerged in water are dissected and aerial leaves are more or less entire.

Heteropolar Pollen polarity in which the two polar hemispheres are different because of displacement of one or more apertures.

Heterospory Adj: *heterosporic.* The formation of two types of haploid spores, microspores and megaspores, within two types of sporangia.

Heterostyly Adj: *heterostylous.* Hercogamy in which the relative lengths or heights of stigmas versus anthers vary among different flowers.

Heterotropous An ovule that varies in orientation.

Heterotypic synonym Syn: *taxonomic synonym.* An unaccepted name based on a type different from that of the correct name.

Hexandrous Six stamens in a flower as in Gynandropsis.

Hilum Funicular scar on the seed coat.

Hirsute With long, rather stiff trichomes.

Hispid With very long, stiff trichomes, often capable of penetrating skin as in *Cucurbita.*

Holoparasite A parasitic plant lacking chloroplasts/photosynthesis.

Holotype The one specimen or illustration upon which a name is based, originally used or designated at the time of publication.

Hornochlamydeous Perianth composed of similar parts, each part termed a tepal.

Homogarny Simultaneous maturation of sex organs, as in *Mirabflis jalapa* and *Vinca rosea.*

Homology Adj: *homologous.* Similarity that is the result of common ancestry.

Homolog/homologue Cf. *homology.* A specific feature that is homologous to another, cited feature.

Homonym One of two (or more) identical names that are based on different type specimens.

Homoplasy Similarity that is not due to homology or common ancestry, but the result of independent evolutionary change.

Homospory Adj: *homosporic.* The formation of one type of haploid spore, by one type of sporangium.

Homotypic synonym Syn: *nomenclatural synonym.* An unaccepted name that is based on the same type is that of the accepted name.

Hood A hood like appendage arising from the gynostegium of some Asclepiadoids of the Apocynaceae.

Hooded Sepals enlarge to form a hood-like structure.

Hooked Syn: *uncinate.* With apical hook like structure.

Hooks Leaf modified into hooks for climbing as in cati, terminal leaflets become claw-like hooks.

Horizontal Syn: *divericafe-, divergent; patent.* More or less horizontally spreading with divergence angle of <15° up or down from horizontal axis.

Horn A hornlike appendage, often associated with a hood, arising from the gynostegium of some Asclepiadoids of the Apocynaceae.

Hyaline cell One of the nonchlorophyllous cells in the specialized leaves of *Sphagnum* moss, having characteristic pores and helical thickenings and functioning in water absorption and retention.

Hybridization Sexual reproduction between different species (interspecific hybridization) or between different populations, infraspecific taxa, or forms within a species.

Hydathode A group of specialized cells that secrete excess, transported water (usually due to root pressure) from leaf margins.

Hydrochory *Adj: hydrochorous.* Dispersal of propagules by water.

Hydroid A specialized cell that functions in water conduction in some mosses.

Hydrophily Adj: *hydophilous.* Pollination by water.

Hygroscopic Absorbing moisture from the air, often resulting in movement.

Hymenopterophily *Melittophily.*

Hypanthium Syn: *floral tube.* A cuplike or tubular structure around or atop the ovary, bearing along its margin the sepals, petals, and stamens.

Hypanthodium An inflorescence bearing numerous flowers on the inside of a convex or involuted compound receptacle as in fig. The fleshy receptacle forms a hollow pear-shaped cavity with a narrow small opening. The flowers are borne on the inner wall of the cavity and are unisexual. Male flowers are borne on upper side (towards mouth) and female flowers are borne at the base of cavity.

Hyphodromous Pinnate venation with only the primary midrib vein present or evident, the secondary veins absent, very reduced, or hidden within the leaf mesophyll.

Hypocotyl A region of the embryo between the root and epicotyl; may function in seedling development and as an anatomical transition between root and shoot.

Hypocrateriform Petals fused with basal portion narrow and tube-like and upper portion bowl-shaped.

Hypogeous Syn: *cryptocoylar.* With cotyledon(s) remaining in the ground during germination. Hypogynous with sepals, petals, and stamens attached at the base of a superior ovary.

Hypogynous A flower with superior ovary as in Cruciferae.

Hypsophylls or bract leaves Special leaves which bear floral buds at their axils.

Hypotropous An ovule in which the micropyle points proximally.

Hypotropous-dorsal A hypotropous ovule in which the raphe is dorsal (abaxial), pointing away from the central floral or ovary axis.

Hypotropous-ventral A hypotropous ovule in which the raphe is ventral (adaxial), pointing toward the central floral or ovary axis.

Hysteranthy Adj: *hysteranthous.* Timing in which leaf and flower development do not coincide.

Identification The process of associating an unknown taxon or other entity with a known one.

Illegitimate name A name that is validly published name but not in accordance with the rules of the International Code of Botanical Nomenclature.

Imbricate (a) Leaves or other structures overlapping. (b) With overlapping perianth parts. Both ends of one sepal are outer and of another sepal are inner while other three sepals have one end outer and another one inner

Imbricate-alternate Outer whorl of perianth parts (sepals or outer tepals) alternating with, along different radii, the inner whorl of perianth parts (petals or inner tepals), a typical perianth aestivation.

Imparipinnate/imparipinnately compound Syn: *odd-pinnate.* A pinnately compound leaf with a terminal leaflet, typically odd-pinnate. Rachis is terminated by an odd leaflet.

Imperfect (a) *Unisexual.* (b) With lateral primary veins covering less than two thirds of the leaf blade area, a sub category of actinodromous and of acrodromous.

In Latin for "in," meaning abbreviation for" in the publication of," referring to a name published within a larger work authored by the person(s) following the "in."

Inaperturate A pollen grain that lacks any recognizable aperture.

Inbreeding Syn. *selfing.* The union of gametes derived from a single individual.

Incanous Covered with dense, fine, grayish white/tetrichomes; whitish pubescent.

Incertae sedis A general term used to describe species in an uncertain taxonomic position.

Incipient speciation Early and incomplete divergence of one lineage into two (or more), such that the terminal entities of those lineages are not fully separated from one another (e.g., exhibit incomplete lineage sorting).

Incised With margins sharply and deeply cut, usually jaggedly.

Incision Indentation of the lamina margin.

Inclined Directed upward, with a divergence angle of 15°-45° from horizontal axis.

Incompatibility reaction The inhibition of pollen germination or pollen tube growth between genetically similar individuals, mediated by incompatibility genes and functioning to promote outcrossing.

Incomplete Lacking one or more of the four major whorls or floral parts: sepals, petals, stamens, carpels.

incomplete lineage sorting Condition whereby a lineage segment at a given point in time contains more than one gene lineage.

Incurved Tip gradually curved inward or upward (adaxially).

Indehiscent legume A secondarily modified legume does not split open as in peanut.

Indehiscent or Achenial Fruits In such fruits, pericarp does not rupture and seeds remain enclosed within it.

Indented phylogenetic classification A classification in which monophyletic groups are ordered in a sequential, hierarchical method.

Indeterminate (a) A shoot that has the potential for unlimited growth, the apical meristem continuing to grow. (b) An inflorescence in which the basal flower matures first; maturation from base to apex.

Induplicate Plicate, with adjacent adaxial sides facing one another, being V-shaped in cross-section.

Indurate Hardened and inflexible.

Indusium Adj: *idusiate.* A flap of tissue that covers a sorus, found in some leptosporangiate ferns.

Ineditus Abbr: *ined.* Not validly published.

Inferior with sepals, petals, stamens, and/or hypanthium attached at the apex of the ovary. Ovary borne below attachment.

Inflorescence An aggregate of one or more flowers, the boundaries of which generally occur with the presence of vegetative leaves below; may be composed of unit inflorescences. The arrangement of flowers and mode of distribution of flowers on the shoot system of a plant.

Inflorescence bract A modified, generally reduced leaf subtending an inflorescence axis.

Inflorescence bud A bud that develops into an inflorescence.

Infraforal selflng *Autogamy.*

Infrafoliar Descriptive of a palm inflorescence that is positioned below the leaves of the crownshaft.

Infrapetiolar bud An axillary bud surrounded by a petiole base, *e.g., Platanus,* sycamore.

Infraspecies Subspecies, varieties, or rarely forms (formae) of a species, generally showing slight and often intergrading morphological differences as well as some geographic, ecological, and/or phytogenetic distinctions.

Infructescence The complete inflorescence at the stage of fruiting.

Infundibular Funnel shaped; with a narrow base and greatly expanded apex.

Infundibuliform Petals fused forming funnel-shaped structure.

Ingroup The study group as a whole in a phylogenetic analysis.

Innerbark Secondary phloem.

Inserted Syn: *cryptantherous.* With stamens included within the perianth. Length of stamens shorter than petals so these remain within corolla tube.

Insignificant Condition in which the differences between two or more groups are due to chance alone; i.e., the probability of the differences being due to chance is calculated as greater than some standard value (generally 5%, or $p > 0.05$).

Integument A sheath or flap of tissue that surrounds the megasporangium (nucellus) of an ovule and develops into the seed coat of the seed.

Intercalary meristem An indeterminate (having potentially continuous growth), basal or sub-basal region of actively dividing cells.

Interfoliar Descriptive of a palm inflorescence that is positioned among the leaves of the crownshaft.

Intermingled With no consistent developmental pattern of sporangia in a fern sorus.

International Code of Botanical Nomenclature Abbr: *ICBN.* The standardized system of rules for naming plants, "algae:' fungi, and organisms traditionally treated as fungi, governing specific names assigned to taxa and the endings that denote taxon rank, and utilized for naming new taxa and determining the correct name for previously named taxa.

Internode (a) The region between two adjacent nodes of a shoot. (b) A cladogram lineage that spaces between two nodes (points of divergence). Syn: *stem.*

Interpetiolar Two stipules lying between the two leaves.

Intine The innermost layer of a pollen grain wall, composed primarily of cellulose and pectins.

Intrapetiolar Present between the leaf and the axis.

Intrastaminal disk A discoid or doughnut-shaped structure arising from the receptacle at the inside of the stamens and/or base of the ovary; may be nectar-bearing. ("nectariferous disk") (flower part)

Intravaonal (axillary) squamules Trichomes found in the axils of sheathing leaves, possibly functioning in secreting a protective mucilage, e.g., many Alismatales. (trichome type)

Introgression Hybridization between two species followed by backcrossing to one or both parents.

Introrse Dehiscing inward, toward the centre of the flower.

Inversion A mutation resulting in the 180° flipping of a segment of DNA.

Involucel A group or cluster of bracts subtending a unit of an. inflorescence.

Involucral bract *Phyllary.* Leaf structural type.

Involucral bract Whorls of green sepaloid bracts occurring around the base of the inflorescence. These form an involucre, as in family Compositae and Umbelliferae. In addition to bracts, there may be smaller involucres round the base of each branch.

Involucre Adj: *involucrate.* A group or, cluster of bracts subtending an inflorescence.

Involute (a) Margins or outer portion of sides rolled inward or upward over adaxial surface (b) Valvate with each perianth part induplicate, folded longitudinally inward along central axis.

Iridoid A secondary chemical compound characteristic of many Asterids.

Irregular *Zygomorphic, bilateral, monosymmetric.*

Isobilateral When both leaf surfaces receive equal light and there is no difference between the two surfaces.

Isolation species Biological species.

Isomerous Having the same number of parts in different whorls.

Isomorphic Appearing identical, e.g.. the gamates of some green plants.

Isopolar Pollen polarity in which the two polar hemispheres are the same but can be distinguished from the equatorial region.

Isotype A duplicate specimen of the holotype, collected at the same time by the same person from the same population.

Iteropary Adj: *iteroparqus.* Referring to plants that reproduce more than one time in the life of the plant, typically in regular cycles.

Jacked layer Antheridial wall, archegonial wall.

Jacknife/jacknifing A method of evaluating cladogram robustness that reanalyzes the data of the original character X taxon matrix by selecting (resampling) characters at random, such that a given character can be selected only once, the resultant resampled data matrix being smaller than the original.

Jactulator Syn: *retinaculum.* Funiculi of the seeds that are modified into rigid, often hook shaped structures that function to disperse the seeds by a catapulting mechanism, characteristic of the Acanthaceae.

Kernel The seed inside the stony endocarp of a drupe.

Key Taxonomic key. An identification device, consisting of contrasting statements used to narrow down the identity of a taxon.

Kranz anatomy A leaf anatomy in which chloroplasts of the bundle sheath cells are typically much larger than those of the mesophyll cells, correlated with C4 photosynthesis.

Labelum A modified, typically expanded, median petal, tepal, or perianth lobe, such as in the Orchidaceae.

Lacerate With sinuses irregularly cut, lobes appearing torn.

Lacinate Cut into narrow, ribbon like segments.

Laevigate Lustrous, polished.

lagenostome A rim or ring of tissue at the apex of the megasporangium, which functioned to funnel pollen grains to a pollination chamber in ancestral seeds.

Lamellar An exine wall structure having stacked, tangentially oriented, planar structures, often constituting the inner wall layer.

Lamina (leaf blade or **Epipodium).** The most important part of the leaf and a seat of gaseous exchange for photosynthesis, respiration, transpiration, etc.

Laminar (a) With a dorsiventrally flattened, leaf like structure bearing the thecae. (b) With ovules arising from the surface of the septae.

Lanate Villous.

Lanceolate Lance-shaped as in Nerium and Polyalthia. Margins curved, widest near base, length:width ratio between 6:1 and 3:1.

Lanceovate Margins curved, widest near base, length:width ratio between 3:1 and 2:1.

Landmarks Defined points of an object that generally correspond to 'homologous features, used in morphometric analyses.

Land plants A monophyletic group of eukaryotes united by an outer cuticle, specialized gametangia-antheridia and archegonia-and an intercalated diploid phase in the life cycle, including the embryo [embryophytes/Embryophyta].

Later homonym A homonym that is illegitimate (unless conserved), being published at a date later than the earliest published homonym.

Lateral (a) *Axillary*. (b) Style arising at the side of an ovary.

Lateral bud Axillary bud.

Lateral mesistem A cylindrical sheath of cells, functioning in secondary growth, that increases width or girth of stems or roots in woody plants; includes the vascular cambium and cork cambium.

Lateral root A root that arises from another root, derived endogenously from the pericycle.

Lateral vein Ventral vein.

Laticifer Cells located in the periphery of some tissues that secrete and store latex, functioning to deter herbivory and to seal and protect, plant tissue upon wounding.

Latrorse Dehiscing laterally relative to the flower center.

Laver One of the ecological criteria of plant communities based on height and plant habit, including the canopy, subcanopy, shrub or subshrub layer, and herb layer.

Lead One of the two contrasting statements in a dichotomous key.

Leaf A generally dorsiventrally flattened organ, usually functioning in photosynthesis and transpiration, either gametophytic (in mosses and some liverworts) or sporophytic (in vascular plants), often variously modified.

Leaf base or hypopodium The point of attachment of the leaf to the stern.

Leaf gap A region of nonvascular parenchyma tissue interrupting the vasculature of the stem at a node, associated with euphylls.

Leaf primordiurn An immature leaf of the shoot.

Leaf scar A mark indicating the former place of attachment of a leaf.

Leaf spine Cf. *prickle; thorn.* A sharp pointed leaf, e.g., cactus spines or glochidia.

Leaf stalk or mesopodium or stipules A pair of lateral outgrowth present at the base of the leaf.

Leaflet A distinct and separate segment of a leaf

Leaflet spine A sharp pointed leaflet, e.g., some palms, such as in *Phoenix.*

Leaf tendrils Tendrils formed by climbing plants which may be modification of entire leaf or its parts.

Laesura PI: *laesurae.* The differentially thickened wall region corresponding to the tetrad attachment scar on each of the four immature spores following meiosis.

Lectotype A specimen that is selected from the original material to serve as the type when no holotype was designated at the time of publication, if the holotype is missing, or if the original type consisted of more than one specimen or taxon.

Legitimate name A name that is validly published in accordance with the rules of the ICBN.

Legume or Pod A dry, dehiscent fruit derived from one carpel that splits along two sutures as in Fabaceae. The fruits develop from monocarpellary superior ovary with marginal placentation. Ovary is unilocular with many ovules.

Lemma The outer and lower bract at the base of the grass floret. In the spikelet, each floret bears a bract called flowering glume or lemma and a bracteole called palea which fits within the lemma.

Lenticel A pore in the bark, generally functioning in gas exchange.

Lenticular Lens shaped; disk shaped with two convex sides.

Lepidote Covered with scales or scale like structures.

Leptoid Specialized cell that functions in sugar conduction in some mosses.

Leptosporangium *Pl: leptosporangia.* The sporangia of the leptosporangiate ferns (Polypodiales), characterized by developing from a single cell and having a single layer of cells making up the sporangium wall.

Liana/liane A woody, perennial vine, in tropical forests often a component of the canopy layer.

Lianas Climbing by twinning woody and stout stems which hang down by thick ropes, e.g. Tinospora, Hiptage, Quisqualis, etc.

Light reactions A series of biochemical reactions of photosynthesis in plants, occurring in the thylakoid membranes of the chloroplast and requiring light as an energy source, during which water is broken down into hydrogen ions, electrons, and molecular oxygen, and producing high-energy ATP and NADPH2, which are utilized in the dark reactions.

Lignin A complex polymer of phenolic compound that impregnates the secondary cell wall of some cells (including tracheary elements and sclerenchyma), functioning to impart strength and rigidity to the wall.

Lignotuber Bur.

Ligulate (a) Strap- or tongue shaped; flattened and somewhat oblong in shape, e.g., ray Powers of some Asteraceae. (b) Having a ligule arising from the top of the leaf sheath, at the junction with the blade. Cf. *eligulate.*

Ligulate Petals fused to form a tube-like structure at the base and broad flap-like structure in the upper part as in ray florets of Compositae.

Ligule (a) A small appendage on the upper (adaxial) side of the leaf, near the leaf base, found in the Selaginellaceae and Isoetaceae of the lycophytes. (b) An outgrowth or projection from the top of a leaf sheath at its junction with the blade, as in the Poaceae.

Limb The expanded portion of the corolla or calyx above the tube, throat or claw.

Lineage Cf: *clade.* A sequence of ancestral descendent populations in which the members are linked or connected by gene flow.

Lineage segment A portion of a lineage, e.g., from one divergence point ("node" of a cladogram) to another.

Lineage sorting The process by which, following evolutionary divergence, several gene lineages inherited from an ancestor converge to a single gene lineage.

Linear (a) With margins straight, parallel, length:width ratio between 12:1 and 6: 1. (b) Stigmas or stigmatic tissue long and narrow in shape. Longer and slightly broader as in grasses and tuberoses (*Polianthes tuberosa*).

Linear regression analysis A statistical procedure for fitting a straight line onto a bivariate plot of two variables and testing the significance of relationship between them.

Linear tetrad A tetrad in which the four pollen grains are arranged in a straight line, e.g., *Typha.*

Lip Cf: *labellum.* Either of two variously shaped parts into which a calyx or corolla is divided, usually into upper (posterior) and/or lower (anterior) lips, each lip often composed of one or more lobes, e.g., Lamiaceae, Orchidaceae.

Lithocyst *Cf. cystolith.* A specialized cell that contains a cystolith.

Lobe A segment of a synsepalous calyx or sympetalous corolla or a divided leaf.

Lobed Having lobes. Sinuses extending (pinnately or palmately) one eighth to one fourth the distance to midrib, midvein or vein junction.

Lobed or incised Margin much dissected or incised as in *Raphanus*.

Locule An ovary cavity bounded by ovary walls and septa. (b) A compartment of the anther, usually the result of two microsporangia fusing within a theca.

Loculicidal capsule A capsule in which longitudinal lines of dehiscence are radially aligned with the locules.

Loculicidal dehiscence Pericarp opens through the middle of locules and placenta is divided as in cotton and okra.

Lodicule One of the (2-3) modified perianth parts of a grass flower, which collectively upon swelling function to open the floret by separating the lemma from palea.

Loment A secondarily modified legume that splits transversely into I -seeded segments.

Lomentum It is a modification of legume. The fruit develops from monocarpellary, unilocular, superior ovary. These are bisutural fruits which are constricted or divided into one seeded mericarps as in groundnut, Mimosa, Tamarindus, etc.

Long-branch attraction A condition in which taxa with relatively long branches

Long branches Having numerous character state changes tends to come out as close relatives of one another in a phylogenetic analysis because of random effects.

Longitudinal Along longitudinal suture.

Longitudinal dehiscence Dehiscing along a suture parallel to the long axis of the thecae.

Longitudinal posture Placement of margins with respect to a horizontal plane.

Lower epidermis The abaxial epidermis of a leaf.

Lumen *Pl: lumina.* The space between muri in a reticulate pollen grain.

Lunate Half moon-shaped.

Lycophyll Adj: lycophyllous. Syn: *microphyll.* The sporophytic leaf of the lycophytes characterized by an intercanary meristem, having a single vein and lacking a gap in the vasculature of the stem.

Lyrate Pinnatifid, but with a large terminal lobe and smaller basal and lateral lobes.

Maculate Spotted; with small spots on a more or less uniform background.

Majority consensus tree A consensus tree in which only those clades that are retained 50% or more of the time are retained (i.e., not collapsed to a polytomy).

Malacophily Pollination carried out by snails and slugs, may be found in subterranean runners.

Male *Staminate.* Individual with male reproductive organs only.

Male gametophyte Syn: *microgametophyte.* A gametophyte that bears only antheridia.

Male sporophyll Syn: *microsporophyll.* A sporophyll that bears one or more microsporangia.

Malpighian Trichomes with two arms arising from a common base, e.g., Malpighiaceae.

Marcescent Ephemeral, but with persistent remains; withering persistent.

Marginal (a) With the placenta along the margin of a unicarpellate (simple) ovary, e.g., Fabaceae. (b) With three or more primary veins diverging from one point and reaching the blade maigin, a subcategory of actinodromous.

Massula PI: *massulae.* A group of fused pollen grains in large, often irregular numbers, but less than an entire theca.

Masticatory A substance that is chewed by humans, usually for its pleasing or euphoric effects, e.g., peyote, *Lophophora williamsii.*

Maturation Syn: *specialization.* Acquisition of the mature structural and functional features of a cell following cell expansion. Relative time of development of parts.

Matutinal In the morning, typically with respect to when flowers open.

Maximum likelihood A method of phylogenetic inference that considers the probability, based on some selected model of evolution, that each tree explains the data.

Mealy Covered with small, fine granules.

Median Cf. *mean.* The middle value of a list of values.

Median vein Dorsal vein.

Mean The numeric average. Cf. *median.*

Megagametogenesis The process of development of the female gametophyte from the haploid product(s) of meiosis.

Megagametophyte Female gametophyte.

Megaphyll *Euphyll.*

Megasporangium *Pl: megasporangia.* Syn. *nucellus.* A female sporangium, within which megasporocytes undergo meiosis to produce haploid megaspores.

Megaspore A female spore, produced via meiosis in the megasporangium and giving rise to a female gametophyte.

Megaspore mother cell *Megasporocyte.* A cell within the megasporangium that undergoes meiosis, forming four megaspores.

Megasporogenesis The process of development of megaspores from the megasporocyte.

Megasporophyll Female sporophyll.

Melittophily Syn: hymenopterophily. Adj: entomophilous. Pollination by bees.

Membranous Thin and somewhat translucent; membrane like.

Mericarp *A* portion of a fruit that separates from the ovary as a distinct unit that completely encloses the seed(s).

Meristem *A* region of actively dividing cells.

Meristic Discrete quantitative characters.

Merosity Number of parts per whorl or cycle.

Mesarch An orientation of xylem maturation in which the protoxylem is surrounded by metaxylem within the vascular tissue, as can occur in siphonosteles.

Mesocarp *A* middle wall layer of the pericarp, if the latter is divided into layers.

Mesophyll The region of a sporophytic leaf between the outer epidermal layers and exclusive of the vasculature, containing the chlorophyllous cells.

Mesophytic Having an intermediate texture, between coriaceous and membranous.

Metaspecies Two or more lineages segments that can be resolved as neither monophyletic or paraphyletic.

Metaxylem The xylem of a group of tracheary elements that matures later, consisting of larger diameter cells.

Microfibril Microscopic fibre like units of intertwined cellulose molecules forming a meshwork within the cell wall.

Microgametogenesis The process of development of pollen grains from microspores.

Microgametophyte Male gametophyte.

Microhair *A* very small trichome, as in the three celled, glandular microhairs of the Commelinaceae.

Microphyll Essentially equivalent to a lycophyll.

Micropylar The distal region of an ovule that is near the micropyle.

Micropyle A small pore at the distal end of the integument of a seed, functioning as the site of entry of pollen grains, or in angiosperms of pollen tubes.

Microsatellite A region of DNA that contains tandem repeats, which are short (usually 2-5) repeats of nucleotides.

Microsporangium *Pl: microsporangia.* A male sporangium, within which micro-sporocytes undergo meiosis to produce haploid microspores.

Microspore A male spore, produced via meiosis in the microsporangium and giving rise to a male gametophyte.

Microsporogenesis The process of development of microspores from microsporocytes.

Microsporophyll A sporophyll bearing one or more microsporangia.

Middle lamella A pectin rich layer formed between the plasma membrane of adjacent cells in land plants, functioning to bind adjacent cells together.

Middle layers Anther wall layers that may occur between the endothecium and tapetum.

Midrib Syn: *costa.* The central, main vein of the blade of a simple leaf or of a compound leaf in some palms.

Midvein The central, main vein of the blade of a leaflet.

Mitochondrion *Pl: mitochondria.* A double membrane bound, cellular organelle with invaginations called cristae that function in the electron transport reactions of respiration.

Mixed bud A bud that produces both flowers and leaves.

Mixed craspedodromous Pinnate venation in which some secondary veins terminate at the margin, but, many terminating away from the margin.

Mixed development A combination of gradate and simultaneous development of sporangia in a fern sorus.

Modifications of the Leaf Lamina The lamina or sometimes the entire leaf undergoes modification to suit various purposes.

Modified adventitious roots Adventitious roots modified for various purposes.

Modified Roots Roots modified to carry out specialised functions of mechanical and physiological nature.

Modified tap roots Tap roots modified for food storage.

Monad A single, unfused pollen grain.

Monadelphous With one group of stamens connate by their filaments but anther lobes are free.

Monandrous Single stamen in each flower as in *Euphorbia*.

Moniliform roots Roots develop alternate swellings and constrictions to give a. beaded appearance.

Monistichous Alternate, with points of attachment in one vertical row/rank, e.g., Costaceae.

Monocarpellary Presence of a single carpel in a flower as in Leguminosae.

Monocarpic A perennial or annual plant, flowering and fruiting once, then dying.

Monochasial or Uniparous cyme A single lateral branch arises from the peduncle of old flower which terminates in a flower. The lateral branch also terminates in a flower.

Monochasium A determinate inflorescence that develops along one axis only.

Monocotyledonous A type of anther wall development in which only the inner secondary parietal cell layer divides to yield the tapetum and a single middle layer.

Monocyclic Uniseriate.

Monoecious Having unisexual flowers, both staminate and pistillate on the same individual plant.

Monograph A detailed toxonomic study of all species and infraspecific taxa of a given taxonomic group.

Monolete Spores with an unbranched, linear laesura that is linear and unbranched.

Monomorphic Being similar to one another, as in leaves.

Monomorphic character A character that is invariable in character state values within an OTU.

Monophyletic/monophyly Referring to a group that consists of a common ancestor plus all descendants of that ancestor.

Monopodial A branching pattern in which a given axis is derived from a single apical meristem.

Monosporic Megasporogenesis in which meiosis of the megasporocyte nucleus results in the formation of four haploid megaspore nuclei, followed by cytokinesis, resulting in four megaspore cells, only one of which contributes to the female gametophyte.

Monosulcate A pollen grain with a single, sulcate aperture.

Monosymmetric Zygomorphic, bilateral, irregular.

Monothecal Cf. *bisporangiate.* Anther with one theca and typically two microsporangia.

Monoulcerate A pollen grain with a single, ulcerate aperture.

Morphocline Transformation series.

Morphologic species *Taxonomic species.*

Morphometrics The study of shape or form, generally using statistical methods.

Movement hercogamy A type of hercogamy involving movement of floral parts, e.g., the rapid closure of the stigmas upon their being touched by a potential animal pollinator.

Mucilage ducts/canals Specialized cells that secrete mucilage.

Mucronate With a stiff, straight apical process, length:width ratio 1:1 to 3:1.

Mucroulate With a stiff, straight apical process, length:width ratio $\leq$ 1:1.

Mucronate Apex broad but with a sharp point as in periwinkle (*Vinca rosea*).

Multifoliate More than four leaflets attached at the top of rachis (petiole).

Multicellular Trichomes having two or more cells.

Multicostate or Palmate From the base of the lamina number of equally strong veins (or costas) which spread like fan, hence called palmate.

Multiple fruit A fruit derived from two or more flowers.

Multiparous More than two branches at each node.

Multiseriate (a) Perianth composed of three or more distinct whorls. (b) Trichomes having more than one vertical row of cells. c) Rays in wood that are made up of many vertical rows of cells.

Multistate character A character with three or more character states.

Muricate Having coarse, radially elongate, rounded protuberances.

Murus P.I: *muri.* The structural element of a reticulate pollen grain.

Mycorrhiza PI: *mycorrhizae.* A symbiotic association between a fungus and roots of a vascular plant, functioning to increase absorptive surface area and mineral uptake.

Mycorrhizal roots Roots of higher plants infested with fungal mycelia which facilitates nutrient absorption specially N, P and K by the higher plants from the soil.

Mycotroph Adj: *mycotrophic.* A plant obtaining nutrition from mycorrhizal fungi in the soil, as in Ericaceae.

Naked bud A bud lacking surrounding protective scales, e.g., *Viburnum,* Caprifoliaceae.

Napiform roots Roots almost spherical at the upper end which abruptly tapers at the lower end.

Narrowly acute Apical margins approximately straight, the intersection angle <45°.

Narrowly cuneate Basal margins approximately straight, the intersection angle <45°.

Narrowly elliptic Margins curved, widest near midpoint, length: width ratio between 6:1 and 3: 1.

Narrowly oblong Margins straight, parallel, length:width ratio between 6:1 and 3:1.

Narrowly triangular Three sided, length:width ratio between 6: 1 and 3:1.

Natural selection The directed and nonrandom genetic modification of a population or species, in which genetic changes that result in an increase in survivorship and/or reproduction are contributed to the next generation.

Neck The distal, narrow extension of the sterile jacket cells of the archegonium, through which a sperm cell must travel to fertilize the egg.

Neck canal cells Cells located within the neck of the archegonium that break down and are secreted from the pore of the neck at maturity.

Nectariferous disk A nectary consisting of a dislike or doughnut shaped mass of tissue surrounding the ovary base or top; may be inner (intrastaminal), beneath (staminal), or outer (extrastaminal) to the androecium.

Nectary A group of specialized cells that secrete sugar- (or protein-) rich fluids to the outside, as a reward for pollination or protection.

Neoteny A type of *paedomorphosis* caused by a decrease in the rate of development of a structure or feature.

Neotropical Distributed in tropical areas of the New World (tropical Mexico, Caribbean, Central/South America).

Neotype A specimen derived from a non-original collection that is selected to serve as the type as long as all of the material on which the name was originally based is missing.

Nerve *Vein.*

Nested clade analysis A methodology that reconstructs the genealogical relationships of individuals using haplotypes.

Nested clade phylogeographic analysis Nested clade analysis correlated with geographic distribution of individuals.

Netted Syn: *anastomosing, reticulate.* With ultimate veinlets forming a reticulum or netlike pattern.

Network Unrooted tree.

Neuter Sex organs (androecium and gynoecium) are altogether absent.

Nexine The inner layers of the exine, including both endexine and the foot-layer of the ektexine.

Nitid *Syn: shining.* Appearing lustrous, polished.

Nocturnal Occurring during the night, typically with respect to when flowers open.

Node (a) The point of attachment (origin) of leaf , leaves or branches on a stem. (b) The point of divergence of one clade into two; the point in time and space of the most common ancestor of the two divergent clades.

Node-based A type of phylogenetic classification in which a node (common ancestor) of the cladogram (and all descendants of that common ancestor) serves as the basis for grouping.

Nomen conservandum Abbr: nom. cons. A conserved name.

Nomen novum Abbr: nom. *nov.* A new name.

Nomen nudum Abbr: nom. *nud.* A name published without a description or diagnosis, making it invalid.

Nomenclatural synonym Homotypic synonym.

Nomenclatural type Syn: type; type specimen. A specimen or illustration that acts as a reference for a scientific name, upon which the name is based.

Nomenclature The formal naming of taxa according to some standardized system; for plants, "algae," fungi, and organisms traditionally treated as fungi, governed by the ICBN.

non Latin for "not."

Nonclathrate Cf: clathrate. Fern scales in which the cell walls of adjacent cells (anticlinal walls) are thin.

Nondecussate Opposite leaves or other structures (e.g., leaflets) not at right angles to the preceding pair; may be superficially the result of stem twisting.

Nondisjunction An irregularity during meiosis in which homologous chromosomes do not segregate, which may result in the production of gametes that are unreduced, i.e., have two sets of chromosomes.

Nonendospermous Exalbuminous.

Nonporous Referring to wood having only tracheids.

Nonvisus Abbr: n. v. Latin for "not seen" meaning that authors did not see a specimen, such as a type.

Normal root A root arising from the radicle.

Nucellar beak A proliferation of cell divisions of the nucellus at the micropylar region of the ovule.

Nucellus Adj: *nucellar.* Megasporangium of a seed.

Nuclear endosperm An endosperm in which the early mitotic divisions are not followed by cytokinesis.

Nucleus A double membrane bound, cellular organelle that contains DNA.

Nude Descriptive of the ovary of a female flower lacking a perianth; thus, ovary position not evident.

Number Whether a Latin name is singular or plural.

Nut A one seeded, dry indehiscent fruit with a hard pericarp, usually derived from 1-loculed ovary.

Nutlet A small nut, e.g., mericarps of the Lamiaceae.

Nutritive tissue Tissue that surrounds or abuts the embryo of a seed and that consists of female gametophyte (in nonangiosperms) or endosperm (in angiosperms).

Obcordate Reverse of cordate with an apical notch as in *Bauhinia.*

Obdiplostemonous Stamens are arranged in two whorls and stamens of outer whorl opposite and inner whorl anther lobes alternate with the petals, as in Rutaceae.

Obhaplostemonous Syn: *antepetalous, anti-petalous* Stamens uniseriate, equal in number to the petals, and opposite the petals

Obkinceelate Margins curved, widest near the apex, length:width ratio 6:1 to 3:1.

Oblance-ovate Margins curved, widest near the apex, length:width ratio 3.1 to 2:1.

Oblate A pollen grain in which the P/E ratio is less than about 0.8.

Obligate hemiparasite A chlorophyllous, parasitic plant requiring a host to survive and reproduce.

Oblique With an asymmetrical apex or base.

Oblong Margin~ straight, parallel, length:width ratio 2-1 to 3:2. More or less rectangular as in banana.

Obovate Margins curved, widest near apex, length:width ratio 2:1 to 3:2. Reverse of ovate as in jackfruit.

Obturator A protuberance of tissue, typically arising from the funiculus or placenta at the base of the ovule as in Euphorbiaceae.

Obtuse Apical or basal margins approximately straight, intersection angle >90°. Apex blunt or broad angled as in hanyan.

Ocrea A specialized, scarious, sheath like structure arising above the node in some members of the family Polygonaceae, interpreted as modified stipules.

Ochreate Stipules which form a hollow tube encircling the stem from the node upto a certain height as in *Polygonum.*

Octastichous or phyllotaxy The ninth leaf is found above first leaf and so on.

Odd-pinnate Imparipinnate imparipinnately compound.

Offset Similar to a runner but shorter and thicker and found in aquatic plants. It runs for sometimes under-water and then surfaces to bear a bunch of leaves and adventitious roots.

Oil A type of triglyceride compound that may function as high-energy storage compounds or secretion products; a type of ergastic substance.

Oil bodies Oil containing structures found within certain cells of most liverworts.

Oil ductstcanals Specialized cells that secrete oil.

Onagrad type Crucifer type.

Ontogenctic sequence The discrete stages of the developmental sequence of a given feature.

Ontogenetic trajectory A plot of developmental change as a function of time.

Ontogeny The developmental sequence of a given feature.

Oogamy A type of sexual reproduction in which one gamete, the egg, becomes larger and non-flagellate and the other gamete, the sperm cell, remains relatively small and flagellate; found in all land plants and independently evolved in many other eukaryaryotes.

Open venation Syn: *free venation.* Fern venation in which the veins arising from the midvein or base of a pinnule do not join back together.

Operational taxonomic unit (OTU) One of the individual, unit taxa of a phylogenetic analysis.

Operculate Having calyx and corolla fused into a cap that falls off as a unit, e.g., *Eucalyptus.*

Operculum An apical lid, as in the capsule of most mosses that falls off during spore release.

Opposite With two leaves or other structures per node, on opposite sides of stem or central axis.

Opposite phyllotaxy Two leaves on each node and opposite to one another.

Opposite decussate Successive pairs of leaves placed at right angles to one another.

Opposite superimposed Successive pairs of leaves exactly on top of one another so that all leaves lie in one plain as in guava and Quisqualis.

Orbicular Adj: *circular.* Circular in outline; margins curved, length:width ratio approximately 1:1.

Ordered Referring to a transformation series in which the character states occur in a predetermined sequence.

Organelles Structural, membrane bound units of the cell that provide some vital metabolic function.

Orientation Referring to the angle relative to a central, usually vertical, axis.

Ornithophily Adj: *ornithophilous.* Pollination by birds.

Ortho-amphitropous An ovule type in which the vasculature is straight, leading from the funiculus base to the middle of the nucellus, with the nucellus bent sharply in the middle along both the lower and upper sides, often with a basal body present.

Ortho-campylotropous An ovule type in which the vasculature is straight. leading from the funiculus base to the middle of the nucellus, with the nucellus bent only along the lower side, with no basal body.

Orthographia conservanda Abbr: orth. cons. A conserved spelling in a scientific name.

Orthotropous type (atropous) The mature ovule in which no curvature takes place during development. It is straight and the micropyle and funiculus are in a straight line and above the hilum.

Outcrossing Syn: allogamy; xenogamy; *outbreeding.* The transfer of gametes from one individual to another, genetically different individual.

Outer bark *Periderm.*

Outgroup A taxon or group of taxa that is not a member of the ingroup.

Ovary The part of the pistil (gynoecium) containing the ovules.

Ovate Egg-shaped as in china rose and banyan. Margins carved, widest near base, length:width ratio 2:1 to 3:2.

Ovule An immature seed, prior to fertilization. Ovule arises from the placenta and are found in the basal part of the carpel having integumented megasporangia, the nucellus containing the embryo sac enclosed by one or two integuments, which after fertilisation and subsequent development becomes a seed.

Ovuliferous scale The modified lateral branch system of a conifer cone that bears one or more seeds.

P/E ratio The ratio of the polar diameter to the equatorial diameter.

Pachycaul An erect, woody, trunk like stem that is swollen basally, the swollen region functioning in storage of food reserves or water.

Padlike nectary Developing as a discrete pad of tissue extending partway around the base of the flower.

Paedomorphosis A type of heterochrony in which the mature or adult stage of the derived ontogenetic sequence resembles a juvenile ontogenetic stage of the ancestral condition.

Palea PI: *paleae.* The inner and upper bract at the base of the grass floret, or the bracts subtending flowers in some Asteraceae, e.g., tribe Heliantheae.

Paleotropical Distributed in tropical areas of the Old World (tropical Africa, Asia, Australasia).

Palinactinodromous With three or more primary veins diverging from one point, the primary veins having additional branching above their main point of divergence.

Palisade mesophyll The usually upper (adaxial), columnar cells of the mesophyll of some leaves.

Palmate/palmately compound A compound leaf with four or more leaflets arising from a common point, usually at the end of the petiole.

Palmate Compound Leaves in which rachis does not differentiate and the lamina appears articulated or attached to a point on the top of the petiole.

Palmately netted, Palmate-netted, Palmately veined With four or more primary veins arising from a common basal point, the ultimate veins forming a fine reticulum.

Palmate-parallel With several primary veins arising from one point, the adjacent secondary veins that are parallel to these having transverse, interconnecting veins as in "fan" palms.

Palmate-ternate Ternate, with the three leaflets joined at a common point.

Palmatifid Palmately lobed to divided.

Palmatisect Palmately divided, almost into discrete leaflets but confluent at the lobe bases.

Palynology The study of spores and pollen grains.

Pandurate Violin-shaped, obovate with the side margins concave.

Panicle An indeterminate inflorescence, consisting of several branched axes bearing pedicellate late flowers.

Pantropical Distributed worldwide in tropical areas.

Pantoporate A pollen grain aperture type in which pori occur globally on the pollen grain surface.

Papilionaceous Petals appearing butterfly like; a zygomorphic corolla with one large posterior standard, two lateral wings and two inner most apically connate petals, the keel as in the Faboideae-Fabaceae.

Papillate Syn: *tuberculate; verrucate.* having minute, rounded protuberance.

Pappus The calyx modified as awns, scales, or capillary bristles as in Asteraceae.

Parallel or Striate With primary or secondary veins essentially parallel to one another, generally converging at the apex, the ultimate veinlets transverse.

Parallelism *Convergence.* parallelodromous *Parallel.*

Paraphyletic/paraphyly A group that includes a common ancestor plus some, but not all, descendants of that common ancestor.

Paraphyletic species A species consisting of lineage segments that form a paraphyletic grade, each lineage segment lacking an apomorphy.

Paraphyses Hair-like structures arising from the receptacle of a fern sorus.

Parasite A plant living on another organism (or plant) and extracts food from it (called host).

Parasitic or haustorial roots The roots that absorb nutrition by sending out haustoria into the tissue of hosts.

Parastichous The condition when it is not possible to establish orthostichy, as on date palm trunk.

Paratype A specimen cited in a publication but that is neither a holotype, isotype nor syntype.

Parenchyma Cells that, at maturity, are alive and potentially capable of cell division, are isodiametric to elongate in shape, and have a primary cell wall only (rarely with secondary wall), forming a solid mass of tissue and functioning in metabolic activities and in wound healing and regeneration.

Parietal Placentae bearing ovules developed on the inner wall of the ovary and their position corresponds to the confluent margins of carpel in which ovary is unilocular as in Cruciferae and Cucurbitaceae.

Parietal cell The outer cell formed if the archesporial cell of an ovule divides to form an inner megasporocyte.

Parietal-axile With the placentae at the junction of the septum and pericarp of a 2 or more loculate ovary, e.g. Brassicaceae.

Parietal-septate With placentae on the inner ovary walls but within septate locules, as in some Aizoaceae.

Paripinnate/paripinnately compound Syn: *even-pinnate.* A pinnately compound leaf without a terminal leaflet, typically even-pinnate as in *Tamarindus indica.*

Parsimony analysis Principle of parsimony.

Parted With sinuses extending (pinnately or palmately) one half to three quarters of the distance to the midrib, midvein or vein junction.

Parthenogenesis Development of an embryo from a cell of an abnormal, diploid female gametophyte, such as a diploid egg.

Partial inflorescence *Unit inflorescence.*

Patent Horizonal; divaricate; divergent.

Peat Fossilized and partially decomposed *Sphagnum* (peat moss).

Pectinate Comblike; pinnately divided with close, narrow lobes. (division)

Pedate A palmately divided leaf in which the lateral lobes are further divided, as in some ferns.

Pedicel A flower stalk. Adj: *pedicellate.* (flower past; inflorescence part)

Pedicillate Flower is produced on a stalk or pedicel.

Peduncle Adj: pedunculate. A stalk which bears inflorescence.

Pellucid Having translucent spots or patches.

Peltate (a) Having a stalk attached away from the margin of a flattened structure, e.g., a petiole attached on the underside of blade. (b) Trichomes with stalk attached on the underside of a disk shaped apical portion.

Pendant/pendulous Hanging downward loosely or freely.

Pendulous Apical.

Penicillate Tufted, like an artist's brush.

Penni-parallel Syn: pinnate parallel. With secondary veins arising along the length of a single primary vein region, the former essentially parallel to one another and interconnected by transverse veins.

Pentacarpellary Gynoecium that consists of five carpels. It may also be called multicarpellary as in Malvaceae.

Penta or polylocular Ovary with 5 or more chambers as in Malvaceae.

Pentamerous The floral whorls contain units five in number or in multiple of five.

Pentandrous. Five stamens in a flower as in Solanum.

Pentastichous or phyllotaxy The sixth leaf is found above the first leaf and seventh above the second leaf.

Pepo A nonseptate, fleshy fruit with parietal placentation and a leathery exocarp derived from a tricarpellary, syncarpous, unilocular inferior ovary, e.g., Cucurbitaceae. These fruits are full of swollen placenta and are many seeded. Epicarp makes the hard rind. Mesocarp and endocarp are fleshy.

Peramorphosis A type of heterochrony in which ontogeny passes through and goes beyond the stages or trajectory of the ancestral condition.

Perennial A plant living more than two years.

Perfect (a) Bisexual. (b) With lateral primary veins covering at least two thirds of the leaf blade area, a subcategory of actinodromous and of acrodromous.

Perfoliate A leaf blade that is sessile with the base completely surrounding the stem.

Perforation plate The region of one or more perforations at the end wall of a vessel member, where one cell makes contact with another; may be compound if composed of several pores or simple if composed of a single opening.

Perianth Syn: *perigonium.* Petals and sepals are of similar appearance and the two can not be distinguished. The outermost, nonreproductive group of modified leaves of a flower, composed of the combined calyx and corolla or of tepals.

Pericarp Syn: *rind.* The fruit wall derived from the mature ovary wall, sometimes divisible into layers: endocarp, mesocarp and exocarp.

Pericycle A cylindrical sheath of parenchyma cells just inside the endodermis, functioning as the site of resumed meristematic growth, forming a secondary root or (in woody plants) contributing to the vascular cambium.

Periderm Syn: *outer bark.* The cork cambium and its derivatives: phelloderm and cork tissue.

Perigonal nectary Nectaries on the perianth, usually at the base of sepals, petals or tepals.

Perigynous Hypanthium present, attached at base of ovary with sepals, petals and stamens attached to hypathium, the ovary superior. A flower with half embedded or sunken ovary as in Rosaceae.

Perine/perine layer A thick, outer layer in the spores of mosses and ferns.

Periodicity Periplasmodial *Arnoeboid.* Periodically repeating phenomena (temporal phenomena).

Persistent (a) Sepals or petals do not fall off rather these remain with the developing fruits as in brinjal and tomato. (b) Leaves last for a long time as in most tropical trees, e.g. Polyalthia.

Perisperm A nutritive tissue derived from the nucellus of the ovule (present in some seeds) or may be called persistent nutritive nucellus (as in members of Nymph-aeaceae). In coffee seed, it is the perisperm that is the source of coffee powder.

Perispermous Having a seed storage tissue in which the chalazal nucellar cells enlarge and store energy rich compounds.

Peristome teeth Hygroscopic, teeth like structures that occur in a whorl along the margin of the opening of a moss capsule and that function in spore release.

Personate Two-lipped, with the upper arched and the lower protruding into corolla throat, e.g., *Antirrhinum.*

Petal An individual member or segment of the corolla, typically (non-green) colored and functioning as an attractant for pollination. (flower part, perianth part) *petalostemonous Epipetalous.* (stamen fusion)

Petaloid Sepals are coloured.

Petaloid bract Bract becomes brightly coloured and appears like a petal in which three small flowers are enclosed within three coloured bracts, e.g. bright red or orange bracts in *Poinsettia pulcherrima.*

Petiolar spine A sharp pointed leaf petiole, e.g., *Fouquieria spp.* Cf: *thorn; prickle.*

Petiole or mesopodium Adj: *petiolate.* Syn: *stipe.* Stalk of the leaf.

Petiolule A leaflet stalk, Adj: *petiolulate.*

Phalaenophily Adj: *phalaenophilous.* Pollination by moths.

Phanerantherous *Enserted* phanerocotylar *Epigeous.*

Phelloderm The inner layers of cells produced by the cork cambium.

Phenetic A classification system in which taxa are grouped by some measure of overall similarity.

Phenogram A branching diagram representing a phenetic classification.

Phloem A tissue composed of specialized sieve elements plus some parenchyma and often some sclerenchyma, functioning in conduction of sugars.

Photosynthesis The series of biochemical reactions in which light energy is used to produce high-energy compounds, in land plants involving reactions of carbon dioxide and water to produce polysaccharides.

Phyllary Syn: *involucral bract.* One of the involucral bracts subtending a head, as in the *Asteraceae.*

Phylloclad Cladode; cladophyll.

Phyllode A leaf consisting of a flattened bladelike petiole.

Phylogenetic A classification that is based on evolutionary history, or pattern of descent; referring to relationships between groups of individuals at a given point in time.

Phylogenetic species A species recognized as the smallest group of populations or lineages diagnosable by a unique combination of character states; a monophyletic group recognized and defined by one or more apomorphies.

Phylogenetic systematics Syn: *cladistics.* A methodology for inferring the pattern of evolutionary history of a group of organisms by grouping taxa based on apomorphies

Phylogenetic tree *Cladogram.*

Phylogeny The evolutionary history or pattern of descent of a group of organisms.

Phylogeography A field of study that assesses relationships among genotypes of a species or closely related species and correlates those relationships with their spatial distribution.

Phyllotaxy The mode of arrangement of leaves on the stem. Cauline Leaves borne directly on the main trunk.

Pilate With a long terete stalk terminating in a globose or ellipsoid apical thickening.

Pilate-glandular Having a glandular cell atop an elongate basal stalk.

Pilose With soft, straight to slightly shaggy trichomes at right angles to the surface. Long, distinct and scattered hairs as in *Grewia pilosa.*

Pin Flowers with a long style and short stamens, found in distylous flowers.

Pinna PI: pinnae. The first discrete leaflets or blade divisions of a fern leaf.

Pinnate/pinnately compound A compound leaf with leaflets arranged oppositely or alternately along a central axis, the rachis. It may be simple pinnate, bipinnate, tripinnate and decompound.

Pinnately veined With secondary veins arising along the length of a single primary vein, the latter a midrib or leaflet midvein.

Pinnate netted Pinnately veined, with secondary veins arising along length of a single primary vein, the latter a midrib or midvein, the ultimate veins forming a fine reticulum.

Pinnate-parallel Penni-parallel.

Pinnate-ternate Ternately compound, with the terminal leaflet arising from the tip of a rachis, evolutionarily derived from a pinnately compound leaf, e.g., some Fabaceae.

Pinnatifid Pinnately lobed to divided.

Pinnatisect Pinnately divided, almost into discrete leaflets but confluent at the midrib.

Pinnule The ultimate divisions or leaflets of a leaf, usually applied to ferns.

Pipered type A type of embryo development in which the zygote divides longitudinally (i.e., parallel to the axis of the female gametophyte and nucellus), thus not forming a basal and terminal cell.

Pistil That part of the gynoecium composed of an ovary, one or more styles (which may be absent), and one or more stigmas.

Pistillate Syn: *female.* Flowers unisexual, with carpel(s) only, lacking fertile stamen(s).

Pit A hole in a secondary cell wall that functions in cell-to-cell communication during development and that may function in water conduction in some tracheary elements.

Pit pairs Adjacent holes in the lignified secondary cell walls of two adjacent cells occurring opposite one another, functioning in allowing communication between cells during their development and differentiation.

Pitcher A pitcher-like structure formed either by entire leaf or by part of it such as by lamina.

Pitcher leaf A leaf shaped like a container that bears an internal fluid and that functions in the capture and digestion of small animals, e.g., leaves of *Darlingtonia, Nepenthes, Sarracenia,* pitcher plants. In Nepenthes, pitchers are modified to catch insects to meet the need of nitrogenous substances and in Dischidia (an epiphytic climber), the pitcher accumulates rain water as well as debris which is absorbed by the adventitious roots formed within.

Pith The central, mostly parenchymatous tissue, internal to vasculature of siphonosteles and eusteles and within the vascular cylinder of some roots.

Placenta The ovule bearing tissue of the ovary.

Placental vein *Ventral vein.*

Placentation Arrangement of ovules within the ovary

Planation Evolutionary flattening, into a dorsiventral, planar posture.

Plane (a) Flat, without vertical curves or bends. (b) With a smooth configuration. Syn: *smooth.*

Plant A group of organisms, defined either by characteristics (photosynthesis, cell walls, stores and a more or less sedentary behaviour), or by phylogenctic relationships to the land plants, Embryophyta.

Plant anatomy The study of tissue and cell structure of plant organs.

Plant habit General form of plant, including aspects of stem duration, branching pattern, development, and texture.

Plant physiology The study of metabolic processes in plants.

Plant press A device used to press and then dry plant specimens, such that they can be effectively used in an herbarium specimen.

Plant sciences The study of plants, which are equivalent to the land plants.

Plaspia membrane A phospholipid bilayer with embedded proteins that envelops all cells, functioning as the cell boundary, in cell-cell recognition and in transport *of* compounds.

Plasmodesmata Minute pores in the primary cell wall through which membranes traverse between cells, allowing for interchange of compounds between cells; an apomorphy for the Charophytes of the green plants.

Plasmodial *Amoeboid.*

Plated Bark split or cracked, with flat plates between fissures.

Pleonanthic An indeterminate shoot that bears lateral flowers but that continues vegetative growth.

Plesiomorphic *Ancestral.*

Pleurotropous An ovule position in which the micropyle points to the side.

Pleurotropous-dorsal A pleurotropous ovule in which the raphe is above.

Pleurotropous-ventral A pleurotropous ovule in which the raphe is below.

Plicate With a series of longitudinal folds: pleated.

Plumose (a) Covered with fine, elongate, ciliate appendages; featherlike, e.g., pappus of some Asteraceae. (b) Stigmas with feathery, trichome-like extensions, often found in wind pollinated taxa such as Cyperaceae, Poaceae.

Plurilocular An ovary with two or more locules.

Pneumatophores Roots that grow upwardly from soil to air, functioning in obtaining additional oxygen for the plant.

Polar axis An extended pollen grain diameter that passes through the centre of the original pollen tetrad.

Polar nuclei In a typical angiosperm female gametophyte, the two haploid nuclei of the central cell that ultimately fuse with a sperm cell (via double fertilization) to form a triploid endosperm cell.

Polar view A pollen grain from the direction of either pole.

Polarity The designation of relative ancestry to the character states of a transformation series/morphocline.

Pollen grain An immature, endosporic male gametophyte of seed plants.

Pollen sac A microsporangium, usually one-half of a theca in an angiosperm anther.

Pollen tube An exosporic process that grows from a pollen grain, functioning as a haustorial organ or to deliver sperm cells to the egg.

Pollen kit A yellowish or orange, carotenoid-like material adhering to the exine, functioning to stick pollen grains in masses.

Pollinarium In an orchid flower, the pollinia plus a sticky stalk (derived from either the anther or stigma), the unit of transport during pollination.

Pollination The transfer of pollen grains from microsporangia either directly to the ovule (in gymnosperms) or to the stigma (in angiosperms).

pollination chamber A cavity formed by the breakdown of cells at the distal end of the megasporangium (nucellus) in gymnosperm seeds.

Pollination droplet A droplet of liquid secreted by the young ovule through the micropyle, functioning to transport pollen grains by resorption.

Pollinium PI: pollinia. Anther in which all pollen grains of both thecae (Orchidaceae) or of adjacent thecae (Asclepiadaceae) are fused together as a single mass.

Polyad A group of pollen grains that are fused in precise units of more than four, e.g., Mimosoideae.

Polyadelphous Stamens with fused filaments arranged in several groups.

Polyandrous Indefinite number of stamens as in Nymphaea.

Polyclave key A key in which all of the known character states that match a specimen are selected in order to narrow down the identity to a smaller subset of the possibilities.

Polygamous A plant with both bisexual and unisexual flowers.

Polygonum type A type of monosporic female gametophyte in which the megaspore nucleus undergoes three sequential mitotic divisions, producing eight nuclei and seven cells, the most common and ancestral type in the angiosperms.

Polymerase chain reaction Abbr: PCR. A procedure for the rapid amplification of DNA using primers, free nucleotides, and DNA polymerase in solution and heating the solution to effect denaturation and replication of the DNA.

Polymorphic character A character that has variable character state values within an OTU.

Polypetalous Apopetalous. Petals are free (not united) as in Cruciferae.

Polyphyletic group A group that consists of two or more, separate monophyletic or paraphyletic groups, each with a separate common ancestor; a group in which the common ancestor of all members is not itself a member of the group.

Polyploidy A mutation in which offspring have an increase in chromosome number by a multiple of some ancestral set.

Polysepalous *Aposepalous.* Sepals are free, i.e., riot united with each other

Polystichous *Spiral.*

Polysymmetric *Actinomorphic, radial, regular.*

Polytomy Three or more lineages arising from a single common ancestor in a cladogram, representing conflicting data or the lack of resolution.

Pome A fleshy fruit with a cartilaginous endocarp derived from a bi- or polycarpellary syncarpous inferior ovary, with the bulk of the edible fleshy tissue derived from the outer, adnate hypanthial tissue as in apple and pear. The thalamus becomes fleshy and swollen and surrounds the fruit. So it is a false fruit. The pericarp is thin and papery.

Population A group of individuals of the same species that is usually geographically delimited and that typically have a significant amount of gene exchange.

Pore (a) A specialized, permanent opening in the upper epidermis of the thallus of some liverworts, functioning in gas exchange. (b) A single hole in the primary cell wall of sieve elements that is lined with callose and through which solutes flow in sugar conduction. (c) Vernacular term for a wood vessel, used in the wood industry.

Poricidal Dehiscing by a pore at one end of the thecae, e.g., Ericaceae.

Porose Vessel cells with more or less circular, pore like perforation.

Porous Referring to wood that contains vessel cells.

Porous or Apical Discharge of pollens through apical pore.

Porous dehiscence Fruit dehisces through the pores developed in the terminal part. Seeds are slowly dispersed by wind through these pores, as in poppy (Papaver).

Porus PI: pori. Adj: *porate.* A pollen grain aperture that is circular to slightly elliptic with a length:width ratio of less than 2:1.

Position (a) Placement relative to other, unlike parts. (b) The placement of a taxon as a member of another taxon of the next higher rank.

Posterior Upper lobe or part, especially in a horizontally oriented structure.

Posture Placement relative to a flat plane.

Praemorse Having a jagged, chewed appearance as in palms.

Prickle Adj: *prickly.* A non-spine, non-thorn, sharp pointed outgrowth from the surface of any organ.

Primary cell wall The first, mostly cellulosic cell wall layer that is secreted external to the plasma membrane during cell growth.

Primary endosymbiosis Endosymbiosis involving the engulfment of an ancestral bacterium by a eukaryotic cell.

Primary growth Growth in height or length of a stem or root, brought about by elongation and differentiation of cells derived from the apical meristem.

Primary inflorescence *Unit inflorescence.*

Primary parietal cells The inner layer of cells in an early stage of anther wall development.

Primary pit field A group of numerous plasmodesmata in the primary cell wall, spatially associated with secondary cell wall pit pairs.

Primary root The root of the sporophyte that develops from the radical of the embryo.

Primary tissue A tissue formed by primary growth as in primary xylem or primary phloem.

Primary vein The major vein or veins of a leaf with respect to size.

Primer A complimentary copy of a short, conserved, flanking region of a region of DNA of interest, used to amplify and sequence the DNA.

Primitive *Ancestral.*

Principal components analysis Abbr: PCA. A multivariate statistical method that transforms numerous data variables into other variables called principal components, in order to study relationships among taxa.

Principle of parsimony Syn: *parsimony analysis.* The principle stating that the cladogram exhibiting the fewest number of evolutionary steps is accepted as the best estimate of phylogeny; a corollary of the general principle of Ockham's Razor.

Priority of publication A principle of the ICBN stating that of two or more competing possibilities for a name, the one published first is the correct one, with some exceptions.

Prismatic A short, prism shaped crystal; a type of ergastic substance.

Procumbent *Prostrate.* Trailing flat on the ground

Proembryo, A very young embryo.

Prolate A pollen grain in which the P/E ratio is greater than 1.2.

Prop root Above-ground, adventitious roots that function in supporting the stem. These strong, woody aerial roots sunken in the soil arise from horizontal branches (stem).

Prophyll/prophyllurn *Bracteolelbractlet.*

Prophylls First few leaves of a branch which differ from other leaves as in wood apple (*Aegle marmelos*) where these are represented by spines.

Prostrate Syn: *procumbent; reclining.* Trailing or lying flat, not rooting at the nodes.

Protandry Adj: *protandrous.* With stamens or anthers developing before carpels or stigma.

Protandrous A dichogamous flower in which anthers mature first as in Lamiaceae, Malvaceae and Rubiaceae.

Prottologue Everything associated with a name at its valid publication i.e., description or diagnosis, illustrations, references, synonymy, geographical data, citation of specimens, discussion and comments.

Proteinoplast *Alaurone grain.*

Protogyny Adj: *protogynuus.* With carpels or stigma maturing before stamens or anthers.

Protogynous A dichogamous flower in which stigma matures before the anthers, as in Magnolia, Michelia and Adhatoda.

Protonema An initial, filamentous form of a gametophyte (e.g., in mosses), prior to its differentiation into parenchymatous tissue.

Protoplasm Everything inside the plasma membrane of a cell.

Protostele A stele with a central, solid cylinder of vascular tissue.

Protoxylem First tracheary elements that develop within a patch of xylem, being typically smaller and with thinner cell walls than the later formed metaxylem.

Proximal Near the point of origin or attachment.

Proximal pole The intersection of the pollen grain polar axis with the grain surface that is near the centre of the microspore tetrad.

Pseudanthium A unit that appears as and may function like a single flower, but that typically consists of two or more flowers fused or grouped together.

Pseudobulb A short, erect, aerial storage stem of certain epiphytic orchids.

Pseudocrassinucellate An ovule type with a second inner layer of nucellar cells by periclinal divisions in the single outer layer, a parietal cell not forming.

Pseudodrupe A nut surrounded by a fleshy, indehiscent involucre, resembling a true drupe, e.g. walnut.

Pseudoelaters Groups of elongate, cohering, non-sporogenous, generally hygroscopic cells that develop within the sporangia of hornworts and function in spore dispersal.

Pseudomonopodial A branching pattern that initiates as dichotomous, but in which one of the two axes becomes dominant and overtops the other, the other axis appearing lateral.

Pseudopetiole Term sometimes used for the petiole like structure arising between a leaf sheath and blade, found in several monocots, such as bananas, bamboos.

Pseudoterminal bud A bud appearing to be apical but is actually lateral near the apex, assuming a terminal position with the death or non-development of the true terminal bud.

Pseudoumbel An inflorescence appearing like a simple umbel, but actually composed of condensed, monochasial cymes, as in the Alliaceae and Amaryllidaceae.

Psilate A pollen grain having a smooth sculpturing.

Psychophily Adj: *psychophilous.* Pollination by butterflies.

Ptyxis Syn: *vernation.* The posture of embryonic structures, such as cotyledons within a seed or immature leaves or leaf parts in a bud.

Puberulent Minutely pubescent.

Pubescent (a) With straight, short, soft, somewhat scattered, slender trichomes or hairs. (b) A general, term, meaning having trichomes.

Pulvinus The swollen base of a petiole or petiolule as in some Fabaceae.

Punctate Covered with minute, pit like depressions.

Pungent *Spinose.*

Pyrene A fleshy fruit in which each of two or more seed is enclosed by a usually bony textured endocarp, or the seed covered by a hard endocarp until itself, regardless of the number.

Pyxide/pyxis *Circumscissile capsule.*

Quadrifoliate Leaflets articulated to the top of the rachis.

Qualitative characters Syn: *categorical characters.* Characters in which the states are not directly measured but are based on defined classes of attributes.

Quantitative characters Cf: *continuous quantitative characters, discrete quantitative characters.* Characters in which the states are measured and based on numbers.

Quincuncial Two sepals are outer and two sepals are inner while one sepal has one overlapped and one exposed margin.

Quincuncial Perianth parts of a single pentamerous whorl having two members overlapping, two being overlapped, and one overlapping only at one margin.

Raceme An indeterminate inflorescence consisting of a single axis bearing pedicellate (stalked) flowers.

Racemose inflorescence It has indeterminate growth of main axis, all flowers are lateral and arranged in acropetal succession, i.e., oldest towards base and youngest towards apex.

Racemose or monopodial branching Main stem (terminal bud) grows indefinitely and lateral branches are produced in an acropetal order (i.e. older towards base), as in Polyalthia.

Rachilla PI: *rachillae.* A lateral or secondary axis of abipinnate leaf. (b) The central axis of a grass or sedge spikelet.

Rachis (a) The main axis of a palmately compound leaf. (b) A major, central axis with in an inflorescence.

Radial *Basal.* (a) *Actinomorphic, polysymmetric,* ***regular***. (b) A longitudinal section of wood, parallel to a stem radius.

Radical Leaves Arise from reduced stem and appear as if arising from the root.

Radicle The primary root. The first organ to appear when a seed germinates.

Ramal Leaves borne on the branches.

Ramet Cf. *gene!.* A clonal unit of a gene that is, at least potentially, independent from other ramets.

Random amplified polymorphic DNA Abbr: *RAPD.* The use of randomly generated primers for the amplification of DNA to identify polymorphic DNA regions of different individuals or taxa.

Range The collective geographic distribution of a taxon; the extent of measurements, from the minimum to the maximum values.

Rank One of the hierarchical taxonomic categories, in which a higher rank is inclusive of all lower ranks.

Raphe A ridge-like structure on seed coat formed by fusion of funicle and body of the ovule.

Raphide A needlelike crystal of calcium oxalate, typically occurring in bundles; a type of ergastic substance.

Ray (a) A secondary axis in a compound umbel (b) A corolla with a short tube and a single, elongate, strap like apical extension as in Asteraceae. Syn: *ligulate.* (c) Radially oriented cells that occur in band like strands in the secondary xylem (wood), functioning in lateral translocation of water, minerals and sugars.

Ray floret These are found at the rim of receptacle (perphery) and have ligulate or strap-shaped showy corolla. Flowers are either sterile (neuter) or female (pistillate) and are zygomorphic.

Recency of common ancestry A measure of phylogenetic relationship, stating that two taxa are more closely related to one another if they share a common ancestor that is more recent in time than the common ancestor they share with any other taxon.

Receptacle Syn: *torus.* The tissue or region of a flower to which the other floral parts are attached.

Reclined Syn: *reclinate.* Directed downward, with divergence angle of I5°-45° from horizontal axis.

Reclining *Prostrate.*

Recognition species Sexually reproducing systems that are maintained by genetically based features that promote reproduction.

Recurved With tip gradually curved outward or downward, abaxially.

Reduced A type of anther wall development in which the secondary parietal cells do not divide further and develop directly into the endothecium and tapetum, respectively.

Reduplicate Plicate, with adjacent abaxiai sides facing one another, A-shaped in cross section.

Reflexed Bent or turned downward.

Regma The fruit develops from tricarpellary (multicarpellary), syncarpous, superior ovary. The fruit is multilocular. On maturation, after splitting, these divide into as many parts as the number of carpels. Each part is known as coccus and has one seed, as in castor and Geranium.

Relative cover A measure of the degree to which each species of a community layer contributes to the total cover of that layer alone.

Remodel/remodeling To change the diagnostic characteristics of a taxon.

Reniform (a) Kidney shaped; wider than long with a rounded apex and reniform base. (b) With two rounded, basal lobes, smoothly concave at intersection of lobes.

Repand Margins wavy in a vertical plane as in Polyalthia.

Repent Creeping or lying flat and rooting at the nodes.

Replum A peripheral rim of a cross wall that persists after fruit dehiscence in silicles and siliques of Brassicaceae.

Reproductive roots Roots that bear adventitious buds which grow into plantlets.

Resealed consistency index (RC) A measure of the relative amount of homoplasy in a cladogram, equal to the product of the consistency index and retention index.

Resin ducts/canals Specialized cells that secrete resin.

Respiratory roots Pneumatophores. These are negatively geotropic vertically growing roots into the air for breathing or oxygen requirement found in mangrove plants growing in saline waters, as in Rhizophora, Heritiera, Avicennia, etc.

Restriction fragment length polymorphism Abbr: *RFLP.* Differences between taxa in restriction sites, and therefore the lengths of fragments of DNA after cleavage with restriction enzymes.

Restriction site A sequence of approximately six to eight base pairs of DNA that binds to a given restriction enzyme.

Resupinate Posture inverted or twisted 180', as leaves of Alstroemeriaceae or ovaries of most Orchidaceae flowers.

Retention index (RI) A measure of the relative amount of homoplasy in a cladogram, equal to the ratio (g - s)l(g - m), where g is the maximum possible number of state changes that could occur on any conceivable tree, m is the minimum possible number of character state changes, and s is the actual number of state changes that occur.

Reticulate (a) Venation in which the veins appear to join back together forming a net-like "reticulum. Syn: anastomosing, *netted.* (b) A pollen sculpturing with a netlike sculpturing, each element termed a murus and the space between termed a lumen.

Reticulation Hybridization of two previously divergent lineages, forming a new lineage.

Reticulodromous (a) Pinnate venation in which secondary veins do not terminate at the margin and branch repeatedly forming a very dense, netlike structure. (b) With three or more primary veins diverging from one point but not reaching the blade margin, a subcategory of actinodromous.

Retinaculum *PI: retinacula. Jaculator.*

Retrorse Bent or directed downward, usually referring to small appendages.

Retroserrate Teeth pointed downwards.

Retuse Having an apical incision, cut up to 1/16 the distance to the midrib, midvein or junction of primary veins. Slightly notched obtuse apex as in Pistia and Clitoria.

Reversal Homoplasy by the loss of a derived feature and the reestablishment of an ancestral feature.

Revolute Margins or outer portion of sides rolled outward or downward over the abaxial surface.

Reward A floral structure, such as pollen, or exudate, such as nectar, that functions to ensure that an animal pollinator will consistently return to transport pollen.

Rheophyte A plant found along streams and river banks (often swiftly flowing).

Rhiphidium A monochasium in which the branches develop on alternating sides of each sequential axis, typically having a geniculate (zig-zag) appearance; sometimes equated with scorpioid cymes.

Rhipidium Sub-umbellate cluster, fan-shaped cyme, lateral branches developing in one plane, alternately in opposite directions with the main axis appearing zig-zag (mainly in Iridaceae)

Rhizoid One of several uniseriate (out cell thick), filamentous processes that function in anchorage and water/mineral absorption arising front the gametophytes of free-living gametophytes and from the underground, sporophytic stems of some ferns (e.g., the psilophytes).

Rhizome (a) Underground, dorsiventral, horizontally growing below the soil surface with distinct nodes and short internodes. Nodes bear scale leaves on the upper surface and adventitious roots on the lower surface. They are usually fleshy due to the storage of food as in ginger and turmeric. (b) A horizontal stem of a fern that grows at ground level.

Rhombic Parallelogram-shaped, widest near the middle, the length:width ratio 2:1 to 3:2.

Rhomboidal tetrad A tetrad in which the four grains are in one plane, with two of the grains separated from one another by close contact of the other two.

Ribosome A cellular organelle that functions in protein synthesis.

Ribulose-bisphosphate carboxylase Abbr: *RuBP-carboxylase.* The enzyme that catalyzes the initial binding (fixation) of carbon dioxided in the dark reactions of photosynthesis.

Rind *Pericarp.*

Ring-porous Wood in which the vessels form only in the spring wood, with summer wood either lacking or having relatively small vessels and usually containing mostly fibres.

Robustness The confidence for which a tree or particular clade actually denotes true phylogenetic relationships.

Root A positively geotropic leafless lower portion of the plant axis. Basically, it arises from radicle and is characterized by non-green smooth surface and not differentiated into nodes and internodes. The cylindrical organ of virtually all vascular plants, consisting of an apical meristem that gives rise to a protective root cap, a central endodermis bounded vascular system, absorptive epidermal root hairs, and endogenously developed lateral roots; usually functioning in anchorage and absorption of water and minerals; initially derived from the radicle of the embryo and typically growing downward.

Root apical meristem Region of continuously dividing cells from which all cells of the root are derived.

Root buttresses Found in some large trees to provide additional support by producing plank-like roots arising from the base of trunk as in Salmalia, Ficus and Terminalia.

Rootcap The outer cell layer at the root tip, functioning in protection and to provide lubrication for the growing root.

Root Climbers Climbing by the formation of adventitious roots arising from the nodes.

Root hair One of several hair like extensions from an epidermal cell of a root, functioning to greatly increase the surface area available for water and mineral absorption.

Root tuber A swollen taproot containing concentrations of high energy compounds such as starch.

Rootstock A general term for an underground stem or shoot, these generally giving rise to aerial shoots either by direct conversion of the terminal apical meristem or via lateral buds.

Rosaceous Petals arranged as in rose.

Rosette Adj: rosulate. An arrangement in which parts, usually leaves, radiate from a central point at ground level, e.g., the leaves of *Taraxacum officinale,* dandelion.

Rotate With a short tube and wide limbs oriented at right angles to the tube. Petals fuse to form a saucer-shaped structure at their upper part.

Rotund Orbicular or circular as in water lily and lotus.

Rounded With apical or basal margins convex, forming a single, smooth arc.

Rugose Syn: *bullate.* Covered with coarse reticulate lines, usually with raised, blister like areas between. Wrinkled surface as in *Rubus rugosus*.

Rugulatetrugulose A pollen sculpturing having irregular to sinuous, tangentially oriented elements, often appearing brain like.

Ruminate Unevenly textured, coarsely wrinkled, looking as if chewed, e.g.,, the endosperm of the Annonaceae.

Runners Develop from the basal internodes of the axillary buds of the lowest leaves and which elongate horizontally along the ground carrying the buds, and after trailing for a short distance along the ground, each runner attaches itself to the soil by means of adventitious roots and develops into a new daughter plant.

Saccate Having a pouch like eyagination.

Sagittate Shape like arrow head with two basal lobes more or less pointed and oriented downward, away from apex.

Salverform Trumpet shaped; with a long, slender tube and flaring limbs at right angles to tube.

Sapromyiophily Adj: sapromyiophilous. Pollination by flies.

Samara A winged, dry fruit, e.g., *Acer,* maple; Ulmus, elm. The fruit develops from bicarpellary syncarpous, superior ovary. It is single seeded. The pericarp becomes flat like wings.

Samaroid Look-like samara fruit as in sal (*Shorea robusta*) as the persistent calyx forms wing.

Saprophyte A plant living off dead organic matter.

Sarcotesta A seed coat that is fleshy at maturity.

Sarmentose Stoloniferous.

Saxicolous Occurring on rock.

Scaberulous Minutely scabrous.

Scabrate *Spinulose.*

Scabrous Rough because of the presence of short rigid points or trichomes, resembling sand paper, e.g. fig leaves.

Scale (a) A small, non-green leaf of a bud or underground rootstock. (b) A bract of a sedge spikelet.

Scaling Adj: *scaled.* Scandent *Clambering.* Assigning a weight in a phylogenetic analysis to correlate characters that is the inverse of the total number of the correlated characters.

Scaly bract Additional scaly bracts found with each individual floret as in capitulum, catkin or cyathium inflorescences.

Scaly or imbricated Scale-leaves arranged loosely with overlapping margins and are not covered by any common tunic.

Scape Adj: *scapose.* A "naked" (leafless) peduncle, generally arising from the middle of a basal rosette of vegetative leaves. It bears a flower, several flowers, or a crowded infloresence as in onion.

Scarious Thin and appearing dry, usually whitish or brownish.

Schizocarp Derived from a two or more loculed compound ovary in which the locules separate at maturity.

Schizocarp of follicles A schizocarp in which the carpels of a pistil split at fruit maturity, each carpet developing into a unit follicle, e.g., Asclepiadaceae.

Schizocarp of mericarps A schizocarp in which (generally two) carpels of a single, unlobed ovary split during fruit maturation, the carpels developing into unit mericarps and attached to one another via a stalk like carpophore, e.g., Apiaceae.

Schizocarp of nutlets A schizocarp in which a single ovary becomes lobed during development, each lobe developing at maturity into a nutlet, e.g., Boraginaceae, Lamiaceae.

Schizocarpic or Splitting Fruits After ripening, these dry and multiseeded fruits split (divide) into one seeded segments, called mericarps. Mericarps do not rupture further.

Scientific name A formal, universally accepted name, the rules and regulations of which (for plants, "algae," fungi, and organisms traditionally treated as such) are provided by the ICBN.

Sclereid A sclerenchyma. Cell that is isodiametric to irregular in shape and often branched, functioning in structural support or possibly aiding in providing protection from herbivory.

Sclerenchyma A tissue composed of nonconductive cells that are dead at maturity, that have a thick, lignified, generally pitted secondary cell wall, and that function in structural support and/or to deter herbivory; composed of fibres and sclereids.

Scorpioid cyme Syn: *cincinnus.* A monochasium in which the branches develop on alternating sides (left and right) of each sequential axis, typically appearing geniulate (zig-zag); may intergrade with helicoid cyme.

Scramblers or Hook Climbers Climbing by the formation of hooks, or curved prickles as in rose.

Scurfy *Farinaceous.* (epidermal excrescence)

Secondary cell wall An additional wall layer, composed of cellulose and lignin, secreted between the plasma membrane and the primary cell wall following cell elongation, found in some cell types (including tracheary elements and fibres).

Secondary contact Whereby two (or more) lineages that had diverged and were likely geographically separated (allopatric) in the past, have come to occupy the same or overlapping ranges.

Secondary endosymbiosis Endosymbiosis involving the engulfment of an ancestral eukaryotic cell by another eukaryotic cell, a possible mechanism of chloroplast exchange between eukaryotes.

Secondary growth. Growth in girth or width by means of cells produced from lateral meristems.

Secondary parietal cells Two layers of cells arising by tangential divisions of the primary parietal cells of the anther wall.

Secondary phloem Syn: *inner bark.* Sugar-conducting tissue produced by the vascular cambium to the outside of a woody stem or root.

Secondary root A root derived endogenously from the pericycle of an existing root.

Secondary tissue A tissue formed by secondary growth, via lateral meristems.

Secondary vein A lateral vein that branches from and is smaller than a primary vein.

Secondary xylem Syn: wood. Water- and mineral-conducting tissue produced by the vascular cambium to the inside of a woody stem or root.

Secretory Syn: *glandular.* An anther tapetum type in which the tapetum remains intact with no breakdown of cell walls.

Secretory structure A collection of cells that secrete compounds, either internally or externally.

Secund Syn: *unilateral.* Flowers, inflorescences, or other structures on one side of axis, often due to twisting of stalks, appearing as if arising from one side.

Sedge spikelet The inflorescence unit of the sedge family consisting of a central axis (the rachilla), bearing spiral or distichous bracts (also called scales or glumes), each subtending a single flower.

Seed An embryo surrounded by nutritive tissue and enveloped by a seed coat; the propagule of the seed plants.

Seed coat Syn: *testa.* The outer protective covering of seed, developing from one or two integuments.

Self-dispersal Syn: *autochory.* Process by which a flowering plant buries its own fruits as in peanut.

Self-incompatibility Outcrossing occurring by the genetic inability for fertilization to occur between gametes derived from a single genotype.

Selfing *Inbreeding.*

Self pollination The pollination between the different flowers or same flower of the same plant.

Semelpary Adj: *semelparous.* Referring to plants that have one episode of reproduction, followed by degeneration and death of the entire plant.

Semi-inferior A condition intermediate to superior and inferior.

Semicraspederomous Pinnate venation in which the secondary veins branch near the margin, one terminating at the margin, the other looping upward to join the next secondary vein.

Seminal root Roots arising from near the base of the radicle as in monocots

Sensu lato Abbr. s.l. Meaning in the broad sense. Designated for a taxon name that is used inclusively for a broad, inclusive taxon circumscription to include other, previously recognized taxa.

Sensu stricto Abbr: *s.s.; s.str.* Meaning in the strict sense. Designated for a taxon name used exclusively to exclude other, previously recognized taxa.

sepal An individual member or segment of the calyx, typically green, leaf like and functioning to protect the young flower.

Sepaloid Petals are green in colour or similar to sepals.

Septal nectary A nectary embedded within the ovary septae, secreting nectar via a pore at the ovary base or top.

Septicidal capsule A capsule in which longitudinal lines of dehiscence are radially aligned with the septa.

Septicidal dehiscence Septa in fruit are ruptured during dehiscence and locules are separated. Pericarp is also separated in the form of valves, as in linseed and Aristolochia.

Septifragal capsule Syn: valvular capsule. A capsule in which the valves break off from the septal as in *lpomoea,* morning glory.

Septifragal dehiscence Placenta remains attached in the centre of undivided fruit during dehiscence and pericarp's segments become separated. Dehiscence can be either septicidal or septifragal as in Datura.

Septum A partition or cross wall of the ovary.

Sequential *Gradate.*

Sericeous With long, appressed, silky trichomes.

Senes Whorl.

Serotinous Opening not at maturity, but generally with the stimulus of fire, e.g., cones of some conifers.

Serrate Saw-toothed. Teeth sharp and ascending, the lower side longer, cut from 1/16 to 1/8 the distance to midrib, midvein or junction of primary veins.

Serrate Margin with teeth pointed upwards as in saw, e.g. china-rose and rose.

Serrulate Diminutive of serrate, teeth cut to 1/1, the distance to midrib, midvein or junction of primary veins.

Sessile (a) Without a petiole or, for leaflets, without a petiolule. (b) Lacking a pedicel. (c) With filament absent, the anther attached directly. (d) Ovary lacking a basal stalk.

Sessile-glandular Glandular cell with a very short or no basal stalk.

Sexine The outer, protruding layers of the exine, including columellae, tecturn and supratectal sculpturing elements.

Sexual reproduction Reproduction involving meiosis with recombination and fusion of egg and sperm in lard plants entailing two "phases" of a haplodiplontic life cycle.

Sheath A flattened leaf base or petiole that partially or fully clasps the stem as in Poaceae and Apiaceae.

Sheathing Attached by a curved or tubular structure that partially or totally encloses the stem.

Shining *Nitid.* Epidermal excrescence.

Shoot A stem plus associated, derivative leaves, initially formed by an apical meristem that gives rise to the stem and external (exogenous) leaf primordia; may be gametophytic or sporophytic.

Short shoot *Fascicle; spur.*

***Spur* shoot**. Stem/shoot type. Shreddy bark coarsely fibrous, often fissured.

Shrub A perennial, short, woody plant without a trunk having several main stems arising at ground level, e.g. Lantana.

Sieve area An aggregation of callose-lined pores in *sieve elements.*

Sieve cell A type of sieve element that has only sieve areas on both ends and side walls, found in nonangiospermous vascular plants.

Sieve element A generally elongated cell that is semi-alive at maturity; has a non-lignified, primary cell wall with specialized, callose-lined pores aggregated into sieve areas and/or sieve plates, and that functions in conduction of sugars.

Sieve plate One or more sieve areas at the end wall junction of two sieve tube members, the pores of which are significantly larger than are those of sieve areas located on the side wall; characteristic of angiosperms.

Sieve tube member A type of sieve element that has both sieve areas and sieve plates, the latter at the end wall junction of two sieve tube members and having larger pores; an apomorphy of angiosperms.

Sieve tube plastid A membrane bound plastid found in sieve tube members, the contents of which (whether starch or protein) can vary between taxa and be systematically informative.

Significant Condition in which the differences between two or more groups are not due to chance alone, i.e., the probability of the differences being due to chance is calculated as less than some standard valve (generally 5% or $p < 0.05$).

Silicle Cf. *silique.* A dry, dehiscent, 2-carpeled fruit that dehisces along two sutures, has a persistent partition (replum), and is as broad or broader than it is long, e.g.. Brassicaceae.

Silicula Resembling with siliqua but its breadth and length are equal. It is wide and flat as in candytuft and shephard's purse.

Siliqua The fruit developed from bicarpellary syncarpous, superior ovary with parietal placentation. The ovary remains unilocular in the beginning but becomes bilocular due to formation of false septum, called replum. Each locule has many seeds. On maturity fruit dehisces from below to upwards as in mustard, radish, turnip, etc.

Simple leaf A leaf not divided into leaflets, bearing a single blade.

Simple cone A cone consisting of an axis bearing sporophylls.

Simple corymb An unbranched corymb, consisting of central axis bearing pedicellate flowers, the collection of flowers being flat-topped or convex.

Simple craspedodromous Pinnate venation in which all secondary veins terminate at the margin.

Simple dichasium A three flowered determinate inflorescence with a single terminal flower and two, opposite lateral flowers, the pedicels all of equal length.

Simple fruit A fruit derived from monocarpellary ovary or multicarpellary syncarpous ovary of one flower.

Simple ovary/pistil An ovary/pistil composed of a single carpel, the gynoecium apocarpous.

Simple perforation plate A perforation plate composed of a single opening.

Simple pinnate or unipinnate Leaflets borne directly on rachis.

Simple sieve plate A sieve plate that is composed of one pore region.

Simple tissue A tissue that consists of only one type of cell.

Simple umbel An umbel with the peduncle bearing pedicellate flowers attached at one point, e.g., Alliaceae.

Simultaneous (a) A type of microsporogenesis in which cytokinesis does riot occur until after meiosis. (b) Development of a fern sorus in which sprorangia mature at the same time. Syn: *sunple.*

Sinistrorse Twining helically like a left-handed screw as in Caprifoliaceae.

Sinuate Sinuses shallow and smooth, wavy in a horizontal plane.

Sinus The indentation of a divided planar structure.

Siphonogamy Cf. *zooidoganty.* Gymnospermous pollen tubes that are haustorial, but deliver non-motile sperm cells to the archegonium or egg.

Siphonostele A type of stem vasculature in which a ring of xylem is surrounded by an outer layer of phloem (ectophloic siphonostele) or by an outer and inner layer of phloem (amphiphloic siphonostele; if dissected, called a dictyostele).

Sister group/sister taxa Two taxa. that are each other's closest relative, representing two descendant lineages from a common ancestor.

Smooth (a) *Plane.* (b) A non-fibrous bark without fissures, plates or exfoliating sheets.

Sobole A special type of underground runner found in many grasses. The grasses send out soboles in all directions and develop numerous offsprings. It is a more effective method of propagation as in *Saccharum spontaneum.*

Softwood Wood derived from a conifer, generally (but not always) softer in texture than a hardwood.

Solanad type A type of embryo development in which the terminal cell divides transversely, the basal cell derivatives forming a suspensor but otherwise not contributing to mature embyro development.

Solenostele Aniphiphloic siphonostele.

Solitary A one flowered inflorescence.

Somatic Vegetative; referring to a cell not directly involved in sexual reproduction.

Sorosis A multiple fruit in which the unit fruits are fleshy berries and are laterally fused along a central axis. The fruit develops from spike, spadix or catkin inflorescence, as in jackfruit, mulberry and pineapple. The fruits are compactly set and appear as one fruit. In jackfruit, stigmas fuse with each other to make rough and spiny rind and bracts, perianth and seeds become edible. In mulberry perianth and the dry achenes is eaten.

Sorus PI: sori. A discrete cluster or aggregation of leptosporangia.

Spadix A spike with a thickened or fleshy central axis covered with large and coloured bract, called spathe, typically with congested flowers as in Araceae.

Spathaceous Having a spathe.

Spathe An enlarged, sometimes coloured bract subtending and usually enclosing an inflorescence as in the Araceae.

Spathulate A shape that is oblong, obovate or oblanceolate with a long attenuate base as in *Euphorbia nerifolia.*

Spathy bract Large, thick. boat-shaped bract and called spathe covering full or a part of a spadix inflorescence as in Colocasia and Amorphophallus. In some palms, the spathe may be very long (upto 6m). Spathes in maize and banana are well developed.

Speciation The formation of new species from preexisting species.

Species A group of individuals related to one another by certain criteria and distinct from other such groups of individuals. Cf: biological (isolation) species, cohesion species, evolutionary species, genealogical species, metaspecies, paraphyletic species, phylogenetic species, recognition species, taxonomic (morphologic) species.

Species complex A group of very similar and presumably closely related species and/or infraspecies.

Species evenness A measure of the number and evenness of distribution of species occurring within a given geographical region.

Species nova Abbr: sp. *nov.* Meaning that a taxon name, at the rank of species, is new to science.

Species richness A count of the number of species occurring within a given geographical region.

Specific epithet The second name of a binomial; may be capitalized or not.

Sperm A haploid, motile or non-motile male gamete that functions to fuse with an egg cell producing a diploid zygote.

Spheroidal Approximately spherical in shape.

Spike An indeterminate inflorescence consisting of a single axis bearing sessile flowers.

Spike It is similar to raceme but flowers are sessile as in *Amaranthus*.

Spikelet It is a secondary small spike with one or a few flowers. Spikelets are arranged in spike, raceme or panicle. Each spikelet bears at its base two minute bracts called empty glumes. Above this is a fertile glume, a third bract or bracteole. The fertile bracteole may be called palea or lemma.

Spine Cf: *thorn; prickle.* Entire or part of leaf modified into pointed hard structure. A sharp pointed leaf (e.g., cactus spines orglochidia) or leaf part, including petiole (e.g., *Fouquieria* spp.), midrib, secondary vein, leaflet (e.g., *Phoenix* spp.) or stipule (e.g., *Euphorbia spp.*).

Spinose Syn: *pungent.* (a) Margins with teeth bearing sharp, stiff, and spine like processes. (b) With a sharp, stiff, spine like apical process.

Spinose surface Surface covered with small prickles as in brinjal.

Spinous Sepals are modified into spines as in water chestnut.

Spinulose Syn: *scabrate.* A pollen sculpturing having spine like elements <1 μm long, each element termed a spinulum.

Spinulum PI: spinuli. One of the spine like elements of a spinulose pollen grain.

Spiny Teeth pointed to form spines as in pineapple and Mexican poppy.

Spiral (a) Alternate, with points of attachment in more than three rows/ranks. Syn: *polystichous.* (b) Perianth parts arranged in a spiral, one per node, not in distinct whorls. (c) Stamens arranged in a spiral, one per node; not whorled. (c) A type of endothecium in which the secondary wall thickenings are spiral or helical in shape.

Spiral or alternate phyllotaxy Leaves are produced alternately or spirally on the long axis or stem.

Spiraperturate A pollen grain having one or more apertures that are spirally shaped.

Spongy mesophyil The usually lower (abaxial), irregular cells of the mesophyll that generally have large intercellular, found in some leaves.

Sporangial wall one or more outer layers of sterile, non-spore producing cells of the sporangium.

Sporangiophore A unit of the strobilus of the Equisetales consisting of a peltate axis bearing pendant (ancestrally recurved), longitudinally dehiscent sporangia.

Sporangium *PI: sporangia.* The spore producing organ of the sporophyte.

Spore A haploid cell that, in the land plants, originates from meiotic divisions of sporocytes within a sporangium, ultimately growing into a gametophyte.

Sporocarp The generally spherical reproductive structure of aquatic ferns, functioning in allowing the sporangia inside to remain dormant and resist desiccation for a long time.

Sporocyte A cell of the sporangium that undergoes meiosis, producing (generally four) haploid spores.

Sporogenous tissue Internal cells of the sporangium that mature into sporocytes, the latter forming spores by meiotic divisions.

Sporophyll A specialized leaf that bears one or more sporangia.

Sporophyte The diploid (2n) phase in the haplodiplontic life cycle of all land plants.

Sporophytic leaf The leaf of a vascular plant sporophyte.

Sprorpollenin A polymer of oxidative carotenoids or carotenoid esters that impregnates the exine of pollen grain walls.

Spring wood The secondary xylem from the first part of the growing season, the cells of which tend to be larger in diameter with thinner walls.

Spur A tubular, rounded or pointed projection from the calyx or corolla, functioning to contain nectar.

Spurred One or more sepals form a long, pointed, spine-like structure, which is called spur.

Spur/spur shoot Fascicle.

Squarrose Sharply curved downward or outward (abaxially) near the apex as in Asteraceae.

Stamen The unit of the androecium; a microsporophyll (typically modified as a filament) bearing generally two pollen bearing thecae, fused into an anther.

Stamen insertion (a) Referring to whether or not stamens protrude beyond the perianth. (b) Denoting the "point of insertion: the point of adnation of an epipetalous stamen to the corolla.

Staminal disk Syn: *nectariferous disk.* A discoid or doughnut shaped structure arising from the receptacle at the base of the stamens; may be nectar bearing.

Staminate Syn: *male.* Flowers unisexual, with stamen(s) only, lacking fertile carpel(s).

Staminode/staminodium A sterile stamen, producing nofunctional pollen; may be modified as a nectary or petaloid structure; nay or may not contain an antherode, a sterile anther.

Standard deviation A measure of the differences of individual values of a sample from the mean of all values, calculated as the square root of the "sum of squares" divided by the sample size minus one.

Standard error A measure of how close the sample mean is to the mean of the entire population, calculated as the standard deviation divided by the square root of the sample size.

Starch A polysaccharide of glucose units (alpha- 1,4-glucopyranoside) in green plants, functioning as the high-energy storage compound.

Starch grain Syn: *amyloplast.* Lamellate deposits of starch in green plant cells, functioning as the high-energy storage compound; a type of elastic substance.

Statistics A branch of mathematics dealing with collecting, organizing, analyzing and interpreting numerical data.

Status novus *Abbr: star. nov.* indication of a change in rank of a scientific name.

Stele The spatial distribution of the primary vasculature of a stem organized into arrangements of xylem and phloem.

Stellate Star shaped trichomes having several arms arising from a common base (stalked or sessile) as in Malvaceae.

Stem Axis of a plant, generally a cylindrical organ, bearing leaves with buds in their axils formed by the plumule of the seed and generally growing upward, functioning in support and elevation of leaves and reproductive structures and in conduction of water, minerals, and sugars in vascular plants.

Stem (cladogram) *Internode.*

Stem-based *A* type of phylogenetic classification in which the stem (internode) region just above a common ancestor plus all descendants of that stem are the basis for grouping.

Stem-based clade *A* clade that includes the "stern"(internode) region just above a common ancestor plus all descendants of that stem.

Stem habit Stem feature based on structure, position, growth, and orientation of above-ground stems.

Stem-tendril A modified axillary or terminal bud into tendril for climbing. Tendrils of *Passiflora* are modified axillary branches whereas those of *Vitis* are modified tips of sympodial branches.

Sterile jacket layer Antheridial wall; archegonial wall.

Sterile segment The photosynthetic blade or lamina component of the leaf of an ophioglossoid fern.

Stigma The pollen receptive portion of a pistil (apex of carpel); maybe a discrete structure or a region of a style or style branch. It may be bilobed, trilobed, bifid, capitate, discoid, dumbell-shaped, feathery, linear, highly branched or striate.

Stilt roots Roots arising from lower nodes of the stem and growing obliquely downwards to anchor the plant into the soil.

Stinging trichome *A* sharp pointed trichome that secretes an irritating fluid upon penetrating the tissue of an animal.

Stipe Syn: *petiole.* (a) A leaf stalk, often used for ferns. (b) *Gynophore.* Adj: *stipitate.* (c) The stalk of the sporophyte of mosses.

Stipel Adj: *stipellate.* Stipule-like appendage at the base of leaflets as in leguminous plants. One of a pair of leaf like structures, which may be modified as spines or glands, at either side of the base of a petiolule as in Fabaceae.

Stipule Adj: *stipulate.* One of a pair of leaf like structures (lateral outgrowth present at the base of the leaf), which may be modified as spines or glands, at either side of the base of a petiole.

Stipular spine Cf: *thorn; prickle.* A sharp pointed stipule, e.g. *Euphorbia* spp.

Stipule sear A mark indicating the former place of attachment of a stipule.

Stolon Syn: *runner.* An indeterminate, elongated, slightly underground or aboveground propagative stem with long internodes, rooting at the tip forming new plants. The special type of runner grows upward like ordinary branch and then arches down to strike roots in the soil and forms a daughter plant as in *Mentha.*

Stoloniferous Syn: *sarmentose.* With elongated propagative shoots (stolons) rooting at the tip producing new plants.

Stoma The opening between two guard cells of a stomate.

Stomate PI: *stomata; stomates.* Specialized epidermal cells consisting of two guard cells which by changes in turgor pressure can increase or decrease the size of the opening (stoma) between them.

Storage root A swollen tap root containing concentrations of high-energy compounds such as starch.

Straight Flat, without vertical curves or bends.

Strap shaped Flat, not needle-like but with length: width ratio greater than 12: 1.

Straphiole An appendage of hilum (found in some seeds).

Striate (a) With fine longitudinal lines. (b) A pollen sculpturing having thin, cylindrical, tangentially oriented elements.

Strict consensus tree A consensus tree that collapses any differences in branching pattern between two or more cladograms to a polytomy.

Strigose Covered with dense, coarse, bent and mostly flat trichomes, often with a bulbous base.

Strigulose Minutely strigose.

Strobile A spike in which each flower is borne in the axil of a persistent membranous bract.

Strobilus Cone.

Students t-test *T-test.*

Style A stalk like portion of the pistil between the stigma and ovary. It arises from the top (apex) of ovary.

Styloid A single, elongated, angular crystal of calcium oxalate; a type of ergastic substance.

Stylopodium A swollen region at apex of the ovary, below the styles/stigmas as in Apiaceae.

Subapical With style arising at one side near the apex of the ovary.

Subbasifixed Anther attached just above its base to the filament.

Subcosmopolitan Having a broad, but tot worldwide distribution.

Suberin A waxy, water-resistant substance, found in stomata, casparian strips of the endodermis and the cell walls of cork cells.

Subglabrous Nearly glabrous with just a few scattered trichomes.

Subopposite With two leaves or other structures on opposite sides of stem or central axis but at different nodes, slightly displaced relative to one another.

Subsessile Having a very short, rudimentary petiole (leaf attachment), petiolule (leaflet attachment), pedicel (flower attachment) or filament (stamen attachment).

Subshrub A short shrub that is woody only at the base and that seasonally bears new, non-woody annual shoots above.

Subsidiary cell An epidermal cell adjacent to a stomate and somewhat different from other epidermal cells in shape and size, functioning in regulating stomatal opening.

Subspecific epithet The third component of a ternary name at the rank of subspecies.

Subtropical fruits Fruits growing in moderate climate, e.g. litchi, ber, pomegranate, loquat, fig, date, phalsa, jamun, grapes, bael, etc.

Subulate Awl shaped; approximately narrowly oblong to narrowly triangular.

Successive A type of microsporogenesis in which cytokinesis occurs after meiosis.

Succulent (a) Fleshy or juicy. Syn: *baccate, carnose.* (b) A plant with fleshy stems or leaves.

Succulent fruits Pericarp is fleshy or fibrous and remains distinguishable into three layers. These are indehiscent.

Sucker An underground runner which after travelling a short distance grows up and forms a daughter plant. These may also develop from the axillary buds.

Suffrutescent Woody basally, herbaceous apically; the habit of a subshrub.

Sulcus Adj: *sulcate.* An elongate aperture similar in shape to a colpus having length: width ratio >2:1 usually occurring at the distal pole.

Summer wood The secondary xylem from the latter part of the growing season, the cells of which tend to be smaller in diameter with thicker walls.

Superior With sepals, petals, stamens and/or hypanthium attached at the base of the ovary. Ovary placed above the point of attachment of corolla and stamens.

Superposed bud Bud(s) occurring above the axillary bud.

Supervolute With one half of a simple leaf coiled tightly around the midrib, the other half coiled (in the opposite direction) around the first half, as in members of the Zingiberales.

Suprabasal With three or more primary veins diverging from one point above the base of the blade, a subcategory of actinodromous and pallnactinodromous.

Suprafoliar Descriptive of a palm inflorescence that is positioned above the leaves of the crownshaft.

Surficial Growing upon the ground.

Suspensor A non-persistent column of cells of the embryo, functioning in transport of nutrients to the mature, persistent embryo during its development.

Syconium A multiple fruit in which the unit fruits are small achenes covering the surface of a fleshy, inverted compound receptacle as in fig.

Syconus Fruit developed from hypanthodiurn inflorescence. The receptacle becomes fleshy and hollow cup-shaped with a narrow apical opening. Unisexual flowers are found on its inner surface, male flowers towards upper side and female (pistillate) flowers towards lower side. Larvae of insects live in gall flowers (female flowers). The fruit is achene. Fleshy receptacles form the edible part as in figs.

Symmetry Referring to the presence and number of mirror image planes of symmetry.

Sympatric Occurring in the same geographic range.

Sympetalous Syn: gamopetalous. With connate petals.

Symplesiomorphy A shared, ancestral features among taxa, which may be used as a basis for grouping in a phenetic classification.

Sympodial A branching pattern in which a given axis is made up of several units, each of which is derived from a separate apical meristem, the units themselves determinate or indeterminate.

Synandrous Both anthers and filaments remain fused as in Cucurbitaceae.

Synangium A fusion product of two or more sporangia, e.g., in the psilophytes and some marattioid ferns.

Synanthy Adj: *synanthous.* Timing in which leaves and flowers develop at the same time.

Synapomorphy An apomorphy that unites two or more taxa or lineages.

Syncarp An aggregrate fruit, typically of berries, in which the unit fruits fuse together as in *Annona.*

Syncarpous With carpels (at least at the base) connate, the pistil or ovary being compound.

Synchronic species A species corresponding to a group of contemporaneous organisms, living at the same general point in time.

Syncolpate A pollen grain in which the colpi are joined at the poles.

Synergid cells The two haploid cells flank the egg and may function in pollen tube entry as in a typical angiosperm female gametophyte.

Syngenesious With anthers connate and filaments distinct as in Asteraceae.

Synonym An unaccepted name, by a particular author or authors, applying to the same taxon as the accepted name.

Synoptic collection A collection of plant specimens that contain one specimen (of all available specimens) of each taxon for a given region.

Synsepalous Syn: *gamesevalous.* With connate sepals.

Syntepalous Syn: *gamotepalous.* With connate tepals.

Syntype A specimen that was cited in the original work when a holotype was not designated, or one of two or more specimens that were all designated as types.

Systematics A science that includes and encompasses traditional taxonomy and that has as its primary goal the reconstruction of phylogeny.

t-test *Syn: Student's t-test.* A statistical test of two samples used to evaluate the probability of their being the same.

Tandem repeat A sequence of DNA that repeats multiple times.

Tangential Referring to a longitudinal section of wood, perpendicular to a stem radius.

Tannin Phenol derivatives in plant cells that may function to deter herbivory and parasites growth, a type of ergastic substances.

Tapering Trichomes ending in a sharp apex.

Tapetum The innermost cell layer of the anther wall, consisting of metabolically active cells that function in the development of pollen grains.

Tap root A persistent, well developed primary root directed vertically downwards bearing small lateral roots giving rise to a root system with a prominent main root characteristic to dicot plants.

Tassel Male inflorescence as in maize.

Tautonym A binomial in which the genus name and specific epithet are identical in spelling, illegitimate in the ICBN.

Taxon (plural, taxa) A group of organisms, ideally monophyletic and traditionally treated at a particular rank.

Taxonomic key *Key.*

Taxonomic revision A change of the definition and delimitation of a taxon.

Taxonomic species Syn: *morphologic species.* The smallest group (or class) of individuals that are similar to one another in one or more features and different from other such groups.

Taxonomy A field of science (major component of systematics) that encompasses description, identification, nomenclature and classification.

Tectate-columellate A common exine structure that consists of an inner foot layer, a middle layer of radially elongate columellae and an outer roof like layer called the tectum.

Tecturn The outer roof like layer of a tectate-columellate pollen exine wall.

Tegmic ovule The ovule with integument. It may be unitegmic (with a single integument as generally found in sympetalae / gamopetalae as in Asteraceae or bitegmic (with two integuments found generally in polypetalae and monocots.

Telomes Ancestral branches, e.g., those surrounding the megasporangium of ancestral seeds.

Temperate Distributed in regions between the tropics and arctic/ antarctic, having a moderate climate.

Tendril (a) A coiled and twining leaf part, usually a modified rachis or leaflet. (b) A long slender coiling branch adapted for climbing.

Tendrillar Stipules modified into tendril or apex forms tendril for climbing.

Tendril Climbers Climbing by formation of special wire like structures twinning around to seek support.

Tentacular leaf A leaf bearing numerous sticky glandular hairs or bristles that function in capturing and digesting small animals as in *Drosera,* sundew.

Tenuinucellate An ovule in which the nucellus is composed of a single layer of cells with no formation of a parietal cell.

Tepal A component of the perianth in which the parts intergrade or in which the perianth is undifferentiated into distinctive sepals and petals.

Terete Cylindrical.

Terminal (a) At or near the top, tip or end of a structure. (b) Entire inflorescence positioned as a terminal shoot relative to the nearest vegetative leaves. (c) Style arising at the apex of the ovary.

Terminal bud Syn: *apical bud.* Bud at the apex or end of a stem.

Terminal bud scar Ring like marks indicating the former place of attachment of the terminal bud scales.

Ternary name A scientific name composed of three names, such as a subspecies or variety name.

Ternate/ternately compound Syn: *trifoliolate.* A compound leaf with three leaflets.

Ternately veined With three primary veins arising from a common basal point.

Ternate Netted/ternately netted: Ternately veined with three primary veins arising from a common basal point, the ultimate veins forming a fine reticulum.

Terrestrial Growing on land.

Tertiary vein A lateral vein that branches from and is smaller than a secondary vein.

Testa Seed coat.

Testa Also called seed coat, and derived from the outer integument and is protective in nature.

Tetrad A fusion product of four pollen grains developing from the four products of microsporogenesis.

Tetradynamous With stamens in two groups of four long and two short, e.g., Brassicaceae.

Tetragonal tetrad A tetrad in which the four grains are in one plane and are equally spaced apart.

Tetrahedral tetrad A tetrad in which the four grains form the points of a tetrahedron, e.g. members of the Ericaceae.

Tetramerous A whorl with four members.

Tetramerous The floral whorls contain units four in number or in multiple of four as in Cruciferae.

Tetraploid A polyploid having four sets of chromosomes.

Tetrasporangiate Cf. *dithecal.* Anther with four microsporangia and typically two thecae.

Tetrasporic Megasporogenesis in which cytokinesis does not occur after meiosis, resulting in a single cell with four haploid nuclei, all of which contribute to the female gametophyte.

Texture Internal structural consistency, sometimes incorporating colour.

Thalloid Referring to or having a thallus.

Thallus The flattened (dorsiventral) gametophyte of some liverworts and all hornworts.

Theca One half of typical anther containing two microsporangia.

Therophyte An annual plant.

Thorn cf: prickle, spine. A sharp pointed stem or shoot.

Throat An open, expanded region of a perianth, usually of a sympetalous corolla.

Thrum Flowers with a short style and long stamens, found in distylous flowers.

Thylakoid A membrane of chloroplasts and photosynthetic bacteria that contains compounds involved in the light reactions of photosynthesis.

Thyrse (or Thyrsus) A densely-branched inflorescence with the central axis racemose (indeterminate), but the lateral branches cymose (determinate) as in grapes.

Tiller A grass shoot produced from the base of the stem.

Tissue A group of cells having a common function or structure; may be simple or complex.

Tokogenetic Relationships between individual organisms with regard to gene flow in ancestral-descendent lineages.

Tomentose Covered with short, dense, interwoven trichomes or cottony hairs as in *Terminalia tornentosa* and *Calotropis procera*.

Tomentulose Minutely tomentose.

Topology The structure of a branching diagram such as a cladogram, including how lineages are connected together.

Torus *Receptacle* (flower part) or *compound receptacle* (inflorescence part).

Total clade A clade that includes a crown clade plus another taxa that share a recent common ancestor with the crown clade but not with other crown clades.

Total cover A measure of the degree to which the total area of the community is covered by members of a given layer.

Tracheary element A generally elongated cell that is dead at maturity has a lignified, secondary cell wall and is positioned end-to-end with other tracheary elements, forming a continuous tube that functions in water and mineral conduction; the major component of xylem tissue.

Tracheid The ancestral type of tracheary element that is imperforate, in which water and mineral nutrients flow between adjacent cells through the primary cell walls at pit-pairs.

Trailor A plant which trails over the ground surface by its weak, prostrate and slender stem.

Transfer in position To reclassify a taxon without a change in rank to place within a different taxon of the same rank.

Transformation series Syn: *morphocline.* The hypothesized sequence of evolutionary change, from one character state to another, in terms of direction and probability.

Transverse (a) Dehiscing at right angles to the long axis of the theca. (b) Referring to a cross-section of wood.

Transverse posture Placement of tip with respect to the horizontal plane of object.

Trap leaf A leaf that mechanically moves after being triggered, capturing and digesting small animals, e.g., *Aldrovanda; Dionaea* venus fly trap.

Tree A generally tall, perennial, woody plant having one well-marked main stem (trunk) arising at ground level, e.g. Eucalyptus, Mangifera, Tectona etc.

Triangular Three sided, length: width ratio 2:1 to 3:2.

Trichome An external, hair like plant structure.

Trichotomosulcate *Triseriate.* A pollen grain aperture type that is three branched tricyclic.

Trifoliolate Tenatelternately compound.

Trifoliate Three leaflets articulated at the top of rachis.

Trigger mechanism A type of movement hercogamy in which an insect pollinator triggers the sudden movement of one or more stamens, dusting the insect with pollen at the point of contact.

Trilete mark A three lined differentially thickened wall (laesura) on the spore of some land plants, being the remnant (scar) of attachment of the adjacent three spores of a tetrad.

Trilete Spores with a 3-branched laesura.

Trimerous The floral whorls contain units the multiple of three as in Liliaceae.

Trimonoecious A plant with pistillate, staminate, and perfect flowers on the same individual.

Trinucleate Having three nuclei,, referring to some angiosperm, pollen grains at the time of release.

Trioecious/trioecy A plant with pistillate, staminate and perfect flowers on different individuals.

Tripinnate/tripinnately compound A compound leaf with three orders of axes, the third (rachillae) bearing leaflets. Incision goes to the third order so that leaflets are found on secondary branches of the rachis.

Triploid A polyploid having three sets of chromosomes.

Triseriate Syn tricyclic. With three whorls of parts.

Tristichous Alternate, with points of attachment in three vertical rows/ranks as in the sedges.

Tristichous or phyllotaxy Leaves are three ranked and the fourth is just above the first leaf.

Tristylly Adj: *tristylous.* Hercogamy in which there are three heights of styles and stamens.

Trisulculate A pollen grain with three elongate apertures on opposite sides of the grain, parallel to the equatorial plane.

Trullate Parallelogram shaped, Widest near base and the length: width ratio 2:1 to 3:2.

Trumpet-shaped Petals fused with basal portion tube-like and upper funnel-like appearance.

Truncate Apical or basal margin abruptly cut straight across, the angle approximately 180'.

Tryma A nut surrounded by an involucre that dehisces at maturity as in pecan.

Tube A cylindrically shaped perianth or region of the perianth, usually of a sympetalous corolla.

Tube cell One of the two initial, haploid cells in the male gametophyte of angiosperms that remains near the tip of the growing pollen tube and may function in its development.

Tuber A thick, underground storage stem, usually not upright, typically bearing outer buds and lacking surrounding storage leaves or protective scales as in potato. The fleshy, swollen, underground stem bear scale leaves with buds (eyes) in their axils. The eyes are spirally arranged, but are more crowded towards the distal end (rose end), each eye develops into a small shoot

Tuberculate papillate.

Tubular Cylindrical. Sepals fuse to form a tubular structure.

Tunicated Scale-leaves arranged in a concentric manner and the whole bulb is enclosed by dry membranous scale leaves, called tunic as in onion.

Turbinate Top or turban shaped.

Twiners Climbing by twinning around its support.

Twining Twisted around a central axis.

Twisted With parts rolled up in such a way that the outer part of each covers the inner part of the one in front of it, while in turn its inner part is covered by the one behind it.

Two-armed Tichomes with two arms arising from a common base as in Malphighiaceae.

Type specimen Nomenclatural type.

Typits conservandus Abbr: sayp. covs. A conserved type specimen.

Typus designates Abbr: syp. *dec,*. Designation of a type specimen.

Ulcerus Adj: *ulcerate.* A circular to slightly elliptic aperture similar in shape to a porus (length width ratio <2:1) occurring usually at the distal pole.

Umbel A determinate or indeterminate, flat-topped or convex inflorescence with pedicels attached at one point. It has a very much shortened main axis and all flowers appear to be arising from the same point. Flowers are bracteate. Younger flowers are arranged towards centre while older flowers towards periphery. The whorl of bracts is called involucre.

Umbellule A unit of a compound umbel.

Uncinate *Hooked.*

Undulate Margins wavy in a vertical plane.

Unguiculate Clawed as in Brassicaceae, Caryophyllaceae.

Unicarpellous With gynoecium composed of one carpel, the pistil or ovary being simple.

Unicellular Referring to a trichome (or other structure) consisting of a single cell.

Unicostate or Pinnate Almost racemose type. One prominent midrib runs from the base to the apex with several secondary and tertiary and so on anastomosing veinlets forming a net-like framework as in banana.

Unifacial growth Growth of a secondary cambium in which secondary xylem (wood) is produced to the inside but with no secondary phloem produced to the outside, as in fossil lycophytes and equisetophytes.

Unifacial leaf A leaf that is isobilateral and flattened side-to-side, having left and right sides, except at the base, where it is often sheathing, e.g., members of the Iridaceae.

Unifoliolate A leaf bearing a single leaflet with petiolule distinct from the petiole, interpreted as the derived reduction of an ancestral compound leaf, e.g., *Cercis,* redbud.

Unifoliate Single leaflet attached to the top of the petiole.

Unilateral *Secund.*

Unilabiate One lipped; with one, generally lower, segment as in Goodeniaceae.

Unilocular Ovary with single chamber (locule), as in Leguminosae and Compositae.

Uninervous Having a central midrib with no lateral veins, e.g., lycopods, psilophytes and equisetophytes, many conifers.

Uniparous Only one branch at each node.

Uniseriate *(a)* With a single whorl of parts. Syn: *monoryclic.* (b) Trichomes having a single vertical row of cells. (c) Rays in wood that are composed of a single vertical row of cells.

Unisexual Syn: *imperfect.* Flowers having only carpell(s) or only stamen(s).

Unistomal Referring to the micropyle of a unitegmic ovule.

Unit Inflorescence Syn: *partial inflorescence; primary inflorescence.* A subunit of the entire secondary inflorescence.

Unitegmic An ovule with a single integument, found in all gymnosperms and derived in some angiosperms.

Univariate plot A graph of a single variable, usually per individual, sample or taxon.

Unordered Referring to a character transformation series in which each character state is allowed to evolve into other character state with equal probability, generally in a single evolutionary step.

Unrooted tree Syn: *network.* A branching diagram in which polarity is not indicated, representing the relative character state changes between taxa.

Upper epidermis The adaxial epidermis of a leaf.

Upward Dehiscing toward the sky in a horizontally oriented flower.

Urceolate Unshaped; expanded at the base and constricted at the apex, e.g., many Ericaceae. Sepals fuse to form urn-shaped structure. Petals fuse to form a pitcher like structure (swollen in middle and narrow at the apex).

Urent With hispid and stinging trichomes.

Utricle A small, bladdery or inflated, one seeded, dry fruit; an achene in which the pericarp is significantly larger than the mature seed as in *Atriplex,* salt bush.

Vacuole A large, internal, membrane bound sac of plant cells that functions in storage of compounds such as pigments, acids or ergastic substances.

Valid name A name that is validly published.

Validly published Referring to the criteria needed for a scientific name to be formally recognized, including effective publication, publication in correct form, publication with a Latin description or diagnosis or with a reference to such, and indication of a nomenclatural type.

Valvate (a) Sides enrolled so that margins touch. (b) Having a whorl of perianth parts meeting at the margins and not overlapping.

Valve A portion of the pericarp of a dehiscent fruit that splits off but does not enclose the seed(s).

Valvular Anther dehiscing through a pore covered by a flap of tissue as in as in Barberis.

Valvular capsule Septifragal capsule.

Variegated With two or more colours occurring in various irregular patterns, generally used for leaves.

Varietal epithet The third component of a ternary name at the rank of variety.

Vascular bundle Syn: *vascular strand.* A strand of associated xylem and phloem tissue.

Vascular bundle scar A mark within the leaf scar indicating the former position of a vascular bundle that extended from stem to leaf.

Vascular bundle sheath A ring of cells, composed of fibre or parenchyma cells, surrounding the xylem and phloem of a vascular bundle, functioning in C4 photosynthesis. ,

Vascular cambium A cylindrical sheath of cells that undergoes primarily tangential divisions, producing secondary xylem (wood) to the inside and secondary phloem (inner bark) to the outside.

Vascular cylinder A central region of vascular tissue (xylem and phloem) in a root or in some stems.

Vascular strand Vascular bundle.

Vasculartissue A tissue made up of xylem and phloem, functioning mainly in conduction of water, minerals and sugars.

Vegetation type An assessment of the habit, habitat and cover of plant species present, forest, woodland, savanna, chaparral, scrub, grassland, meadow, strand marsh, swamp, pond and vernal pool.

Vegetative bud A bud that develops into a vegetative shoot, bearing leaves.

Vein Syn: nerve. The vascular bundle of a leaf or leaf homologue such as a sepal, petal, stamen or carpel, containing the conductive tissues.

Velamen A specialized, multilayered epidermis of some roots, functioning in protection, prevention of water loss or water and mineral absorption as in Araccae, Orchidaceae.

Venation The pattern of arrangement of vascular system in the lamina (veins and vein branching).

Venter The swollen, basal portion of an archegonium, containing the egg.

Ventral Adaxial.

Ventral canal cell A second cell within the swollen base of the archegonium located just distal to the egg,

Ventral vein Syn: *lateral vein, placental vein.* One of the two veins of a carpel near the carpel margins.

Vernal Appearing in spring.

Verruca PI: *verrucae.* One of the short, warlike elements of a verrucate pollen grain.

Verrucate (a) *Papillate.* (epidermal excrescence) (b) A pollen sculpturing having short, warlike elements, each element termed a verruca.

Versatile Filament attached to the back of anther at a point only, so that anther can swing freely as in grasses.

Verticillaster A secondary inflorescence with an indeterminate central axis bearing opposite, lateral, sessile cymes, the flowers appearing congested. The leaves are borne on opposite side on a node, each node bears flowers arranged in dichasial cyme. After the formation of three flowers, each dichasial cyme passes on to monochasial cyme of scorpoid type. Flowers are condensed, sessile, bisexual and bilipped. The inflorescence is characteristic to family Labiatae (e.g. *Ocimum*).

Verticillate Having three or more leaves or other structures per node.

Verticillate phyllotaxy Whorled phyllotaxy. Three to five leaves form a whorl in each.

Very widely obovate Margins curved, widest near the apex with a length: width ratio ca. 1: 1.

Very widely ovate Margins curved, widest near the base, with a length: width ratio ca. 1: 1.

Vernation Ptyxis.

Vespertine In the evening, typically with respect to when flowers open.

Vessel Syn: pore. Term for several vessel members attached end-to-end, forming a continuous, conductive tube.

Vessel member A type of tracheary element that is perforate in which continuous holes or perforations in the cell wall occur through which water and mineral nutrients flow between cells.

Vestiture Tachome cover, a combination of trichome type, length, strength, shape, density, and colour.

Vexillary See description under corolla.

Vicariance The splitting of one ancestral population into two (or more) populations, such as by continental drift or the formation of a geographic barrier.

Video To cite a reference. Abbrvid.

Villose Hairs long, soft, closely set as in *Leucas aspara.*

Villous Syn: *lanate.* Covered with very long, soft, crooked trichomes.

Vine A plant with elongate, weak stems, supported by means of scrambling, twining tendrils, or roots; may be annual or perennial, herbaceous or woody.

Viscid Syn.: glutinous. Having a shiny, sticky surface.

Viscin thread One of the long strands of carbohydrate material that function in sticking pollen grains together.

Viscose Sticky because of the some secretion as in leaves of Cleome and *Nicotiana tabacum*, sometime called glutinous.

Voucher specimen An herbarium specimen in a plant collection serving as reference material for a named taxon or as part of research project.

Wax A type of triglyceride compound that may function as high energy storage compounds or secretion products; a type of ergastic substance.

Webbing Evolutionary development of thin tissue between the axes of the branches, thought to have occurred in euphylls.

Weight The specific assignment of taxonomic importance when weighting a character in a phylogenetic analysis.

Weighting/character weighting The assignment of greater or lesser taxonomic importance to certain characters over other characters in determining phylogenetic relationships.

Whorl Syn: *series.* A cyclic group of floral parts, e.g., of sepals, petals, stamens, or carpels.

Whorled (a) With three or more leaves or other structures per node. (b) Perianth parts in distinct whorls or series, with parts arising from the same nodal region. (c) With stamens in one or more whorls or series.

Widely elliptic Margins curved widest near the midpoint, with a length:width ratio ca. 6:5.

Widely obovate Margins curved, widest near the apex, with a length: width ratio ca. 6:5.

Widely ovate Margins curved, widest near the base, with a length: width ratio ca. 6:5.

Widely triangular Three sided, with a length: width ratio ca. 6:5. (shape).

Wood Secondary xylem.

Woody Having a hard, woodlice texture.

Wrinkled With irregular, fine lines or deformations.

Xenogamy Adj: *xenogamous. Outbreeding.* Pollination between flowers of different genetic constitution.

Xerophyte A plant adapted to live in a dry, generally hot environment.

Xylem A tissue composed of tracheary elements plus some parenchyma and sometimes sclerenchyma, functioning in conduction of water and mineral nutrients.

Zig-zag Referring to the micropyle of a bitegmic ovule in which the micropylar pore of the outer integument is spatially displaced relative to the inner integument.

Zonoaperturate Syn: *stephanoaperturate.* Descriptive of pollen grains with apertures occurring in the equatorial region.

Zonocolpate A pollen grain with colpi occurring in the equatorial region.

Zonoporate A pollen grain with pori occurring in the equatorial region.

Zoochory Dispersal of propagules by animals.

Zooidogamy Cf - siphonogamy. Gymnospermous pollen tubes that are haustorial and deliver motile sperm cells to a fertilization chamber, where the sperm swims to the archegonium or egg.

Zygomorphic Syn: *bilateral, irregular monosymmetric.* Bilaterally symmetrical, with one plane of symmetry, e.g. a flower capable of equal division by only one plane of symmetry.

Zygote A diploid cell that results from the fusion of two haploid cells (egg and sperm) and that ultimately matures into a new sporophyte in the land plants.

SUBJECT INDEX